Skeletal Anatomy of the Newborn Primate

Although much is known about the anatomy of adult primates, the same cannot be said for the anatomy of young primates, especially non-hominoid primates such as lemurs and marmosets. This is the first book dedicated to newborn skeletal and dental anatomy and how it varies across primate species, which is important for interpreting adult primate skeletal form, as well as for comprehending primate and human evolution.

Structured according to anatomical regions, the book includes hundreds of detailed anatomical illustrations, a color atlas illustrating entire skeletons in representative taxa, and boxes at the end of each chapter providing further detail on key aspects covered in the main text. While the book is primarily a guide to comparative anatomy, it also highlights the links between development and behavior. It is an indispensable resource for students and researchers in the fields of biological anthropology, anatomy, primatology, growth and development, dental biology, and veterinary medicine.

Timothy D. Smith is Professor at the School of Physical Therapy, Slippery Rock University, Pennsylvania, USA. He has researched growth and development in primates and other mammals for over two decades, with a particular focus on the perinatal period. He is also an illustrator and has produced anatomical illustrations for numerous journal articles and books. He is currently an Associate Editor for *The Anatomical Record.*

Valerie B. DeLeon is Associate Professor in the Department of Anthropology, University of Florida, Florida, USA. She has studied craniofacial development and growth in humans, non-human primates, and other mammals for over 20 years. She currently serves as President-Elect of the American Association for Anatomy.

Christopher J. Vinyard is Professor in the Department of Anatomy and Neurobiology, Northeast Ohio Medical University, Ohio, USA. He has studied mammalian craniofacial biology for two decades, with a special emphasis on feeding adaptations in primates. He is the co-editor of *Primate Craniofacial Function and Biology* (Springer, 2008).

Jesse W. Young is Associate Professor in the Department of Anatomy and Neurobiology, Northeast Ohio Medical University, Ohio, USA. He has nearly two decades of experience conducting research on postcranial growth and locomotor development in humans, non-human primates, and other mammals.

Skeletal Anatomy of the Newborn Primate

TIMOTHY D. SMITH
Slippery Rock University, Pennsylvania

VALERIE B. DELEON
University of Florida

CHRISTOPHER J. VINYARD
Northeast Ohio Medical University

JESSE W. YOUNG
Northeast Ohio Medical University

CAMBRIDGE
UNIVERSITY PRESS

CAMBRIDGE
UNIVERSITY PRESS

University Printing House, Cambridge CB2 8BS, United Kingdom

One Liberty Plaza, 20th Floor, New York, NY 10006, USA

477 Williamstown Road, Port Melbourne, VIC 3207, Australia

314–321, 3rd Floor, Plot 3, Splendor Forum, Jasola District Centre, New Delhi – 110025, India

79 Anson Road, #06–04/06, Singapore 079906

Cambridge University Press is part of the University of Cambridge.

It furthers the University's mission by disseminating knowledge in the pursuit of education, learning, and research at the highest international levels of excellence.

www.cambridge.org
Information on this title: www.cambridge.org/9781107152694
DOI: 10.1017/9781316591383

First published 2020

Printed in Singapore by Markono Print Media Pte Ltd

A catalogue record for this publication is available from the British Library.

Library of Congress Cataloging-in-Publication Data
Names: Smith, Timothy D., 1966– author.
Title: Skeletal anatomy of the newborn primate / Timothy D. Smith, Valerie B. DeLeon, Christopher J. Vinyard, Jesse W. Young.
Description: New York : Cambridge University Press, 2020. | Includes bibliographical references and index.
Identifiers: LCCN 2019038225 (print) | LCCN 2019038226 (ebook) | ISBN 9781107152694 (hardback) | ISBN 9781316606643 (paperback) | ISBN 9781316591383 (epub)
Subjects: LCSH: Primates–Anatomy. | Skeleton.
Classification: LCC QL737.P9 S63 2020 (print) | LCC QL737.P9 (ebook) | DDC 599.8–dc23
LC record available at https://lccn.loc.gov/2019038225
LC ebook record available at https://lccn.loc.gov/2019038226

ISBN 978-1-107-15269-4 Hardback

Contents

The color atlas can be found at the end of the book.

Foreword

Perinatal Anatomy of Primates - A Neglected Ontogenetic Stage

I discovered the importance of perinatal ontogenetic stages rather by chance: When I began – more than 45 years ago – to build up a collection of serial sections in my own histo-lab, I had no easy access to early embryonic material. Instead, we had a rich influx of aborted late fetuses and dead-born neonates from the Zoological Garden of Frankfurt. *Faute de mieux*, I started working on the heads of these specimens. Because primates are precocial, their generally well-ossified bodies are difficult and time-consuming to process histologically, even at birth. However, I discovered soon that especially in the ethmoidal region I could study structures that were quite unknown in the published craniogenetic studies (de Beer, 1937).

I became more and more interested in these late fetal and early postnatal stages, because I realized that there existed a gap of anatomical knowledge, a scientific "terra incognita," so to speak. These perinatal stages also represent one of the neglected size dimensions, namely that between macroscopy and microscopy: the specimens are too small for macroscopic dissection and preparation, but too large and untractable for conventional histology. Normally the heads had to be prepared by the time-consuming celloidine technique, in which it usually takes 3–4 years to produce a good sectional series! We further developed 3D modeling on the basis of 2-mm sheets of Styrofoam instead of wax plates, and that speeded up the production of models enormously (Maier, 1980). Nowadays, 3D reconstructions based on microCT-scans help to visualize complex structural systems more rapidly and easily – but hand-made models may still be advantageous for understanding.

Neglected dimensions are normally connected with technical innovations. For example, improved histological and light-microscopic techniques as developed in the second half of the nineteenth century helped researchers to explore the small size dimensions of early embryonic structures, i.e., in the sizes of a few tens of micrometers. The invention of microtomes and serial sectioning allowed reconstructions of 3D models with the wax-plate technique (Born, 1883), and this approach to morphogenesis speeded up embryological research as testified by the handbook of Hertwig (1906).

I gradually began to grasp the research potential of the "mesoscopic" structures of perinatal ontogenetic stages: they still provide information from embryonic and fetal states of differentiation, but they already show the outlines of the anatomy of the adults. Moreover, I began to realize that the millimeter size range was also of specific functional importance, at least in organs in which extensive exchange with blood capillaries is optimized. Later on, I coined the term "capillary organs" for these interesting structures of microscopic anatomy (Figure 1).

The intensive and systematic use of the new technological facilities produced the modern disciplines of histology, microscopic anatomy, and embryology within a few decades. There was much to discover and a largely new concept of comparative anatomy was founded, as demonstrated by the first modern text- and handbooks of

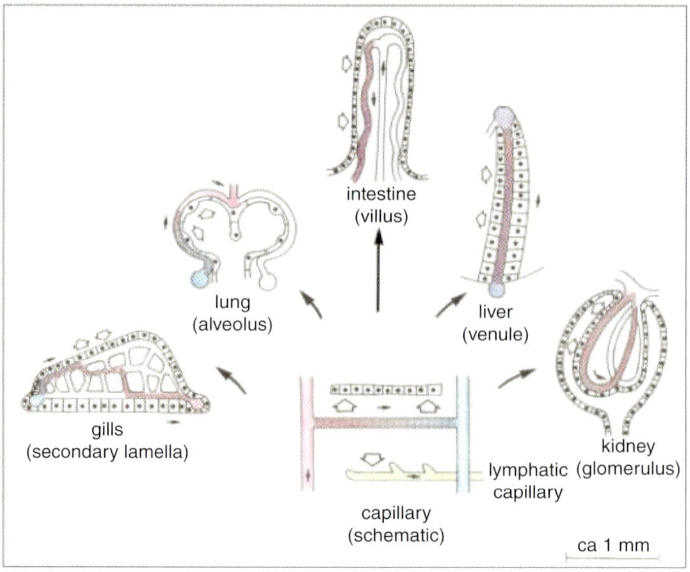

Figure 1 "Capillary organs" in mammals. Capillary exchange occurs in the size range of 1–2 mm. Quite a number of organs, which are specialized for extensive exchange of metabolites or gases with the capillary blood, are thereby showing very regular organization patterns in the millimeter dimension (original).

histology and embryology (Kölliker, 1879; Hertwig, 1888, 1906 – to name just the most important pioneering authors). A much-ignored highlight of modern comparative anatomy extensively using microscopic anatomy is the handbook of Bolk et al. (1931–1939). This period was followed by the predominance of electron microscopy (TEM, REM), but this technique had its most fascinating results in cytology and histology rather than in microscopic anatomy (Krstic, 1984).

We are presently witnessing a further methodical step into molecular biology, classical descriptive and functional embryology often being considered outdated. Certainly, modern developmental biology and developmental genetics open many new aspects of ontogeny underlying histology and microscopic anatomy (LeDouarin, 1982; Gilbert, 2014; and many others). Their application on problems of homology and phylogeny resulted in the concepts of EvoDevo, which still need a broader empirical basis, however.

Traditional histology and light microscopy still allow specific insights which are not reached by the new approaches. MicroCT and new microscopic techniques allow us to study anatomical structures, sized a few millimeters, very effectively. However, I am convinced that in the future only intelligent combinations of all available techniques will enhance morphological research. Before this background, it is to be very welcomed that the authors of the present volume fill a deplorable gap in our knowledge of primate anatomy.

Wolfgang Maier
Professor Emeritus, Department of Biology,
University of Tübingen

Preface

We currently lack a broad comparative perspective on skeletal anatomy of newborn primates. With this book we aim to bring together new and existing anatomical information on this fascinating life stage of primates. This goal is challenging because the age since conception is frequently unknown. Thus, another aim is to describe samples in a diversity of primate species in order to recognize patterns of morphological maturity in newborn primates. Most living primates are less frequently available to study compared to most other mammals. This is because of their longevity and sometimes based on their conservation status. Still, the slowest-reproducing primates, the hominoids (humans and apes), are the best studied at the newborn age. We begin most chapters by reviewing existing information on newborn hominoids. To avoid repetition in subsequent chapters, we typically do not cite the most comprehensive sources, the works of Scheuer and Black (2004, on humans; and see Cunningham et al., 2016) and Schultz (e.g., 1940, 1941, 1944, 1969, regarding great and lesser apes). Subsequently, a primary goal of each chapter is to expand our knowledge base to other, less well-studied primates.

We alert readers to special topics at the end of each chapter in which a more detailed discussion is provided (the "Box" series). At the end of the book is a color atlas providing views of the entire skeleton for six primate species. Throughout the book, original illustrations are provided to convey information on bone morphology, usually based on multiple specimens. Except where another source is identified, all drawings are by Tim D. Smith.

The reader should note certain conventions throughout the text. When possible, we use common names for groups of primate species (e.g., families). However, when anatomical details apply to specific taxa only, we use Latin names for primate species or genera. In addition, in an abbreviated reference to age, we make reference to postnatal age in days. For example, specimens known to die one day after their day of birth are denoted as "P1" individuals, and so on.

This project is only possible because of the specimens saved and preserved by zoological and research institutions over two decades. Below, Table I summarizes the primates referred to throughout the text by common name and genera. The book references prior studies and describes new findings based on 156 newborn primate specimens (details in Table 9.4). Numerous other specimens consulted in selected passages are at earlier (fetal) or later subadult stages. Finally some museum samples were used to develop figures.

The rarest specimens we used were obtained after natural deaths in captivity, and we thank the veterinary and research staff of the following institutions for their great care in saving the samples: Cleveland Metroparks Zoo, Dallas Zoo, Dallas World Aquarium, Duke Lemur Center, Dumond Conservancy, Gladys Porter Zoo, and Philadelphia Zoo. We are also grateful to the collection managers and curators of the American Museum of Natural History, Carnegie Museum, and National Museum of Natural History for allowing us to CT scan specimens. Finally, our work benefited from CT scans of a 5-day-old specimen chimpanzee (*Pan troglodytes*) from the Digital

Table I Common names of primates with corresponding Latin nomenclature

Common names	Corresponding genera referenced in text
Hominoids (apes and humans)	
hominids (humans and great apes)	*Gorilla, Pan, Pongo, Homo*
lesser apes	*Hylobates*
Old World monkeys	
baboons	*Papio*
colobus	*Colobus*
guenons	*Cercopithecus*
langurs	*Trachypithecus*
macaques	*Macaca*
mangabeys	*Cercocebus, Lophocebus*
New World monkeys	
capuchins	*Cebus, Sapajus*
howling monkeys	*Alouatta*
marmosets	*Callithrix, Cebuella*
owl monkeys	*Aotus*
saki monkeys	*Pithecia*
squirrel monkeys	*Saimiri*
tamarins	*Leontopithecus, Saguinus*
titi monkeys	*Callicebus*
Tarsiers	
tarsiers	*Tarsius*
Lemurs	
bamboo lemurs	*Hapalemur*
dwarf lemurs	*Cheirogaleus*
mouse lemurs	*Mirza, Microcebus*
ring-tailed lemurs	*Lemur*
sifakas	*Propithecus*
sportive lemurs	*Lepilemur*
true lemurs	*Eulemur*
Lorises	
bush babies	*Galago, Galagoides, Otolemur*
lorises	*Loris, Nycticebus*
pottos	*Potto*

Morphology Museum of the Primate Research Institute of Kyoto University (hereafter, KUPRI). This work would not have been possible without grant support to each of the coauthors from the National Science Foundation (BCS-0959438; BCS-1231717; BCS-1231350; BCS-1728263; BCS-1830894; BCS-1830919). We thank numerous collaborators for working with us on previous manuscripts that provided much of the newer information discussed in this book. We also thank Christopher Bonar, Anne Burrows, Brigitte Demes, Leanne Nash, and Kristen Prufrock for making specimens available for study. We thank our many student collaborators, too numerous to mention, but especially Andreana Cunningham, Jayna Cunningham, Serena Engel, Ellen Fricano, Gladys Galdamez, Brianna Harvey, Naomi Levin, Catherine Llera, Matt McMahon, Michelle Millen, Hannah Rader, Gabriel Hughes, Kelsey Paddock, Dominic Violi, Sarah Zaleski, and Larissa Zeigler, who joined us in our efforts to understand specific skeletal and dental regions in newborn primates. A special thanks to Mark Mooney, Magdalena Muchlinski, Liza Shapiro, and Holly Smith for their comments on earlier versions of some of the chapters. We thank Ilaria Tassistro and Jenny van der Meijden at Cambridge University Press for guidance throughout the preparation of this book, and Wolfgang Maier for writing the Foreword. Finally, we are thankful for our families. In particular, our children provided an inspiration for this book. We often miss those early days.

1 Introduction

Many of life's greatest challenges occur at an early age. Subadult years are not just challenging – they are costly and hazardous. It is a costly time because growth has metabolic costs, and faster growth or the development of highly specialized anatomical systems incurs additional metabolic expenses (e.g., Arnott et al., 2006). The hazards of the subadult years are abundantly clear when looking at age-biased mortality rates of prey species (e.g., Adams et al., 1995). Predation provides an obvious explanation for why subadulthood is so very hazardous, but there are other factors. Life expectancy is age-dependent: even captive mammals have a notable increase in life expectancy by surviving 1 year (Kohler et al., 2006). In this setting, with predation controlled for, those within the first week of life are especially at risk of death regardless of whether they are frequent prey species (e.g., small primates) or predators (e.g., carnivorans). Subadulthood, and especially infancy, is inherently risky.

As helpless as infants might appear to be, they have adaptations to intense selection pressures. Not all infants are entirely helpless, but it is the norm for primates on their day of birth. However, the degree and duration of dependency subsequent to birth vary immensely according to behavioral studies (see Chapter 11). Yet anatomical variation, which may reflect this, is understudied.

1.1 Development: Assessing Stages and Ages

Broadly, newborn mammals vary in maturity. Some species are notably altricial, that is they are born underdeveloped, with eyes closed, insufficient fur to regulate body temperature, and insufficient muscle mass to locomote. Marsupial pups are the most extreme examples of altricial infants, and are born at an embryonic stage in which placental mammals remain in the uterus (Smith and Keyte, in press). Other species are precocial. These mammals can locomote soon after birth, if imperfectly at first. Ungulates and cetaceans are excellent examples. Primates are generally considered precocial based on certain advanced somatic traits (especially neural), relatively long gestations, and other contrasts to altricial mammals (Portmann, 1941; Martin, 1990). However, primates vary in the pace of somatic maturation and locomotor development and some are born less somatically mature than others (Crompton, 1983; Nicolson, 1984; Grand, 1992; Smith et al., 1994; Atzeva et al., 2007; Rosenberg and Trevathan, 2015).

Multiple approaches have been used to define subadult stages of development, some of which are observational, and some strictly physiological or morphological. "Juveniles" are of interest to primatologists because humans, like all primates, have a prolonged period of dependence on parental care. "Infants" and "newborns" are notably more dependent and less mobile than older subadults. A careful study of the way in which living non-human primates care for their young may suggest the ancestral patterns of care-giving. While some definitions of juvenility are behavioral in nature, we are left with morphology alone to assign developmental status to extinct primates. Morphologically, the infant and juvenile stages were discretely defined by Adolph Schultz (1960, 1969) based on dental criteria (details of these stages are discussed in Chapter 2).

The term newborn, or "neonate," would seem easy to apply if one knows the date of birth. Neonates are often assigned based on chronological age that falls within a range. For humans, the first 4 weeks are commonly considered the neonatal period (e.g., Scheuer and Black, 2004). For non-human primates, there are no well-established conventions. Because primate species vary in the duration of life stages (Schultz, 1969), it cannot be assumed that the duration of the neonatal period is similar across species, even between species of similar sizes. Because of vast differences in longevity among humans and other primates, we adopt a shorter interval of postnatal time: the time from day 0 (P0) to 1 week old (P7); this is a similar timeframe to some recent studies of non-human primates (Godfrey et al., 2004; Smith et al., 2015; 2017a).

In one important way, the duration of the neonatal stage is irrelevant. We should not expect perfect morphological similarity among primates because rates of development are potential adaptations. In other words, we expect that selection may act on primates with differing diets, environments, or even differing social systems, perhaps in the service of a somewhat more precocious locomotor competence, or as a means for diverting metabolic costs toward a particular specialized organ or morphology. The significance of the term "neonate" relates to the earliest postnatal morphology, regardless of how long the infant remains at this stage.

One final consideration is that in the practice of identification, a neonate can be a moving target of sorts. Two neonates may not be at similar *stages* of development even if precise postnatal *age* is known. One reason has a basis in development. Even if postnatal age is known, age since fertilization may or may not be known. Moreover, neonates may be born prematurely, or preterm. Thus, an infant may be of unknown precise age, and could appear to be either late fetal or neonatal. Perinatal specimens (or "perinates") can provide useful clues of the adaptions of a species at birth, but such information must be assessed with caution.

1.2 Special Challenges of Studying the Neonate

Research on juvenile primates, especially prenatal and newborn specimens, entails special challenges. First, the precise age of a perinate, that is, the duration in days since fertilization, is often unrecorded. Perinatal deaths may be due to complications in pregnancy, premature birth, and trauma during parturition, among other factors. Thus, in our endeavor to describe the neonate, we must exclude specimens that are born prematurely. In primates the following may signify a fetal maturational status: specimens with too little body fur (bearing in mind that some species are born poorly furred), small heads compared to other perinatal specimens of the same species, or small overall body length (sitting height) compared to other perinatal specimens of the species (Smith et al., 2015). Amniotic membranes also suggest that stillborn animals may have been aborted or removed for veterinary reasons prior to full gestation. Most species described in this book were available in samples of two or more specimens. In response to small sample sizes, we frequently defer to description at the generic taxonomic levels. And although both sexes were examined in most cases, there is no attempt to identify sexual dimorphism in somatic dimensions. Although neonatal dimorphism occurs in at least some primates, it has only been identified using very large samples (e.g., Blomquist and Williams, 2013).

A second factor is the relative rarity of a slowly reproducing species, many of which are rare and endangered. Study of the neonatal stage adds additional complications because cadavers must be acquired and staged (as discussed above). In sum, this book reviews data on specimens studied over the course of many decades (Schultz, 1935; Sirianni and Swindler, 1985; and see the comprehensive survey by Smith et al., 1994), as well as newly available material (Smith et al., 2015).

A final factor is the small size of neonates of the smallest species. As such, the smallest forms are difficult to study via gross observation or conventional radiography. This creates methodological challenges which have traditionally been answered using histology, and more recently micro-computed tomography (μ-CT).

1.3 Advantages and Disadvantages of Methods for Studying the Newborn

Gross observation and conventional radiography have previously provided detailed knowledge on newborn osteology and dentition of relatively large primates (Schultz, 1935; Sirianni and Swindler, 1985), and even some smaller species that are relatively advanced in terms of ossification (Niemitz and Sprankel, 1974). For small specimens (e.g., fetuses or small primate species), histological methods such as serial sectioning or whole-mount preparations have been used to study bone ossification patterns or tooth cusp mineralization (e.g., Tarrant and Swindler, 1973; Phillips, 1976; Newman and Hendrickx, 1984), while gross observation has usually sufficed for detecting dental eruption and gingival emergence (see Smith et al., 1994; Godfrey et al., 2001).

Radiological methods are known to underestimate the amount of mineralized tissue to some extent. This relates to limits in resolution, but radiographic detection may be especially problematic with subadults if structures (bones, teeth) have only just initiated mineralization. Thus radiographic techniques have tended to underestimate the amount of mineralized tissue (Garn et al., 1959; Kraus and Jordan, 1965; Winkler, 1995; Swindler, 2002). In the past two decades, the ability to study bones and teeth via radiographic methods has improved to allow researchers to examine microanatomical features of adult and subadult bone, even in small samples (e.g., Rossie, 2006; Ryan and Krovitz, 2006). In particular, μ-CT has far finer resolution (i.e., the capability to allow close objects to be distinguished from one another) than conventional radiography, and provides the raw data that allow three-dimensional reconstructions with specialized software.

Histology and μ-CT each have advantages. The level of detail on bone that can be gleaned from histology far exceeds that which can be detected by radiographic observation, although the discrepancy in resolution is narrowing (Reinholt et al., 2009). However, interpretations from histology must be made with the knowledge that histological methods usually introduce distortions or artifacts (Box 1.1). Dehydration in order to stain and otherwise prepare slides shrinks tissues, and can produce artifactual tears or folds (Rolls and Farmer, 2008). Bone is typically less affected by shrinkage than other tissues, but thin bony projections may shrink more than more robust regions (e.g., cortices), and thinner bone is more prone to artifactual bending following decalcification (DeLeon and Smith, 2014). Teeth are especially difficult to study histologically after decalcification due to fragmentation of enamel (Box 1.1). These drawbacks can be minimized by a careful comparison of histology and μ-CT from the same individual specimens (see Box 1.1), allowing identification of specific artifacts (e.g., DeLeon and Smith, 2014; Smith et al., 2016). Effects of shrinkage and distortion are far reduced when using μ-CT or gross observation. However, most fixatives introduce some shrinkage, and some distortion can occur as subadult bones or dental arcades dry for gross examination. Nonetheless, μ-CT has proved successful for identification of perinatal and even fetal bone morphology (Box 1.2; e.g., Ryan and Krovitz, 2006). In the context of fetal and perinatal specimens, an important consideration is selecting the correct parameters for scanning, because immature bone is less mineralized, and best visualized with lower energy levels. However, this also means that variation in the rate of ossification can produce different results among species. Therefore, identification of optimal CT parameters for studying subadult samples is difficult and essential prior to a study.

1.4 The Newborn Body and Tissues

The newborn primate bears superficial resemblance to adults of the species, while differing in proportions (e.g., head to body size, or forelimb to hindlimb). In the terminology used to refer to anatomical parts and locations within parts, all non-human primates are treated as quadrupeds (Box 1.2).

The "body plan" concept pertains to developmental mechanisms, which are briefly considered in the next chapter. Through observation, one recognizes the body plan as tissues begin to differentiate; for most of the skeleton the morphology is first recognizable in the form of cartilage. By the newborn stage, the cartilaginous and other components (membranous bones and teeth) of this body plan are for the most part differentiated (although chondrogenesis and osteogenesis continue postnatally – Chapter 2). In all mammals, the skeletal system has more separate components in subadults than in adults (Figure 1.1). This is because joints that normally ossify or may fuse as adulthood approaches have a transient function in growth. As such, epiphyses and sutures are more prominent in our discussion than in traditional osteology books.

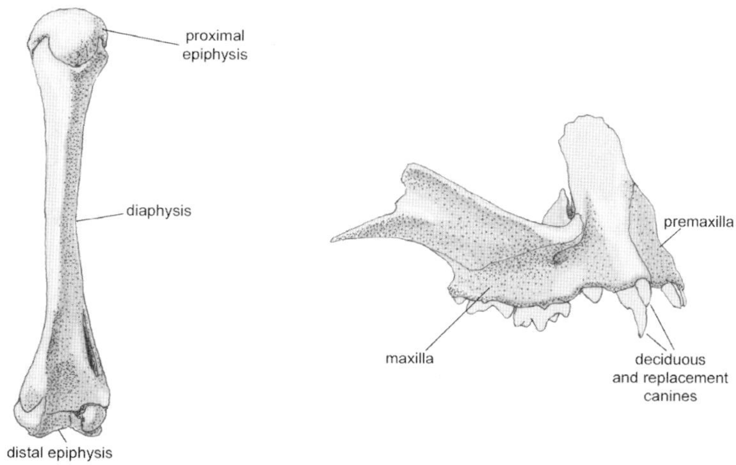

Figure 1.1 (Left) Long bone of an immature cat, showing three separate elements. The diaphysis (or shaft) is the central osseous element, and the bone is capped at either end by epiphyses. The elements are joined by transient cartilaginous joints (synchondroses), the epiphyseal growth plate. (Right) Midfacial skeleton of an immature cat, showing the mixed dentition. The transient function of the deciduous teeth is emphasized by the incipient replacement teeth.

In addition, a more nuanced discussion is merited for subadults. Most mammals form two complete sets of dentition, with different functional types. Subadult mammals rely for a portion of their life on deciduous, or milk, teeth (Figure 1.1). In this book, the variable rates at which the primate skeleton and dentition develop (e.g., ossify, erupt) are explored and discussed in a functional context.

Box 1.1 Methods for Studying Skeletal and Dental Anatomy in Perinatal Specimens

Historically, the study of the skeleton or dentition in late prenatal and perinatal primates has been possible using gross or radiographic observations, but only histology provided highly detailed information. The most limiting factor for the former was one of resolution. In recent decades, micro-computed tomography (μ-CT) has provided increased resolution to allow visualization of a microscopic level of detail.

Below, a newborn tarsier (*Tarsius syrichta*) is shown in a cross-sectional view at the level of the first molar (M1). The specimen was CT scanned first, and subsequently histologically sectioned. Note that the CT slice (a) is from a "resliced" series, produced by rotating a reconstructed skull volume in the same plane as histology (b), producing matching views. Note that μ-CT can faithfully provide many details of skeletal and dental cross-sectional morphology, and can yield detailed 3D reconstructions (c). However, histology still remains a gold standard for tissue-level observations. The limitations of CT relate to degree of mineralization and whether mineralized elements reach at least the level of resolution in their thickness (d).

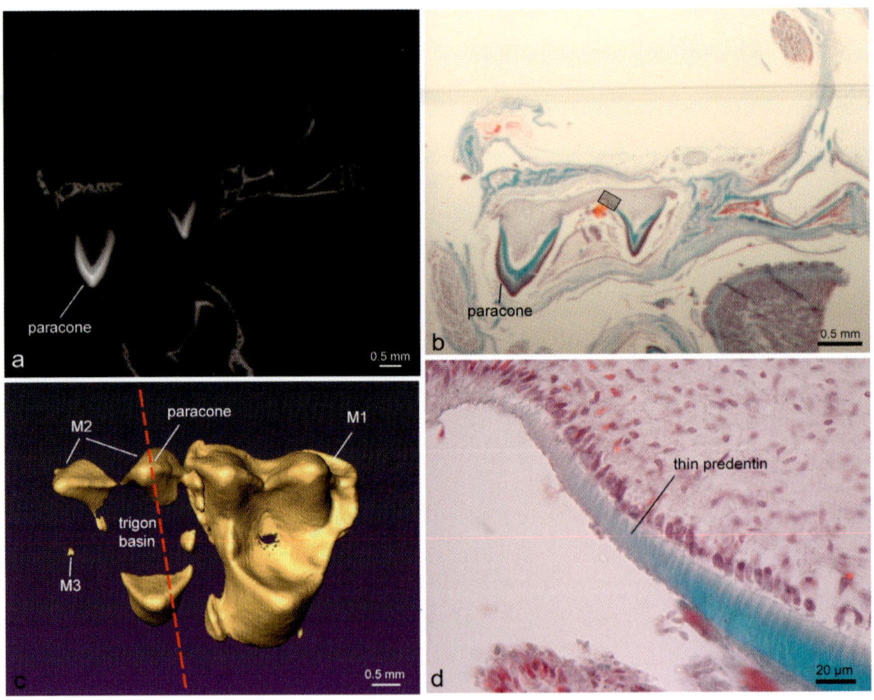

However, any histological method produces some degree of distortion. Below, a newborn lemur (*Eulemur collaris*) is shown in the same manner as the tarsier. In this case, we can show certain artifacts that may occur. First, note the wider cervical dimensions of the crown of dp2 in the CT slice (e) compared to the matching histology (f). Higher magnification views reveal that elements within the confines of the dental follicle collapse inward. Specifically, the mineralized tissue separates from the internal enamel epithelium, creating a large gap around the shrunken crown (g). Also note that enamel, once fully mineralized, shatters easily in decalcified specimens (h), creating an artifactual "enamel space" (Nanci, 2007). mt, maxilloturbinal; ns, nasal septum.

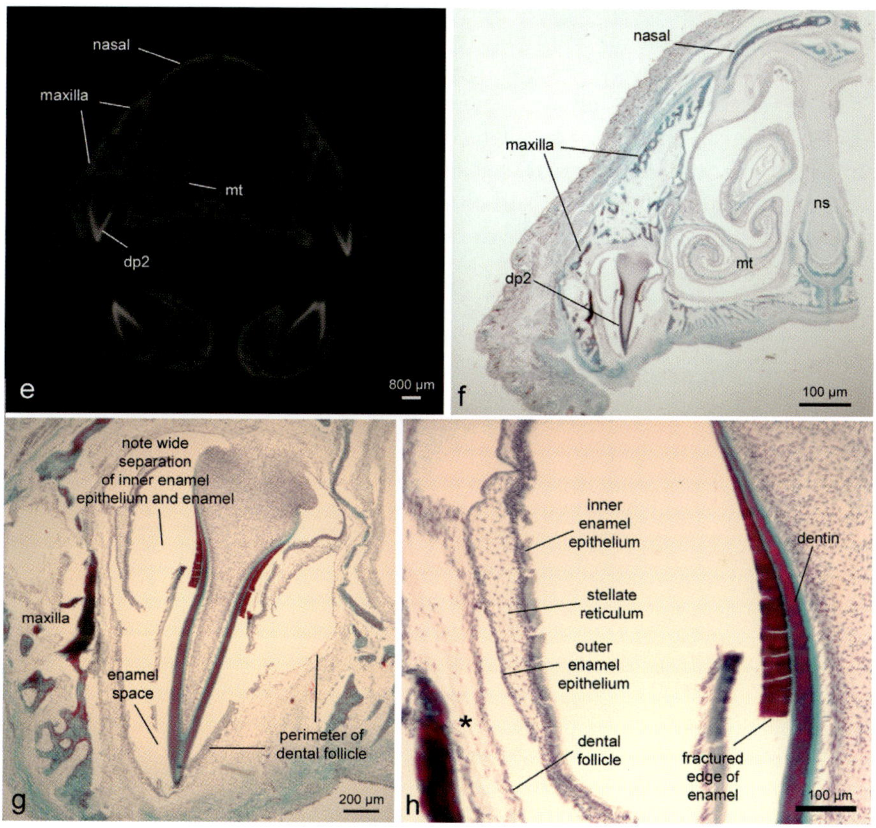

Box 1.2 Fetal Loris

Below, a fetal slender loris (*Loris tardigradus*) is shown grossly (left) and in a CT reconstruction of the skeleton (right). Fetal primates emerge from the embryonic period with certain characteristics that contrast our Order with other mammals: such as relatively large brains compared to their faces, relatively (or extremely) large eyes, and well-developed digits on hands and feet.

Anatomy of non-human primates may be discussed with a terminology that applies broadly to mammals. They have cranial (toward skull) and caudal (tail) ends of their body (regardless of whether they have tails). They have ventral and dorsal surfaces. These terms take precedence over the more human-specific terms superior and inferior, and other terms (anterior, posterior) that change meaning when comparing quadrupeds to bipeds.

The skeleton is divided into axial (skull, vertebrae, ribs) and appendicular (limb, limb girdles) parts. The CT reconstruction of the fetal loris reveals similarities to other primates at the same stage and even at birth: epiphyses of long bones remain unossified and sutures remain widely patent. In the perinatal period, the "fetal" growth trajectory may continue for some tissues/structures.

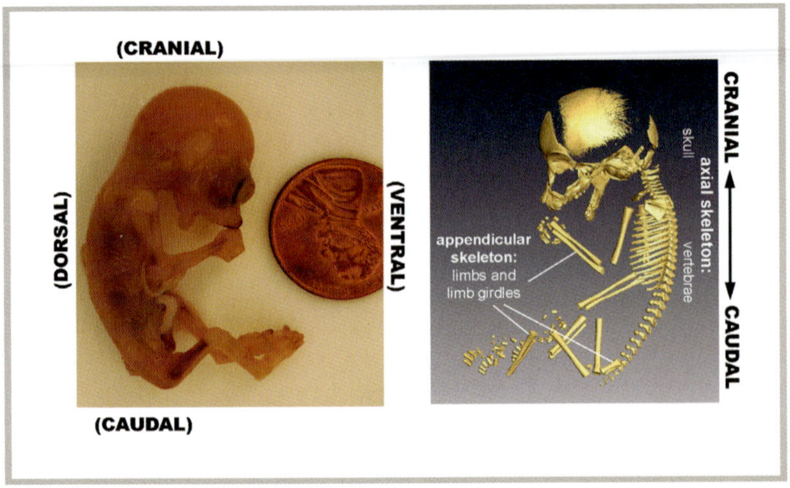

2 Primate Development and Growth

Observations on primate growth and development have advanced our understanding of the evolutionary roots of what may be the most important human adaptive trait – an enlarged brain (Jerison, 1975; Gould, 1977; Lieberman, 2002; Leigh, 2004). Over the past five decades, the manner in which we evolved our brains and our unique craniofacial profile has been explicitly linked to evolutionary trends in ontogeny. The newborn is an important stage for understanding developmental trajectories, because it represents a transitional point in time between prenatal and postnatal development. It is a static moment that borders a time of complete dependency and a time in which the subadult either rapidly or gradually achieves independence.

Strategies for rearing young primarily rest on primate mothers. In a real sense, primate mothers have evolutionary means for enhancing adaptive traits of their young, with two crucial strategies: evolutionary alteration in duration of gestation or in the timing of weaning. The anatomy of the newborn primate is at once a potential measurement of maternal prenatal investment as well as an indication of the extent and duration of infant dependency. Throughout this book, the maturational state of the newborn is discussed in both macroscopic and microscopic (tissue-level) perspectives. This chapter provides certain basic concepts of vertebrate development and histology (readers are referred to Hall, 2015 for further details).

2.1 The Prenatal Body, Whole and Parts

The terminology for prenatal stages reflects major developmental processes: differentiation and growth. The two phases of prenatal development, embryonic and fetal periods, differ in the balance of these two processes. The embryonic period is dominated by differentiation events, with some growth (Carlson, 2014). Less growth occurs as we track time backwards toward fertilization. Indeed, in the first mitotic event (cleavage) the volume of the one-cell zygote is more or less parceled into a two-cell zygote of the same volume. In contrast, in the fetal period there is far more growth and relatively less differentiation of tissues/organs. In a broad sense, vertebrate embryos have strong similarities in form, while fetal-staged vertebrates have many defining features of an Order (Figure 2.1). To be certain, superficial distinctions are visible among embryos, thought to reflect differential growth of body regions; even more differences exist in the developmental mechanisms that produce body regions and parts (Richardson et al., 1997). Yet the basic body plan of vertebrates includes prominent neural structures (brain and sensory placodes or their derivatives), pharyngeal (or branchial) arches, somites, and limb buds (in tetrapods) that can at least be recognized across disparate taxonomic groups (Figure 2.1). These structures render the embryo quite alien in relation to the postnatal appearance of the species, at least until the embryonic period nears its end.

Another consideration bearing on prenatal form concerns anatomical terminology. Whether referring to the prenatal form of humans, non-human primates, other mammals, or other vertebrates, the terminology becomes restricted compared to the system used to refer to postnatal form. This is because some anatomical terminology prioritizes language that relates to positional behavior. Anterior is synonymous with cranial in quadrupeds, not so in humans or birds (because the latter are bipeds). Such terminology is of limited use with the embryo or fetus. Thus, prenatal locational terms emphasize primitive body surfaces, while "directional" terminology (anterior–posterior; superior–inferior) is best avoided, with the exception of the skull in primates. For quadrupeds, dorsal indicates the back of the animal, while ventral indicates the underside. These are easily located in the embryo as the side to which the spinal column (dorsal) and the abdominal wall (ventral) are positioned. All quadrupeds have

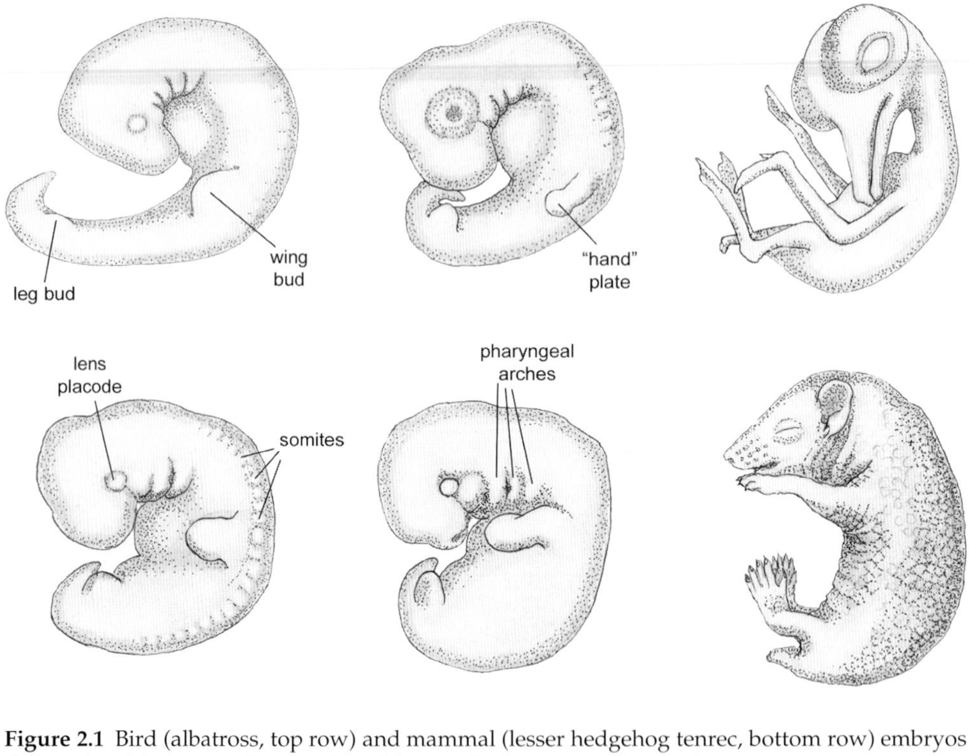

Figure 2.1 Bird (albatross, top row) and mammal (lesser hedgehog tenrec, bottom row) embryos (left and middle, top and bottom) and fetuses (right, top and bottom).

a "head" (cranial or cephalic) and "tail" (caudal) end. These parts are also clearly recognizable throughout most of prenatal development (see Box 1.2). Terminology relating to the limbs is identical in the fetus and postnatal tetrapod of any age (proximal and distal; medial and lateral); however, it will be important to know "primitively ventral" surfaces of limbs (Chapters 7 and 8) to fully understand postnatal form and function.

2.2 Morphogenesis of Prenatal Form

Embryogenesis (also known as organogenesis) shapes the body. Maturation of tissue types occurs in this phase, such that not just overall form but basic functionality of structures (muscles, bone, brain, etc.) is recognizable once embryogenesis is complete. After a series of stages, the embryonic body plan is complete, and the fetal period of development ensues. In the fetus, growth produces more observable changes in the size of the fetus and its body parts. Growth may occur by an increase in cell number, or hyperplasia. This occurs at all stages of development, but while mitosis produces two daughter cells from one cell, the daughter cells themselves are smaller than the parent cell during the early embryonic period. The embryo does eventually grow in size, but the extent of growth pales compared to fetal trajectories. Hypertrophy, or increase in cell size, is a far more potent growth mechanism at the cellular and tissue level, and this mechanism amplifies fetal growth trajectories (Tanner, 1990). Hypertrophy occurs as cells differentiate, and acquire cytoplasmic elements related to cell function. A final mechanism for size increase occurs as connective tissues mature and produce extracellular matrix (see below).

When growth trajectories *diverge* among body regions, this can produce changes in overall form – a major mechanism for evolutionary change. Change that is realized through a shift in growth rates is referred to as heterochrony, a process that can be recognized prenatally (Figure 2.2) or postnatally.

Because animals develop at drastically differing rates, numerous systems have been devised for prenatal staging of embryos, some specific for birds or mammals (e.g.,

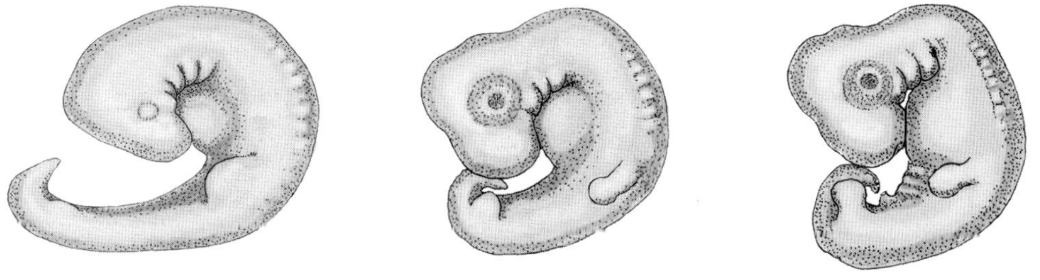

Figure 2.2 Embryonic limb growth in a bird, an example illustrating heterochronic growth *within* a species during limb formation (albatross; based on Rehorek et al., 2016). In the albatross, the embryonic leg bud appears first (preceding the above stages). Subsequently, growth of the wing bud overtakes that of the leg bud, shown in an intermediate stage above. In a still later stage shown at the right, growth and morphogenesis of the leg bud has accelerated such that limb sizes and shape are more similar.

Werneburg et al., 2013). Even among mammals, specializations (e.g., wings of bats) require a special consideration in staging (Nolte et al., 2009). Fortunately, many non-human primates have been staged using criteria for human embryos, as developed by Streeter (1920) and modified by O'Rahilly and Müller (1987, 2010). External features of the limbs and face appear in a similar sequence across a diversity of primates.

2.3 Prenatal Stages of Development and Morphogenesis of Body Tissues, Parts, and Regions

In human development, the end of the eighth week (56 days post fertilization) marks the end of the embryonic period. There is no one crucial characteristic that differentiates a 56-day embryo and a 57-day fetus, but the two periods of development differ in terms of the preponderance of events at the cellular and tissue level (i.e., differentiation versus growth, as discussed above). Before reflecting on the two processes as they relate to the skeleton and dentition, we should look at the continuum of development. In early prenatal development, skeletogenesis and odontogenesis proceed in a similar manner across mammals. First, some embryonic features indicate that skeletogenesis is imminent, such as a grossly observable swelling, like a pharyngeal arch or limb bud. At the outset, the appearance of a limb bud or facial prominence reflects mitosis within the "core tissue" of the structure, and an influx of migratory cells; this tissue proliferation results in the grossly visible outgrowth.

The genesis of bony elements begins after the initial appearance of limb buds or facial prominences. Skeletogenesis occurs in a sequence that spans time and morphological space. For example, limb bones form in proximal to distal direction over developmental time. This appears to be conserved among tetrapods. Primitively, the ossification sequence may also follow suit. However, ossification of limb elements is not tightly coupled to the same spatial sequencing in amniotes (Fröbisch, 2008).

Another basic similarity among tetrapods is the basic body plan, or organizational template, in which components of the axial skeleton (head, vertebral column, and other skeletal elements of the neck and thorax) begin to differentiate first, and components of the appendicular skeleton follow, with some overlap. We have gleaned important details of the mechanisms that govern the earliest phases of morphogenesis from experiments on animals. Many mechanisms are broadly conserved among most tetrapods. Birds are similar to mice in limb formation, especially in early processes, e.g., in the role of apical limb ectoderm in limb outgrowth (Hall, 1995). During fetal development, mechanisms of growth (e.g., cellular hypertrophy) become more potent, and the proportions of the body gradually begin to resemble a particular species, although not in its adult proportions (Figure 2.1).

2.4 Postnatal Stages of Development

An increasing number of field studies of living primates over the past half century have raised interest in primate ontogeny (e.g., Pereira and Fairbanks, 1993; Bolter and Zihlman, 2007). Accordingly, life stages may partially be based on developmental milestones that may be observed. For example, primates may be regarded as infants prior to weaning. Juvenility follows, the time between weaning and sexual maturity (Pereira and Altman, 1985; Janson and Van Schaik, 1993). The phrasing can vary. To Pereira and Altman, an infant is a subadult that is "directly dependent on its mother and other possible caretakers for survival" (1985, p. 225) and juveniles are "prepubertal individuals that are no longer dependent on their mothers" (or other adults) for survival (1985, p. 236). In other words, these life stages reflect the level of subadult dependency.

For our purposes, life stages reflect maturational status. Dental staging is commonly used to arrange primates into maturation groups (Schultz, 1960, 1969). The earliest stage is prior to alveolar eruption of permanent teeth, and Schultz grouped them as "infants." Schultz further elaborated that infants have at least some and at most all deciduous teeth (1960, p. 342). The juvenile period spans from eruption of first to eruption of last permanent teeth (see Schultz, 1960, who also subdivided juveniles into two substages). The adult period is the last and longest, from eruption of last permanent teeth through senescence.

Dental stages create imperfect comparisons because they encompass a range of chronological age across primates. For example, all known strepsirrhines have at least some erupted teeth at birth, whereas many newborn anthropoids have none (see Chapter 5). And yet, because chronological age of skeletal samples is frequently unknown, there may be no better option. When chronological ages *are* known, a comparison of staged samples highlights differing rates of maturation (e.g., Rossie, 2006). This relates to a central question explored in this book: how do primates compare to one another at birth in maturational status?

2.5 Differentiation and Growth of Mineralized Tissues

2.5.1 Skeletogenesis

When skeletal or dental structures are first identifiable, they are composed of a primitive tissue called mesenchyme (Box 2.1), which is most abundant during the embryonic period (e.g., Carlson, 2014). It is conceptually helpful to regard embryogenesis as a developmental process that is not tied to our concept of "staging" the prenatal organism as a whole. Morphogenesis of many structures occurs in a sequential, iterative manner such that late phases of differentiation are also occurring at the same time that other tissues are beginning to differentiate. Because the early development is dominated by differentiation events, most differentiation events do indeed occur during what most would call the embryonic period. However, mesenchyme persists even after the body "template" is established (see Chapter 4, e.g., regarding nasal capsule morphogenesis). In this way, mesenchyme serves as a pool for continued differentiation events throughout prenatal development.

The first indication of skeletogenesis occurs internally, in the form of a "mesenchymal condensation." For example, during outgrowth of limb buds the bony elements of the limb develop as condensations (i.e., areas where mesenchymal cells associate more closely in space) for individual bones, in a proximal to distal sequence. Mesenchymal cells within the condensation become more adherent to one another, and narrow their morphogenetic potential as they differentiate into chondroblasts (Hall,

2015). Cranial bones and cartilage undergo similar transitions (Box 2.1). Each mesenchymal condensation precedes a chondrified element for a particular bone. Cranial skeletogenesis is complicated in vertebrates by the novel mesenchymal cell population derived from neural crest, the "ectomesenchyme" (Gans and Northcutt, 1983; McBratney et al., 2008). This additional source of chondrogenic and osteogenic mesenchyme fuels the complex sequence of skeletogenesis.

As a body tissue, mesenchyme is not a subcategory of one of the four mature tissue types (i.e., epithelial, nervous, muscle, and connective tissue). In its components, mesenchyme most closely resembles certain connective tissues in that it is composed of cells and cellular products (in this case, a fluid called ground substance, but few visible fibers). In a sense, mesenchyme is a partially differentiated tissue (ready to differentiate into many different mature connective tissues or other tissues). The mesenchymal cells, not yet fully functional, are nonetheless responsive to cellular signals. When looking at mesenchyme through a microscope, condensed mesenchyme is easily located because the cell nuclei are grouped in dense clusters or islands (and nuclear stains emphasize this in histological preparation – see Box 2.1). The aggregation occurs in response to chemical signals from adjacent tissues (Thorogood and Hinchliffe, 1975). At the physiological level, the cell behavior change is matched by different activities: cells within the condensation communicate more frequently. Mesenchymal cells that will form bone (osteogenic mesenchyme) or cartilage (chondrogenic mesenchyme) exhibit specialized proteins in the cell membrane called gap junctions, which facilitate cell–cell exchanges. Experimentally blocking these exchanges prevents differentiation from occurring (Loty et al., 2000).

There are four major types of mineralized tissue of the body: bone, cartilage, enamel, and dentin (odontogenesis is discussed in Chapter 5). Mesenchymal condensations that precede bone progress by direct or indirect routes. Intramembranous bone forms within mesenchymal tissue, and in association with a "membrane" (i.e., a collagenous fascia). Mesenchymal cells differentiate into bone-producing cells (osteoblasts) at these sites, producing calvarial bones, bones of the superficial facial skeleton, clavicle, and distal tips of distal phalanges.

In "indirect" forms of ossification, bone *replaces* an existing tissue, such as cartilage, marrow, tendons, or ligaments (Hall, 2015). Replacement of cartilage by bone, endochondral ossification, is the most common of these. At the outset, "templates" for bones or parts of bones are produced via differentiation of mesenchyme into a pool of chondrogenic (cartilage-producing) cells. The result is hyaline cartilage, an avascular supporting tissue (Box 2.1), and the cartilaginous elements that are formed resemble the individual bones that will replace them, if not in the proper proportions.

An important distinction to make is that intramembranous and endochrondral ossification are osteogenic processes, and the tissue produced may be named dermal and chondral bone, respectively. This distinction allows us to refer to individual bones based on origin, while acknowledging that they may be of *dual* origin. In addition, Hall (2015) distinguishes a subtype of chondral bone, "perichondrial" bone. This forms within the capsular membrane of cartilage, the perichondrium. The perichondrium, as well as the capsular membrane of bone (the periosteum), are more than protective and nutritive (general functions of membranous capsules of other body structures). They are also chondro- or osteogenic, in that immature connective tissue cells reside within them. These are a pool for the chondroblasts or osteoblasts, respectively. Denoting the perichondrial subtype of chondral bone will clarify two concepts below.

Most chondral bone in the body replaces cartilage templates, and such bones include most long bones, the bones of the hands and feet (excluding the distal phalanges), the vertebrae, sternum, and the bones of the cranial base. This replacement is a major event, although not the first event, in endochondral ossification. First, the embryonic mesenchymal condensation for an individual bone (or part thereof) differentiates to hyaline cartilage. Prenatally, a part of most chondral bone templates (e.g., the diaphyseal portion in long bones) is replaced by bone; this portion of the bone is termed the primary ossification center. The initial bone production precedes cartilage breakdown, occurring on the peripheral surfaces of cartilage. In a long bone, this bone tissue is produced by the perichondrium surrounding this diaphysis. This perichondrial bone forms a bony collar around the diaphysis (Box 2.2), and the perichondrium may now be called a periosteum. The periosteum transmits blood vessels deep within the cartilage of the primary ossification center, and in this vascularized environment, the cartilage matrix becomes mineralized and the chondrocytes become hypertrophic (enlarged or swollen) and die. The destruction of cartilage spreads toward both ends of the primary ossification center, replaced by bone and marrow tissue. In most chondral bones, a region of the bone remains cartilaginous: this represents a secondary ossification center (in long bones this is the epiphysis, one located proximally and one distally to the diaphysis). Replacement of the secondary ossification center takes place postnatally in most bones. In long-lived organisms, such as humans and many of the non-human primates, endochondral ossification takes years to complete. And even when the secondary ossification center is replaced by bone, a disk or bar of cartilage (e.g., the epiphyseal plate) separates the different ossification centers. By retaining this cartilage, chondral bones can harness a growth potential which bone (as a tissue) does not have.

Additional elaboration of bones occurs even after bones are formed. Generally, this bone is produced on the external surfaces of existing bone, and the process is termed "appositional growth." Bone is produced by osteoblasts that migrate to the bone's surface from the periosteum, and it is laid down on external surfaces. This is the way in which long bones grow in circumference. Mundy (1999) notes this is "probably analogous" to intramembranous bone formation. An additional activity occurs to balance this new bone production, osteoclast-mediated resorption of bone tissue in another region. When coordinated, these additive and subtractive processes can enlarge individual bones (Box 2.3). The addition of new bone can also result in elaborate new parts that extend beyond the initial chondral bone template. In the German literature this growth process is termed *Zuwachsknochen* (Maier, 1987, 1993; Sánchez-Villagra and Maier, 2002); it has also been called appositional bone (Starck, 1967). The new bone produced begins in the perichondrium of existing, developing bone, and then extends outward as a "new" bone surface. For example, the sphenoid bone is mostly endochondral in origin, but the alisphenoid is produced via *Zuwachsknochen* (Maier, 1993), when perichondrial bone emanates outward from the cartilaginous "root" of the alisphenoid (see Chapter 4). Even after a chondral bone ossifies,

Zuwachsknochen is an apt description for novel parts of bones (e.g., some accessory lamellae of nasal turbinals) that emanate from the perichondrium or periosteum to elaborate on the bone that formed from the template. Smith et al. (2016) liken the elements of the cartilaginous template to the foundation and frame of a home, while those parts forming via *Zuwachsknochen* are the additions to the home. This mode of bone production deserves special consideration because it may be especially prevalent in late fetal or early postnatal ontogeny, and because the elements produced in this manner might be neomorphic (Sánchez-Villagra and Maier, 2002), or in other cases epigenetic (Smith et al., 2016).

2.5.2 Properties of Mineralized Tissues

The mineralized tissues of the body are remarkable in ways beyond the mineralization process itself. The type of minerals and extent of mineralization (least in hyaline cartilage) differ among the types, but in all cases the mature tissue has a highly specialized function. In each case, the tissue cannot be produced rapidly (i.e., within days as in epithelial repair), and thus these tissues might properly be regarded as metabolically expensive to produce and maintain. Enamel (the hardest tissue) is not replaced at all – it is a permanent tissue. The only route to maintaining enamel throughout postnatal life is to continuously generate "fresh" enamel near the base of the tooth (as in rodent incisors) or to generate a new generation of teeth (i.e., replacement teeth).

Both bone and hyaline cartilage can be maintained throughout life. In bone, maintenance is achieved via localized wholesale replacement, or remodeling. Whether in maintenance or growth, this process involves the same cells: osteoblasts generate new bone, and osteoclasts resorb existing bone. Once bone cells are trapped in matrix, they are termed osteocytes, and they are no longer capable of creating matrix. In other words, bone as a tissue is not produced internally; it is only added to existing surfaces.

Hyaline cartilage is a remarkable tissue which is generally specialized for support and shock absorption, in a different manner than bone. Another highly beneficial characteristic of cartilage is its potential for *interstitial* (within tissue) growth and repair, in additional to matrix deposits on the periphery. Unlike osteocytes, chondrocytes are capable of manufacturing new matrix within the tissue. The interstitial growth potential of hyaline cartilage is very instrumental in the directional growth observed in the epiphyseal growth plate. It is here that longitudinal growth of bone occurs. Distinct zones of cells reflect both growth and the life cycle of the chondrocyte (Box 2.2). Chondrocyte proliferation and hypertrophy propel the epiphysis away from the diaphysis at both ends of a typical long bone.

2.6 Prenatal and Postnatal Growth

2.6.1 Growth Centers

Our brief introduction to tissue-level distinctions of bone and cartilage allows us to turn our attention back to the whole organism, and its modes of growth. Above, we have emphasized the different growth characteristics of cartilage and bone. Either tissue can generate extracellular (i.e., matrix) volume, but the intrinsic growth potential of cartilage relates to the concept of a "growth center," which in recent years has been mostly relegated to dental and orthodontic literature. Growth centers have an inherent and internal capacity for expansion, although hormonal and epigenetic influences may affect the actual rate of growth. Growth centers (e.g., the epiphyseal plates of long bones or cranial synchondroses) have been contrasted to other locations where the growth potential of hyaline cartilage is lacking. For example, the growth that occurs at sutures is a passive separation in response to soft-tissue expansion or other active growth processes (Baume, 1953; Moss and Young, 1960).[1] Thus, one factor influencing growth is tissue composition: skeletal tissues (bone, cartilage, dentin, enamel) have different growth potentials.

There is a fine contrast between a discussion of how individual bones grow, and how skeletal tissues grow. As a tissue, bone only increases in volume by deposition along existing bone surfaces (appositional growth). Cartilage has the capacity to grow in this way, but also by proliferation and matrix production within existing cartilage (interstitial growth; Box 2.2). Many individual bones harness both growth potentials by maintaining cartilaginous parts of epiphyses after most of the bone has ossified. During the process of secondary ossification, the epiphyseal cartilage is mostly ossified and what was a rather broad zone of chondrocyte proliferation and hypertrophy (Box 2.2b–d) becomes compressed into a cross-sectionally slim epiphyseal plate (Box 2.2e,f). A small reserve zone remains, where chondrocytes divide and provide daughter cells for the proliferating zone. Cells in the proliferating zone divide very rapidly, and this division is at first perpendicular to the long axis of the bone. Through unknown means, the two daughter cells then re-orient along the long axis (Hunziker, 1994). The cells swell as they enter the hypertrophic zone. The basis for longitudinal growth, then, includes hyperplasia, hypertrophy, and increased matrix production. Growth in the epiphyseal plate is regulated by hormones (e.g., growth hormone, insulin-like growth factor 1) and cytokines (e.g., fibroblast growth factor), and is subject to numerous other influences such as temperature (Serrat, 2013) and mechanical forces (Rot et al., 2014). The reader is referred to recent works that discuss the epiphyseal plate over a more extended postnatal span of time (Burr and Organ, 2017; Kjosness and Reno, 2017) to understand the complex interactions among such factors.

Heritable disorders, such as achondroplasia, demonstrate early fusion of synchondroses globally, affecting the linear dimensions of endochondral bones. They also reveal the profound interdependence of development of the cranial base (in which synchondroses reside during development) and the cranial vault and face (Perlyn et al., 2006; Parsons et al., 2015). In addition, growth of individual bones of the skull can be profoundly influenced by the evolutionary loss of synchondroses. This is likely the case with the human pisiform (Kjosness et al., 2014) and may explain extreme midfacial reduction in humans (Baume, 1968; Smith et al., 2017b).

[1] Growth centers are contrasted to "growth sites," such as sutures, which are considered passive locations of expansion during growth (Baume, 1953).

2.6.2 Cranial Growth Mechanisms, With Special Reference to Craniofacial Growth Theory

The capacity of cartilage to generate new cellular and matrix volume *within* the tissue (interstitial growth) is an important contributor to craniofacial form. As with growth plates, the capacity of cranial synchondroses to drive adjacent bony elements apart is the basis for midline elongation across the cranial base; this "pacemaker" role in growth (as contrasted with passive separation across sutures) is experimentally demonstrable (see further discussion in Hall, 2015). Although the midline synchondroses have been studied for many decades (e.g., Scott, 1958; Baume, 1968; Melson, 1974; Wealthall and Herring, 2006; Gonzalez et al., 2013), bilateral synchondroses have received far less scrutiny. Interstitial cartilaginous growth is also implicated in the downward/forward growth of the midface in humans (Scott, 1958; Latham, 1970; Mooney et al., 1989). Cartilage has two potential means of generating movement across space. The interstitial expansion is one potential mechanism. However, during endochondral ossification, chondrocyte mitosis occurs directionally. This, with eventual enlargement of the chondrocytes (hypertrophy), is the driving force that expands bone directionally. Wealthall and Herring (2006) demonstrated that both the intrinsic capacity of cartilage to expand (interstitial growth) and the directional growth that occurs during endochondral ossification are at play in facial growth of the mouse. In the minipig, Dayeh and Herring (2014) found that in the septal cartilage there are regional areas of mitosis that relate to the ossified ethmoid (perpendicular plate) dorsally, and the body of the presphenoid ventrally. They found relatively more mitoses in the ventral region, the portion more in line with the basicranial axis. Because endochondral ossification had already occurred in the perpendicular plate, these observations imply that the septum itself drives facial elongation (although the authors note a small area of septal ossification anteriorly).

The regional complexity in septal growth, implied by mouse and minipig observations, may be a factor that influences variable results in surgical methods to understand the influence of septal cartilage growth on midfacial dimensions. As a driver of midfacial growth, a ligament that connects the anteroinferior base of the nasal septum to the premaxillary bone is posited to pull the midfacial skeleton anteriorly in tandem with growth of the septum ("septal traction model": Latham, 1970; and see Mooney et al., 1989; Foster and Holton, 2016). Resection of this ligament reduces the relative anterior dimensions of the midface in some mammals such as the rabbit (Sarnat and Wexler, 1966). Yet, in other mammals resection does not reduce anterior growth of the midface (e.g., ferrets: Cupero et al. 2001; chimpanzees: Siegel and Sadler, 1981). If the influence of the septum affects facial growth differently among mammals, differential growth rates in different portions of the septum may provide an explanation.

In absence of a growth center, many bones may enlarge by modeling, a selective depositional and resorptive process that causes bone as a tissue to "drift." The result can be envisioned as a gradually expanding circumference of bone (e.g., long bone diaphysis). The process presumably ensures bones do not become too heavy for an animal to support, but could also easily be understood as a means of strategic conservation of a tissue that must be continually remodeled throughout life (Karsenty and Oury, 2010). Both ideas are hypotheses for why bones pneumatize via mucosal invasion (see below). Modeling is not simply a means of bone enlargement. During growth, as later, bone adapts according to internal deformational forces imposed based on biomechanics. Thus, the ultimate bone architecture reflects locomotor and postural demands (Ryan and Krovitz, 2006).

Another factor affecting bone growth is its unique capacity to accommodate adjacent tissues (Box 2.3). Bone modeling rapidly shifts osseous tissue in response to external demands such as compression, for example, from growing muscle masses (creating depressed surfaces, or fossae). Additional complex processes are involved in craniofacial morphogenesis. There, bone may be displaced, that is, one region can grow faster than an adjacent region (again, differential growth), resulting in a change in the relative position of one to the other (Enlow, 1990; discussed further in Chapter 3). Similarly to sutural growth, growth via displacement is not *dependent* on modeling. In a sense it may be regarded as a "passive" means of enlargement, although bone modeling is concurrent to ensure bones maintain proper form individually.

Bone's capacity to adapt to surrounding soft tissues is complex, because in many locations, bone serves as a shared support tissue between two adjacent soft-tissue regions (Smith et al., 2011, 2014). This renders craniofacial growth complex due to sometimes competing influences of adjacent soft-tissue regions. Functional matrix theory, developed by Melvin Moss (e.g., Moss and Young, 1960; Moss and Greenberg, 1967) has provided a framework for untangling this process. Functional matrices are soft-tissue regions (e.g., contents of the orbit) or even organs (e.g., brain) that are housed by bone. Individual bones form "skeletal units" that support functional matrices; a single bone or combinations of articulated bones may serve as a *shared* skeletal unit supporting neighboring functional matrices. A prime example would be the orbital plate of the frontal bone, a skeletal unit that simultaneously serves as the roof of the orbit and the floor of the anterior cranial base in anthropoid primates. The shared skeletal units are theorized to be affected by both functional matrices. One or the other (in this case, the frontal lobe of the cerebrum or the contents of the orbit) may have primacy (Smith et al., 2010, 2014).

2.6.3 Growth Trajectories

Tissue-level differences in the manner in which bone and cartilage grow explain the complex trajectories of growth across ontogeny. For example, if sutures can be conceptualized, in part, as passive "gates" that facilitate expansion of neurocranium (Box 2.4), skull growth can be linked to brain growth. Differences in the extent of skull growth achieved by birth become a crucial part of an index (neonatal neurocranial volume/ adult neurocranial volume), which provides insight into prenatal versus postnatal investment in brain growth. The capacity of epiphyseal growth sites to increase the length of long bones explains the changing proportion of limbs. The contrasting factors that drive growth in different regions of the body explain the different timing of postnatal deceleration and cessation of growth in different parts of the body (e.g., Schultz, 1969). In the chapters that follow, additional growth concepts will be introduced to expand on the basic biological properties of bone and cartilage. According to the focus of our book, these concepts are discussed regarding their importance to functional anatomy and adaptation of the newborn primate.

The nature of growth during embryonic and fetal periods has been studied grossly and for parts of the growing embryo or fetus. In mammals, growth rates are not uniform across time, as indicated by velocity curves of body length data on developing humans (Tanner, 1990; Bogin, 1999a, 1999b). Fetal growth is more rapid than embryonic growth for increasing body length (or height, as we like to refer to it postnatally). In humans, the peak velocity (measured in the number of cm added per unit of time) occurs at about 18 weeks postmenstrual age (~16 weeks after fertilization). Subsequently, there is a rapid velocity decline during the remainder of gestation, followed by a very gradual decline postnatally, until growth stops (Tanner, 1990; Bogin, 1999a, 1999b). Because velocity is measured based on periods of time (days, weeks, or months) it is a poor indicator of rate changes, such as pubertal growth spurts, which also characterize primate growth.

Increase in body dimensions means the body tissues are growing as well. In humans and non-human primates, volumetric increase of cartilage and bone is aptly described using logarithmic regression lines (e.g., Kimes et al., 1988; Mooney et al., 1991). This prenatal growth curve closely tracks tissue-level changes. For example, centers of chondrification are densely cellular, with relatively little matrix, during the late embryonic period (Box 2.1). Fetal cartilage is further differentiated with more matrix, although chondrocytes may appear relatively small. Differentiation of connective tissues such as cartilage is likely related to early phases of exponential fetal growth – matrix production, rather than mitosis alone, multiples growth potential. Postnatally, mature cartilage has an even greater matrix to cell ratio (Box 2.1). Organs grow too, of course, and not all grow at the same pace. The primate brain has accelerated growth compared to most other tissues, with some variation in the extent of prenatal versus postnatal growth (Leigh, 2004). Many other mammals have rapid growth of the brain (although not necessarily prenatally), but primates distinctly delay the growth of reproductive organs, deferred during the prolonged juvenile period.

It should be emphasized that the manner in which primates grow is not the same way that all mammals grow (Kilborn et al., 2002). Primates are distinct from most other mammals in how uneven their growth is across postnatal time. The most distinctive common feature for all primates is a juvenile rate decrease, preceding sexual maturation, as compared to a more even, continuous increase in most other mammals (Bogin, 1999a, 1999b). Not all primates grow in precisely the same way (Leigh and Park, 1998). Humans generally follow the primate pattern, but they defer growth through an even more prolonged juvenile period. The distinctions among non-human primates are discussed in Chapter 11.

Box 2.1 Modes of Ossification

Osteogenesis

There are two primary modes of ossification, both beginning with a primitive connective tissue called mesenchyme. Most simply, mesenchymal cells can directly differentiate into osteoblasts, the cells that synthesize bone matrix. This mode occurs within or deep to collagenous membranes, and is thus termed intramembranous (or membranous) ossification. Most bones of the body, however, form by replacing cartilage templates (endochondral ossification). Most bones throughout the body are formed by one of these modes (but see Hall, 2015, for discussion of other mechanisms).

In the head skeleton, both modes of ossification play prominent roles. Image a (below) shows the facial region of a rodent embryo, in which an island of intramembranous bone (double arrow) is forming adjacent to nasal capsule cartilage (NC). Most parts of the nasal capsule are replaced (endochondral ossification) by parts of the ethmoid bone. Some parts of the nasal capsule disappear, and small portions remain as hyaline cartilage postnatally. Small portions of the nasal septum (NS) remain cartilaginous.

Intramembranous Bones

Relatively few bones of the appendicular skeleton are intramembranous (e.g., clavicle). In contrast, numerous cranial bones form by direct differentiation of mesenchymal cells to osteoblasts. One such bone is the mandible. Images b and c show the body of the mandible in an embryonic and a late fetal rodent, respectively. The mandible forms around, but does not replace, Meckel's cartilage (MC). Note the dense, cellular region around MC in the embryo (b), and the spongy bone of the mandibular body (MB) in the fetus (c). The first evidence of any bone is a condensation of mesenchyme, that is, dispersed cells become more closely "clumped" together (and thus create a densely stained region, see g). This occurs following an inductive interaction with epithelium

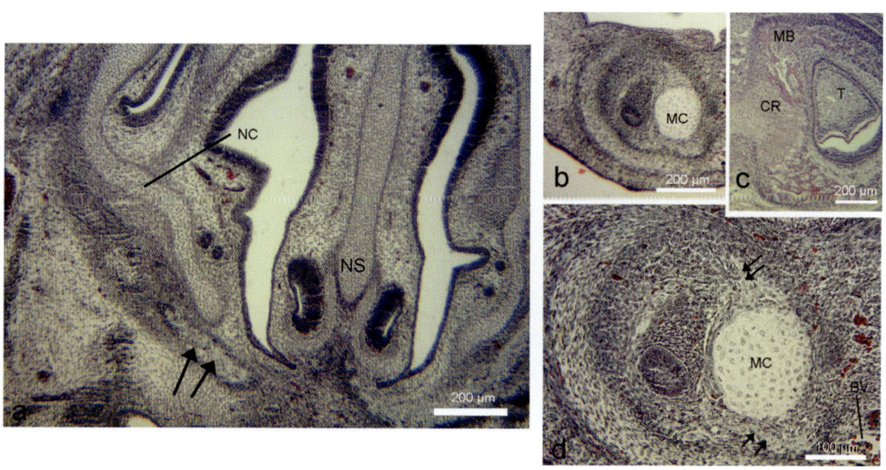

(in this case with that of the mandible). In this rodent embryo (b, and see enlarged view in d), differentiation to osteoblasts has begun, as evidenced by the light green struts of bone matrix (double arrows). BV, blood vessels; CR, cartilage remnant.

Chondrogenesis and Chondral Bones

Endochondral ossification is an indirect mode of osteogenesis. It begins with cartilage formation. Chondrogenesis (e) progresses via differentiation of mesenchymal cells to cartilage-producing cells. At first, the embryo has a core of dispersed mesenchyme, as shown in the embryonic facial primordia of a rodent embryo (f). A more advanced embryo (g) shows the critical stage of mesenchymal condensation (see box for enlarged view) presaging the formation of nasal capsule structures (open arrows). In image h, in a late embryo, cartilage structures are formed of densely packed newly differentiated chondroblasts. In a late fetal rodent (i), chondrocytes (C) are beginning to disperse within a hyaline cartilage matrix.

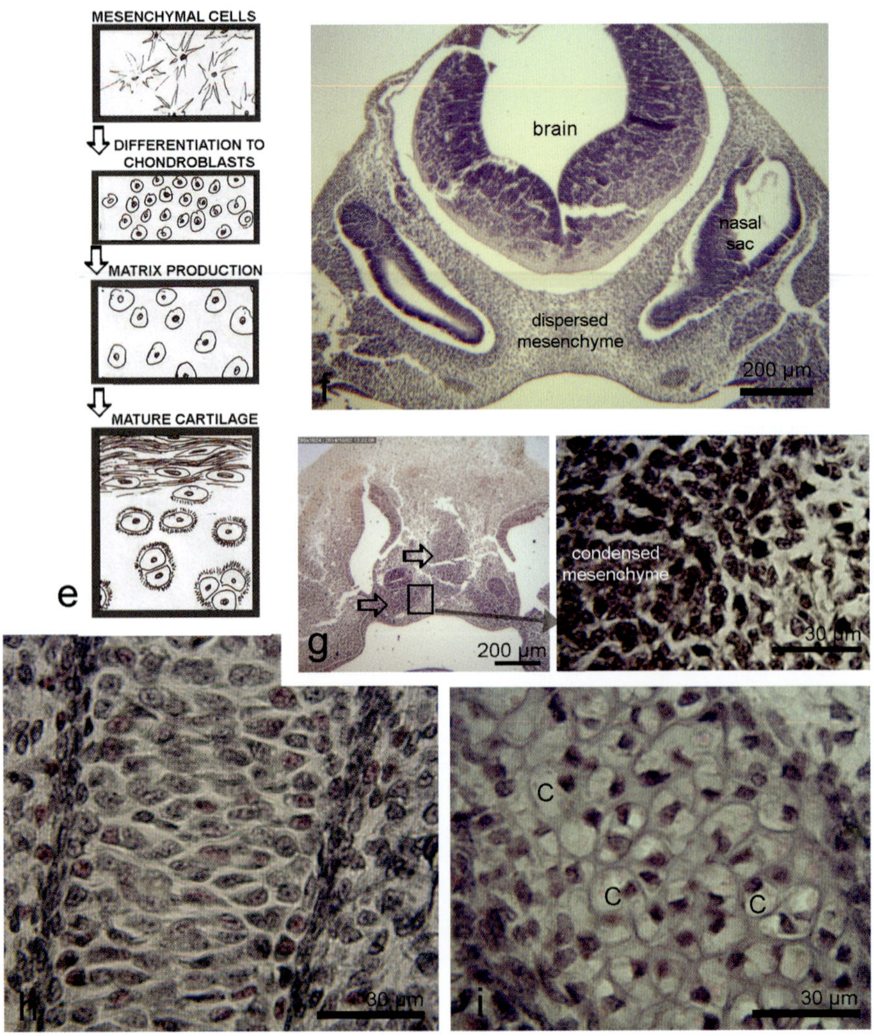

Box 2.2 Epiphyseal Growth Plates

Below (a): a parasagittal section through the upper limb of a late fetal mammal is shown. As in humans at the same stage, the proximal (PE) and distal epiphyses (DE) remain cartilaginous, and the diaphysis (D) is ossified. Each epiphysis possesses five zones located next to the diaphysis. Growth occurs at proximal and distal ends, such that proliferating and enlarging chondrocytes literally propel each epiphysis away from the diaphysis.

This occurs at both ends (below: b, c). The zones are: (1) reserve (or resting) zone of cartilage in which typical hyaline cartilage is maintained (a, b); (2) proliferating zone in which chondrocytes are stacked parallel to the bone's long axis (c); (3) hypertrophic zone in which the chondrocytes enlarge (c); (4) calcified cartilage zone in which the matrix accumulates mineral deposits and the chondrocytes are dead/dying (c); and (5) ossification zone in which new bone matrix is being deposited along remnants of calcified cartilage (c).

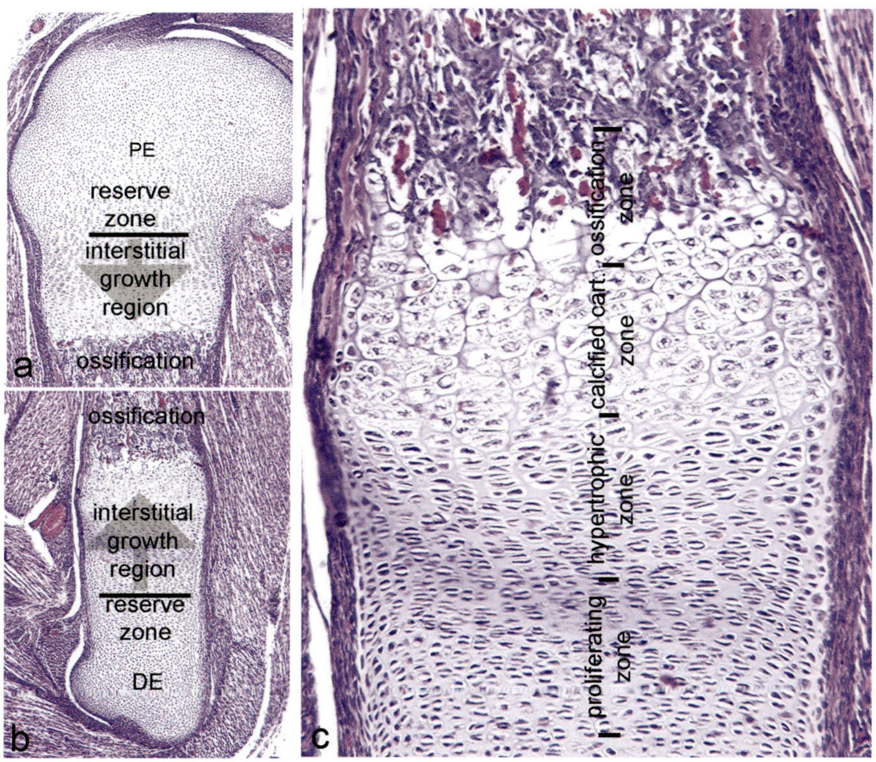

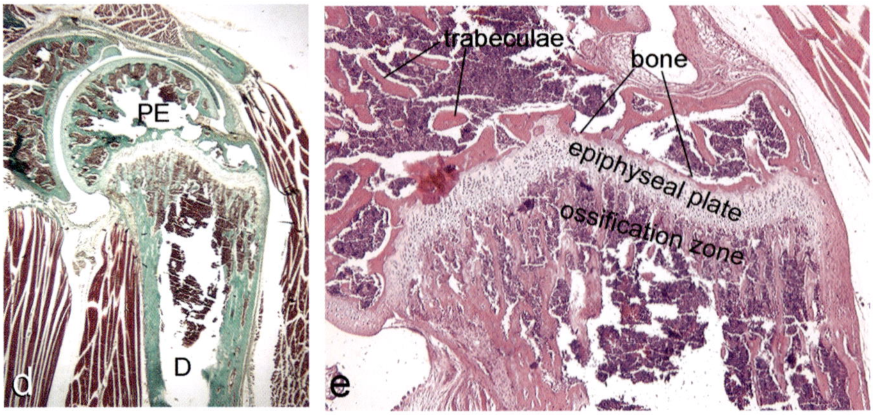

After ossification of the epiphysis, the region of epiphyseal growth is reduced to an irregular disk, the epiphyseal growth plate (d, e). The reserve zone is now adjacent to a thin layer of bone in the epiphysis which connects to multiple trabeculae (e). The same zones are seen in this relatively compressed region of growth (f, g). Note the changing, denser staining of the cartilage matrix approaching the ossification zone, reflecting mineralization (f). In the ossification zone itself, chondroclasts (CCl) are at work breaking down the mineralized matrix (g).

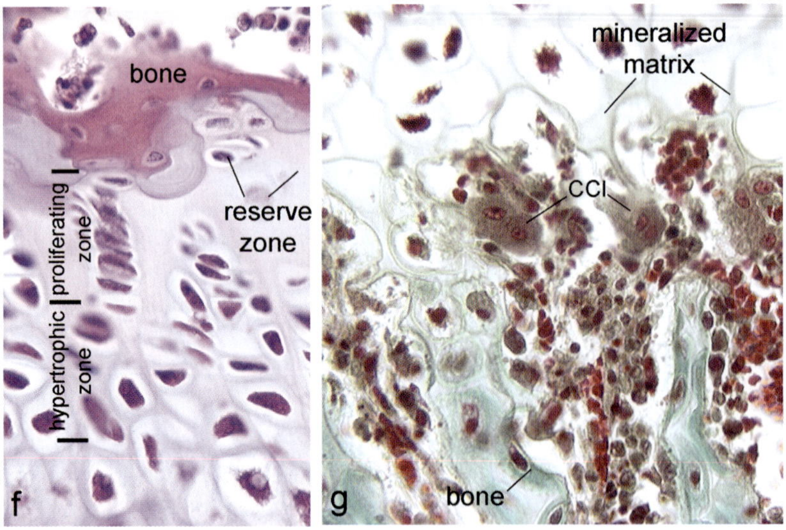

Box 2.3 Growth Trajectories in the Newborn Face

Growth of the facial skeleton is dependent on spatial relationships to soft-tissue structures, such as the eyes and teeth (see below). Some monkeys are very similar to humans in patterns of growth of the face. Based on their small size, monkeys such as tamarins and marmosets offer convenient models for understanding these influences. In image a, the cranium of a newborn *Saguinus* (tamarin) is shown from the right side, and the midface as well as most of the upper face is boxed to indicate the region shown on the right. On the right, two major elements that effect midfacial growth are shown superimposed on the bone: the eye and deciduous teeth.

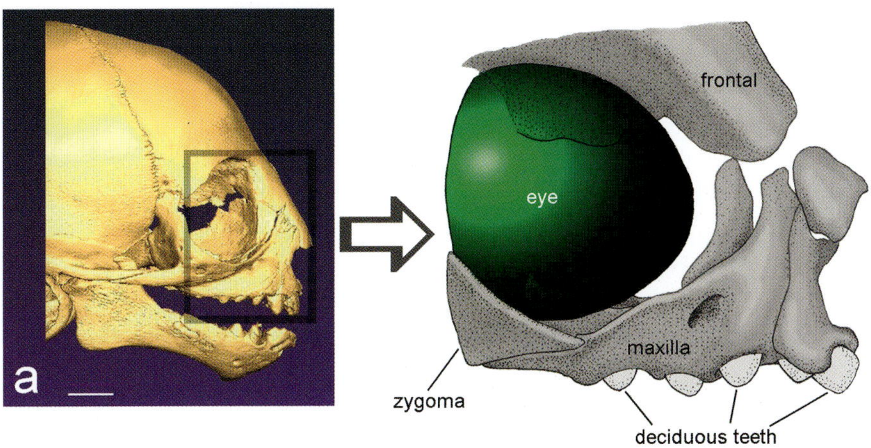

In figure b, a coronal section of the entire upper and middle facial regions is shown, revealing the spatial relationships of the eye, nasal cavity, and the second deciduous premolar (dp3) at birth. Detailed mapping of the location of osteoclasts and osteoblasts along the surfaces of the frontal bone (upper inset) and maxillary bone (lower inset) reveal modeling patterns. The superior–medial orbit is resorptive on the orbital side and depositional on the endocranial side. The inferior surfaces of the orbit are mostly depository in these monkeys (as in humans). Notice that opposing surfaces have opposing activity (deposition vs. resorption) in many bony plates that serve as shared boundaries between two regions (e.g., orbit vs. dental).

Taken together, such observations reveal trajectories of growth that characterize changes in the facial skeleton (figure c), such as an upward drift of the orbital floor due to selective bone deposition (green arrows, figure d) and resorption (red arrows).

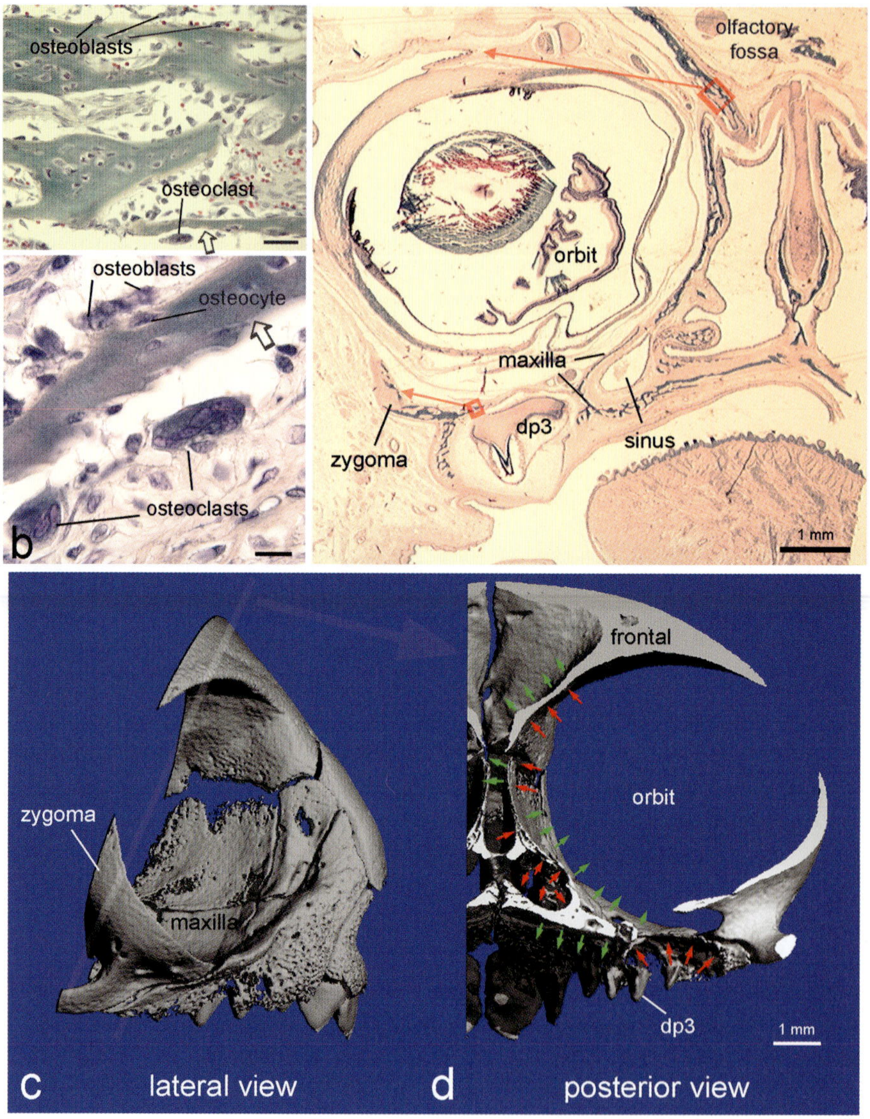

Box 2.4 Sutural Growth

Most cranial joints are sutures. Image a shows a coronal section of a fetal loris (*Loris tardigradus*) at the level of the metopic suture. Here the soft-tissue regions of the head are clearly visible, especially the eyes (E) and brain (B). In a close-up view of (b), the metopic suture is visible, a joint between the right and left frontal bones (F). The fibrous nature of this suture is evident (open arrow). As this is a midline gap between bones, it is considered a growth site, a region at which expansion occurs. In this case, the growth is a passive expansion of the skull in transverse dimensions (see large arrows). One can picture the calvarial bones as islands of tissue that are carried away from one another, at sutural interfaces, for the duration of brain growth. Generally, sutures fuse after brain growth ceases, but the rate of fusion among different sutures can vary. (Other abbreviations: NA, nasal airway; T, tongue.)

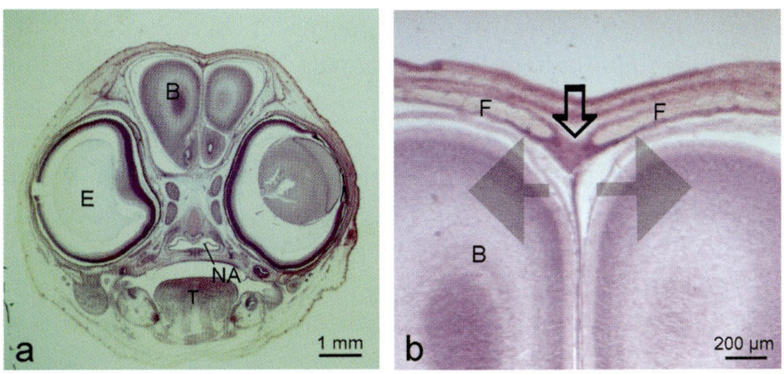

The role of sutures is also understood via a contrast to endochondral portions of the growing skull (c). Endochondral ossification occurs at multiple locations within the skull, and various bones beneath the brain are prime examples. In sagittal sections of the contralateral side of the same loris (c), the brain is surrounded by the membranous neurocranium (MN), and below by the chondrocranium (CC). Calvarial bones ossify within the collagenous membrane that initially surrounds the growing brain. A higher magnification view of an adjacent section (d) shows the collagenous fontanelle (small arrows) that intervenes between the margins of the frontal and parietal bones (P). This particular suture allows brain expansion in a longitudinal direction. Image e is an enlargement of the basicranium in the same specimen. Unlike calvarial sutures, when basicranial bones first ossify they are separated by synchondroses (growth centers; see hypertrophic chondrocytes (HC) in panel e).

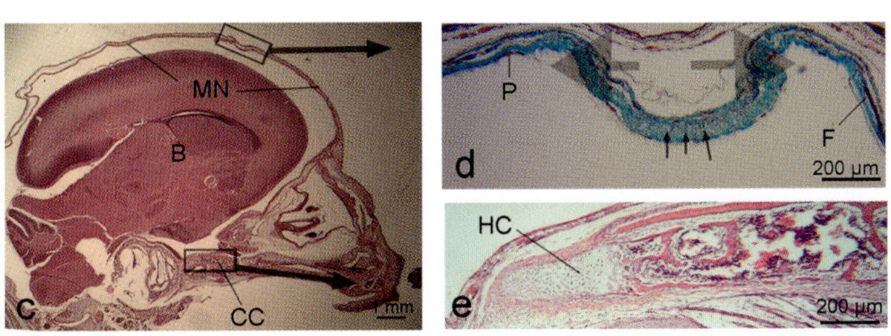

3 Why Ontogeny Matters

Development evolves, unless it is constrained. (*Hall, 2015, p. 271*)

Hall's quote, attributed to an anonymous source, reminds us that our understanding of natural selection was first applied only to animals or plants that had already hatched or sprouted (Darwin, 1902). In his 1859 monograph, Charles Darwin devotes a large section to "Development and Embryology." He observed the great importance of embryonic stages to classification, and discussed animals whose larvae display adaptations allowing independent living (i.e., without parental provisioning or protection). However, he also observed that adaptive features (e.g., a "much-curved beak") are of little utility to the young that are provisioned by their mother, and this is especially so while they are in the egg or womb. Thus, to Darwin, the embryo or fetus was to a great extent "buffered" from the influence of selection.

Mendel's discoveries were unfolding nearly in parallel time, and thus the link of phenotype to the heritability of genes was, as yet, unavailable to Darwin as he wrote his most influential book. Now we more clearly understand the means by which selection affects the prenatal animal. The reproductive suppression model (Wasser and Barash, 1983) offers an explanation for why some pregnancies may fail: "females can optimize their lifetime reproductive success by suppressing reproduction" (p. 513) pending better environmental conditions. In this sense, natural selection may have a very immediate impact on survival of embryonic or fetal animals. We also presume there are longer-term effects of selection, still mediated through the mother. For example, animals vary in their maturational state at birth (see Chapter 1). Although altricial mammals are highly dependent, there is a great range in the duration of their dependency (Smith and Keyte, in press). On the other hand, some mammals that are born somatically precocial (such as most primates) vary greatly in degree of dependency at birth. The degree of precociality and dependency has a complex relationship to prenatal and postnatal development.

3.1 Pace and Pattern of Development

Because natural selection is at work on prenatal animals, we are able to discern certain adaptive features in the newborn. Using a developmental timeframe, rather than relying on a static endpoint (such as the adult stage), has helped to fully recognize adaptive traits that define lower-level taxonomic groups. For example, indriids are not just characterized by robust teeth to process a tough, largely leaf diet, they accelerate mineralization and eruption of permanent molars for that purpose (Godfrey et al., 2004; see Chapter 5); this rapid pace of dental development (but not all other somatic development) differentiates indriids from all other large-bodied lemuroids with the exception of *Lepilemur*.

At another extreme, extant hominoids delay dental eruption the longest of all living primates, at least in absolute time (Smith et al., 1994). However, the pace of dental development in hominoids varies postnatally, and considerable debate has centered on whether fossil hominids (*Homo* and *Australopithecus* spp.) are primitively ape- or human-like (e.g., Beynon and Dean, 1988; Simpson et al., 1990; Lampl et al., 1993; Smith, 1994; Cofran and Walker, 2017).

Developmental observations have also been used to inform us of convergent paths of development (Morimoto, 2018), which may allow us to recognize homoplasy in adults. For example, wrist development among knuckle-walking apes involves distinct ossification patterns (Kivell and Schmitt, 2009). Postorbital septation in tarsiers and anthropoids also derives from very distinct growth patterns, and we have argued elsewhere that this reflects different functional demands during development (Smith et al., 2013; DeLeon et al., 2016).

Thus, a goal in modern ontogenetic studies is to recognize developmental differences in pace and pattern among species (Smith, 1992; Bromage and Janal, 2014). Temporal discrepancies in appearance of features between or within groups of vertebrates imply different rates of development (the pace) for specific features. The term for change in the *relative timing* of appearance of morphological features is heterochrony (see Figure 2.2). The phenomenon can be inferred by comparing ancestors to descendants, or it can be an observed phenomenon within an individual species (Hall, 2015). In practice, heterochrony is also detected by comparing extant species (Figure 3.1).

The ultimate explanation for heterochrony as a phenomenon, whether observed within a species or inferred by comparative morphology, is the same: developmental mechanisms are subject to natural selection. The agents of natural selection that influence rates of development are able to affect major morphological distinctions between ancestors and descendants. Within the Order Primates, such distinctive features include relative size of the neurocranium, face, special sense organs, limbs, and many other features. Studies of the newborn primate have the potential to determine the extent to which prenatal as opposed to postnatal processes influence these traits.

3.2 Neoteny

It has been argued that adult humans resemble juveniles more than is the case for any other primates (Gould, 1977), a phenomenon related to a change in the rate of maturation, such that adults are "juvenilized" (paedomorphosis). If one observes that all apes are more similar as infants than as adults (Figure 3.1), we might hypothesize the infant form is a common ancestral state, what Gould referred to as the "prototype."

When adults evolve to more closely resemble the prototype, they are said to be paedomorphic, and in this particular case, *neotenous* (Figure 3.1).

Gould discussed the long history for the term neoteny, and the observations that underlie it. He defined it as a "retention of formerly juvenile characters by adult descendants," produced by deceleration of somatic development. Shea (1989) cautioned against conflation of neoteny with other forms of paedomorphosis. Discussing Gould's "clock" models of development, Shea noted that in neoteny, although the morphology is juvenilized in a descendant, the duration of growth remains unchanged. Yet, the result (juvenilization) may also be produced via reducing magnitude of growth per unit time or by a reduced duration of growth, these being two other

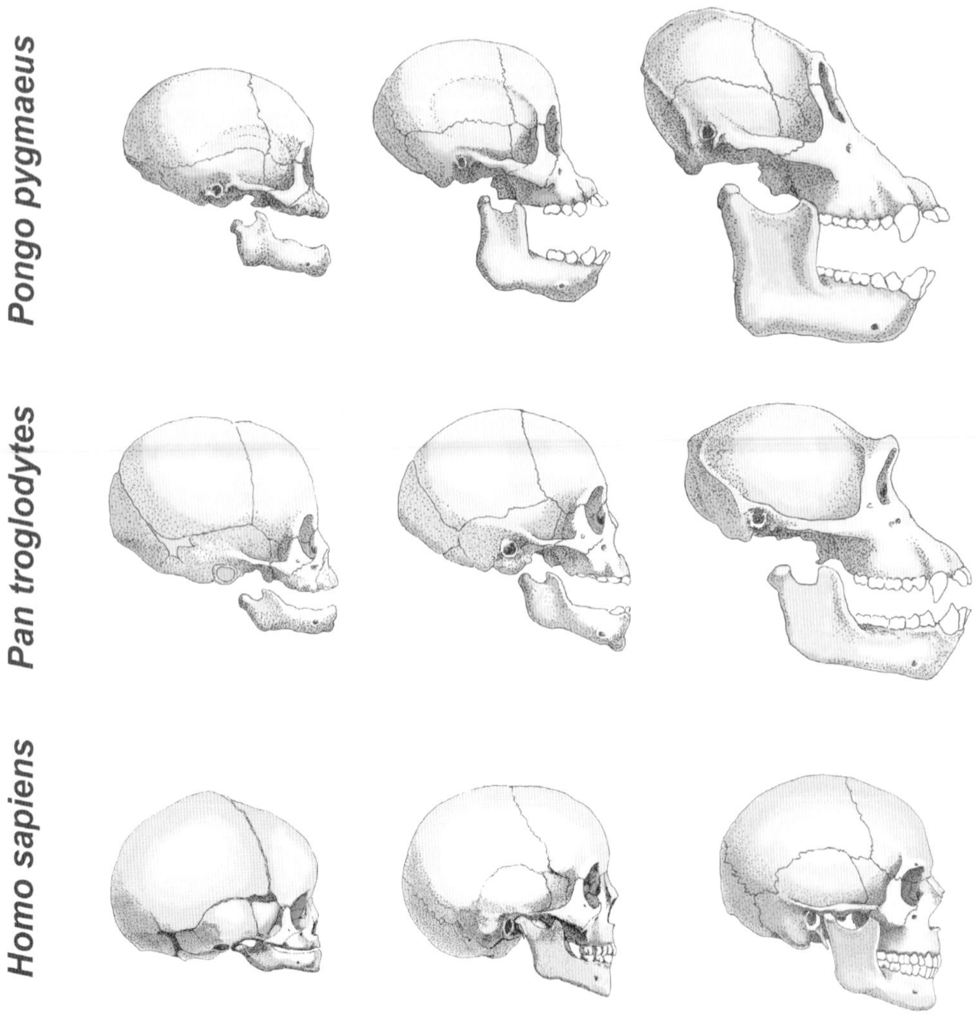

Figure 3.1 Newborn (left column), full deciduous tooth eruption (middle), and adult (right) stages in great apes and humans. Ape series redrawn from Schultz (1940, 1941); human series redrawn from Schwartz (2007). Gould (1977) contends that adult humans resemble subadult humans more so than any other living hominoids.

forms of paedomorphosis. In neoteny, it is the allometry of growth – rather than growth rate or magnitude – that changes (Shea, 1989).

The developmental mechanisms that produce human neoteny have never been fully explained. Molecular studies consistently report a closer genetic affinity between humans and chimpanzees than among all great apes (e.g., Mann and Weiss, 1996; Kaessmann and Pääbo, 2002). How does this square with the common trajectory of great apes for a greatly extended midface and more robust teeth compared to humans (Figure 3.1)? Mitteroecker et al. (2004) assert that the timing of developmental differences has a profound influence on morphological disparity between humans and great apes. The smaller genetic difference between humans and chimpanzees impacts an earlier developmental period, while genetic differences among apes have effects on a later developmental timeframe.

This interpretation of hominoid development shows the importance of recognizing differences in pace and pattern. One exercise is then to apply this information to the fossil record, to draw an inference on the antiquity of what we see as a novel pattern – does it characterize all extinct hominids (Smith, 1992)? Subadult morphology provides a unique opportunity for understanding the life history of a species. "When to be born, when to be weaned, when to stop growing, when to reproduce, and when to die are basic elements of mammalian life history" (Smith, 1992, p. 135), and developmental information is a key to recognizing adaptive life-history patterns.

Our focus is not to revisit such questions on hominoids in particular, but to identify a broader primate pattern of development. Gould's assertion that humans, among all primates, most closely resemble the "young stages of our forebearers" (1977, p. 387) appears well-founded based on comparative knowledge of hominoid postnatal development. However, the postnatal growth trajectories of non-hominoids remain very poorly studied, especially in the case of non-anthropoids (Figure 3.2).

3.3 What Does Anatomy of the Newborn Primate Tell Us?

Schultz's work demonstrated that an understanding of ontogeny facilitates interspecific as well as higher-level comparisons. Common stages may be recognized even when chronology differs starkly, e.g., short- versus long-lived animals. Also, the timing at which these stages are recognized can have important adaptive meaning because, again, *development evolves*. With this in mind, Bolter and Zihlman (2007) emphasized the importance of development to phylogeny and life history. Subadult morphology has phylogenetic significance because specific patterns of growth produce adult morphologies. In the context of life-history theory, which posits a link between natural selection and the pace at which an animal develops (e.g., Promislow and Harvey, 1990; Janson and Van Schaik, 1993), we are able to identify adaptive features of a species across their lifespan.

The newborn stage offers a unique opportunity to recognize adaptive features. At parturition, metabolic costs for growth commence a transition from the mother to the infant. This shift is relatively long in primates compared to many mammals, but primates do vary significantly in the relative duration of the gestational period, time before weaning, or age at first reproduction (Harvey et al., 1987; Ross, 2003a). The

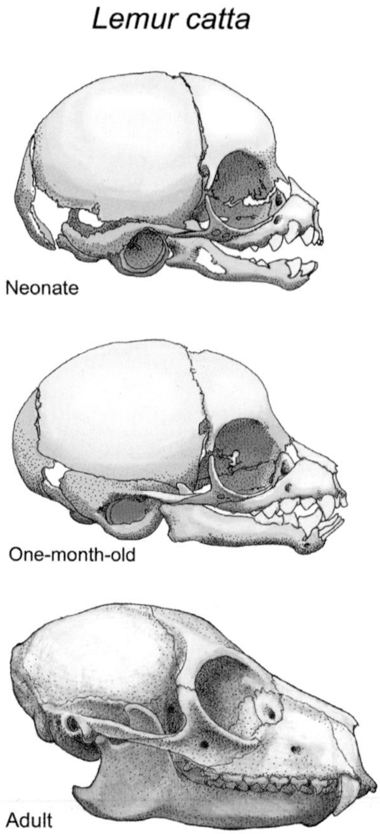

Figure 3.2 The head skeleton of *Lemur catta*, at newborn (top), infant (middle), and adult ages (bottom). The most profound ontogenetic change takes place in the viscerocranium.

maturational state of the neonate offers a particularly revealing indication of maternal adaptations to environmental or social selective factors.

The most intuitively understood morphology may be that of teeth – feeding competence is directly dependent on the degree to which the teeth of the subdult are mineralized or erupted. Newborn primates vary widely in their state of dental maturation (Smith et al., 1994; Godfrey et al., 2004; Smith et al., 2015), with some having most deciduous teeth erupted and others having few or no erupted teeth (Figure 3.3). Primates also vary broadly in the extent of adult brain size achieved by parturition (Harvey et al., 1987; Leigh, 2004; and see Chapter 4). And while all primates are initially physically helpless, there are subtle differences in the pace at which species achieve locomotor competence (see Chapters 9 and 11) and how long they are carried by the mother (Ross, 2001), and these differences likely correlate with differences in somatic maturity of the limb skeleton (Young et al., in press). Given the many challenges for young primates, we aim to understand how newborn anatomy of the skeleton and teeth reflects life-history patterns (Chapter 9).

3.4 Regional Skeletal Anatomy of the Newborn Primate: Modularity of Development

3.4.1 Craniodental Anatomy

This book adopts a regional approach to skeletal anatomy in order to allow ease of comparison to existing texts on dental or skeletal anatomy of the adult (e.g., Ankel-Simons, 2007; Gebo, 2014). For functional interpretation, it is revealing to compare individual body parts among mammals and among primates. For example, primates exhibit a large range of variation in extent of dental eruption achieved at birth (Figure 3.3). Some other orders of mammals may exhibit similar broad variation, but

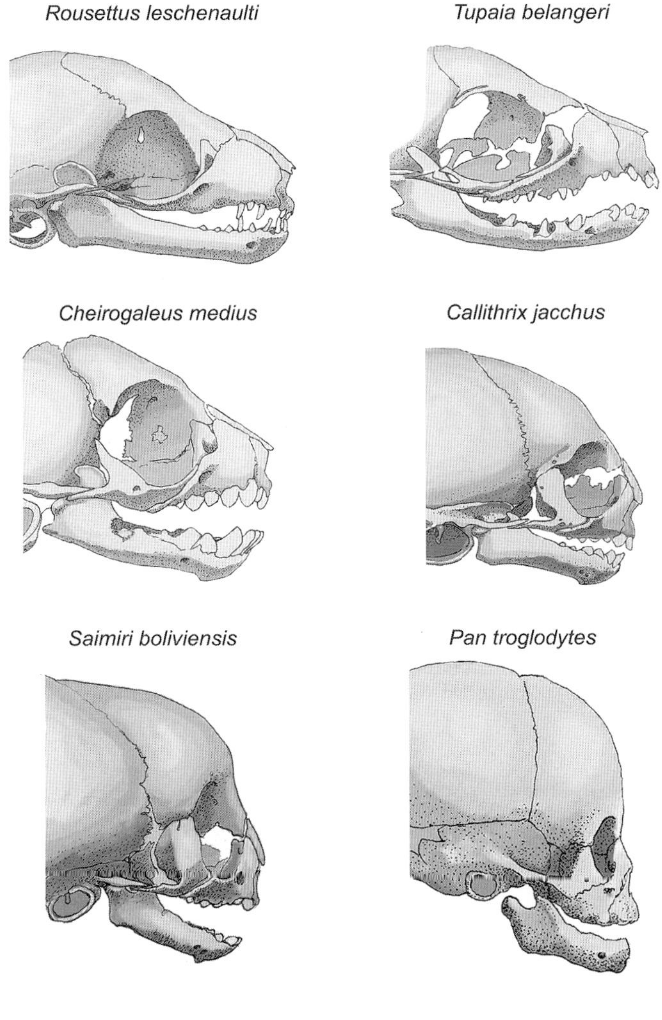

Figure 3.3 Alveolar eruption status of the teeth in a perinatal bat (*Rousettus leschenualti*), tree shrew (*Tupaia belangeri*), and four primates. Many mammals are slow to mineralize and erupt deciduous premolars (top row). Primates vary profoundly in this regard, with some quite rapidly erupting these teeth.

the tree shrews and insectivores that have previously been used to model ancestral primates are uniformly slow to erupt their teeth, especially their deciduous premolars (Figure 3.3; Asher and Lehman, 2008).

While such comparisons are often revealing, another goal is to understand the integration of parts within skeletal regions of the head or postcranium, and between skeletal regions. Much of this synthesis occurs within the last three chapters of the book. Within the head skeleton itself, the different parts have different rates of growth and development; that is, growth is modular (Cheverud, 1996). For example, the basicranium is thought to be more evolutionarily conserved, and the morphology is more phylogenetically stable within higher taxonomic groups of primates (Cardini and Elton, 2008; Goswami and Polly, 2010; Neaux et al., 2018). Rates and patterns of neurocranial or viscerocranial growth are far less stable, even within primate families or superfamilies. Based on comparative growth of cranial modules in humans and great apes (Figure 3.1), human neoteny evolved through discordance between growth of the viscerocranium and neurocranium, and *within* the viscerocranium. Our knowledge on non-hominoids is scant by comparison. For certain, brain growth rates vary substantially within other primate families and superfamilies. For example, some tamarins (*Saguinus* spp.) defer over half of the total brain growth to the postnatal timeframe, while squirrel monkeys (*Saimiri sciureus*) accomplish more than half of brain growth prenatally (Leigh, 2004).

Because our understanding of variation of neonatal cranial form has been biased toward hominoids, the patterns of cranial integration that are common to all anthropoids, or all primates, remain poorly understood. It is our goal to enable the reader to see phylogenetic patterns in a common early stage of development with the goal of contextualizing adult morphologies. To some extent, the newborn can substitute for Gould's prototype, and thus allow us to see whether neoteny, or alternatively hypermorphosis, occurs in phylogenetic patterns across all primates. Although we do not endeavor to discuss the entirety of growth and development, we aim to provide sufficient information for the reader to do so by revealing neonatal and early infant anatomy, which can then be compared to adults (Figure 3.1). A broader comparative understanding of overall skull ontogeny may lead to a more nuanced view of neoteny. To what extent are other primates neotenic? Such a discussion may require separate considerations of brain and dental ontogeny. While trends in primate dental development show immense variation (Figure 3.3; Chapter 5), head size in newborn primates is generally large (Trevathan, 2015), either due to rapid prenatal brain growth or relatively slow somatic growth compared to most mammals (Halley, 2017). Accordingly, we may expect that neoteny of the neurocranium relates in part to the degree of postnatal hypermorphosis. In addition, a degree of neurocranial neoteny may occur even if viscerocranial development yields a large projecting face (Figure 3.2). In this sense, aspects of adult morphology are hypothesized to be epiphenomena of selective events during ontogeny, such as relatively enhanced primate encephalization.

3.4.2 Postcranial Anatomy

The work of Schultz (1940, 1941, 1944) suggests that the postcranial axial skeleton has a slower trajectory of epiphyseal fusion than the skull in hominoids. This is in keeping with what appears to be a broader trend among infant anthropoids. Both the trunk and the limbs grow for a longer duration of postnatal ontogeny than the cranium in primates that have been well-studied (mainly catarrhines, and few platyrrhines). In part this reflects precocious growth of the brain, resulting in infants with disproportionately large heads (Grand, 1977).

After the brain growth is complete, the limbs and trunk continue to grow in length and mass (Turnquist and Wells, 1994; Bolter and Zihlman, 2003; Shapiro and Raichlen, 2006; Young, 2012a, 2012b). Juvenile primates may have a brain grown to nearly adult proportions (e.g., 98% of adult cranial capacity in vervets) while the femur (87% of adult length) and trunk (84% of adult length) are less grown (Bolter and Zihlman, 2003). As a result of these discordant growth trajectories, the center of mass of the primate shifts caudally during postnatal development (Young, 2012a, 2012b; Young and Shapiro, 2018).

Thus far a pattern has emerged for anthropoid primates. The brain achieves most of its adult mass before most structures that function in a locomotor capacity, reproductive structures, or even teeth and masticatory muscles. However, hands and feet provide an interesting exception, being proportionally large in subadults (Bolter and Zihlman, 2003; see also Chapters 9 and 11).

These are examples of whole body mosaicism across development (Bolter and Zihlman, 2007). Our understanding of development of the whole primate body at a comparative level is currently very narrow, again heavily biased toward hominoids.

Figure 3.4 Illustration of neonatal tree shrew (left, *Tupaia belangerii*) and lemur (right, *Eulemur flavifrons*). This illustration presents certain dichotomous features between altricial mammals (e.g., *Tupaia* is born furless, with eyes closed) and precocial mammals (e.g., *Eulemur* is born furred with open eyes). In addition, some features presage specializations in the precocial primate, such as the large head relative to the trunk, reflecting enhanced encephalization that typifies all primates. Also note the relatively large hands and feet, with well-formed joints evident at birth. This has been documented quantitatively in some primates (see text), and may be a primate characteristic more broadly.

There is at least some published evidence that strepsirrhines follow a similar pattern of relatively large hands and feet (Lawler, 2006), and descriptively, all primates have relatively large heads at birth (Figure 3.4). A clearer understanding of which traits typify all or only some newborn primates will put the developmental trajectory of hominoids, and our own species, in a clearer context.

4 The Skull

4.1 Introduction

The skull is the rostral-most part of the axial skeleton and includes both the cranium and the mandible. The cranial diversity of adults has been documented extensively in a broad spectrum of primates (e.g., Tattersall & Schwartz, 1974; Fleagle et al., 2010), including ontogenetic studies (e.g., Baume, 1953; Lieberman et al., 2000; Mitteroecker et al., 2004). The elements of the skull are highly conserved across primates and retain most of the features that characterize the mammalian skull. The primate skull is distinct from other mammals for trends toward encephalization and emphasis on visual systems over olfactory systems (Cartmill, 1980; Smith et al., 2007). Because many bones of the skull interface with more than one of these soft-tissue regions, growth of one organ (e.g., the brain) can directly affect one skull region and indirectly affect another region. Thus, the relatively large brains and eyes of primates require specialized osseous support (Box 4.1).

The complex structure of the skull leads us to subdivide this integrated structure into component parts. In this chapter, we divide the skull into three parts (face, cranial vault, and cranial base), but anatomists and other researchers have used a diverse range of frameworks to describe the component parts of the skull. The dichotomous terms *neurocranium* and *viscerocranium* are used to partition the skull based on embryology and the evolutionary history of these regions. Comparative anatomical texts traditionally use these terms. The neurocranium surrounds the brain, and the viscerocranium is the remaining part of the cranium that surrounds and supports the eyes, mouth, nasal cavity, and other rostral gut tube derivatives. In most texts, the term viscerocranium is used to describe all pharyngeal arch derivatives. One source of confusion is that some pharyngeal arch derivatives do not contribute to our lay understanding of the "face": for example, the bones of the middle ear, the styloid process of the temporal bone, and the hyoid bone. In another source of confusion, human anatomical texts commonly describe the "face" in terms of three regions: upper face, midface, and lower face, which correspond to the distribution of neurovascular structures to these regions (e.g., divisions of the trigeminal nerve). It is a common source of confusion that the "face" described in this manner does not correspond directly to the region defined as the viscerocranium. The "upper face" (as described in human anatomy texts) refers to the region of the forehead in humans and related taxa, and from a developmental and comparative context is more properly considered as part of the neurocranium.

The terms *chondrocranium* and *dermatocranium* are applied to distinguish regions of the cranium based on their type of ossification. For example, these terms subdivide the neurocranium into two parts based on whether the bone forms endochondrally (chondrocranium) or intramembranously (dermatocranium) (Sperber, 2000; Figure 4.1a). The chondrocranium forms from numerous chondrification centers including (from caudal to rostral) a midline basal plate, paired otic capsules, paired parachordal cartilages, paired trabeculae cranii, and paired olfactory capsules (de Beer, 1937; Starck, 1975). The centers closest to the midline give rise to endochondral portions of the occipital and sphenoid bones, as well as the midline part of the ethmoid bone. The otic capsule forms the largest portion of the temporal bone and the olfactory capsule yields the lateral masses of the ethmoid bone and the maxilloturbinal. Midline centers are joined to the cartilaginous sensory capsules by cartilaginous bridges called commissures. Presence or absence of various commissures in embryos of primates and other mammals has been documented (e.g., de Beer, 1937; Spatz, 1964). After ossification, the chondrocranium is *mostly* synonymous with the anatomical term basicranium, or cranial base. However, because the rostral portion of the developing

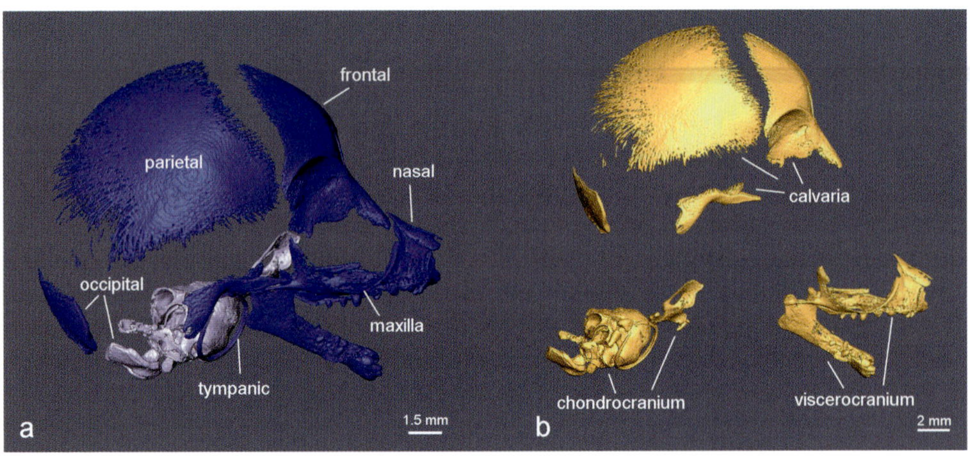

Figure 4.1 Skull of an early fetal (39.6 mm crown–rump length, CRL) specimen of *Loris tardigradus*. (a) The skeletal elements that form from ossification of cartilage templates are shown in light gray, and comprise most of the cranial base. Bones forming via intramembranous ossification are shown in blue. Note that some bones (e.g., occipital) form from both modes of osteogenesis. (b) On the right, regions of the skull are emphasized by "exploding" its parts. The cranial base supports the brain; the neurocranium is completed by the calvarium, which surrounds the sides and dorsum of the brain. Note that endochondral ossification is far from complete at this stage. Thus, the ossified facial skeleton at this age represents the membranous part of the viscerocranium.

chondrocranium projects further rostrally into the midface and gives rise to many bones of the nasal cavity, the facial skeleton incorporates endochondral bone from the chondrocranium (see below). Similarly, the presphenoid appears to expand anteriorly via endochondral ossification in many anthropoids and contributes to the interorbital septum (Ashley-Montagu, 1943; Smith et al., 2017b). Additionally, in all primates, the frontal bone cups the frontal lobes ventrally, articulating posteriorly with the lesser wing of the sphenoid. As such, the anterior cranial fossa, unlike the other fossae, is formed more by the intramembranous frontal bone than by the endochondral bone of the orbitosphenoid or the ethmoid.

At the junction of neurocranium and viscerocranium, bones perceived as individual elements in the adult skull are formed through the fusion of ossification centers of diverse origin. For example, the temporal bone represents a junction of facial skeleton, calvaria, and cranial base. The petrous portion of the temporal contributes to the cranial base. It is derived from endochondral ossification and supports components of the auditory system, including inner and middle ear. The middle ear (tympanic) cavity surrounds the ear ossicles. These small bones – the malleus, incus, and stapes – are formed through endochondral ossification and are derived from the first and second pharyngeal arches. In both embryological and evolutionary contexts, they are part of the viscerocranium. The squamous portion of the temporal contributes to the calvaria and is derived from intramembranous ossification. The ectotympanic portion of the temporal is often considered part of the cranial base, and yet it is formed by intramembranous ossification.

In this chapter, we partition the head skeleton in terms of the functional matrices that they encompass (see Chapter 2): facial skeleton, calvaria, and cranial base. The facial skeleton includes the orbits, nasal cavity, upper and lower jaws, and masticatory and pharyngeal structures. The majority of bones in the facial skeleton form intramembranously, including the paired nasal, maxillary, lacrimal, zygomatic, and palatine bones. A single median bone, the vomer, forms intramembranously (from paired ossification centers) and articulates with the perpendicular plate of the ethmoid (which is still at least partly cartilaginous at birth). The ethmoid bone is developmentally part of the chondrocranium; this is also true for the maxilloturbinal (inferior nasal conchae). Thus, the chondrocranium contributes rather significantly to the facial skeleton. Skeletal elements of the facial skeleton articulate with bones of the cranial base and incorporate membranous extensions from endochondrally derived bones (*Zuwachsknochen*, Chapter 2), such as the alisphenoid and orbitosphenoid contributions to the orbit and the pterygoid plates in the pharynx. In addition, the hyoid (although not part of the skull) and styloid process of the temporal are included here in descriptions of the facial skeleton, based on its embryological derivation from pharyngeal arch structures.

The calvaria (cranial vault) is defined spatially as the sides and roof of the neurocranium. Bones of the calvaria are formed entirely of intramembranous bone. This is an assemblage of bones or parts of bones that form without a cartilaginous precursor. Bones of the calvarium include the frontal (2), parietals (2), squamous parts of temporals (2), upper portions of the alisphenoids (greater wings of the sphenoid) (2), and the squamous part of the occipital (1). Some of the calvarial bones present at birth (e.g., parietals and squamous occipital) form from the prenatal aggregation of multiple centers of ossification. Four fontanels may be open: the anterior fontanel at the intersection of the coronal, sagittal, and metopic sutures; the posterior (occipital) fontanel at the intersection of the occipito-parietal and sagittal sutures; the anterolateral (pteric) fontanel at the intersection of the alisphenoid with the frontal and possibly parietal bones; and the posterolateral (mastoid) fontanels at the intersection of the temporal, occipital, and parietal bones (Figures 4.2 and 4.3).

The cranial base is defined spatially as the floor of the neurocranium. Much of the cranial base is formed from chondrocranium, but the anterior cranial fossa is formed substantially by intramembranous ossification of the orbital plates of the frontal bone. Bones of the cranial base include the occipital (supraoccipital, exoccipital, and basioccipital parts), temporal (petrosal and ectotympanic parts), sphenoid (presphenoid, basisphenoid, orbitosphenoid, ventral portions of alisphenoid), and ethmoid (crista galli, cribriform plate). These structures are spatially integral to the cranial base, and serve also as a developmental bridge to the facial skeleton.

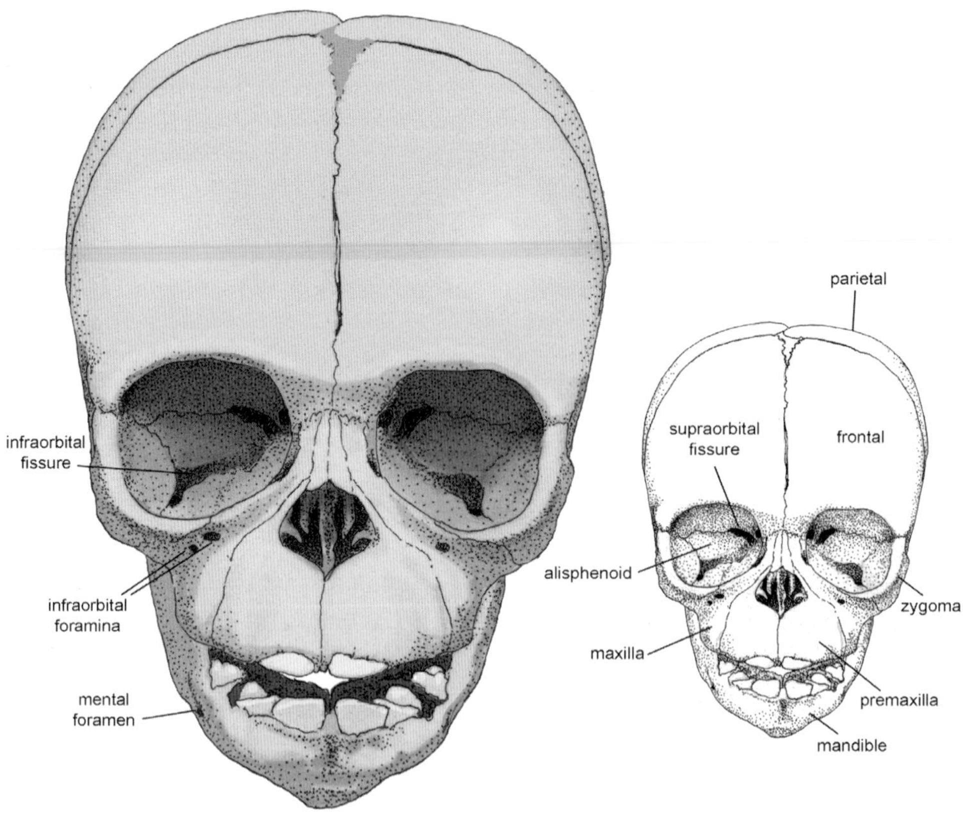

Figure 4.2 Anterior view of a chimpanzee (*Pan troglodytes*) skull identified as "fetal." Many characteristics of this skull match descriptions of newborn chimpanzee skulls described by Schultz (1940), suggesting this skull is a near-term fetus (from the collection of the University of Pittsburgh, Department of Anthropology).

4.2 The Skull of Newborn Hominoids

Extensive literature exists documenting the anatomy of the newborn human skull (reviewed in Scheuer and Black, 2004). To discuss apes, we refer to existing accounts of the bones of the skull in newborns. Skull osteology has been described in 0- and 10-day-old chimpanzees (*Pan troglodytes*; Schultz, 1940), in two newborn orangutans (*Pongo pygmaeus*) who lived a "few days" (up to 8; Schultz, 1941), a newborn gorilla (age in days not provided) figured by Diogo et al. (2015), and a newborn gibbon (*Hylobates lar*; Schultz, 1944). As in previous chapters, the works of Schultz (above) and Scheuer and Black (2004) inform our descriptions of hominoids. We also refer to published CT volumes of newborn chimpanzees (*Pan troglodytes*) at Kyoto University Primate Research Institute (KUPRI) Digital Morphology Museum and skeletal specimens from The Field Museum (Chicago, IL).

4.2.1 The Facial Skeleton

The midface appears small relative to the large proportions of the neurocranium in hominoids (Figure 4.2). A comparison to a perinatal strepsirrhine provides a stark contrast (Figure 4.3). Apes are born with orthognathic faces, while the facial protrusion of strepsirrhines is already established at birth, albeit to a varying degree. A basal view (Figure 4.4) emphasizes the degree to which newborn ape faces are positioned beneath the neurocranium compared to the precociously prognathic strepsirrhine face. The overall form of the newborn ape facial skeleton is notably similar to humans (see Gould, 1977, discussing the work of Naef and others), especially relative to the eventual divergence.

The bony orbit is already fully formed at birth in apes (Figures 4.2 and 4.3, and see Schultz, 1940, 1941), as is the case in all extant anthropoids (Smith et al., 2013; DeLeon et al., 2016). One of the most variable regions in primates generally, the orbital "mosaic" (i.e., the assemblage of skeletal elements that form the eye socket – Cartmill, 1980) is composed of the same bony elements in apes and humans. In apes and humans, the orbital surfaces of the frontal, maxillary, ethmoid, zygomatic, alisphenoid (greater wing of the sphenoid), and orbitosphenoid (lesser wing of the sphenoid) are the largest contributors to the orbit. The lacrimal forms a small (inferomedial) part of the wall, and the palatine has a small contribution posteriorly near the apex of the orbit.

The orbital plates of the frontal make up most of the roof and are inclined approximately 25–30° relative to the visual and occlusal planes. The orbitosphenoid has very little contribution to the orbital roof, projecting only minimally rostral to the optic canals. The floor of the orbit is composed primarily of maxilla and zygomatic. The zygomaxillary suture is widely patent, and obliquely oriented across the floor of the orbit.

The lateral wall of the orbit is already ossified in the neonate. The postorbital septum, a bony wall separating the orbit from the temporal fossa in anthropoids (also see *Tarsius*, below), is precociously complete in newborn apes. In the chimpanzee, the zygomatic and the alisphenoid contribute equally to the lateral wall. In addition, the zygomatic process of the frontal extends away from the neurocranium and contributes to the upper third of the lateral wall. The zygomatic, which forms the lateral and

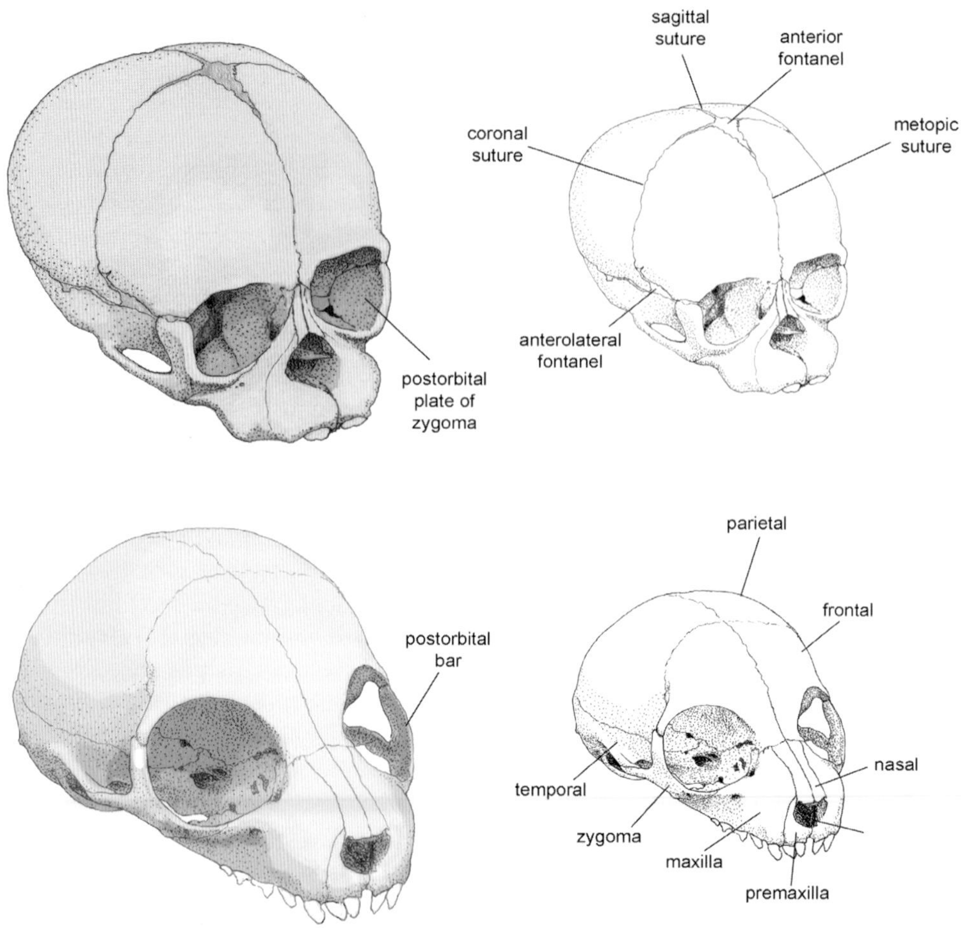

Figure 4.3 This anterolateral view of the late fetal *Pan troglodytes* skull clearly indicates features and joints of individual skull bones. Below, anterolateratal view of an infant potto (*Perodicticus potto,* specimen CMNH # 69182). Based on dental eruption status, the potto is likely older than a neonate, but better ossified to show morphology.

inferior parts of the orbital rim, already has formed its posterior lamina, which extends from the body of the zygomatic posteriorly, cupping the orbital contents and articulating with the alisphenoid. The orbital plate of the alisphenoid has a flange of appositional bone extending into and contributing to the postorbital septum that separates orbit from temporal fossa.

The medial wall of the orbit is substantially ossified in humans and we observe the same in the perinatal chimpanzee. The frontal bone contributes substantially to the medial wall. The lacrimal is well ossified and closely apposed to the frontal and the frontal process of the maxilla. At least some ossification is apparent in the ethmoid and palatine bone contributions to the medial wall. Again, the orbitosphenoid extends only minimally rostral to the optic canals and has almost no contribution to the medial wall of the orbit.

The nasal cavity of hominoids is bounded anteriorly by the nasal aperture, nasal bones, and frontal processes of the maxillae and extends posteriorly to the presphenoid and the choanae, which are open and continuous with the nasopharynx. The nasal cavity extends superiorly within the ethmoid to the cribriform plate. The floor of the nasal cavity is bounded by the palate. Laterally, the nasal cavity expands into the surrounding bone of the ethmoid and maxilla as paranasal sinuses. Nasal conchae are the scrolls of bone that separate the nasal airway from the openings of the paranasal sinuses, increasing the mucosa-lined surface area. Hominoids and other haplorhines lack the complex nasal structures observed in strepsirrhine primates and have only up to three nasal conchae in the lateral wall of the nasal cavity. There is at least some ossification in all nasal conchae in the newborn chimpanzee, and ossification of the middle nasal concha is the most developed. The nasal septum is mostly cartilaginous at birth. The vomer, which forms through intramembranous ossification, exists as a trough of bone extending rostrally from the presphenoid along the posterior and inferior border of the nasal septum and articulating with the palate inferiorly.

The hard palate separates nasal and oral cavities. Well-ossified palatal processes from the maxilla and palatine are apposed in the midline at a widely patent midpalatal suture. The incisive foramina are incomplete medially, and the palatine foramina are complete. The maxillary alveolar ridge is U-shaped, and its length and width at birth are approximately equivalent in hominoids. A major distinction between humans and apes is the early disappearance of the premaxillary suture on its facial aspect in the former. In humans, the premaxillary suture fuses, or perhaps the site is simply overgrown by an expanding maxilla, during fetal development (Sperber, 2000). At birth, a trace of this suture may be observed on the palatal surface, but the premaxilla and maxilla are completely united on the facial surface. In great apes, at least some neonates of some species have a patent premaxillary suture at birth. In the newborn chimpanzee, the maxillary–premaxillary suture was visible only in dry specimens, but may be obscured in virtual reconstructions from CT data.

The body of the mandible is significantly larger than the ramus and has a billowed appearance anteriorly, reflecting the deciduous tooth germs of anterior teeth. The ramus is relatively short and inclines posteriorly as it ascends, in contrast to the very vertical orientation in adult hominoids (see Chapter 2). The coronoid process is relatively small and backwardly curved; the condylar process has an anteroposteriorly compressed appearance. The genial tubercle anteriorly and the gonial angle posteriorly are both well-developed. The mandibular symphysis is widely patent in the newborn chimpanzee, and there is a small secondary ossification center at the inferoanterior border of the symphysis, comparable to that described in humans. The body and greater horns of the hyoid are ossified. The body of the hyoid is a mediolaterally elongated oval, with a midline projection anteriorly. There is a wide, non-ossified gap between the greater horns and the body in the newborn chimpanzee. The styloid process of the temporal did not appear to be ossified in our observations of the newborn chimpanzee.

4.2.2 The Calvarium

The frontal bone forms bilaterally, and right and left frontal bones meet in the midline at the metopic suture, which remains patent in humans until approximately 2–3 years of age and is patent in the neonatal chimpanzee. The parietal bones form bilaterally from multiple ossification centers, which quickly coalesce into a single center. The squamous portion of the occipital also forms from multiple ossification centers and is apparent as a single center at birth. The squamous portion of the temporal forms laterally. The lateral portion of the alisphenoid (greater wing of the sphenoid) also contributes to the wall of the vault in the temporal fossa. In the newborn chimpanzee, the cranial vault is composed almost entirely of frontal, parietal, and squamous occipital elements. The squamous temporal and alisphenoid have only minor contributions to total vault surface area, which is reflected in the small temporal fossa and may relate both to differential brain growth and to the small volume of the temporalis muscle at birth.

In our observations of the newborn chimpanzee, all cranial vault sutures are widely patent. In *Pan* and *Pongo*, the fontanels were described as small at birth by Schultz (1940, 1941). Schultz noted that soon after birth the fontanels have "practically disappeared" (1940, p. 51) in the chimpanzee, although we observed that they remained open in our dry specimen and the virtual reconstruction. Both Schultz's figures (figure 11 in 1940, figure 22 in 1944) and our observations (Figure 4.4) indicate the posterolateral fontanel is the largest at birth in apes. In *Hylobates*, Schultz describes the bones bordering the anterolateral and posterior fontanels to be completely merged at birth (1944).

4.2.3 The Cranial Base

The cranial base in newborn hominoids already reflects the profound influence of encephalization and partitioning of the endocranial base into anterior, middle, and posterior cranial fossae. The sphenoid is positioned at the center of the cranial base and contributes to anterior and middle cranial fossae in all primates.

The anterior cranial fossa consists mostly of the paired frontal bones. The orbital plates of the frontal extend posteriorly to the orbitosphenoid (lesser wings of the sphenoid) in the newborn chimpanzee, although the suture appears open. Compared to the area of the orbital plates, the olfactory fossa and cribriform plate are relatively small. Crista galli and the cribriform plate in our virtual reconstruction of the chimpanzee do not appear to be ossified.

In the midline at the sella turcica, the presphenoid and basisphenoid are ossified and closely apposed, although the intrasphenoidal synchondrosis may be fusing perinatally (see below). The orbitosphenoid appears already fused to the presphenoid. The bilateral alisphenoids are anchored to the basisphenoid at birth and extend laterally to surround the temporal poles of the brain, projecting rostrally beyond the limits of the presphenoid. The lateral margins of the alisphenoid are already fully apposed to the squamosal portion of the temporal bone in the middle cranial fossae of the newborn chimpanzee.

Immediately posterior to the alisphenoid and squamous temporal, the petrous part of the temporal forms from endochondral ossification. It houses the middle and inner ear and allows passage of the internal carotid artery. The squamous temporal and petrous temporal are closely apposed, and the suture between them remains visible in some specimens. The bony labyrinth of the inner ear (including cochlea and

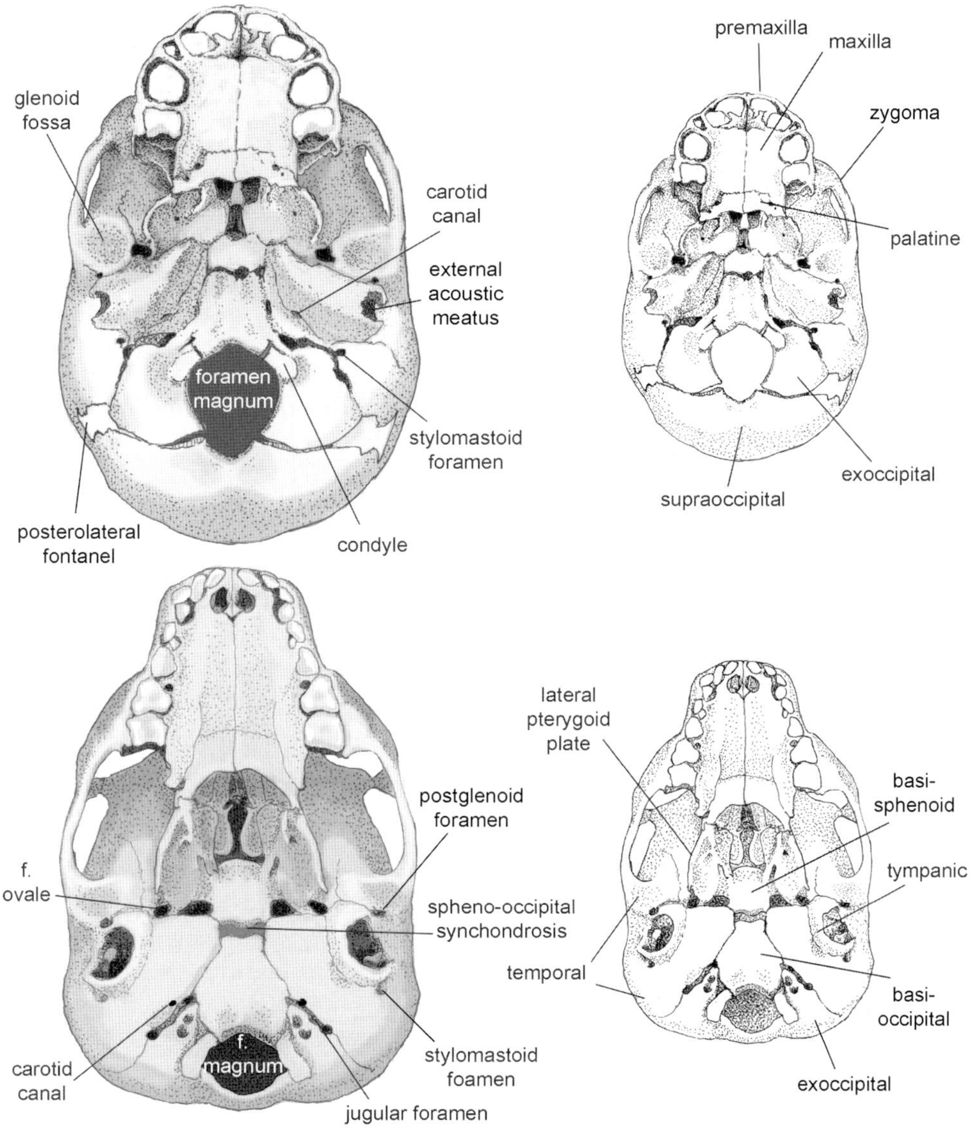

Figure 4.4 Basicranial view of late fetal *Pan troglodytes* and infant *Perodicticus potto*. Note certain differences in the position of passageways, such as the carotid canal, and foramen magnum.

semicircular canals) does not undergo modeling, and so the size and morphology of these structures is determined at the time of initial ossification and will remain static for the lifetime of the individual, although the capsule itself moves relative to other structures. The same is true for the ossicles of the middle ear (malleus, incus, and stapes). As a result, these inner- and middle-ear structures appear relatively large in the newborn primate compared to other cranial structures which do continue to grow. The tegmen tympani is well-ossified in newborn hominoids. The middle-ear cavity of primates is formed primarily by the petrous temporal.

The ectotympanic bone forms as an incomplete ring of intramembranous bone on the external wall of the skull, lateral and/or inferior to the petrous temporal and middle ear. The ectotympanic serves as the anchor for the tympanic membrane and in many cases contributes to the external auditory canal. In hominoids, other catarrhines, and tarsiers, the ectotympanic eventually grows laterally to form a bony ectotympanic tube. In platyrrhines and strepsirrhines, the ectotympanic retains its ring-like structure, and the external ear canal remains entirely cartilaginous. In the newborn chimpanzee, the ectotympanic is still ring-like and closely apposed to the auditory bulla, with some evidence for anterior and posterior tubercles forming at the margins of the meatus (see Figure 4.4). This is similar to humans, in which the tympanic annulus is still ring-like at birth (Scheuer and Black, 2004; Schwartz, 2007; Fricano, 2018).

The posterior cranial fossa houses the brainstem and cerebellum. Its anterior wall is formed by the clivus in the midline, extending from sella to foramen magnum, and the petrous temporal laterally. The floor and posterolateral walls are formed by the occipital, which is represented by four separate elements in the neonate: the supraoccipital, the bilateral exoccipitals, and the basioccipital. The intersections of the four occipital masses form the margin of the foramen magnum (Figure 4.4). The occipital bone has a relatively similar degree of ossification and similar features in all anthropoids, with some differences in the orientation of the foramen magnum. The occipital condyles are well-formed but bifurcated by the basioccipital–exoccipital synchondroses, which also appear patent.

Synchondroses of the cranial base have a profound effect on perinatal growth. By far, most of the literature has centered on midline synchondroses and their effect on anteroposterior elongation and cranial base angle (e.g., Lieberman et al., 2000; Smith et al., 2017b). The sphenooccipital synchondrosis is easily visible in dry skulls and the most frequently studied. Several other midline synchondroses are potentially highly relevant to the newborn stage of development: the intrasphenoidal (midsphenoidal), sphenoethmoidal, presphenoseptal, and ethmoseptal synchondroses. Two of these midline joints, the intrasphenoidal and presphenoseptal synchondroses, are not growth centers in humans at birth (Baume, 1968). Smith et al. (2017b) suggest this may be true for all hominoids.

Bilateral synchondroses have received little consideration in the literature, even concerning humans. In humans, such joints exist at birth only in the posterior cranial fossa. They include the jugular (petro-exoccipital) synchondrosis and the anterior and posterior intraoccipital synchondroses. The jugular synchondrosis of humans remains patent until adult ages in humans (Scheuer and Black, 2004). Within the middle cranial fossa, the alisphenoid and basisphenoid are separated by "a mixture of cartilage and ligaments" at birth, according to Bosma (1986, p. 105). Whether this site is a growth center at an earlier point in ontogeny has not been discussed.

The basicranium of the newborn skull possesses the following apertures: foramen magnum, foramen ovale, foramen rotundum (best visualized in endocranial view), jugular foramen, stylomastoid foramen, postglenoid foramen, and the optic foramen (Figure 4.4). In the newborn chimpanzee, the optic canals are fully ossified. The superior orbital fissure is oriented close to vertically and located close to the presphenoid. The foramen rotundum is fully ossified, but foramen ovale appears open laterally, and the foramen spinosum is not ossified. The carotid canal is incomplete in most cases, indicating some boundaries are as yet cartilaginous at birth. The condylar canal is present in the exoccipital. The exoccipital also has a deep notch for the hypoglossal nerve, but it remains open anteromedially.

4.3 The Skull of Newborn Old and New World Monkeys

A number of traits distinguish catarrhine and platyrrhine primates, including external structure of the nose, dental formula, and the sutural pattern at pterion. Our descriptions are based on CT image volumes of newborn primates representing five families of platyrrhines and a narrower sample of non-hominoid catarrhines.

4.3.1 The Facial Skeleton

The face of Old and New World monkeys is proportionately larger than that of hominoids, consistent with the expectation of a lower degree of encephalization. Nevertheless, the neurocranium is still wider than the face (specifically the width across the zygomatic arches) in newborn monkeys. The maxillary, premaxillary, zygomatic, lacrimal, and nasal bones all have facial surfaces which form the external midfacial skeleton, and the external sutures between these bones are well-established at birth. Although Old and New World monkeys display a range of orthognathic and prognathic types as adults, the newborn specimens we have observed are predominantly orthognathic. Monkeys displaying some prognathism include *Papio* and *Alouatta*, predictive of their characteristic adult morphology (Figure 4.5).

The general structure of the orbit is similar to that in hominoids. Proportionally, the orbits differ greatly among anthropoids at birth. The eyes of anthropoids are grown to a greater degree at birth compared to strepsirrhines (Cummings et al., 2012; unpublished data). Furthermore, the eyes of anthropoids achieve more than half of the adult axial diameter at birth, irrespective of head size. An allometric consequence is that the eyes of small-bodied species are proportionally larger, relative to head size, than larger species such as apes. The specific bones that form the postorbital septum are precociously in articulation (Schwartz, 2003; Smith et al., 2013; DeLeon et al., 2016, 2017). The contour of the orbit varies from rounded to slightly flattened in newborn anthropoids. *Saimiri* represents an extreme with a completely rounded orbit. Most bones are well ossified along the margin of the orbit. The degree of ossification in the orbit of some newborn monkeys (e.g., *Saimiri*) is comparable to that observed in the chimpanzee. In others (e.g., *Allenopithecus*, *Colobus*, and *Alouatta*), ossification is even more advanced.

The orbital plate of the frontal bone extends posteriorly to articulate with the orbitosphenoid, and the roof is more inclined than that in hominoids, ranging approximately 30–45° relative to the visual and occlusal planes. The frontal–orbitosphenoid suture is apposed, but still visible. The contribution of the orbitosphenoid to the orbital roof is highly variable, ranging from a minor contribution comparable to that of the chimpanzee in *Alouatta* and *Colobus*, to a more extensive contribution in *Allenopithecus* and *Saimiri*. The relative contribution of maxilla and zygomatic to the floor of the orbit is variable across monkeys, and the zygomaxillary suture is visible and clearly patent

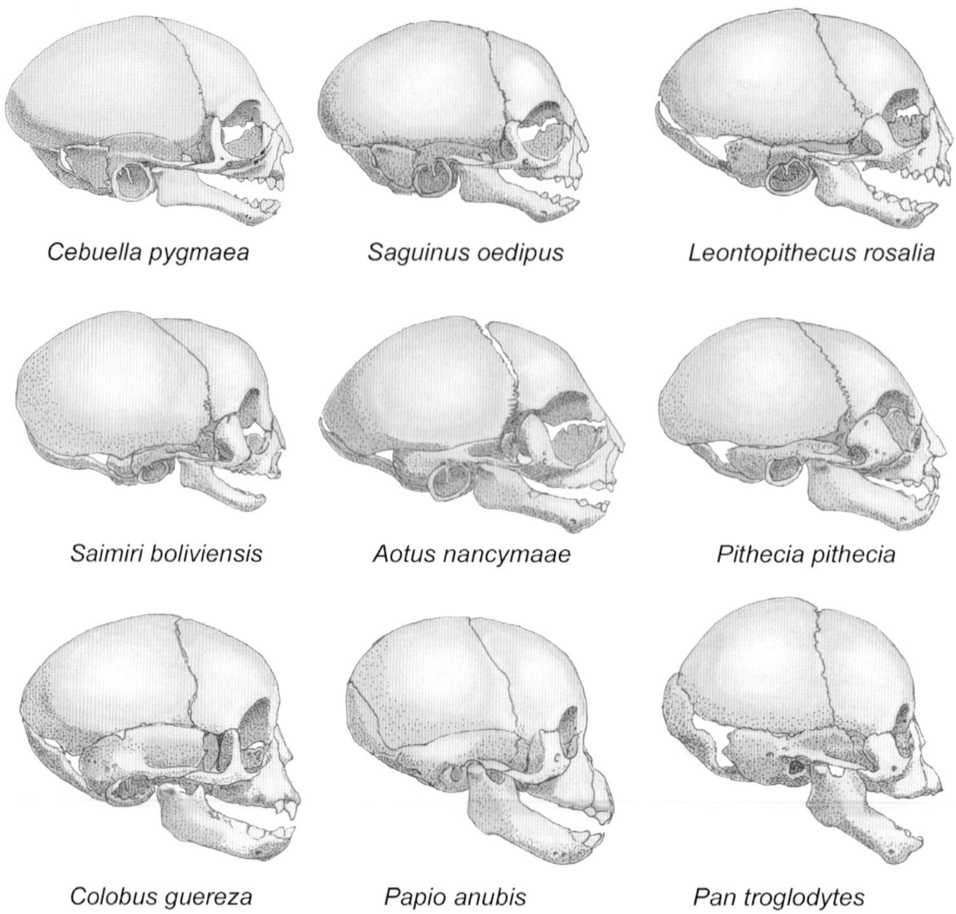

Figure 4.5 Lateral view of newborn skull in nine species of anthropoid primates. All skulls drawn to the same scale. Note varied shape of newborn calvarium. Also note relatively diminutive facial profiles, even in species that eventually develop greatly projecting midfaces (e.g., *Papio*).

in all specimens observed. The maxilla and zygomatic contribute almost equally to the orbit floor in *Allenopithecus*, and the zygomatic contributes to a slightly lesser extent in *Colobus*. In contrast, the floor is made up almost entirely of the maxilla in platyrrhines. The palatine also contributes to the inferomedial floor of the posterior orbit in platyrrhines (Box 4.2), but to a lesser extent in catarrhines. Some platyrrhines (e.g., *Alouatta* and *Saimiri*) show a characteristic cupping in the floor of the orbit, where the inferior orbital rim (including maxilla and zygomatic) is elevated above the floor of the orbit.

The lateral wall of the orbit is formed by the posterior lamina of the zygomatic and the orbital plate of the alisphenoid. As in hominoids, the orbital plate of the alisphenoid is a shared wall between the orbit and the middle cranial fossa, and the posterior lamina of the zygomatic is a shared wall between the orbit and the temporal fossa. The zygomatic process of the frontal bone also contributes to the upper portion of the lateral wall, although this is highly variable across monkeys. In some platyrrhines (e.g., *Alouatta*, and to a lesser extent in *Saimiri*), flanges from the alisphenoid extend rostrally to articulate with the zygomatic. In *Alouatta*, a second, similar flange extends rostrally just lateral to the foramen rotundum. It is continuous with the posterior maxilla and appears to contribute to the roof of the posterior alveolus.

The ethmoid, lacrimal, frontal, maxilla, and presphenoid all contribute in variable amounts to the medial wall of the orbit in monkeys. In *Saimiri* spp., the eyes are so large relative to the skull that the orbital plate of the ethmoid is resorbed during development (Maier, 1983). The result, postnatally, is an interorbital fenestra, or an open communication between the right and left orbits. However, the septum is still intact at birth (Maier, 1983) and even juvenile *Saimiri* can still possess a complete bony partition between the orbits. One little-emphasized aspect of the orbit in anthropoids is the varying extent of the presphenoid contribution to the orbit. Maier (1983) noted that perinatal *Saimiri* possesses an interorbital septum composed of the presphenoid. Recently, is has become clear that presphenoid projection into the midfacial region is a broader trend in monkeys (Smith et al., 2017b). The level at which the presphenoid projects is both between the posterior orbit and *superior* to greatly elongated choanal passageways.

The articulated right and left nasal bones are tall and slender and form a concave upper boundary of the piriform aperture in all newborn anthropoids. In shape, platyrrhines have a broader, more rounded superior border of the piriform aperture compared to catarrhines. In particular, the nasal bones of cercopithecoids may be described as nearly triangular, with the apex directed superiorly. Non-human anthropoids lack a recognizable anterior nasal spine at birth. The maxilla has a tall, vertical nasal process and a short zygomatic process. The premaxilla is relatively broad with alveolar open sockets anteroinferiorly. In cercopithecoids (e.g., *Allenopithecus* and *Colobus*), the nasal process of the premaxilla is tall and slender, extending superiorly between the maxilla and nasal bones. In contrast, in platyrrhines the premaxilla is shorter.

The nasal cavity in New and Old World monkeys is similar in most respects to that of hominoids, but the proportional size of the eyes limits interorbital portions of the nasal fossa in some species more than others, and may also limit maxillary pneumatization (Rossie, 2006; Smith et al., 2011). Medially, the perpendicular plate of the ethmoid remains cartilaginous in most monkeys, although we observed ossification in *Allenopithecus*. The cartilaginous nasal septum articulates with the presphenoid. Histological observations of the septal cartilage that abuts the presphenoid indicate active endochondral growth with zones resembling the epiphyseal plate of long bones, but with an anteriorly convex organization (Smith et al., 2017b). The alae of the vomer are closely apposed to the palatines and the presphenoid in platyrrhines, and expanded laterally to contribute to the medial orbit in newborn *Saimiri*. In the catarrhines, the vomer is ossified only as a trough along the inferior border of the cartilagenous septum.

Laterally in the nasal cavity, all newborn monkeys have tall, plate-like ethmoturbinals, while the maxilloturbinal is a simple horizontal fold (Figures 4.6 and 4.7). The nasoturbinal, if present, remains cartilaginous and is of rudimentary size in anthropoids compared to that of strepsirrhines. In tamarins and *Saimiri*, the root of

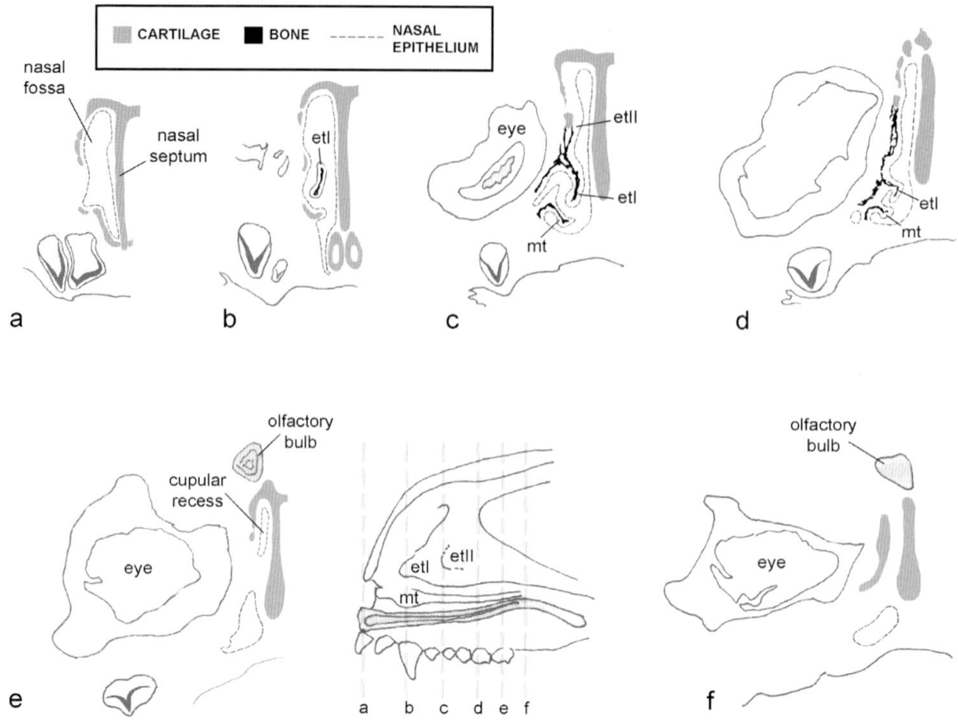

Figure 4.6 The right nasal fossa of a neonatal *Callithrix jacchus* is shown in schematic views, drawn from histology. The sections shown in a–f run from anterior to posterior through the midface, with cartilage and bone highlighted (see key, above left). Note that cartilage of the lateral wall occurs in small nodular masses superiorly. Where the ethmoid forms the medial orbital wall (c,d) it is undergoing endochondral ossification in newborn *Callithrix*. Turbinals are extensively ossified in *Callithrix* and many other anthropoids at birth. Note the nasal fossa ends as a vestigial space superiorly, the cupular recess (e), which lacks turbinals; a respiratory passageway continues inferiorly toward the pharynx (f). Approximate anteroposterior positions of the cross-sections are indicated using a schematic hemisected view of an adult marmoset (bottom center).

ethmoturbinal I may be partially unossified at birth. In *Saguinus* spp., a small, superiorly situated portion of the orbital plate of the ethmoid may also be unossified. The variability in ossification of these latter two foci suggests the ethmoid begins ossifying late fetally and that many elements are completing the process perinatally. It should be recognized that some portions of the cartilaginous nasal capsule of primates are *not* replaced by bone (Box 4.3) (Maier, 2000; Smith et al., 2008, 2012).

Homologues of the olfactory recess and paranasal spaces are easier to identify early in ontogeny. The only paranasal space that pneumatizes (invades bone) prenatally is the maxillary recess (Figure 4.7). Thus, many anthropoids have a true maxillary sinus at birth (exceptions are most catarrhines and *Saimiri* and *Cacajo*: Rae and Koppe, 2004; Rossie, 2006). In many platyrrhines, there remains a vestige of the olfactory recess, the cupular recess – two blind-ended culs-de-sac on either side of the septum. In newborn *Aotus* and *Alouatta*, cartilage surrounds the recess like a thimble at

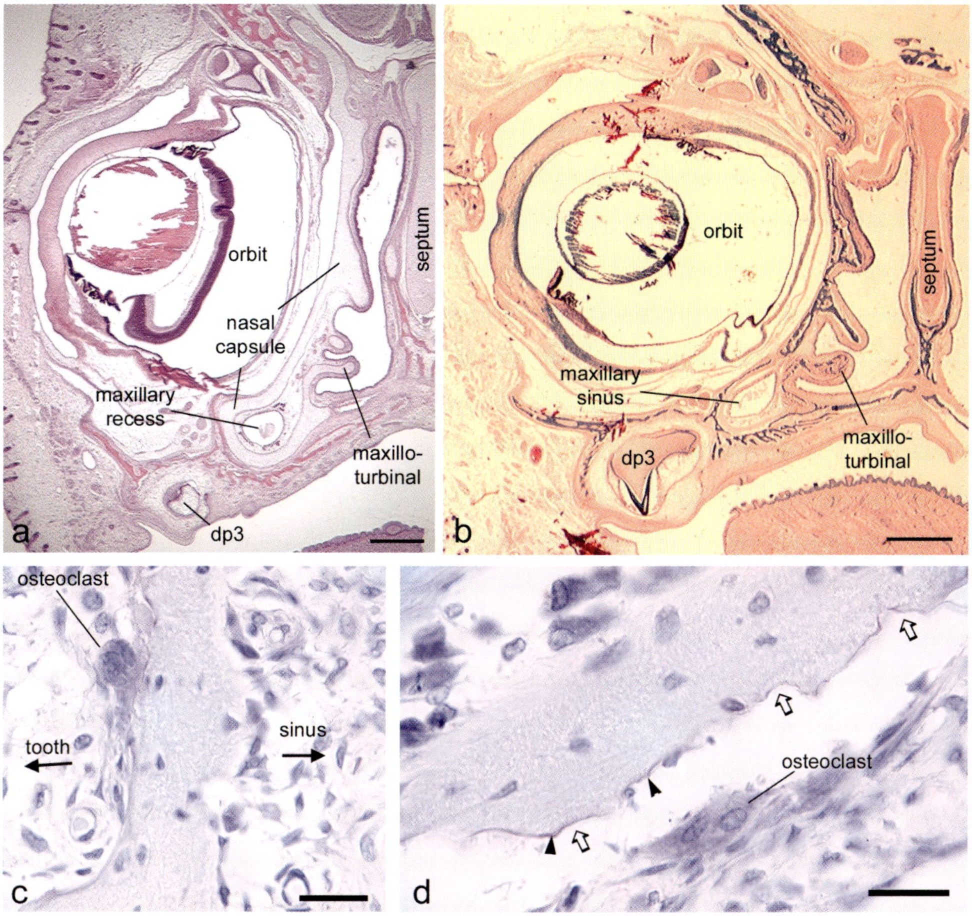

Figure 4.7 Pneumatization of the skull occurs in two phases. The first is termed "primary pneumatization," a result of folding of the fetal nasal capsule in tamarin (a, *Saguinus geoffroyi*, stained coronal section). At this stage the sinus is sometimes referred to as a recess. Once the cartilage of the capsule disappears (see Box 4.4) a paranasal recess may expand by osteoclastic activity, becoming a "true" paranasal sinus. The maxillary sinus of an older *S. geoffroyi* (b–d, neonate) is just commencing "secondary pneumatization." Sinus expansion can be identified where osteoclasts are present, or "scallop-shaped" marks indicate their activity (c,d). Expansion is opportunistic, and other adjacent structures, such as developing tooth germs (e.g., dp3, above) may inhibit sinus growth locally (c).

its back end (Box 4.3). As in hominoids, a small frontal recess persists in some platyrrhines (Rossie, 2006). This recess undergoes pneumatization postnatally (into the ethmoid and/or frontal bones) (see Box 4.4).

The palate of monkeys is highly variable in proportions, but similar in general structure to that observed in hominoids. Sutures of the palate are established at birth, but vary in patency across monkeys, ranging from fully apposed in some (e.g., *Alouatta*) to widely patent in others (e.g., *Saimiri*). The palatine foramina are complete in most specimens, but not, e.g., *Saimiri*. The incisive foramina are incomplete medially in *Alouatta* and *Saimiri*, comparable to chimpanzee, but complete in *Allenopithecus* and *Colobus*. The maxillary alveolar ridge is highly variable across monkeys, and shape is greatly influenced by the dentition. More details on relative development and eruption of teeth are provided in Chapter 5.

The general form of the mandible is similar to that of hominoids in many respects (e.g., short, posteriorly inclined ramus). The volume of the body corresponds to a great degree with the volume of the alveolar sacs for developing teeth (see Chapter 5). The condylar shape varies, being anteroposteriorly flattened in catarrhines, similar but more rounded in *Alouatta*, and rounded in many small-bodied platyrrhines. Both the coronoid and condylar processes are relatively low, and the coronoid slopes posteriorly. The mandibular symphysis is closely apposed and may already be fused in some newborn catarrhines (e.g., *Allenopithecus*), but in platyrrhines (with the exception of marmosets) it remains unfused at birth. No secondary ossification centers comparable to the chimpanzee were observed in our monkeys, but some specimens (e.g., *Alouatta*) showed a great deal of complexity in the symphysis. The mental spines are visible in some monkeys, and the muscular attachments on the gonial angle are well-developed. Subchondral bone on the mandibular condyle is thinly ossified. Characteristic morphology is already apparent in some specimens. In *Alouatta*, the flange along the inferior border of the symphysis is present and the gonial angle is expanded with the rounded contour observed in adults. In contrast, in *Saimiri*, the entire mandible is notably gracile, but already shows the characteristic flange at the inferior border of the symphysis.

The temporal fossa is small, and the articulation of temporal and zygomatic approximately halfway along the zygomatic arch is already complete. *Alouatta* has a pronounced postglenoid process at birth. The pterygoid plates are also small but fully ossified, including the pterygoid hamulus, which is notably elongated in *Alouatta*. For the specimens where the hyoid was present, the body and greater horns of the hyoid are ossified. The body of the hyoid is already superior–inferiorly elongated in *Allenopithecus* relative to that observed in hominoids. There is a wide non-ossified gap between the greater horns and the body. The styloid process in most specimens does not appear to be ossified, although small ossification centers inferior to the stylomastoid foramen were observed in some (e.g., *Alouatta*).

4.3.2 The Calvarium

The bones that participate in the calvarium are identical across primates, but the sutural interfaces vary, especially laterally. The cranial vault of newborn monkeys is composed almost entirely of frontal, parietal, and squamous occipital elements. The squamous temporal and alisphenoid have only minor contributions to total vault surface area. This is particularly pronounced in many platyrrhines (e.g., *Alouatta* and *Saimiri*), where the parietal is extensive, and the squamous temporal has an extremely small contribution to the lateral vault. In all anthropoids, the frontal bone contributes to the region of the "upper face" (Figure 4.5) and retains a metopic suture separating right and left frontals at birth. It is prominent in part due to orbital "frontation" typical of anthropoids, both in adults and subadults. Yet, there is also an ontogenetic effect. In monkeys at early ages, as in hominoids, the proportionately diminutive size of the midface accentuates the frontal region. However, the extent to which any part of the frontal bones is upright (in the frontal plane) varies. In all newborn cercopithecoids examined here, including *Trachypithecus*, *Colobus*, *Macaca*, *Papio*, and *Allenopithecus*, the frontal bone bulges above the orbits, to form at least a small, upright "forehead" (Figure 4.5). The remainder of the bone slopes posteriorly. In most newborn platyrrhines, the frontal bone begins to slope posteriorly immediately above the superciliary ridges (Figure 4.5). *Saimiri* is an exception.

The patency of calvarial sutures is highly variable across newborn monkeys. Bones are closely apposed in *Alouatta*, *Callithrix*, *Cebuella*, *Pithecia*, *Papio*, and *Macaca*, but more broadly patent in *Saguinus* spp. and *Aotus*. For calvaria that were undamaged in our observations, midline anterior and posterior fontanels were found in all species, and most species possess an anterolateral fontanel and a relatively large posterolateral fontanel (Figure 4.5), which is discussed further below in light of temporal bone ossification.

The pattern of sutures at pterion that distinguishes platyrrhines and catarrhines is already established at birth, although the anterior lateral fontanel remains open in many specimens that we observed. In catarrhines (e.g., *Colobus*), the frontal articulates with the alisphenoid (Ankel-Simons, 2007; Fleagle, 2013). In most platyrrhines, the frontal and alisphenoid are excluded from contact because the parietal and zygomatic bones have a sutural interface (Ashley-Montagu, 1933). There are exceptions to this rule. *Alouatta* has a frontoalisphenoid suture, such that the coronal suture ends inferiorly at the alisphenoid, as observed in catarrhines. Some *Saguinus* specimens have a small bony bridge connecting the two bones. In one *Colobus*, the squamous temporal extends anteriorly to close the anterior lateral fontanel, and as a result, the coronal suture ends inferiorly at the temporal. Interspecific variations have been noted in adults as well (Ashley-Montagu, 1933; and see Ankel-Simons, 2007), and there is evidence for strong heritability of sutural patterns (Wang et al., 2006).

4.3.3 The Cranial Base

In all newborn anthropoid monkeys, the bones of the basicranium are advanced in ossification and have established most synchondroses or sutural articulations. As in hominoids, the anterior cranial fossa consists mostly of the frontal bones. The orbital plates of the frontal extend posteriorly to the orbitosphenoid, and form a squamous suture, with the orbitosphenoid superior to the frontal margin. The relative contribution of the orbitosphenoid is variable across monkeys, being smaller in some (e.g., *Colobus*) and more extensive in others (e.g., *Saimiri*). In many specimens of both platyrrhines and catarrhines (e.g., *Colobus*, *Alouatta*, and *Saimiri*), each orbit protrudes into the anterior cranial fossa, creating a convexity. Relative to the area of the orbital plates, the olfactory fossa in monkeys is small. Crista galli and the cribriform plate remain unossified in newborns. In many platyrrhines, the olfactory bulbs are inferiorly displaced, resulting in an olfactory fossa that is deeply distinct from the contour of the anterior cranial fossa.

In the midline, basicranial bones (presphenoid, basisphenoid, and basioccipital) are substantially ossified in all primates. The orbitosphenoid is well-ossified and shows at least partial attachment to the presphenoid. The optic canal is fully surrounded by bone of the orbitosphenoid, comparable to other anthropoids. In some specimens, we observed a bony bridge from the posterior margin of the orbitosphenoid to the posterior margin of the presphenoid contributing to the lateral border of the optic canal, which differs from the caudally projecting anterior clinoid processes in, e.g., humans. The chiasmic sulcus is positioned at the posterior margin of the presphenoid. The intrasphenoidal synchondrosis (between presphenoid and basisphenoid) is closely apposed in most catarrhines (e.g., *Allenopithecus, Macaca*). The synchondrosis appears to have some initial bony bridging laterally in *Colobus*, indicating that fusion may be imminent. In contrast, the synchondrosis is more widely patent in newborn platyrrhines (e.g., *Alouatta* and *Saimiri*).

The bilateral alisphenoids, which make up a large proportion of the middle cranial fossae, are fused to the basisphenoid at birth (Box 4.5), except in some perinatal *Saguinus*. Thus, growth centers in the middle cranial base are lacking in nearly all newborn monkeys known to us (unpublished histological observations), as is apparently the case in humans (Bosma, 1986). To be sure, a joint may remain – a slight groove between the basisphenoid and alisphenoid. Maier (1993, citing Gaupp, 1910) called this joint the "junctura basipalatina." The lack of an actively growing synchondrosis at this joint represents a contrast to at least some newborn strepsirrhines and other mammals (see Section 4.5.3). In catarrhines, the alisphenoids project rostrally, such that the middle cranial fossa extends farther rostrally than the deepest part of the orbit. In contrast to the catarrhines, the alisphenoid wings in platyrrhines (e.g., *Alouatta* and *Saimiri*) do not extend more rostrally than the presphenoid. The lateral margins of the alisphenoid are already fully apposed to the squamosal portion of the temporal bone. The foramen rotundum is fully ossified, but the foramen ovale is open posteriorly, and the foramen spinosum is not present. Both of the more posterior foramina are known to form without the benefit of an endochondral template (Presley and Steel, 1976), so their contents traverse soft tissue in the newborn.

The petrous portions of the temporal bones contribute to the posterior part of the middle cranial fossae. The otic capsule and the anterior portions of the petrous temporal are well-ossified in almost all monkeys. The petrous part is incompletely ossified posteriorly, indicating a small amount of cartilage remains; the posterolateral fontanel is thus enlarged in dried crania. The mastoid region of the petrous temporal is more variable and not ossified or poorly ossified in some specimens (e.g. *Saimiri* and *Colobus*). The squamous temporal and petrous temporal are closely apposed, although the articulation between them remains visible. The mastoid process displays only a slight expansion. The bony labyrinth of the inner ear (including cochlea and semicircular canals) and the ossicles of the middle ear (malleus, incus, and stapes) appear relatively large in the newborn primate compared to other cranial structures which do continue to grow. The tegmen tympani is well-ossified in most specimens, but the anterior wall of the middle ear cavity remains cartilaginous in many specimens (e.g., *Colobus, Alouatta*, and *Saimiri*). A prominent carotid canal penetrates the petrous bone at its inferior surface near the tympanic ring. In numerous monkeys, the stylomastoid foramen is large and closely adjacent to the tympanic annulus, but it is poorly visible in some monkeys. In such cases, the facial nerve likely exits through a gap between the tympanic bone and the petrous temporal. The postglenoid foramen, when present, empties dural sinus blood into the external jugular vein (Conroy, 1982), representing a second major drainage route for cranial blood. A taxonomic pattern to its presence or absence in newborns is not clear at present. A prominent postglenoid foramen is present posterior to the glenoid fossa in most monkeys available for study and also in the late fetal chimpanzee (Figure 4.4). However, no postglenoid foramina are visible in some monkeys (e.g., it is lacking in *Saguinus*, *Alouatta*, and in single specimens of *Pithecia* and *Colobus*).

As described above, the ectotympanic is distinct between adult catarrhines and platyrrhines. Old World monkeys and hominoid adults have a bony ectotympanic tube. New World monkeys have a less-extensive bony ectotympanic ring. In newborn platyrrhines, the ectotympanic ring is ossified, but positioned variably at an oblique angle. At birth, the ring is incomplete superiorly. The anterior and posterior arms are closely apposed to the squamous temporal superiorly, but the inferior articulation between ectotympanic ring and petrous bulla remains visibly patent. This may allow for continued lateral expansion of the floor of the petrous bulla, eventually bringing the ectotympanic closer to a parasagittal plane. In newborn catarrhines, the ectotympanic has the general proportions of a ring and is positioned obliquely and relative to the squamous temporal and petrous bulla comparable to platryrrhines. However, the bony extensions that will eventually form the ectotympanic tube in catarrhines are already beginning to form by birth (Fricano, 2018).

In the midline of the posterior cranial base of all newborn monkeys, the spheno-occipital synchondrosis is widely patent. Bilateral synchondroses of the posterior cranial base (jugular, interoccipital) resemble those described above for hominoids. These are patent in all newborn anthropoid skulls and may thus remain active. In all monkeys, the occipital condyles are apparent, and extend anterior to the basioccipital–exoccipital synchondroses. Articulations of the exoccipitals with basioccipital and squamous occipital are variable in their patency, remaining widely patent in most monkeys but more closely apposed in some (e.g., *Colobus*).

4.4 The Skull of *Tarsius*

Tarsiers possess some soft-tissue similarities in common with other haplorhines, such as the lack of a hairless rhinarium and disassociation of the external nose from the vomeronasal system (Pocock, 1918; Rossie and Smith, 2007). Osteology of the adult tarsier skull is unlike any other living primate in the degree of ocular hypertrophy, with skeletal support that has engendered some controversy (e.g., Cartmill, 1980; Simons and Rasmussen, 1989). Compared to living anthropoids, their eyes are extra-ocular (ectopic) to a degree that more closely resembles nocturnal strepsirrhines (Rosenberger et al., 2016), and in size the eyes are suggested to rival brain size in adults (Sprankel, 1965). Subadult tarsiers, however, lack some of the skeletal features in the adult orbits (e.g., postorbital plate of zygomatic), and also have eyes that appear not to be as proportionally large as those of adults (see Box 4.6). The eyes of tarsiers are grown to a lesser degree at birth than in other living primates, only approximately one-third of adult axial diameter (Cummings et al., 2012). Our observations are based on image volumes of four *Tarsius syrichta* in the perinatal period (late fetal to 6 days postnatal; one newborn is featured in the atlas section of the book).

4.4.1 The Facial Skeleton

The newborn tarsier face is notably prognathic relative to anthropoids, with midfacial prognathism similar to strepsirrhines described in the following section. The nasal process of the frontal bone slopes anteriorly, very similarly to newborn lorisoids. As a result, the lateral contour suggests a bulbous calvarial part while the nasal process projects to emphasize the shape of a "snout" in the newborn. As observed in anthropoids, external midfacial sutures are well-established at birth.

Certain features that have been attributed to eyeball hypertrophy, such as the circumorbital flanges, are lacking at birth. These flanges, as well as an outward rotation of the ascending process of the zygomatic, gradually emerge during the first week of infancy, presumably in association with a rapid spurt of eyeball growth leading to increasing ectopy (DeLeon et al., 2016; Rosenberger et al., 2016). Elsewhere we have asserted that the partial-to-complete postorbital closure in tarsiers and anthropoids may be evolutionarily convergent, due to compensatory osteogenic support of the rapidly enlarging eyes (Smith et al., 2013; DeLeon et al., 2016).

The structure of the orbit in newborn tarsiers is distinct in many respects from anthropoid primates, primarily in the lateral wall and floor of the orbit. This unique structure is apparent even before the massive postnatal expansion of the eyeballs that defines the tarsier's unique cranial characteristics. The zygomatic process of the frontal is quite small and more similar to that in anthropoids, as opposed to the frontal spur observed in strepsirrhines. In the roof, the orbital plate of the frontal bone extends posteriorly to articulate with the orbitosphenoid, and the roof even at birth is more inclined that that in anthropoids, in excess of 45°. The frontal–orbitosphenoid suture is apposed, but still visible. The contribution of the orbitosphenoid to the orbital roof at this stage is comparable to that observed in some monkeys.

The floor of the orbit is distinct from that in anthropoids. The floor of the orbit in tarsiers consists almost entirely of maxilla anteriorly, similar to platyrrhines. The palatine and pterygoid plates of the sphenoid contribute substantially to the posterior floor of the orbit. The relative volume of the molar tooth crowns is notably large, and the maxilla which houses them extends almost to the lateral edge of the orbit. The zygomatic bone is remarkably flat (planar). It and the zygomaxillary suture are oriented close to a parasagittal plane, and as a result, the zygomatic does not contribute to the floor of the orbit. The tarsier does not show any cupping of the floor of the orbit as observed in some anthropoids.

One of the most striking differences in the tarsier orbit relative to other haplorhines is the lack of postorbital septation and the existence of a postorbital membrane at birth (Smith et al., 2013; DeLeon et al., 2017). The lateral wall of the orbit has relatively little bony support (see Figure 9.5). The frontal process of the zygomatic extends superiorly in a paddle-like shape, but there is no posterior lamina articulating with the alisphenoid. Smith et al. (2013) used histology to demonstrate the existence of a fan-like, fibrous membrane extending from the frontal process of the zygomatic to the frontal and most superior edge of the alisphenoid. Ossification of a vertical orbital plate of the alisphenoid observed in anthropoids is not apparent in the newborn tarsier, and only the rostral edge of the alisphenoid contributes to the lateral wall of the orbit.

The lacrimal, ethmoid, frontal, maxilla, and presphenoid all contribute to the medial wall of the orbit in newborn tarsiers, as in some monkeys. The frontal bone makes up approximately the superior half of the medial wall. The rostral extension of the presphenoid into the medial orbit as the interorbital septum observed in platyrrhines is also evident in tarsiers. Tarsiers show a notable contribution of the palatine bone to the medial wall, where the vertical plate extends superiorly and closely approximates the presphenoid. The region between presphenoid, frontal, ethmoid, and palatine remains cartilaginous through birth, but ossifies rapidly soon after birth. In a 6-day-old tarsier, free edges of a small apical interorbital foramen are apparent in the ossified medial wall. The osseous medial orbit is primarily the frontal bone, with a small contribution of presphenoid to the posterior margin.

In the nose, the piriform aperture is rounded superiorly and tapers inferiorly. The nasal process of the maxilla is slender and pointed; the same is true for the premaxilla. The articulated right and left nasal bones are relatively narrow, more similar to anthropoids than to strepsirrhines. The interpremaxillary junction is rounded on the superior aspect, at the floor of the piriform aperture.

The newborn tarsier nasal cavity is internally simple, as in anthropoids. The nasal septum is primarily cartilaginous, and the vomer is present as a shallow trough along the inferior border of the septum, approximating the vertical plates of the palatines posteriorly and extending anteriorly only as far as the premaxillary–maxillary suture. Laterally, the newborn tarsier has a rudimentary semicircular crest, two small ethmoturbinals, a single scroll maxilloturbinal, and no smaller turbinals. There is no frontal recess, and a very small maxillary recess that shows no signs of pneumatic expansion at birth. Ossification advances rapidly postnatally and is far more progressed in a 6-day-old compared to a P0 or P1 neonate, with only the anterior tips of the ethmo- and maxilloturbinals remaining cartilaginous.

The maxillary alveolar ridge is distinctive and influenced primarily by the dentition. As described in Chapter 5, the permanent molars of tarsiers are significantly larger than the other dentition. As a result, the tarsier dental arcade has a distinct "ham"-like shape apparent even at birth; narrow anteriorly and flaring dramatically wider in the molar region. The maxilla and palatine are closely apposed on the palate, although all sutures are clearly patent. The maxilla–palatine suture is obliquely oriented in the tarsier and the palatine foramina are relatively small and widely separated along that suture. These distinctive characteristics are likely related to bone modeling as a product of growth of the large molars. The incisive foramina are complete, and the maxillary–premaxillary suture is apposed but clearly visible.

The mandibular corpus is relatively deep dorsoventrally, with a similar state of dental eruption comparable to the upper jaw. Overall, the alveolar bone is thicker than in most other primates at birth; it is thicker medially than laterally. The lingual surface of the mandible is smooth, while externally the corpus bulges at the intersection of dp4 and M1. The symphysis is open with rounded edges approaching the joint, and no secondary ossification centers are visible. The ramus leans posteriorly to a similar degree as seen in galagids, but less than in lemurids. The condylar process is low, and the subchondral bone is thinly ossified. The coronoid process is low, hooked, and directed posteriorly. The mental spines are well-developed.

The temporal fossa is small. The zygomatic process of the temporal is well-developed as an extension from the squamous temporal, and the articulation of temporal and zygomatic approximately halfway along the zygomatic arch is already complete, but remains a simple, obliquely oriented suture at birth. The pterygoid plates are fully ossified. The lateral pterygoid plate is anteroposteriorly elongated and overlaps extensively with the petrous temporal deep to it. In contrast, the medial pterygoid plate is small, although the pterygoid hamulus is already present. We did not observe styloid processes or hyoids in these specimens.

4.4.2 The Calvarium

Similar to the anthropoids, the cranial vault of newborn tarsiers is composed almost entirely of frontal, parietal, and squamous occipital elements. The parietal is most extensive, followed by the frontals, with a smaller contribution from the squamous occipital. The squamous temporal and alisphenoid have minimal contributions to total vault surface area. Calvarial bones are well-ossified with most sutures closely articulated, but patent. The anterior fontanel is small at birth and absent in a 6-day-old infant. The posterolateral fontanel is smaller than that observed in many anthropoids and strepsirrhines, primarily due to a more fully ossified temporal bone in the tarsier. The anterolateral fontanel is the largest of the fontanels in newborn tarsiers, but closes very rapidly. It is nearly closed in the 6-day-old *T. syrichta*.

The pattern of sutures at the pterion is distinct from other primates in the newborn tarsier. The zygomatic is not in close approximation to the frontal, parietal, or alisphenoid. Instead, the arc of the coronal suture is roughly in line with the anterior border of the alisphenoid, although the large anterior lateral fontanel intervenes. Eventually, the frontal and alisphenoid will make direct contact.

4.4.3 The Cranial Base

As in anthropoids, the orbital plates of the frontal bones make up most of the anterior cranial fossa. The relative contribution of the orbitosphenoid is comparable to the lower end of the range observed in anthropoid primates. Relative to the area of the orbital plates, the olfactory fossa in tarsiers is small and rostrally displaced, similar to that observed in *Saimiri*. The orbital roof protrudes into the anterior cranial fossa, creating a pronounced convex surface. Derivatives of the mesethmoid, including crista galli and the cribriform plate, remain unossified in the perinatal specimens observed.

In the middle cranial fossa, the presphenoid and basisphenoid are ossified and notable for being narrow and mediolaterally compressed relative to the width of the cranial base. In contrast, the orbitosphenoid is wide and fully fused to the presphenoid. The intrasphenoidal synchondrosis remains widely patent, and the chiasmic sulcus is positioned dorsal to the synchondrosis. The sella turcica is formed by the basisphenoid. The bilateral alisphenoids are anchored at synchondroses lateral to the intrasphenoidal synchondrosis via lateral (alar) processes off the rostral margin of the basisphenoid (Box 4.5), comparable to the spatial relationship observed in *Saimiri* and more rostral than observed in most anthropoids. The alisphenoids are also distinct in being rostrocaudally compressed relative to the width of the cranial base. This may be related to crowding from the expansive petrous temporals immediately caudal. Similar to the platyrrhines we observed, the alisphenoid wings only extend as far as the caudal border of the presphenoid. The suture between alisphenoid and squamous temporal is closely apposed. As in anthropoids, the optic canals are fully ossified at birth. The superior orbital fissure is widely open, due to lack of ossification in the alisphenoid. Foramen rotundum is fully ossified, but foramen ovale and foramen spinosum are not encased in bone as yet.

The otic capsule and the petrous temporal are well-ossified in the tarsier, including the mastoid region which showed delayed ossification in some anthropoids. The squamous temporal and petrous temporal are closely apposed, although the articulation between them remains visible. The mastoid process displays only a slight expansion. The bony labyrinth of the inner ear (including cochlea and semicircular canals) and the ossicles of the middle ear (malleus, incus, and stapes) appear relatively large in the newborn primate compared to other cranial structures which do continue to grow. The tegmen tympani is well-ossified, but the anterior wall of the auditory bulla remains cartilaginous at birth.

The ectotympanic ring in newborn tarsiers is ossified, and positioned closer to a parasagittal plane than that observed in many other primates. The diameter of the ectotympanic ring appears small relative to the volume of the otic capsule. Inferiorly, the ring is positioned just anterior to the carotid canal. At birth, the ring is incomplete superiorly. The anterior and posterior arms are closely apposed to the squamous temporal superiorly, but the inferior articulation between ectotympanic ring and petrous bulla remains visibly patent (Fricano, 2018).

4.5 The Skull of Newborn Strepsirrhines

Strepsirrhines are distinct from haplorhines in a lesser degree of encephalization, lateral orbits enclosed only by a postorbital bar, wider interorbital breadth, and longer snouts. In adult strepsirrhines, relative orbit size varies according to activity pattern (Kirk, 2006), but in newborn strepsirrhines such differences (e.g., larger eyes in nocturnal species) are not yet realized (Cummings et al., 2012).

4.5.1 The Facial Skeleton

Among primates, strepsirrhines have the largest facial skeleton relative to neurocranial size, again consistent with the expectation of lower encephalization relative to haplorhines. The maxillary, premaxillary, zygomatic, lacrimal, and nasal bones all have facial surfaces which form the external midfacial skeleton, and the external sutures between these bones are well-established at birth (Figure 4.8). In newborn strepsirrhines the orbits are moderately frontated and the frontal bone exhibits slight "bossing" at birth. This is a sharp contrast to adults, in which the orbits and especially the calvarial portion of the frontal bone lean posteriorly (Figure 3.2). In addition, because newborns lack the superciliary ridge of adults, the superior orbital margin is more integrated with the calvaria. Even though most newborn strepsirrhines possess a "snout," this latter aspect represents a parallel to anthropoid ontogeny. In both groups, the primate develops a brim to the orbit, the superciliary ridges, as the midface and orbital rim project away from the calvaria (Figure 4.8). Newborn strepsirrhines are distinct from anthropoids in the rostral positioning of the nasal cavity relative to the anterior cranial fossa, and the inclined orientation of the cribriform plate.

The frontal and zygomatic bones form the majority of the circumorbital perimeter. The maxilla forms the inferomedial margin until its articulation with the lacrimal. The lacrimal fossa and opening to the nasolacrimal canal occurs on or slightly anterior to the circumorbital margin. Along the orbital margin, the zygomaticomaxillary and frontolacrimal sutures are in close articulation in newborn strepsirrhines. Laterally, the zygomaticofrontal suture is the least closely articulated of the circumorbital sutures.

Unlike all known haplorhines, the frontal bone of strepsirrhines contributes to the lateral margin of the orbit with a distinct zygomatic process. The process is present even prenatally, as an inferiorly and laterally projecting "spur," prior to the formation of the zygomaticofrontal sutural articulation (Ramaswami, 1957; Cummings et al., 2012). The precocious, prenatal presence of a frontal spur is a contrast to extant anthropoids and tarsiers (Smith et al., 2013) and is suggested to be a structural response to the bilateral width of the eyes in strepsirrhines relative to the width of the brain (DeLeon et al., 2016). The ends of the opposing bones are sharply tipped in galagids and most lemurids. In some large lemurids and *Propithecus*, the orbital processes of the zygomatic and sometimes the frontal bones are more blunt-ended. In *Nycticebus* they are blunt-ended and closely approximated. In cheirogaleids the tips of the frontal and zygomatic barely touch at the lateral orbit, and there may even be a gap between the bones. Sutural characteristics have not been histologically examined in many species, but in *Mirza*, the frontal and zygomatic approach one another within a ligament, suggesting that ossification of the orbital processes follows this path during formation of the postorbital bar (Smith et al., 2013).

The floor of the orbit is distinct from that in anthropoids and more similar in many respects to that of the newborn tarsier. The maxilla makes up almost all of the floor of the orbit, and extends all the way to its lateral boundary, although it is thinly or minimally ossified around the developing first molar tooth crown. The narrow maxillary process of the zygomatic contributes to the inferior margin of the orbital rim, extending medially as far as the infraorbital foramen (e.g., *Varecia*). Posteriorly, the palatine contributes to the floor of the orbit, but mostly the bony orbit is open to the temporal fossa laterally and inferiorly. Most strepsirrhines (e.g., *Varecia*) show a cupping of the floor of the orbit, just posterior to the orbital rim, as observed in some anthropoids.

The lacrimal, ethmoid, frontal, maxilla, and sphenoid bones all contribute in variable amounts to the medial wall of the orbit in strepsirrhines (Box 4.2). Similar to tarsiers, the frontal bone makes up approximately the superior half of the medial wall, related to the ventral position of the olfactory fossa relative to the orbital roof. The rostral extension of the presphenoid into the medial orbit observed in platyrrhines and tarsiers is also evident in strepsirrhines (e.g., *Varecia*). Also similar to tarsiers, strepsirrhines show a notable contribution of the palatine bone to the medial wall, where the vertical plate extends superiorly and closely approximates the presphenoid (Box 4.2). However, in contrast to tarsiers, the pterygoid plates of the sphenoid do not make a substantial contribution to the orbit. The region between presphenoid, frontal,

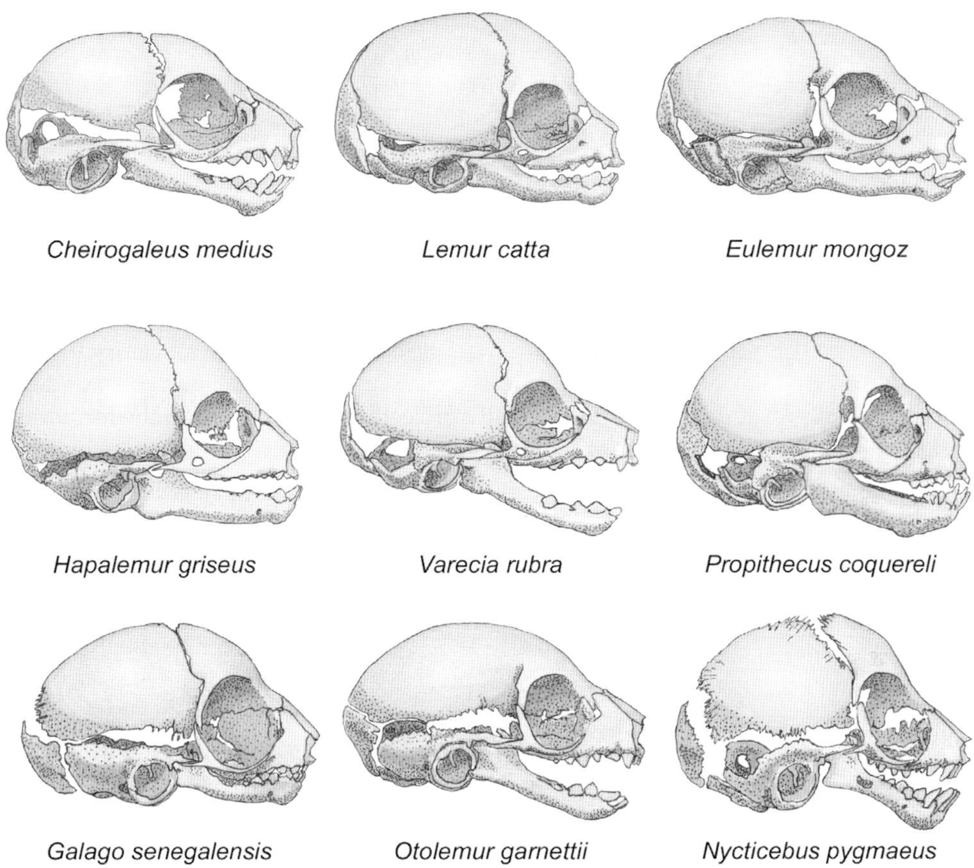

Figure 4.8 Lateral view of newborn skull in nine species of strepsirrhine primates. All skulls drawn to the same scale. These views emphasize variation in "snout" length and relative orbit size at birth. Note more erupted teeth in small species (*Cheirogaleus*), lorisoids, and *Propithecus*. Also note the relatively robust mandible in the folivores (*Hapalemur* and *Propithecus*). The *Nycticebus* specimen was the smaller of two specimens recorded at P0, and may represent premature birth, thus explaining the exaggerated appearance of sutures and fontanels.

ethmoid, and palatine is well-ossified and the bones are apposed in many of our specimens (e.g., *Varecia*).

The nasal process of the maxilla is tall and rounded in lorisoids, but low and rounded in lemuroids. In all the lemurids examined, a triangular gap exists at the intersection of the frontal, nasal, lacrimal, and maxillary bones, an area of exceedingly thin or poorly mineralized bone (Figure 4.8).

In strepsirrhines, as in most mammals, the hindmost part of the developing nasal capsule gives rise to the ethmoturbinal complex (Maier, 1993; Smith and Rossie, 2008). The presphenoid body ends at the prespheno-septal synchondrosis, which is nearly parallel to the end of the nasal capsule (in *Varecia* spp.) or fragments thereof (Box 4.4). Bony extensions from the orbitosphenoid grow around the posterior end of the nasal cavity; thus, the posterior nasal capsule support is taken over by the sphenoid bone in strepsirrhines. All strepsirrhines possess at least four ethmoturbinals, and one or more of them are sequestered within a cul-de-sac that is surrounded by capsular cartilage – the olfactory recess (Figure 4.9; Box 4.3; Maier, 1993; Smith and Rossie, 2008). The posterior end of the nasal capsule varies at birth in strepsirrhines, being completely absent (*not* ossified) in lorisoids, and fragmented in most lemuroids (with the exception of *Varecia*, in which it is still cartilaginous to the posterior end at birth). Newborn primates are unusual among mammals in the fragmentation of the posterior-most portion of the nasal capsule, and this may be regarded as one reflection of a broad trend in nasal reduction (Smith et al., 2017b). As in most mammals, strepsirrhines have a horizontal separation of the posterior-most nasal chamber from the respiratory passageway. This partition, the transverse lamina, separates the olfactory recess, above, from the nasopharyngeal duct(s), below. The transverse lamina of adults is partly endochondral (via ossification of the lamina transversalis posterior) and partly intramembranous (choanal process of the vomer).

The nomenclature for nasal turbinals has varied greatly (see Smith et al., 2007, for further discussion). We adopt a terminology rooted in ontogeny, as advocated by Maier (1993). The only distinction of our terminology from that of Maier is that the first two medially positioned scrolls of the ethmoid are treated as independent

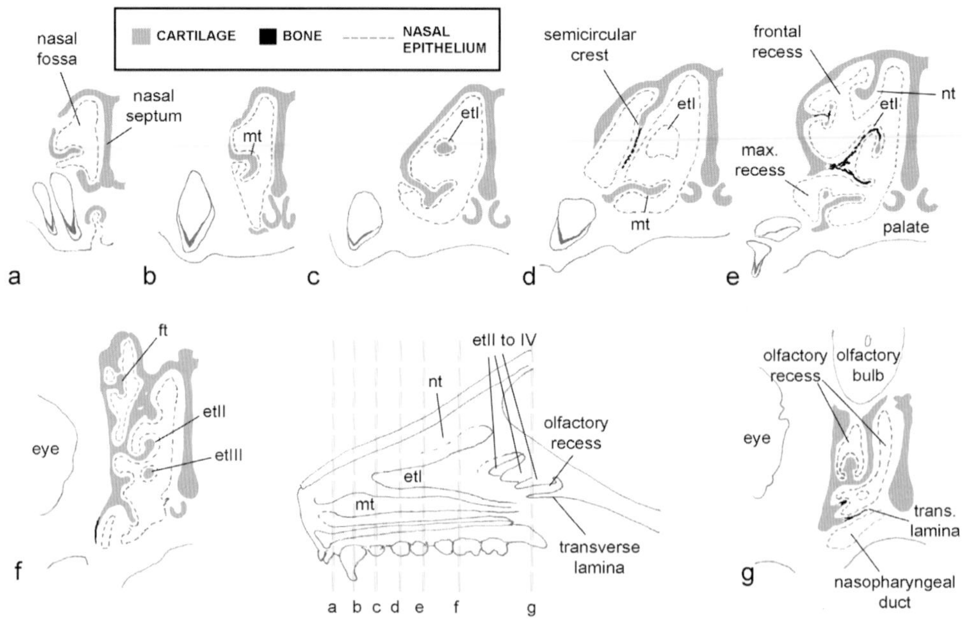

Figure 4.9 The right nasal fossa of a neonatal *Mirza coquereli* is shown in schematic views, drawn from histology. The sections shown in a through g run from anterior to posterior through the midface, with cartilage and bone highlighted. Note that much more cartilage of the lateral wall remains compared to anthropoids. The turbinals also have more persisting cartilage in this and some other strepsirrhine species (see text). Approximate anteroposterior positions of the cross-sections are indicated using a generalized hemisected view of an adult cheirogaleid (bottom center). et, ethmoturbinal; ft, frontoturbinal; mt, maxilloturbinal; nt, nasoturbinal.

turbinals rather than anterior and posterior lamina of the same turbinal (for rationale see Smith and Rossie, 2008). With this in mind, most strepsirrhines have four turbinals within or projecting anteriorly from the olfactory recess located in the medial row; these are the ethmoturbinals (often called "endoturbinals" because they occupy the space closest to the nasal septum). An exception is *Daubentonia*, which has six ethmoturbinals (Maier and Ruf, 2014). Strepsirrhines are greatly variable in the number of turbinals that nest far away from the midline. These form from separate regions of the developing nasal capsule (Box 4.3). Those that develop between ethmoturbinals are called interturbinals. Frontoturbinals develop within the frontal recess (Smith and Rossie, 2008). The turbinals of strepsirrhines are morphologically more complex than those of haplorhines, forming elaborate scrolls or double scrolls rather than simple folds/plates or single scrolls. The number and complexity of turbinals in newborns may not correspond to adults in all cases, for two reasons. First, newborn strepsirrhines have some intraspecific variation in the number of smaller turbinals (interturbinals, frontoturbinals). Second, turbinals become more elaborate postnatally via secondary bony outgrowths after the turbinals have ossified (Smith et al., 2016).

The ethmoid bone, as a whole, is less ossified at birth in strepsirrhines than in anthropoids. This is particularly so in lemuroids (Box 4.2). The result of this is that CT reconstructions of infant strepsirrhines show an incomplete posterior separation of the nasal fossae, and an incomplete medial orbital wall. In infant *Propithecus*, small islands of bone may be observed within the septal cartilage just anterior to the presphenoid. This suggests mesethmoid ossification may be imminent near birth in sifakas. In other respects all strepsirrhines are broadly similar. The first ethmoturbinal is the only turbinal to commence ossification before birth, although subtle differences in rate of its ossification may exist among lemurids (Martell et al., 2016). The most posterior turbinals remain unossified at birth. This may also be true of *Daubentonia*, based on a late fetal specimen (Maier and Ruf, 2014). The transverse lamina is in a state of partial ossification at birth. The vomer is present in most strepsirrhine specimens as a bony trough along the inferior margin of the cartilaginous nasal septum. The alae are widely separated, and each is in close approximation to the orbitosphenoid/presphenoid and palatine at the roof of the nasal cavity.

The alveolar process of the maxilla is low and has a "scalloped" appearance where the incisors, canine, and all or most deciduous molars are erupted through the alveolar socket in newborn strepsirrhines (Smith et al., 2015). In lemuroids the interpremaxillary junction is a depressed region following the rounded contour of the piriform aperture. However, in galagids the bones unite to form a rostrally projecting keel. The premaxilla is narrow with small alveolar open sockets anteroinferiorly. The nasal process is slender, partially overlapping the nasal bone on its lateral side, but only minimally in large-bodied lemurids, and not at all in galagids. The piriform aperture is oval or rounded, tapering inferiorly. The facial surface of the lacrimal bone is larger than the orbital surface and intervenes between the frontal and maxillary bones. The articulated right and left nasal bones are relatively broad in newborn strepsirrhines, especially so in galagids.

All newborn strepsirrhines that we observed had a rostrocaudally elongated hard palate relative to anthropoids, consistent with the elongated snout in these animals. The palatal shelves of the maxilla, rather than that of the palatines, appear to be the major source of this proportional variation. Nevertheless, there is a great deal of variation in the shape of the maxillary–palatine suture.

The mandibular corpus is relatively slender in all strepsirrhines, and the bone is thin. Both the external and lingual surfaces of the mandible are relatively smooth, with no discernible mylohyoid line, spine of the lingula, or mental spine. The symphysis is open with rounded edges approaching the joint. Interdental septa are prominent slats of bone separating the crypts of anterior teeth and dc from dp2 crypts, but are incomplete between crypts of postcanine teeth. The ramus leans posteriorly, such that the angle of the mandible is obtuse, although less so in cheirogaleids. The condylar process is relatively low, and the coronoid is relatively narrow anteroposteriorly and upright in most strepsirrhines, but more rounded and anteroposteriorly broad in cheirogaleids.

Similar to haplorhines, the temporal fossa in newborn strepsirrhines is small relative to its eventual adult size. The medial pterygoid plate and the pterygoid hamulus are well developed, but the lateral pterygoid is less pronounced.

4.5.2 The Calvarium

Calvarial bones are fully formed except in several respects. First, in all strepsirrhines the metopic suture is fully patent at birth, and thus two frontal bones are found. Second, all fontanels are large openings in the skull except for the anterolateral fontanel, which is relatively small or absent in all strepsirrhines. The posterolateral fontanel is exceedingly large as a bony aperture (Figure 4.8). However, a relatively large extent of this opening is occupied by unossified portions of the endochondral part of the temporal bone.

Calvarial sutures are closely articulated except adjacent to fontanels. The parieto-temporal suture appears to be more widely patent compared to haplorhines. In the region of the pterion, the parietal extends farther rostrally than the alisphenoid, and the frontal and alisphenoid are widely separated. The zygomatic makes no contribution to the braincase in strepsirrhines and is limited to the inferior portion of the postorbital bar.

4.5.3 The Cranial Base

As in other primates, bones of the cranial base are advanced in ossification and have established most synchondroses or sutural articulations. Strepsirrhines are distinct from haplorhines in the relatively large size of the olfactory fossa and proportionately reduced contribution of the frontal bones to the anterior cranial fossa. The frontal bones slope posteriorly to the orbitosphenoid, which is expansive in most strepsirrhines, overlapping the high end of the range observed in haplorhines. The orbital plates of the frontal bones also slope medially to contribute to the walls of the expansive olfactory fossa, which cups the relatively large olfactory bulbs and is deeply recessed relative to the orbital plates. The olfactory fossa is often triangular in shape, with the base anteriorly and the apex posteriorly. Crista galli and the cribriform plate remain unossified in newborn strepsirrhines.

In the midline, basicranial bones (presphenoid, basisphenoid, and basioccipital) are substantially ossified in all primates. Well before birth, the presphenoid body and orbitosphenoid are fused. The optic canal is complete, and the orbitosphenoid extends posteriorly as a bony bridge to the posterior margin of the presphenoid, contributing to the lateral border of the optic canal, similar to that observed in many monkeys. The chiasmic sulcus is positioned at the posterior margin of the presphenoid. A possible foramen of Rauschke was identified at sella in some specimens (e.g., *Varecia*). The intrasphenoidal synchondrosis (between presphenoid and basisphenoid) is widely open in all strepsirrhines, and there is variation in the relative width of the presphenoid and basisphenoid.

Between the basiocciptal and basisphenoid is the spheno-occipital synchondrosis, and the intrasphenoidal synchondrosis separates the basiphenoid and presphenoid; both are active, bidirectional growth centers at birth. More anteriorly, there is variability in the degree of ossification of the presphenoid body and the ethmoid bone. In newborns, the presphenoid body typically has a blunt end posteriorly, adjacent to the intrasphenoidal synchondrosis; anteriorly, it may be round or oval and blunt-ended.

Bilateral to the basisphenoid are paired alisphenoids, closely apposed in most cases, yet still visible (Figure 4.4). Fusion may occur in the perinatal timeframe, but at least some newborn strepsirrhines still possess cartilage at the basi-alisphenoid synchondrosis (Box 4.5). This growth center has been described in other mammals during development (Spatz, 1964; Maier, 1993), and we therefore suspect anthropoids (if not all haplorhines) are derived in lacking it. In most specimens, the alisphenoid extends anteriorly, closely apposed to the lateral side of the intrasphenoidal synchondrosis and the presphenoid. Further characterization of growth that results in apposition of alisphenoid against the lateral side of the presphenoid is needed. The alisphenoid is associated with the temporal pole of the brain in strepsirrhines, as in haplorhines, but the "cupping" effect of the middle cranial fossa noted especially in anthropoids is not observed in most strepsirrhines (with the exception of *Otolemur*). The wings of the alisphenoid do not extend more rostrally than the presphenoid or orbitosphenoid. Foramen rotundum remains incomplete. In contrast, the posterior margin of the alisphenoid is more completely ossified in many strepsirrhines (e.g., *Varecia* or *Galago*). In these specimens, foramen ovale had a substantial bony margin (open posteriorly). In addition, thin bone extended posteriorly to overlap the apex of the petrous temporal.

As in all primates, the petrous portions of the temporal bones contribute to the posterior part of the middle cranial fossae. The otic capsule and the anterior portions of the petrous temporal are well-ossified in almost all monkeys. A large posterior portion of the petrous temporal, in the region external to the subarcuate fossa, remains unossified in most strepsirrhines, least so in *Eulemur* and galagids. In the newborn skeletal form, this creates a "subarcuate foramen" and effectively widens the posterolateral fontanel (Figure 4.8). Histology reveals this gap is mostly cartilage in the mastoid region. The squamous temporal and petrous temporal are closely apposed, although the articulation between them remains visible. The bony labyrinth of the inner ear (including cochlea and semicircular canals) and the ossicles of the middle ear (malleus, incus, and stapes) appear extremely large in the newborn primate compared to other cranial structures. The tegmen tympani is well-ossified in most specimens, but the anterior wall of the middle ear cavity remains cartilaginous in many specimens (e.g., *Varecia*). A prominent carotid canal is visible on the posterior side of the petrous bulla in some species. The stylomastoid foramen is poorly defined, and in most species, the facial nerve likely exits through a gap between the tympanic bone and the petrous temporal. A postglenoid foramen is visible in all newborn strepsirrhines.

The ectotympanic is ring-like in newborn strepsirrhines and expected to remain so through ontogeny. It is separated from the auditory bulla by a narrow gap in lemurids. The bulla expands laterally to cover some or most of the ectotympanic in *Lemur* and *Hapalemur*, but less so in *Varecia*, where the annulus remains externally exposed. In cheirogaleids and lorisoids, the ectotympanic is fully exposed at birth, and is attached to the petrous bulla by fibrous tissue (MacPhee, 1979).

In all newborn strepsirrhines, the posterior part of the chondrocranium is well-ossified at birth and synchondroses or sutural articulations are formed. The spheno-occipital synchondrosis is fully patent. As in all other primates, the floor and posterior walls of the posterior cranial fossae are composed primarily of occipital bone, represented by four separate masses surrounding foramen magnum: squamous occipital, paired exoccipitals, and basioccipital (Figure 4.4). A small gap between the supra- and infranuchal parts of the occipital is visible in some lemurs (e.g., *Hapalemur*), but this is restricted to the margins, a remnant of the sutura Mendosa that separates them during fetal development. The plane of the squamous occipital takes an angular turn at the level of attachment of nuchal muscles, i.e., at the superior nuchal line. Thus the nuchal line may or may not form a low ridge. The condylar canal is present in the exoccipital. The exoccipital also has a deep notch anteriorly, indicating the position of the hypoglossal nerve.

4.6 Early Postnatal Transformations

Parturition is followed by rapid morphological changes in all primates, and yet the rate of morphogenesis of skull regions is not identical. Skull growth is modular (Cheverud, 1996), with some regions being more "preformed" while others progress through rapid transformations. Studies on diverse mammals suggest the basicranium is phylogenetically more stable than other cranial regions (e.g., Cardini and Elton, 2008; Goswami and Polly, 2010). Major developmental distinctions of perinatal chondrocranial development in primates may be summarized as follows. Nearly all primates are characterized by prenatal loss of the posterior nasal cupula, before birth in most species (Box 4.3). All anthropoids may be characterized by early (prenatal) loss of bilateral synchondroses between the ossified basisphenoid and alisphenoids. Hominoids have an earlier fusion of the intrasphenoidal synchondrosis than monkeys or strepsirrhines, perhaps by perinatal loss of the presphenoseptal synchondrosis (Smith et al., 2017b). Note that synchondroses may fuse or undergo "fibrous degeneration" (Baume, 1968; Box 4.5). If the latter, they may exist as suture-like joints, at least transiently allowing passive separation driven by growth of functional matrices. A broader comparative ontogenetic perspective is needed regarding the postnatal angular changes in the cranial base. Non-hominoids remain poorly studied during early postnatal development.

The remainder of the neurocranium undergoes more dramatic postnatal changes, and it is well known that the brain grows at different rates postnatally among primates (e.g., Leigh, 2004). This is also true of primate eyes (see above) and thus these two

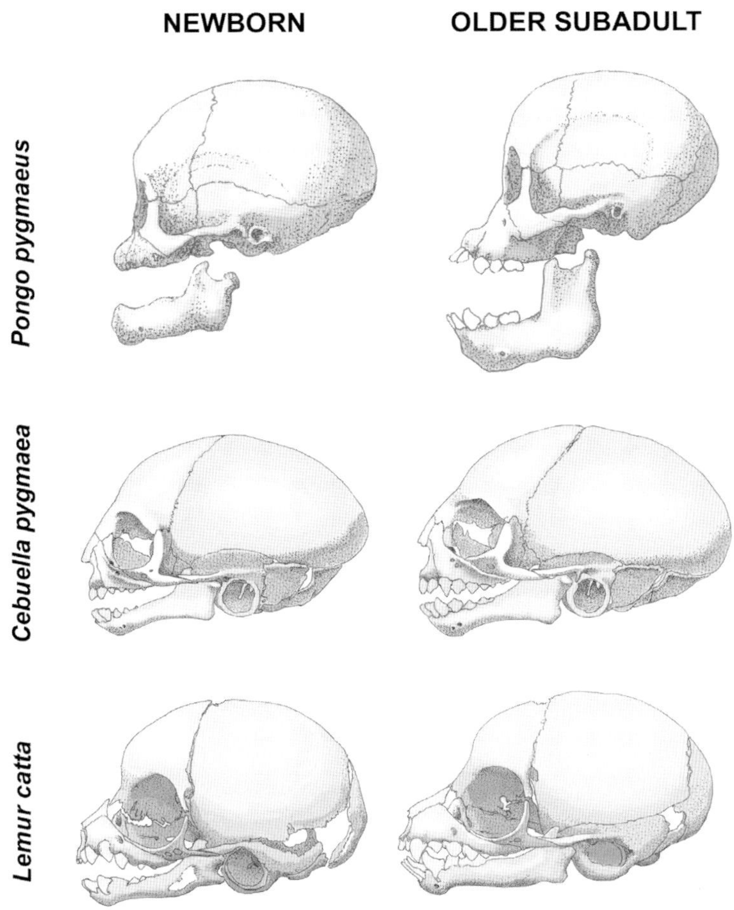

Figure 4.10 Postnatal transformations of the skull in an ape (*Pongo*), monkey (*Cebuella*), and lemur (*Lemur*). Neonates are on the left. Infants with all deciduous teeth at least partly erupted on the right. *Pongo* redrawn after Schultz (1941).

neighboring functional matrices together dictate form of the calvaria and upper face. Features such as superciliary ridges do not appear in early infancy in most primates (Figure 4.10), but tarsiers have more rapid circumorbital changes after birth than most primates. Later postnatal morphogenesis results in sexually dimorphic features (e.g., muscular attachment sites).

Early changes in the mid- and lower face are the most profound changes. Infancy represents a period of great proportional changes in the majority of primates, such that the facial skeleton outpaces the neurocranium in growth (Figure 4.10). At the end of the infant period (full deciduous eruption), the facial skeleton is proportionally enlarged compared to neonates, but this is perhaps most pronounced in larger primates, and strepsirrhines generally. The midfacial region of most haplorines remains inferiorly and anteriorly displaced relative to the neurocranium; in strepsirrhines displacement is anterior. There are vast differences in the pace of midfacial projection. In the species

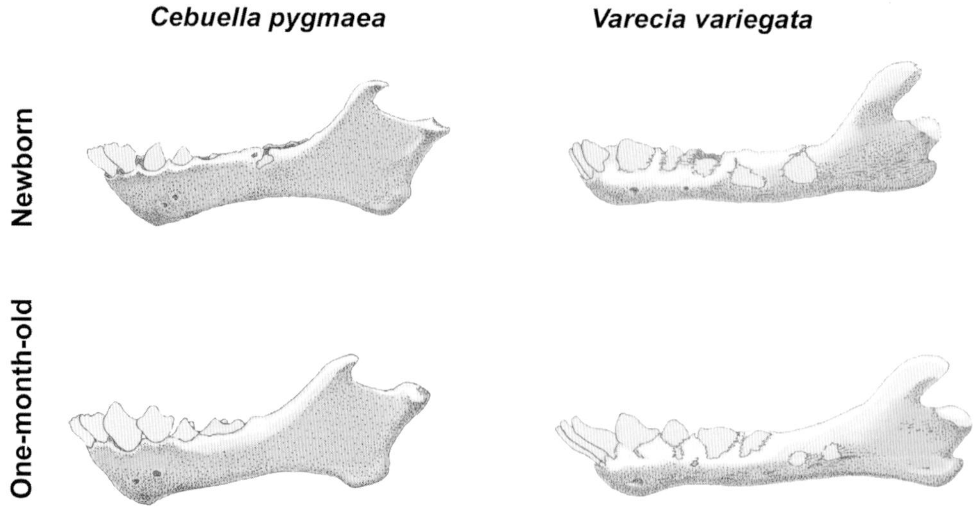

Figure 4.11 Postnatal transformations of the isolated mandible in a monkey (*Cebuella*) and lemur (*Varecia*).

shown in Figure 4.10, the infant monkey and lemur are at about 1 month of age, and full deciduous eruption appears imminent. However, in *Pongo*, the full deciduous eruption occurs at or after about 1 year of age (Fooden and Izor, 1983).

Early postnatal changes in the mandible are equally profound in strepsirrhines and most anthropoids. During infancy, the ramus becomes more vertically inclined (and at a more acute angle with the body in apes). This is more exaggerated than in humans due to the greater robusticity of the teeth. Angular changes of the mandible occur in most other primates, especially large-bodied species, but these may mostly await the eruption of permanent teeth. Two exceptional cases are illustrated in Figure 4.11. Some small-bodied primates have especially advanced eruption of mandibular teeth at birth (e.g., *Cebuella*). It may be for this reason that early postnatal changes are rather subtle, involving increased robusticity of the body. On the other hand, some primates that are born in a dentally altricial state (e.g., *Varecia*) appear to close the disparity with other primates very rapidly, and thus undergo a remarkable morphological transition during the same time period.

Box 4.1 Big Eyes and Brains – Challenges of Growing Primate Heads

The cranial skeleton forms around and between soft tissue elements of the head, such as the eyes, mucosa-lined nasal airways, and brain. The spatial relationship of these soft-tissue regions, or functional matrices (Moss and Young, 1960), is apparent before the bones of the face are fully ossified, and primates are distinctive from most non-primates, even in late embryonic/ early fetal stages (see a–d).

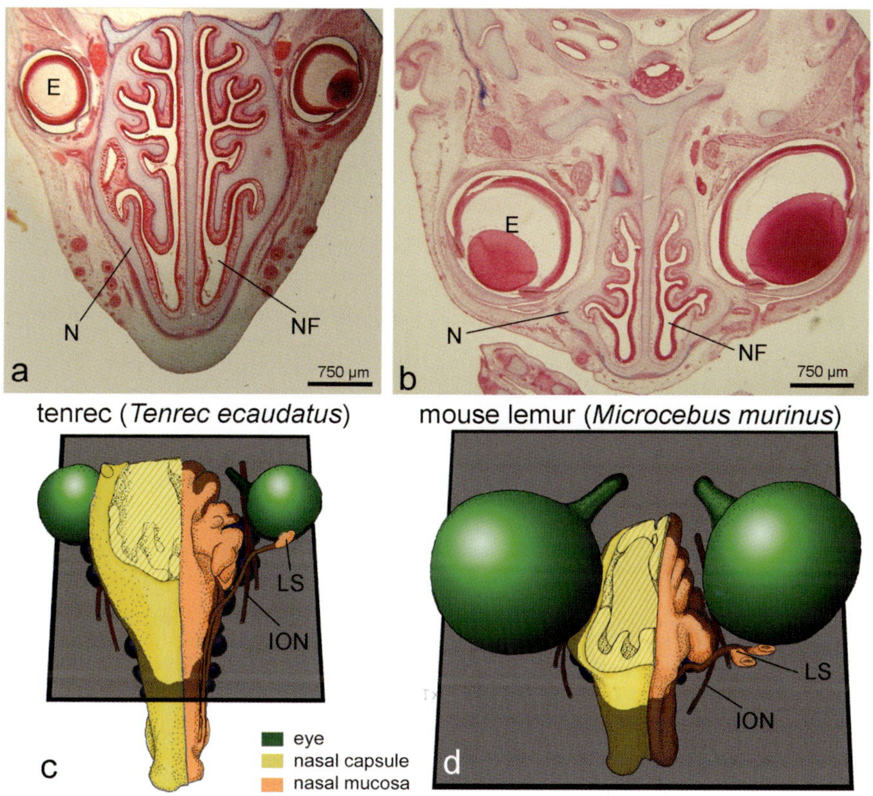

Above: Histologically sectioned and stained fetal heads of a non-primate (a, the shrew-like tenrec) and mouse lemur (b) show the greater spatial overlap of the eye (E) with the cartilaginous nasal capsule (N). Below (c,d) are three-dimensional reconstructions of cartilaginous and soft tissue structures from the same specimens. The cut-planes for a and b are indicated in gray (LS, lacrimal sac; ION, infraorbital nerve. Scale: 1 mm). Portions of figures a–d are from Smith et al. (2014), used with permission. NF, nasal fossa.

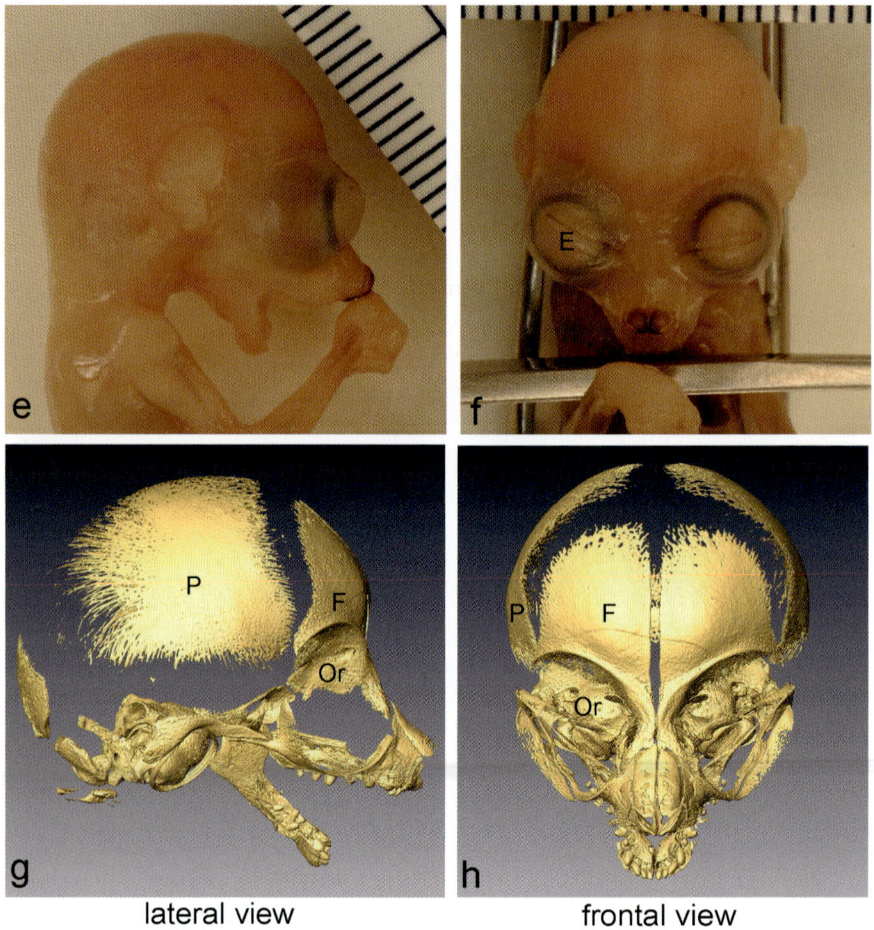

Above: Gross (e,f) and skeletal (g,h) views of the head in a slender loris (*Loris tardigradus*) fetus. Note the relatively small midface compared to the eyes (E) and neurocranium. Calvarial bones (e.g., F, frontal; P, parietal) are widely separated. Or, orbit.

Eyes may be viewed as having a constraining influence on the nasal region. Even in strepsirrhine primates, those with the most primitive nasal cavities, the relatively large eyes reduce interorbital space posteriorly (d). However, in some primates the eyes may extrude beyond the confines of the orbit to a great degree (e,f). In all primates, the brain is also proportionally large compared to other mammals. This is easier to recognize prenatally (above), as growth of the facial skeleton outpaces that of the neurocranium postnatally. The spatial relationship of the eyes and brain is another factor affecting the complexity of the nasal cavity (DeLeon et al., 2017).

Box 4.2 The Medial Orbit

Configuration of the orbit is essentially established at birth in most primates, except for bones that remain relatively poorly ossified, such as the orbital plate of the ethmoid in cheirogaleids, or the postorbital plate of the zygomatic in tarsiers.

Equally well-developed in all primates is the orbital plate of the frontal bone, which forms an extensive part of the orbital roof in all newborn primates (a–c, see *). The frontal forms a large portion of the medial orbit in strepsirrhines that lack ethmoid exposure (c,d). The lacrimal bone and orbitosphenoid also form portions of the medial orbit (see text). Exposure of the ethmoid bone is variable. It forms a large portion of the medial orbit in newborn anthropoids (a, ethmoid coded green) and will go on to form part of the medial orbit in cheirogaleids and tarsiers after ossifying. The palatine bone (b,c, coded red) appears to form a portion of the orbit in most (if not all) primates. As in adults, it forms a small part of the posteromedial orbit in anthropoids (b) and a large part of the medial orbit in large lemurs (c).

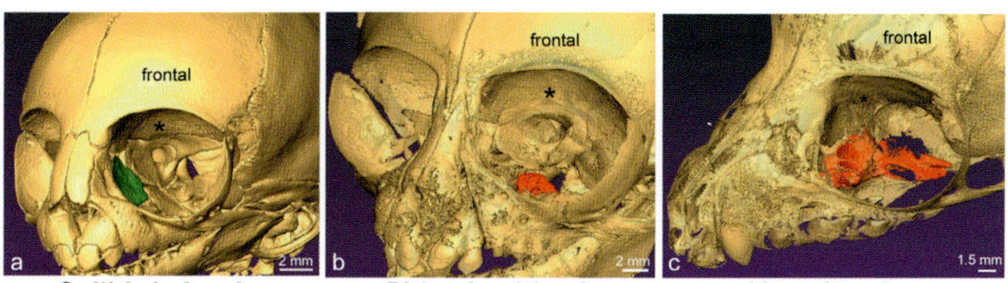

Callithrix jacchus ***Pithecia pithecia*** ***Varecia rubra***

In large lemurids the palatine and frontal approach one another with some established sutures at birth (d: frontal is blue, palatine is red, maxilla is green). Histology of the same perinatal lemur shows these bones meet over the unossified nasal capsule (e).

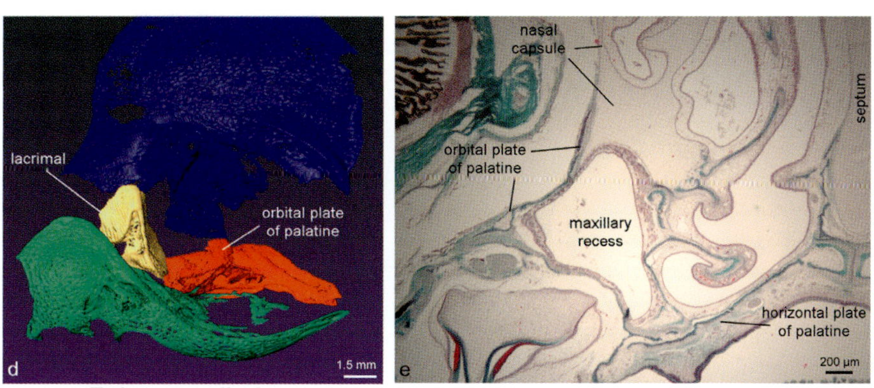

Eulemur collaris

Box 4.3 Fate of the Nasal Capsule

The cartilaginous nasal capsule forms an incomplete enclosure for the nasal fossa in the fetal mammal, leaving communications for the respiratory passageways and olfactory nerves. It is tripartite, forming from three portions that support the main nasal chamber, paranasal recesses, and the olfactory recess (if present). Only certain portions of the nasal capsule actually remain cartilaginous postnatally, such as a portion of the septum nasi and small elements that support the external nose. The remainder of the nasal capsule ossifies or disappears.

Pars anterior. The anterior part of the nasal capsule in newborn primates (a, Alcian blue reveals cartilage matrix of the anterior nasal capsule in a newborn *Saguinus geoffroyi*) covers the dorsal side (via tectum nasi), lateral sides (paries nasi), and partitions the nasal chamber into two halves (via septum nasi). This region is relatively complete in many newborn strepsirrhines, but lateral parts are virtually absent in newborn catarrhines and exist in fragments in platyrrhines. In some specimens, chondroclasts are seen near the cartilage fragments (see b and c, high-magnification views of sections closely adjacent to the previous section; see box in a for location).

As these cells are immediately adjacent to cartilage and not bone, it would appear likely that these fragments are being isolated and gradually resorbed by the chondroclasts. In newborn platyrrhines, chondroclasts have also been observed along the outer margins of the tectum nasi, and between the paries nasi and the maxilloturbinal (Smith et al., 2008, 2012). Thus, chondroclastic activity appears to reduce the pars anterior and isolate the maxilloturbinal, which is initially continuous with the nasal capsule. Elements that ossify (e.g., semicircular crest) are few.

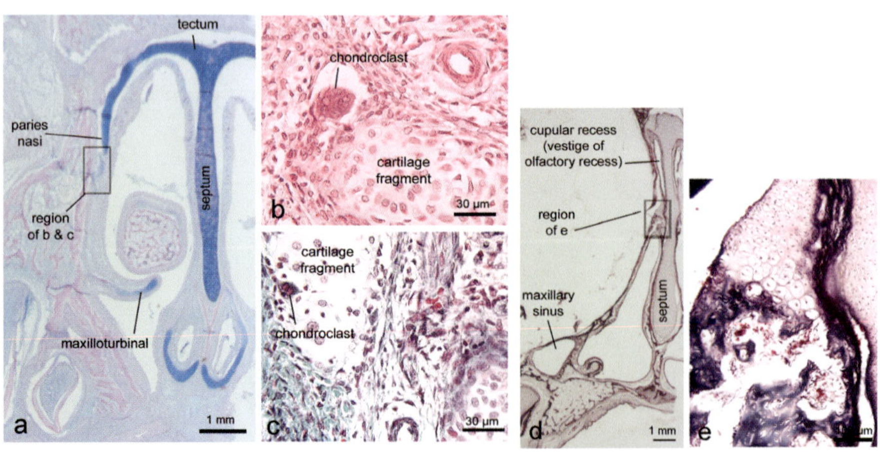

Pars intermedia. The pars intermedia is reduced in anthropoids (Maier, 1980), and at birth may exist as fragments surrounding the maxillary recess or sinus. However, these fragments are already absent in many newborns (d, *Pithecia pithecia*). The means by which this part of the pars intermedia disappears is unclear, but there is no evidence that it ossifies. The more dorsal part of the pars intermedia is more developed in newborn strepsirrhines (see Figure 4.11), although much of the perimeter may disappear. In strepsirrhines, a variable number of frontoturbinals are ossified products of the pars intermedia.

Pars posterior. The posterior portion of the nasal capsule yields the most endochondral bone, all of which becomes part of the adult ethmoid bone (which also includes the more anterior ossifications, except for the separate maxilloturbinal). Ethmoturbinals extend as scrolls or plates near the midline septum, and between them some primates form smaller interturbinals. In most anthropoids, the ethmoturbinals and lateral portions of the pars posterior between the orbits have at least initiated (see, e, enlarged from d).

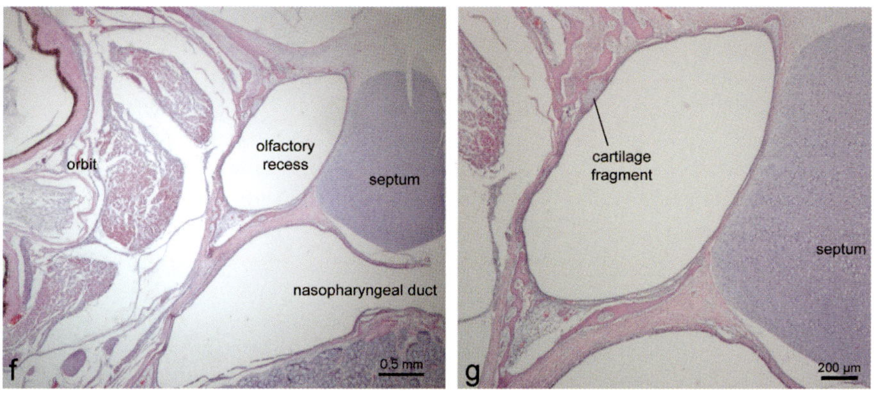

The fate of the most posterior limit of the nasal capsule is poorly described in mammals generally. In most newborn mammals, the nasal capsule ends in a thimble-like cartilage enclosure, called the posterior cupula. In most newborn primates, the posterior cupula only exists as fragments, if present (f,g: newborn *Lemur catta*). Thus far, there is no evidence that the posterior cupula ossifies, and no chondroclasts have been described in the vicinity. However, vestiges of the posterior cupula are seen in some perinatal strepsirrhines, and Alcian blue preparations indicate the matrix is poor in proteoglycan content compared to median cartilage (h, late fetal *Eulemur*) suggesting the matrix may be degrading. Some haplorrhines have a vestigial posterior cupula at birth (i, *Alouatta seniculus*), the fate of which is not known.

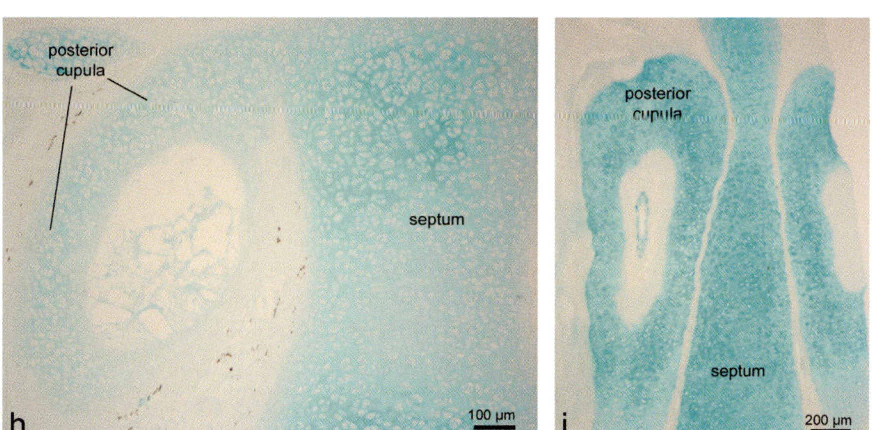

Box 4.4 Paranasal Spaces

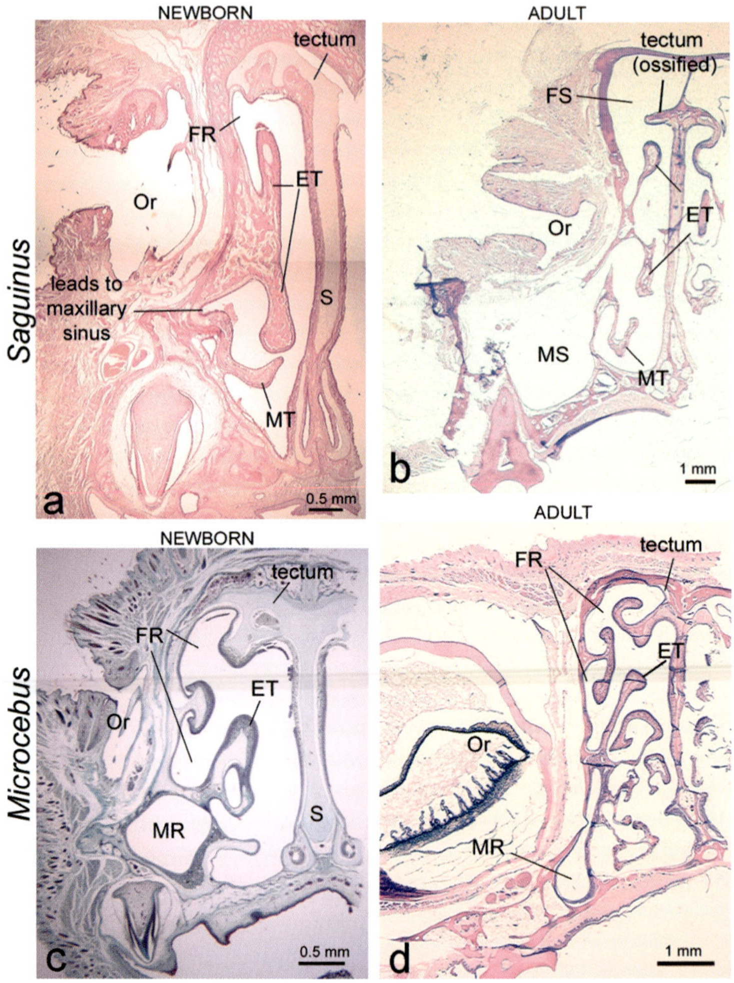

Above: Histological sections of newborn and adult tamarin (top row) and mouse lemur (bottom row) at a cross-sectional level through the frontal recess (FR) or sinus (FS) to illustrate the invasive and expansive nature of pneumatization. Note that in the tamarin, the frontal recess expands beyond the level of the nasal tectum, the cartilaginous "roof" of the nasal fossa, during prenatal and early postnatal ontogeny. In contrast, note that the FR of the mouse lemur does not expand dorsal to the nasal tectum. ET, ethmoturbinal; MR/MS, maxillary recess/sinus; MT, maxilloturbinal; Or, orbit; S, nasal septum.

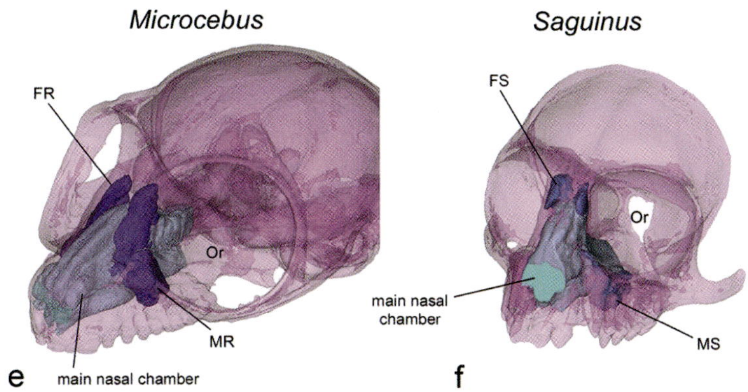

Above: Crania of adult mouse lemur (e) and tamarin (f) showing the spatial relationship of the paranasal spaces (dark blue) to the main nasal chamber and orbit (Or). FR/FS, frontal recess/sinus; MR/MS, maxillary recess/sinus. Courtesy of Tom Eiting.

Pneumatization of paranasal spaces occurs in two phases. The "primordial sinuses" are the products of primary pneumatization, which occurs due to the folding of the cartilaginous nasal capsule during fetal development (Sperber, 2000). These primary sinuses, also called recesses, are initially completely confined by cartilage (Figure 4.7a). After ossification or breakdown of the nasal capsule, these paranasal spaces may remain spatially "static," growing only in stable proportions. For example, the primordial maxillary sinus of squirrel monkeys (*Saimiri* sp.) does not expand during early postnatal ages, that is, it remains proportionally similar in infants, juveniles, and adults (Rossie, 2006). In contrast, the maxillary sinus of many other anthropoids expands inferiorly as the dentition erupts, and also posteriorly as new molars develop. Similarly, the frontal sinus of some but not all primates progressively invades the skull space dorsal to the primitive "roof" of the nasal fossa (b, above). Thus, a "true sinus" expands beyond the space of the primitive nasal fossa via secondary pneumatization.

Box 4.5 Synchondroses of the Cranial Base in Newborn Primates

Basicranial elements of the skull ossify within a cartilaginous template. Midlines bones such as the basioccipital, basisphenoid, and presphenoid ossify within an initially continuous cartilaginous plate, but synchondroses persist between them at birth (a, showing a perinatal fruit bat, *Rousettus leschenaulti*). These are active growth centers, which function just as epiphyseal plates of long bones (Box 2.2) to propel two bones apart via interstitial growth. Most midline synchondroses have a "bipolar" appearance (b), with chondrocyte proliferation and hypertrophy (H) at caudal and cranial sides (i.e., growth at both ends). Although this may be expected to primarily affect longitudinal growth of the cranial base, it has been suggested that differential growth at endo- and ectocranial sides of the synchondrosis could influence the degree of basicranial bending, or flexion (e.g., Scott, 1958). However, this has not been tested with a quantitative ontogenetic study. Moreover, a comparative perspective on midline synchondroses at birth is almost completely lacking in the literature.

The midline growth centers include the spheno-occipital (SOS), intrasphenoidal (ISS), and prespheno-septal (PSept) synchondroses. The SOS is patent in most or all newborn mammals (e.g., a,b: *Rousettus*). The SOS and the ISS are also present in most newborn primates (d,e: *Callithrix jacchus*; f,g: *Macaca nemistrina*). The ISS is fused perinatally in humans and probably in some or all apes. Although a recent study (Smith et al., 2017b) suggests that these two bipolar growth centers are ubiquitously present in mammals, their growth characteristics should not be assumed to be identical.

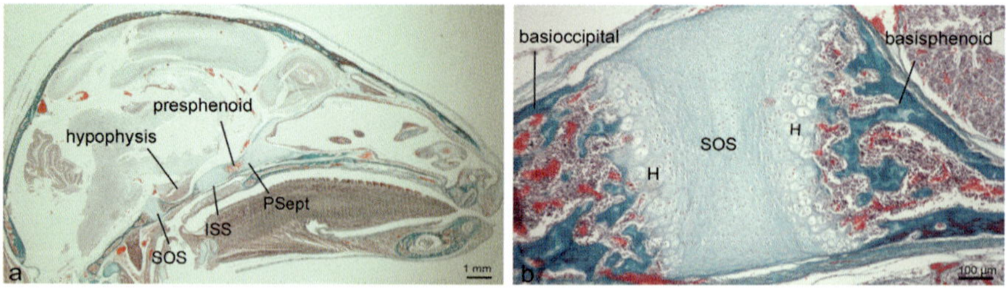

For example, one may note that they do not exist in uniform thickness in all cases (e.g., *Galago moholi*: c–f). Moreover, they cannot all be described as flat "plates," perpendicular to the long axis of the cranial base (e.g., *Saimiri*, below). Thus, it is possible they contribute to basicranial angulation.

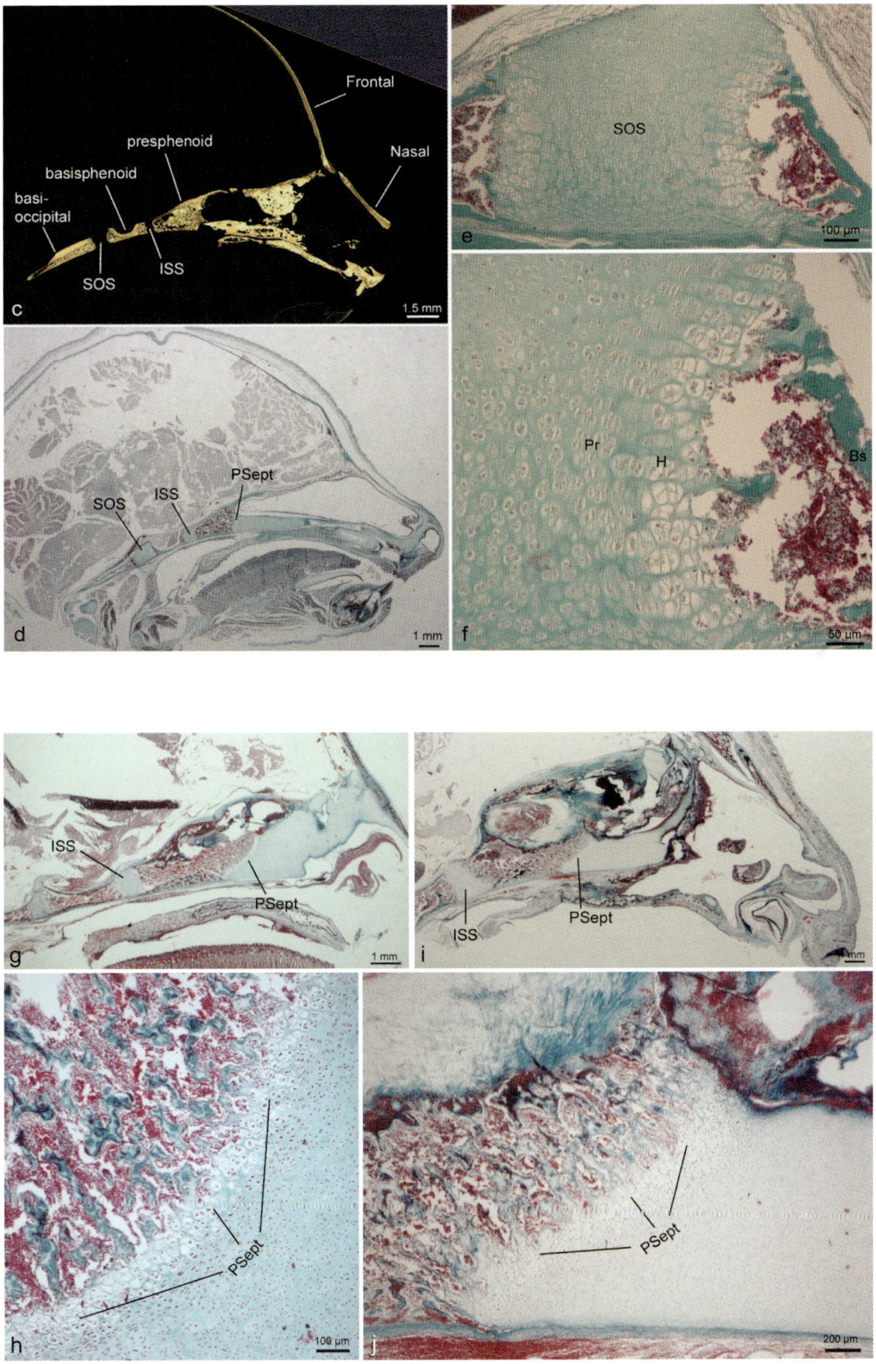
Frontal
presphenoid
Nasal
basisphenoid
basi-
occipital
SOS
ISS
c
1.5 mm
SOS
e
100 µm
PSept
SOS
ISS
d
1 mm
Pr
H
Bs
f
50 µm
ISS
PSept
g
1 mm
i
ISS
PSept
PSept
h
100 µm
PSept
j
200 µm

The PSept has notable morphological variations among extant primates. It has a longitudinal orientation in most mammals (e.g., a), including strepsirrhine primates (c). In anthropoid monkeys, PSept has a distinctively radial appearance in newborn monkeys (e.g., g,h: *Callithrix jacchus*; i,j: *Saimiri boliviensis*), which may influence perinatal growth trajectories (see text).

Bilateral synchondroses have attracted even less attention in the literature. One of these growth centers exists between the basisphenoid and alisphenoid. This synchondrosis is present for a longer duration in strepsirrhines (e.g., k,l: perinatal *Otolemur crassicaudatus*) than anthropoids (e.g., m,n: *Saimiri boliviensis*). In the former it may be an active growth center at birth (l), whereas in monkeys no trace of cartilage remains at birth (n).

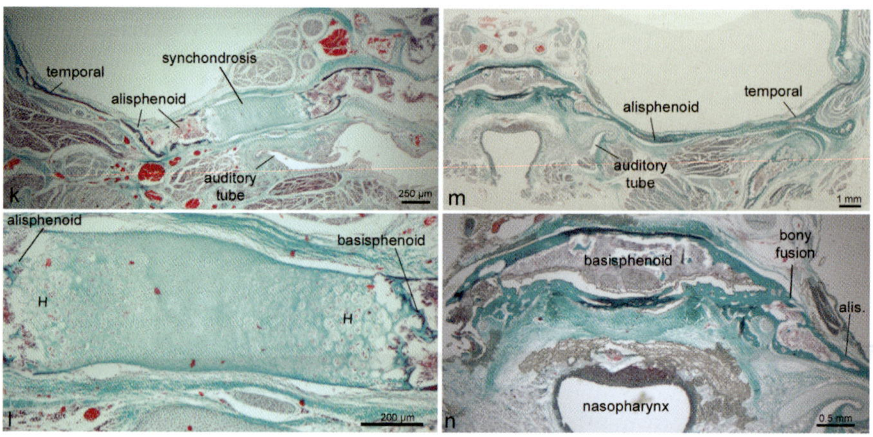

Box 4.6 Cranial Soft Tissues in Tarsiers

One notable feature of adult tarsiers is the extreme hypertrophy of their orbits, relating to greatly enlarged eyes. Compared to other primates, their eyes are large relative to their brain (see a) and body size (Rosenberger, 2010), and eye volume has been said to rival brain volume (Sprankel, 1965). Although both the eyes and brain are reported to have rapid prenatal growth rates in tarsiers (Jeffery et al., 2007), the two organs may have quite different growth trajectories. Below, prenatal tarsiers that are estimated to be in late embryonic (b), early fetal (c), and mid-fetal (d) stages are shown using histology sections. Note that the late embryonic and early fetal tarsiers differ in proportional mid–cross-sectional dimensions of the eyes. In the fetal specimen (c), the eyes appear slightly larger and more convergent toward the midline. However, note that in neither case does the eye protrude away from the cranial contour as they do in adults (a). Indeed, even in a later fetal specimen (d) in which soft-tissue structures are isolated in cross-sections and three-dimensionally reconstructed, the eyes are proportionally smaller compared to the adult, if brain size is taken into account (e). Note that the eyes are also partly "recessed" beneath the frontal lobes (f), in the same manner that is typical of anthropoids. Taken together, these age comparisons suggest that growth of the eye accelerates late in prenatal development, and may be especially rapid postnatally.

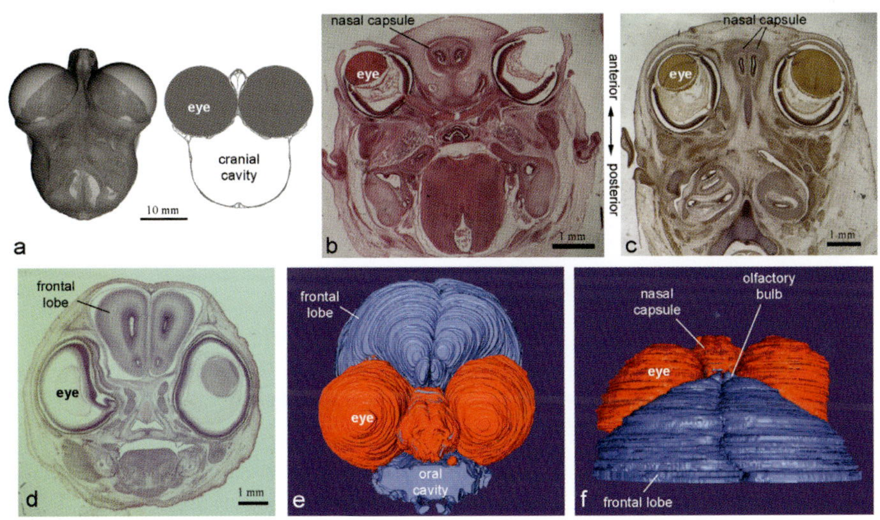

Above: a, an adult *Tarsius spectrum* (from Rosenberger et al., 2016, used with permission). A 3D reconstruction of the eyes and skull is shown next to a horizontal cross-section emphasizing that eye size rivals brain size, and that the eyes extrude out of their sockets in adults. Two prenatal specimens of *T. bancanus* are shown in similar cross-sectional planes using histology: (b) 13 mm CRL *T. bancanus* and (c) 20 mm CRL *T. bancanus*. In (d), a 25 mm CRL fetal *T. bancanus*, a coronal section through the brain and eyes at the level of the frontal lobes is shown. Note slight shrinkage of the brain indicated by the size of the space surrounding the frontal lobes. When the soft-tissue structures from this specimen are segmented and stacked using Amira software (e), the eyes are shown to be proportionally smaller compared to adult tarsiers, and more tucked or "recessed" beneath the frontal lobes (f).

5 Dentition

5.1 Introduction

Dental development must be discussed at both the tissue level – how mineralized the teeth are at a given time – and in terms of their functional status – how rapidly the cusps transition out of alveolar bone (eruption) and then through the gingiva (emergence). Even well-mineralized teeth that have pierced the gingiva are not immediately functional, at least not as masticatory organs, until they are in full occlusion. The dental status at birth is further complicated by the development of two sets of teeth, one of which is shed (the deciduous, or milk, teeth). The deciduous teeth may develop as a temporary, fully functional set which suits masticatory needs throughout subadult ontogeny. Indeed, many definitions of life stages rely on the status of eruption and replacement of the deciduous teeth (Chapter 1; Schultz, 1969).

We discuss newborn teeth with respect to how progressed they are toward their future role for subadult and adult mastication of food (Figure 5.1). The newborn, after all, is first fully dependent on and later supplemented by maternal milk for a variable amount of time. Hence, mammals may or may not have very mineralized teeth at birth and may or may not have erupted teeth. The potential for the mother to provision the newborn with food, for however long, arguably gives natural selection an unusually light hand on the dental battery of the newborn mammal.

Some inferences on the evolution of dental ontogeny are possible even without direct observations on dental development. Primates lack a permanent third incisor when compared to the primitive mammalian dental formula of I3/3, C1/1, PM 4/4, M 3/3. In addition, living primates lack at least the first premolar while early euprimates of the Eocene maintain the primitive number of four (e.g., *Teilhardina* sp.). There was a subsequent evolutionary loss of one premolar in one or more primate clades. There was the loss of an additional premolar following more recent divergence within the Suborder Anthropoidea. All extant catarrhines (including humans) lack two dp2. Because primitive fossil catarrhines (parapithecids) possess three premolars, it seems likely the additional loss of a premolar occurred after catarrhines and platyrrhines

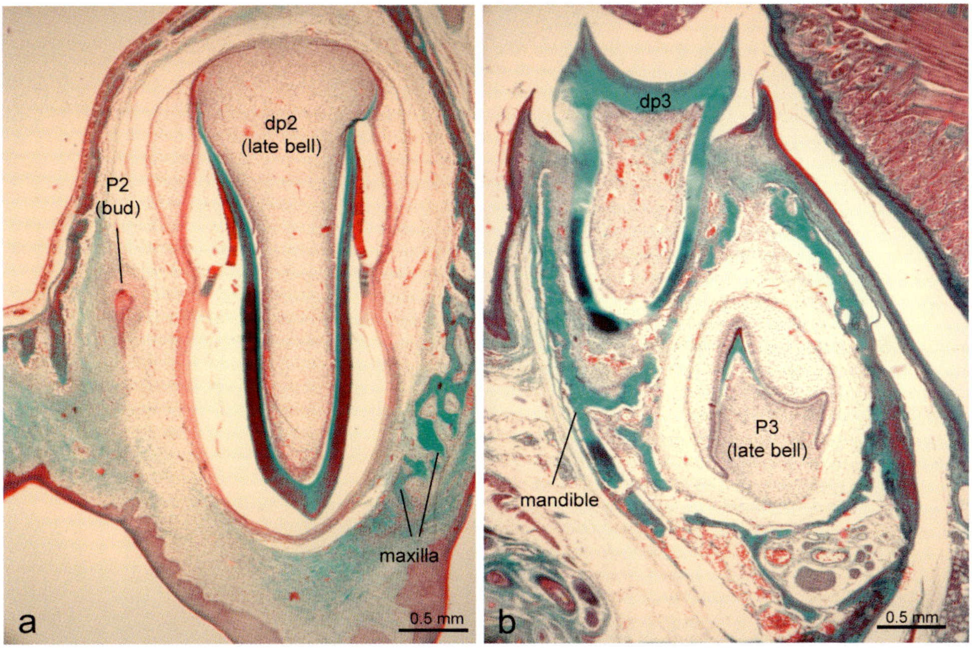

Figure 5.1 Different stages of dental maturation and eruption in subadult primates (newborns of *Eulemur mongoz*, at left; *Cebuella pygmaea*, at right). (a) Deciduous premolar (dp^2) beginning to erupt, but not yet emergent through the gingiva; successional P^2 is at the bud stage. (b) Erupted deciduous premolar (dp_3) that has pierced the gingiva; successional P_3 is at the late bell stage.

diverged. We know from extant primates that a deciduous premolar may form even if its replacement tooth fails to develop.

Our understanding of the evolution of dental development is tenuous because we have detailed information on the dental ontogeny of very few extant primates (e.g., Tarrant and Swindler, 1973; Luckett and Maier, 1982; Godfrey et al., 2004; and see Smith et al., 1994; Swindler, 2002). In this chapter, we review and add to the information on the status of dental maturation, eruption, emergence, and morphology in the newborn primate (e.g., Smith et al., 1994; Godfrey et al., 2001; Guthrie and Frost, 2011). In particular, we intend this chapter to partially resolve the paucity of available morphological observations on the deciduous teeth of non-human primates, noted in 2002 by Daris Swindler in his book, *Primate Dentition* (Cambridge University Press). As of that date, detailed observations accompanied by pictorial accounts were only available on hominoids, and selected anthropoids (mainly catarrhines). By providing such a resource, we intend this to serve as a point of comparison to information on deciduous teeth of extinct forms, which will continue to accumulate (e.g., Franzen et al., 2009).

5.1.1 Ontogeny: General Considerations

The terms deciduous and primary teeth are sometimes used interchangeably. However, primary is better used as a "generational" term: deciduous teeth are the first wave of teeth formed. These teeth derive from the dental lamina, an epithelial bridge to the oral epithelium of the developing jaw. As the first permanent molar also derives from the dental lamina, it too can be considered a primary tooth (Ooë, 1979). During early juvenility, one may detect a continuous transition of morphological features of primary teeth (Figure 5.2).

Secondary (or successional) teeth are the second generation of teeth that derive lingually from the tooth germ they will replace (Luckett, 1993). Deciduous teeth are the primary teeth that are shed, and permanent teeth are simply all teeth that are not shed.

For simplicity, our terminology here divides teeth into three groups, deciduous teeth (i1, i2, dc, dp2–dp4), "replacement" teeth (I1, I2, C, P2, P3, P4, i.e., the secondary teeth), and permanent molars. (Note that P1/dp1 are absent in all extant primates.) To facilitate a discussion of tooth location in the developing jaw, we use "deciduous premolars" as the nomenclature for postcanine deciduous teeth. However, it should be kept in mind that the term "molar" refers to occlusal morphology, and in this sense

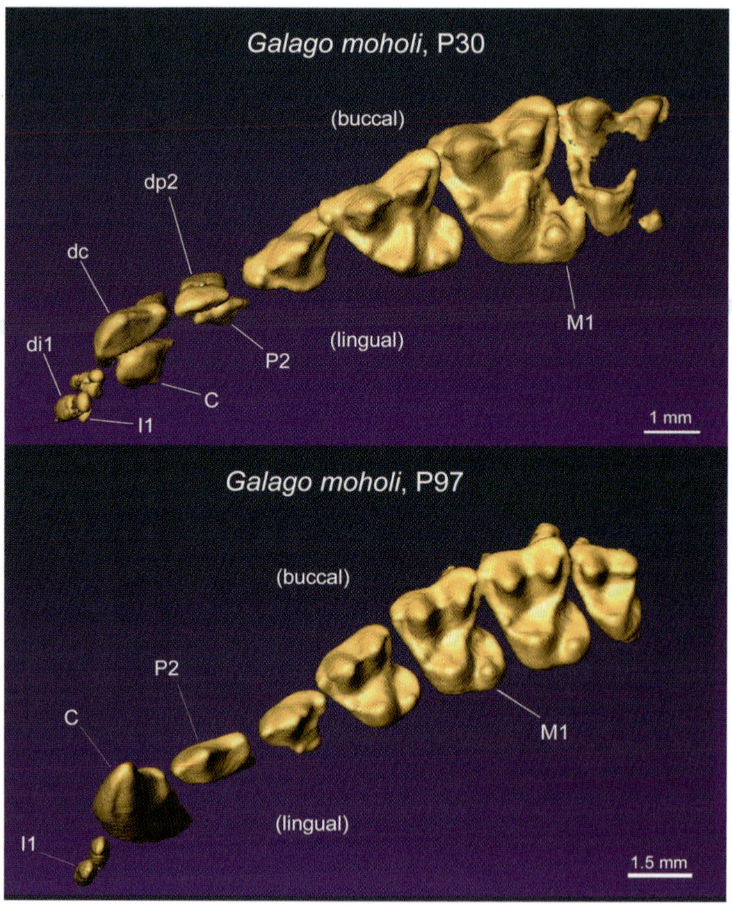

Figure 5.2 Mixed dentition of the upper jaw in a juvenile and weanling *Galago moholi*. C, canine; dc, deciduous canine; di, deciduous incisor; dp, deciduous premolar; I, incisor; M, permanent molar.

another term, "deciduous molar," is a more appropriate term for the primary teeth replaced by the adult premolars (e.g., Schwartz, 2007).

Another dynamic aspect of perinatal dental development is the degree to which the tooth crown projects beyond the margins of the maxilla/mandible (i.e., dental eruption) or pierces the gingiva (i.e., gingival emergence). A practice related to these processes is "dental aging." Primates are commonly sorted into dental ages with respect to which tooth classes are predominantly in occlusion. For example, the infant period may be said to end with the eruption of permanent teeth (Schultz, 1935, 1960). As we shall see, the practice of dental aging provides a means of comparing maturational classes, and is not necessarily expected to reflect the same actual or *chronological* age across all species.

5.1.2 Anatomy of the Tooth: General Considerations

The crown is the protruding and "working" part of the tooth, and is typically swollen to a greater diameter compared to the root of the tooth. The crown surface is complicated by protrusions (cusps) and ridges, or crests that enable mastication by the way in which they interact, or occlude, with surfaces of the teeth in the opposing jaw (Avery, 2002; Nanci, 2007; Schwartz, 2007). This is the occlusal surface of the tooth. The crown also has surfaces that face the tongue (lingual) and cheek (buccal). Buccal surfaces are thus outward-facing; the term labial (referring to the lips) may be substituted for the outer-facing surface of the anterior-most teeth. The crown also has surfaces facing adjacent teeth. Because the dental arcade is arranged in an arc in which right and left sides are joined anteriorly, one of the surfaces is closer to the midline – the mesial surface (Figure 5.3). The opposite side of the tooth crown faces away from the midline – the distal surface. The mesial surface may face medially (e.g., incisors), or anteriorly

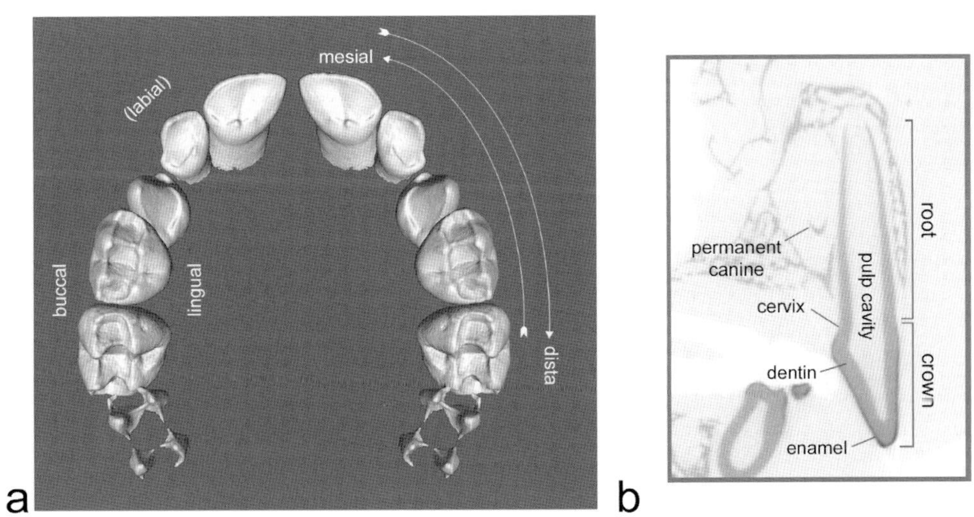

Figure 5.3 Spatial terminology and anatomical parts of the tooth using an occlusal view of the dental arcade in newborn *Allenopithecus nigroviridis* (a) and CT slice through the deciduous and permanent canine in a juvenile *Saguinus geoffroyi* (b).

(e.g., M1). The root is anchored to the alveolar bone of the maxilla and mandibles by a host of short collagenous bundles (collectively, the periodontal ligament). The external junction between the root and the crown is termed the cervix of the tooth.

Primate teeth, as in other mammals, have three mineralized tissues that form the bulk of the volume comprising each tooth: enamel, dentin, and cementum. Enamel surrounds the apical part of the tooth, the crown (Osborn, 1981; Nanci, 2007). The mineralized tissues are microanatomically distinct, and exist as adjacent layers that are grossly or histologically visible in cross-sections. Crown enamel is apposed to dentin at the amelodentinal junction. Dentin (and not enamel) is a continuous core of tissue that extends to the root. The root is covered by cement, the third mineralized dental tissue, creating a cementodentinal junction. Enamel and cementum are not coextensive. They meet at the cervix, which is therefore also considered the amelocemental junction. These tissues are discussed microanatomically below.

Within the jaw, the root of the tooth is usually covered by the alveolar ridges of the maxilla or mandible, thus protecting the root while exposing the enamel. A small portion of the enamel near the cervix may be covered by the gingiva (gums), which are ridges of the oral mucosa (Figure 5.1). In the neonate, this arrangement may be altered in that the roots of a deciduous tooth may also be adjacent to the crown of its replacement tooth, and may be partially resorbed. If they are not already shed, deciduous teeth may be restricted to the gingiva while in the process of shedding.

5.1.3 Terminology of Dental Crown Morphology

Terminology referring to the cusps, crests, and depressions of the occlusal surface of teeth is specific to maxillary and mandibular teeth (here, we follow Swindler, 2002). Cusps of maxillary teeth end in the suffix "-cone"; for mandibular teeth the suffix "-conid" is used. Secondary cusps are denoted by -conules for maxillary teeth and -conulids for mandibular teeth. The crown crests, which may bridge the distance between cusps, are referred to as "-cristae" for maxillary teeth and "-cristids" for mandibular teeth. A band of enamel may form a ledge around the perimeter of the crown: in maxillary teeth this is a cingulum, while in mandibular teeth this is a cingulid. These may form structures that resemble cusps, styles (maxillary), or stylids (mandibular), but are usually not as prominent as primary cusps (Swindler, 2002). However, an expansion of a cingulum may develop robustly into a cusp, as is the case with the hypocone of upper deciduous premolars or permanent molars.

The prefixes for the above terminology have a long history of usage (e.g., Gregory, 1934; Osborn, 1981; Swindler, 2002). The terms for maxillary teeth, regardless of tooth class, include at least one cusp located on the buccal side. In maxillary teeth this cusp is the paracone; in mandibular teeth this is the protoconid. The canine is a single-cusped tooth matching this description; in some cases, incisors assume this "caniniform" morphology. In most primates, at least one of the incisors has an incisal ridge instead of a pointed cusp; the ridge itself may have several tubercles (or mamelons). Such a tooth has a spatulate or shovel-like appearance, a morphology referred to as incisiform.

Posterior teeth (premolars and molars) are the most complex teeth. Premolars may be single-cusped (caniniform), bicuspid, or may possess three to five cusps (as more

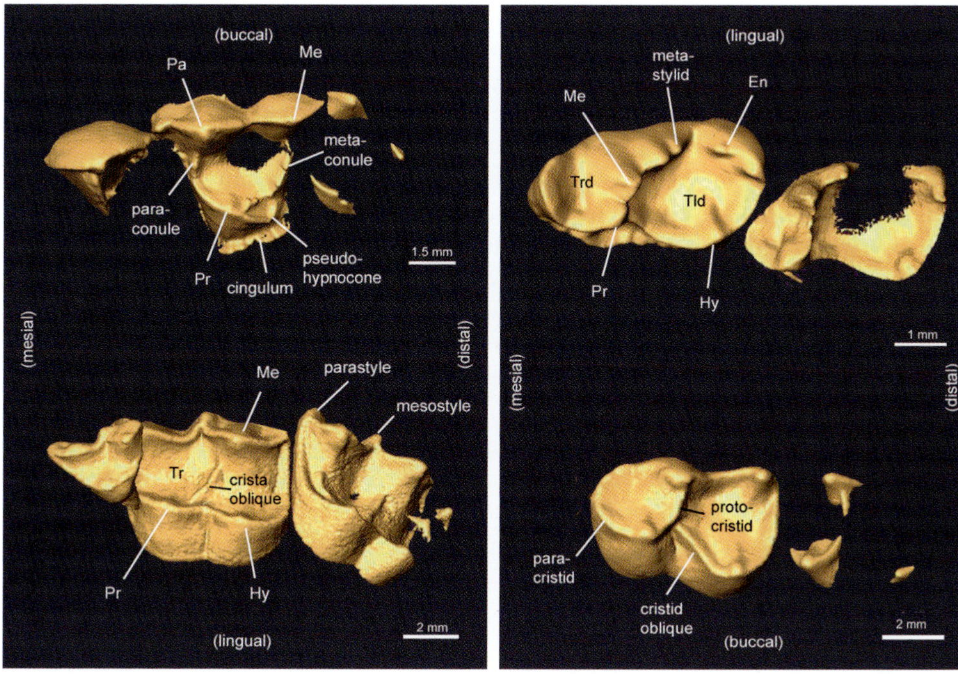

Figure 5.4 Parts of the tooth crown, shown using maxillary teeth (left) and mandibular teeth (right). Top left: dp^4 of *Eulemur rubriventer*; bottom left: M^1 and M^2 of *Propithecus coquereli*; top right: dp_4 of *Hapalemur griseus*; bottom right: dp_4 of *Alouatta seniculus*. Maxillary tooth primary cusps: Pa, paracone; Pr, protocone; Me, metacone; Hy, hypocone. Maxillary tooth basin: Tr, trigon basin. Mandibular tooth primary cusps: Pr, protoconid; Me, metaconid; Hy, hypoconid; En, entoconid. Mandibular tooth basin: Trd, trigonid basin; Tld, talonid basin.

cusps are present, the tooth is described as more "molariform"). In many primates, the deciduous premolars exhibit an increasing gradient of complexity, becoming more molariform as the series progresses distally (Swindler, 2002). Thus dp4 is very similar to M1 (Figure 5.4) and may be more complex, as will be explained.

In teeth with multiple cusps, the second cusp is located to the lingual side of the paracone/protoconid: the maxillary protocone or the mandibular metaconid. The maxillary dp4 has at least three cusps: the paracone, protocone, and the metacone (the latter is buccally situated, distal to the paracone). In some primates the distolingual cingulum is adorned with an additional cusp – the hypocone. The mandibular dp4 has four cusps in its simplest form. Distal to the protoconid is the hypoconid; distal to the metaconid is the entoconid. At the leading edge (i.e., mesially) of molariform teeth, a fifth primary cusp may be present: the paraconid. In M1 this cusp is only present in one extant primate genus – *Tarsius*. However, its presence on deciduous dp4 has been described in some strepsirrhines (Tattersall and Schwartz, 1974). At the distal edge of the tooth (between the hypoconid and entoconid) a final cusp may be prominent: the hypoconulid.

Crests or ridges may be present on any tooth. Even single-cusped teeth may have marginal ridges (marginocristae) occurring on the mesial and distal limits of the tooth.

In premolars and molars marginocristae may create fovea, usually one mesially (anterior fovea) and one distally (posterior fovea). In molariform teeth, cristae/cristids also occur between cusps. Detailed terminology exists for crests (see Swindler, 2002 for a list of these terms, including synonyms). Here, we will focus on a limited number of crests. In the most complex maxillary deciduous premolars and M1, the mesial and distal pairs of cusps may be connected by crests; these are the anterior and posterior transverse cristae, respectively. In maxillary molarized teeth, an addition crest may pass obliquely between the protocone and the metacone or transect the anterior transverse crista; this is called the crista oblique. On the most complex mandibular deciduous premolars, a secondary cusp, the hypoconulid, may form near the center of the crest between the distal cusp pair. The name for a crest between the protoconid and metaconid is the protocristid. The protocristid, on its distal side, may intersect with a cristid obliquid; when present, this crest passes from the hypoconid to the protocristid, or possibly ends at the base of a mesial cusp. A final crest flares mesially away from the protoconid and may form an arc to the metaconid. This crest is called the paracristid; the paraconid, when present, is located at the mesial-most edge of this crest. Prominent crests render the basins deeper in their appearance.

5.1.4 Dental Microanatomy

Enamel, dentin, and cementum are mineralized tissues, like bone, comprised in part by hydroxyapatite. Enamel is the hardest mineralized tissue of the body, and harder than many inorganic materials (Swindler, 2002). Mature enamel is densely organized with mineral content, with scant ground substance (Box 5.1). It is acellular, produced by ameloblasts within the tooth germ. Dentin, which forms a greater bulk of the tooth overall, is likewise an acellular tissue produced by certain cells of the tooth germ, odontoblasts. However, unlike enamel, dentin has a dense collagenous component, which becomes mineralized as the tissue matures. In cross-section, one can easily detect the less-mineralized part of dentin because it is less radio-opaque in imaging that relies on x-ray radiation, and has a more fibrous appearance and contrasting staining density compared to mature dentin. The immature portion of the tissue exists as a discrete region called predentin; this layer is found closest to the odontoblasts (Box 5.1). Mature dentin is softer than enamel but harder than bone (Swindler, 2002). As the tooth forms roots, dentin will be found underlying the enamel in the tooth crown region, and extending from there into the roots, where it is covered by cementum. Cementum is produced by cementoblasts, which produce a mineralized collagenous matrix during root formation. Unlike dentin and enamel, some cementum (that closer to the root) is cellular (Nanci, 2007).

Once a tooth erupts, enamel has reached its full thickness. During mastication, enamel gradually erodes. Thus, morphological descriptions of deciduous teeth can be based, in part, on artifacts of tooth use. For example, incisors of hominoids and some other primates erupt with minute cusplets along the incisal margin, the mamelons. These are the first enamel structures to erode in hominoids. Swindler (2002) describes such artifacts in his description of deciduous tooth morphology in anthropoids. Our descriptions of newborns will provide full details of cusp morphology whenever resolution allows it.

Final enamel thickness reflects the disruption and death of the epithelium that secretes and orchestrates maturation of enamel. In contrast, dentin continues to be produced postnatally. The initial framework of matrix produced is referred to as primary dentin (which *is* a mature matrix; i.e., the meaning is not the same as in

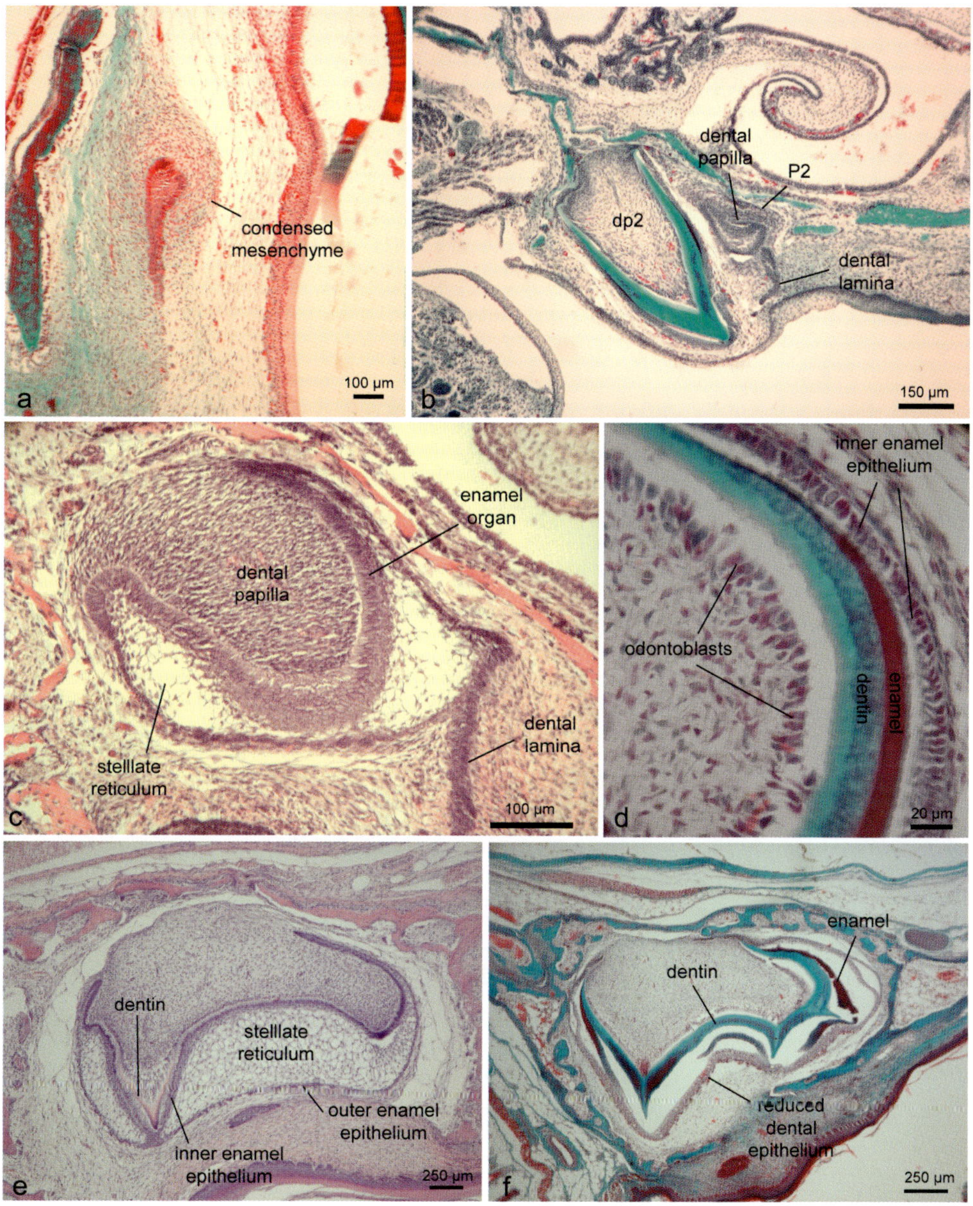

Figure 5.5 Tooth microanatomy and staging. In most stages of dental development, such as the bud (a), cap (b), and early bell (c) stages, the tooth germ lacks any mineralization. In the late bell stage the epithelia of the enamel organ are fully differentiated (d). (e) M2 of an infant marmoset, with dentin only. (f) M1 of the same specimen, with an advanced degree of enamel secretion.

primary bone). Subsequently, dentin continues to be produced during growth and as a reparative process, resulting in secondary dentin. Gradually throughout life, the enamel erodes and dentin continues to accumulate (Avery, 2002).

5.1.5 Tooth Maturation: Stages and Mineralization

5.1.5.1 Presence or Absence of Mineralized Cusps

Developing teeth have been studied using a variety of histological procedures. Our focus in this chapter concerns light-microscopic study techniques using hematoxylin–eosin or trichrome procedures on serially sectioned jaws (Smith et al., 2015). "Staging" of teeth is based on changes in cross-sectional appearance of the tooth germ as viewed using microscopy (Osborn, 1981; Table 5.1; Figure 5.5). The tooth germ is first seen when mesenchyme condenses locally around a thickening of the dental lamina. The result is a spherical/oval mass that remains connected to the dental lamina via a thin epithelial stalk – the tooth bud. As the mesenchymal mass (or papilla) becomes centralized and pushes deeply within the epithelial portion of the developing tooth, the tooth germ acquires a "capped" appearance. As the epithelium over the papilla differentiates, it acquires a bilaminar appearance: two opposed simple epithelia called the inner and outer enamel epithelium. Stellate reticulum, identifiable as a loose mesenchymal tissue between the inner and outer enamel epithelia (Nanci, 2007), forms during the late cap stage. The early bell stage is signified by a change in shape from the cap stage (Table 5.1). Mineralization of the first cusp signifies the beginning of the late bell stage. Cells on the convex side of the dental papilla differentiate into odontoblasts, and secrete dentin matrix toward the inner enamel epithelium. Later, the inner enamel epithelium manufactures enamel on it deep side, facing the existing dentin. During the late bell stage, di1–dp4 extend a secondary lamina that give rise to the permanent tooth buds. These do not form in synchrony. Hence, some permanent teeth may delay their appearance considerably, while others could appear and even begin to mineralize prenatally. Another significant event in the bell stage is the breakdown of the stalk of dental lamina connecting the tooth germ to oral epithelium: the tooth now has autonomy. Later the stellate reticulum disappears, ending the earliest phase of amelogenesis, and bringing the inner enamel epithelium in closer proximity to blood supply (Nanci, 2007).

Disappearance of the stellate reticulum signifies the late bell stage, at which time the enamel is nearly mature. Prior to eruption, the inner and outer enamel epithelium (now nonfunctional) fuses into a thin "reduced dental epithelium." The reduced dental epithelium remains until eruption; the roots are beginning to form at this time as well. This is the so-called "pre-eruption phase" (Avery, 2002). The most advanced stage of dental ontogeny, of course, is "erupted" or "shed."

The rate of crown mineralization is of profound interest to anthropologists because of a potential link to body mass and life-history characteristics. Understanding this relationship in extant primates gives us a better informed view of fossil primates. Measurable features of enamel growth are observed cross-sectionally as lines in the enamel. The lines travel concentrically if the crown is viewed in a transverse plane, parallel to the tooth cervix. In a longitudinal cross-section, the lines are roughly parallel to the amelodentinal junction. These lines, called the striae of

Table 5.1 Stages of tooth development (from Smith et al., 2015)

Thickening of dental lamina[a]		Prior to formation of a bud, the dental lamina thickens
Bud[b]		The bud is a spherical or ovoid mass, still bearing a connection to the dental lamina, which is surrounded by condensed mesenchyme
Cap[b]		The mass is now "cap-shaped," due to an indentation on its deep surface by mesenchyme forming the dental papilla. Later, the cap-shaped mass now bears a more distinct internal enamel epithelium of columnar cells; stellate reticulum first appears
Bell[b]	Early	The developing tooth now has a "bell-shaped" appearance; four identifiable layers surrounding the dental papilla: (1) external enamel epithelium, (2) stellate reticulum, (3) stratum intermedium, (4) internal enamel epithelium
	Late	Associated with formation of dental hard tissues; extension of dental lamina into permanent buds

[a] From descriptions in van Nievelt and Smith, 2005; [b]from description in Osborn, 1981.

Retzius, reflect a single cycle of enamel maturation; the line itself occurred near the end of the period, when the enamel band is at the maturational stage of enamel mineralization (Nanci, 2007). The production of the striae is periodic, occurring in whole days, and hypothesized to relate to life-history variables (Bromage et al., 2012). At birth an especially dark stria is produced, the so-called "neonatal growth line" (darker lines are an indication of stress such as that produced during parturition). There is a great practical value in counting the striae in known-age individuals. It allows one to calculate the duration of time required to produce tooth crowns (Reid et al., 1998). The duration of time to complete the enamel of the crown, as measured by counting these lines, varies among primates.

Hominoids seem to share a very long period of amelogenesis, extending years postnatally. There may be great similarities in the total time period of crown formation, as evidenced by the best-understood comparison of *Pan* to humans (Anenome et al., 1996; Reid et al., 1998). The similar state of cusp mineralization at birth in hominoids may reflect a much conserved developmental plan of amelogenesis. There may be considerably more plasticity in the formation of the tooth root compared to the crown formation; taxa may have great similarities in cusp development, but great variation in tooth eruption patterns. Much is still unknown about the nature of periodic production of enamel. It is suggested that periodicity of deciduous and permanent teeth differs, for example (Reid et al., 1998).

5.2 Morphology of the Dentition in Newborn Hominoids

The deciduous teeth of humans have been described extensively in the literature, including some comprehensive works (Kraus and Jordan, 1965). The great apes have also garnered much attention, especially as models to help us draw inferences on our common ancestors (e.g., Mann, 1988; Smith et al., 1994). In this section we will summarize the information of dental morphology in newborn apes, as published by Swindler (2002) and others.

5.2.1 Status of Dental Maturation, Eruption, and Gingival Emergence

According to Seibert and Swindler (1991), no deciduous teeth of the chimpanzee (*Pan troglodytes*) are erupted "in the intact oral cavity" at birth (i.e., there is no gingival emergence). Incisors may be visible in intact skulls of late fetuses (see Figure 4.2).

The maxillary and mandibular di's of *Pan* are described as "fully formed" at birth, although this is a reference to the crown in particular (Seibert and Swindler, 1991). The dc is described as "lagging" in mineralization and eruption compared to the di's; only the tip of the cusp is mineralized at birth. Mandibular dp_3 has three (of four) cusps mineralized in the newborn (the authors do not specifically mention the maxillary dp^3). The maxillary dp^4 is described as partially mineralized. Neonates may have only the mesial two cusps mineralized, or the two mesial cusps plus the hypocone. The crista oblique shows initial calcification late prenatally. However, the basin of maxillary dp^4 remains unmineralized at birth, and continues to mineralize in older infants (4 months). The mandibular dp_4 has four mineralized cusps, with a more quadrate appearance compared to dp_3 (thus more closely resembling M_1). In the maxillary M^1, the two distal cusps are unmineralized at birth. The mandibular M_1 is also incompletely calcified at birth, although all primary cusps are at least partially calcified. Some teeth (e.g., dp's) are described as having "gelatinous" (i.e., poorly mineralized) roots. No portions of replacement teeth are mineralized at birth. In newborn (stillborn and 2-day-old) *Pongo*, all deciduous teeth are mineralized (Winkler et al., 1991). Radiographs indicate deciduous crowns are well-formed, including mineralization of tooth basins. Winkler et al. also dissected the permanent molars of a stillborn *Pongo* and stained them with Alizarin red, revealing isolated cusps. In the maxillary M^1, only the hypocone is unmineralized. In the mandibular M_1, the entoconid is unmineralized. Oka and Kraus (1969) examined one specimen in which the entoconid of M_1 had begun to mineralize as well. No studies at this level of detail exist on newborn samples of other apes.

Seibert and Swindler (1991) describe the sequence of cusp mineralization in *Pan* as follows. For the maxillary dp's and M^1: paracone, protocone, metacone, then hypocone. For the mandibular dp's and M_1: protoconid, metaconid, hypoconid, entoconid, and hypoconulid. This sequence of mineralization corresponds to reports for other primates, including humans (Kraus and Jordan, 1965). Moreover, the largely incomplete state of calcification of the crowns, except perhaps that of the incisors in some individuals, is similar. More work on the state of mineralization of newborn ape teeth, especially for lesser apes, is needed to provide a fully balanced view.

5.2.2 Maxillary Dentition

Swindler (2002) provided some information based on casts of worn deciduous incisors of subadult common *Pan*, *Gorilla*, and *Pongo*. He confirmed the spatulate form (convex labially, concave lingually) of the maxillary di^1, a morphology in common with humans. Mesial and distal marginocristae are present, as well as a lingual cingulum.

The deciduous incisors of the *Pongo* he illustrates (Swindler, 2002, figure 4.5) possess a lingual longitudinal ridge, and this is described to typify the "unworn" morphology in apes generally (p. 55).

The di^2 is narrower than di^1 in apes, but similar in form in *Pan* and *Gorilla*. Swindler refers to di^2 in *Pongo* as caniniform. All great apes have pronounced dc's, nearly conical, but with mesial and distal marginocristae that give them a somewhat triangular appearance. A longitudinal ridge is present on the lingual surface. Swindler (2002) describes a mesolingual fossa in *Gorilla* and some *Pan*. In *Pongo* he describes an additional depression, the distolingual fossa.

The dp^3 possesses two cusps, the paracone and protocone (Figure 5.6). A longitudinal groove runs between them, described as deep in *Pan* and *Gorilla*, and shallower in *Pongo* (Swindler, 2002). In *Pongo*, dp^3 is distinguished by a transverse crest between the protocone and paracone. In the gorilla, dp^3 has a small lingual cingulum. *Pongo* is distinguished by a wrinkle pattern on both deciduous premolars, as described for permanent molars. In all apes, the dp^4 has four cusps and a crista oblique connecting the metacone and protocone. The dp^4 of *Gorilla* and *Pan* has a lingual cingulum. The trigon basin is described as "spacious" and there is a distal fovea (Swindler, 2002). It is restricted lingually in *Pan* and more extensive (wrapping mesial to the protocone) in *Gorilla*.

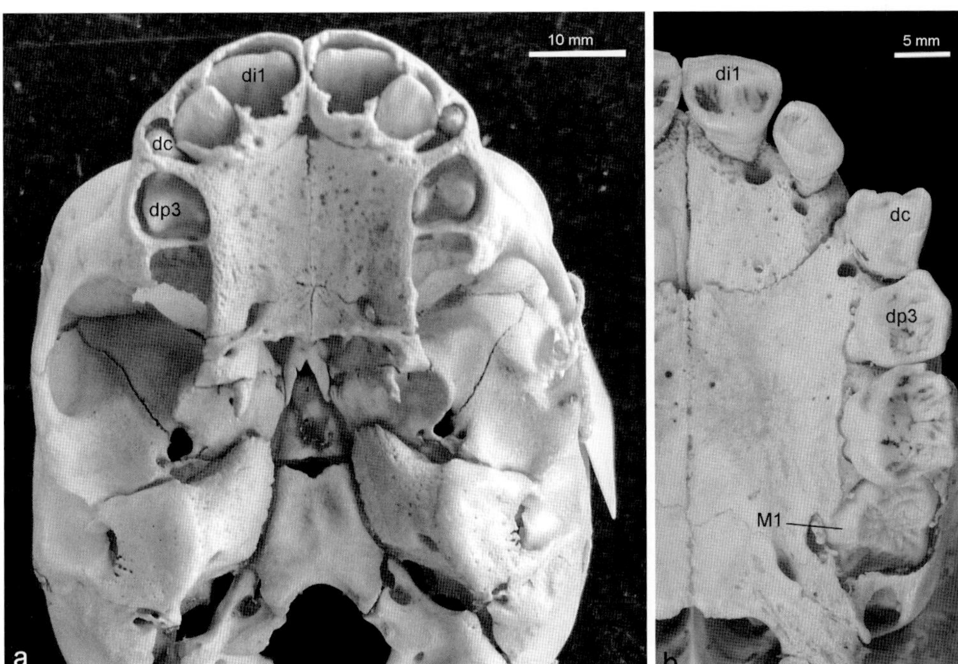

Figure 5.6 (a) Perinatal zoo specimen of *Pan troglodytes* (FM122719, precise age not recorded) showing the deciduous maxillary dentition mostly still within alveolar crypts. (b) Fully erupted deciduous maxillary dentition of *Pongo pygmaeus* with an unerupted M^1 (CM61335). The cusps are all somewhat worn, but the "wrinkled," or crenulated, enamel on deciduous premolars and M^1 is still evident.

In all species, M^1 is morphologically similar to dp^4, although lagging in rate of mineralization.

5.2.3 Mandibular Dentition

Lower deciduous incisors resemble permanent incisors, being spatulate in form. The di_2 is described as narrower than di_1, although in Swindler's figures (2002: figures 4.5 and 4.6) this is a more apt description of *Pongo* than the gorilla. In the gorilla the tooth apices are relatively flat, although Swindler notes the samples he examines are worn. In *Pongo*, di_2 has a more rounded incisal border than di_1. In *Pongo* and *Pan*, there are mesial and distal longitudinal ridges (presence/absence uncertain in the gorilla), and Swindler's figure 4.5 indicates a central lingual longitudinal ridge in *Pongo*.

All great apes have mesial and distal marginal ridges on the deciduous canine. The distal ridge descends from the cusp to a distal ledge. There is a diastema between dc and di_2.

The dp_3 possesses four cusps in all great apes according to Swindler. Swarts (1988), however, finds absence of the entoconid to be a unique trait of *Pongo*. No hypoconulid

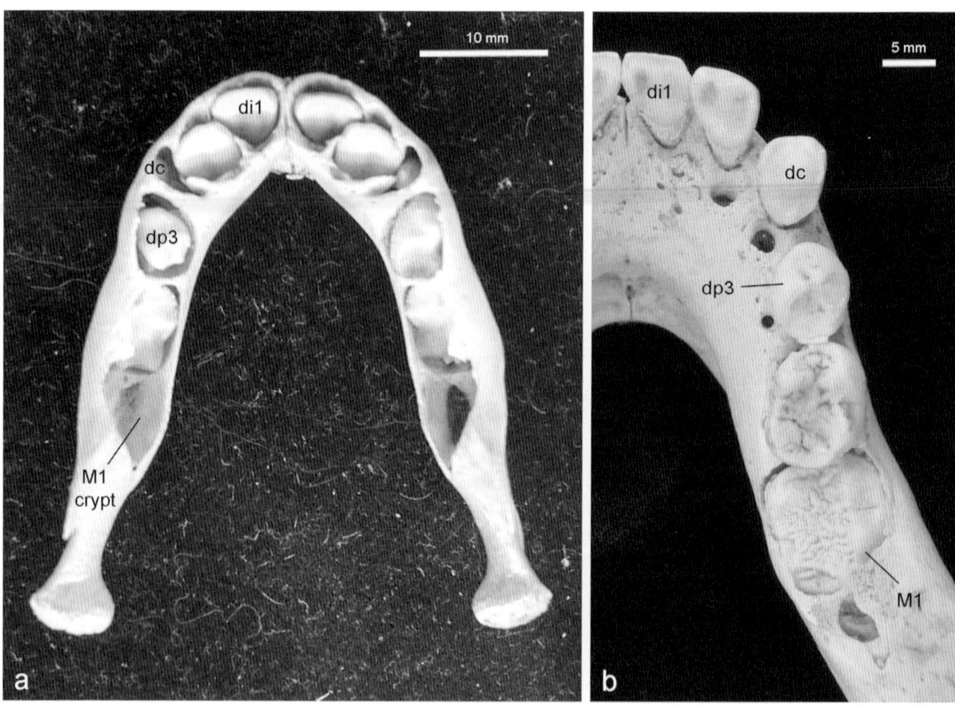

Figure 5.7 (a) Perinatal zoo specimen of *Pan troglodytes* (FM122719, precise age not recorded) showing the deciduous mandibular dentition mostly still within alveolar crypts. Note the M_1 crypt is quite large (and probably contained mineralized cusps prior to clearing the skeleton). M1 develops in the mandible more rapidly than in the maxilla (see Figure 5.6). (b) Fully erupted deciduous mandibular dentition of *Pongo pygmaeus* with an unerupted M_1 (CM61335). The "wrinkled" or crenulated enamel on deciduous premolars and M1 is apparent.

was observed by Swindler (2002); Swarts (1988) observes that it is "indistinct" from the distal marginal ridge, when present. A protocristid connects the protoconid and metaconid. A paracristid extends mesially and creates a border for a small anterior fovea, or trigonid basin. Swarts describes this basin as small (~1/3 the mesiodistal length of the tooth) in *Pongo*, compared to about half the mesial distal dimension in all other teeth. The trigonid basin may even occupy more than half the mesiodistal length of dp_3 in some specimens (Figure 5.7). The paracristid may have a small stylid (Swindler makes note of Jorgensen's opinion that it is a leap to consider this stylid as a homologue of a paraconid). There is a slight buccal cingulid in the gorilla, but not in the other great apes.

There are five cusps in the dp_4 in all great apes. The added hypoconulid, and the contact of the metaconid and hypoconid, together help to create the *Dryopithecus* ("Y-5") groove pattern on the occlusal surface. Swarts (1988) notes that the protocristid may or may not be present in *Pongo*, and thus the basins can appear confluent. In *Pongo*, the wrinkle pattern is seen on both deciduous premolars, as observed in permanent molars. There is a trigonid basin between the protocristid and the mesial marginal ridge. This basin is slightly higher than the plane of the broader talonid basin. No anterior or posterior fovea are seen (Swindler, 2002).

Subsequent descriptions of other primate taxa are based on small samples, sometimes on individual specimens. We strove to examine at least two of every species when possible. Our knowledge of presence and absence of cingula and secondary cusps should be assumed to be incomplete except for species with exceptional crown mineralization at birth. When possible, our samples are supplemented by older specimens to ascertain the progression of crown mineralization.

5.3 Morphology of the Dentition in Newborn Catarrhine Monkeys

5.3.1 General Considerations

The morphology of the crowns of all deciduous teeth as well as M1 is clear in catarrhine monkeys even at birth, because crown mineralization is at an advanced state. The di1 is wider than di2. In upper and lower jaws, di2 is caniniform to a varying extent (Swindler, 2002). Thus, there is a varied gradation of morphology from di1 to dc. A striking characteristic of the postcanine deciduous teeth is the bilophodont form, that is, there are pronounced transverse crests between mesial and distal cusp pairs, seen on dp3 and dp4, and beginning to mineralize on M1. At least four primary cusps are present on deciduous premolars. However, in the lower jaw, an additional cusp may be present at the mesial end of the paracristid, a possible paraconid. This cusp, when present, is more buccally positioned than the paraconid in *Tarsius*. It is typically more mineralized than the paracristid, suggesting earlier mineralization. Thus, putative paraconids may represent an additional primary cusp, but more developmental work is needed to understand whether these form similarly to the other cusps. Tables 5.2 and 5.3 summarize the status of crown mineralization in deciduous premolars and M1 across primates. A discussion of dental eruption is deferred until later in the chapter.

Box 5.2 shows the fully formed crowns of deciduous dentition of *Trachypithecus francoisi*. The crowns of all deciduous teeth are well-mineralized, including transverse crests and even the marginal ridges; the state of mineralization is more advanced compared to newborn hominoids.

5.3.2 Maxillary Dentition

5.3.2.1 Subfamily Cercopithecinae

Newborn *Allenopithecus nigroviridis* (Figure 5.8), *Macaca mulatta*, and *Papio anubis* are described based on new specimens. The di^1 has a slightly rounded incisal border. The incisal margin of di^2 reaches a central rounded tip. The dc has a nearly conical cusp and slight distal and mesial marginal ridges. Swindler (2002) previously described the deciduous teeth of several cercopithecines (*Chorocebus aethiops*, *Erythrocebus patas*,

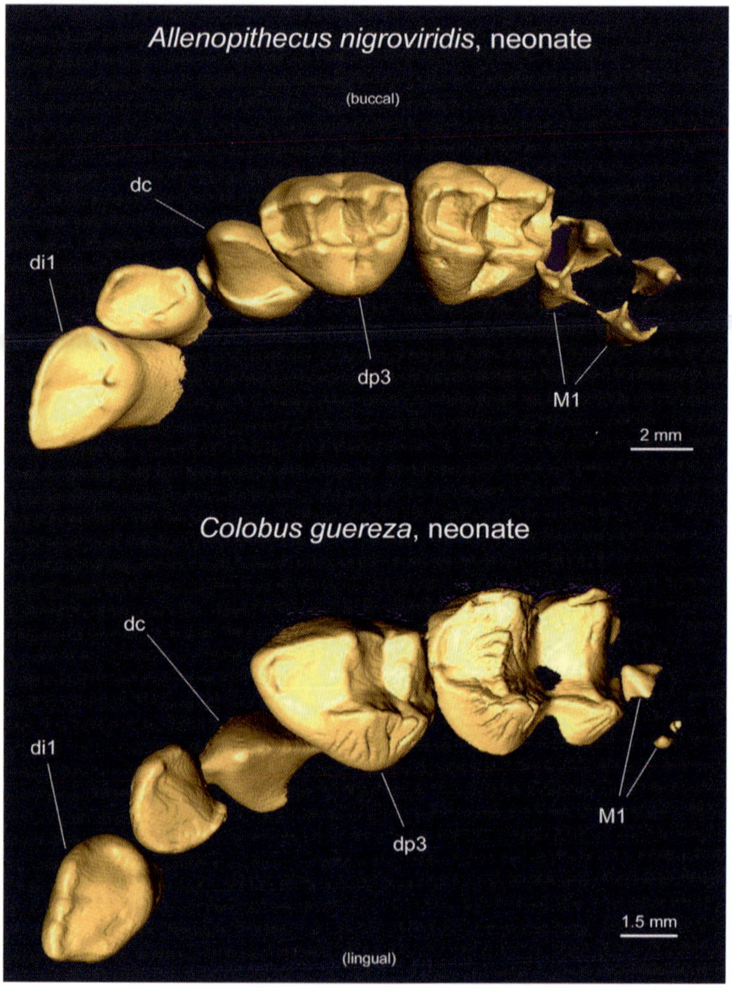

Figure 5.8 Maxillary dentition of a newborn cercopithecine (top) and colobine (bottom).

Cercocebus torquatus, and several *Cercopithecus* and *Lophocebus* spp.). All of these species are described as possessing a lingual cingulum or at least a "slight lingual ledge" (in the case of *Cercocebus*) on di's. A lingual tubercle on the cingulum of di^1 is described in most of these species. The di^2 is described and illustrated as being more rounded at the tip compared to di^1. In some species, Swindler states di^2 is caniniform. Swindler notes a broad triangular shape (at the base) of the dc in several mangabey species, various *Cercopithecus* spp., *Chlorocebus*, and *Erythrocebus*. He describes mesial and distal marginal ridges and a lingual cingulum. *Allenopithecus* is similar in all of these respects. However, the lingual longitudinal ridge that Swindler notes for *Cercopithecus* spp., *Chlorocebus*, and *Erythrocebus* is lacking in the dc of *Allenopithecus*.

In *Allenopithecus*, *Macaca*, and *Papio*, dp^3 and dp^4 are roughly quadrate in shape. The dp^3 is slightly compressed in an ovoid contour on the mesial side. A trigon basin is present. Both maxillary dp's also have the following features: the mesial and distal pairs of cusps are connected by transverse crests; well-demarcated marginal ridges border anterior and posterior fovea; parastyles are present. M^1 has four well-mineralized cusps and transverse crests that are almost mineralized, but the trigon basin is, as yet, not mineralized. M^2 is not mineralized (and has not progressed to even the bud stage – Smith et al., 2015). Swindler's descriptions of deciduous premolars of *Cercopithecus* spp., *Chlorocebus*, and *Erythrocebus* are largely similar to the above. However, in contrast to our findings, he considered styles to be quite rare in cercopithecines. Swindler describes a "small cusplet" on dp^3 in *Cercocebus torquatus* that resembles a parastyle. Swindler reviews studies on cercopithecines and concludes that the crista oblique is sometimes present, although rare.

5.3.2.2 *Subfamily Colobinae*

In newborn *Colobus guereza* (Figure 5.8) and *Trachypithecus francoisi* (Box 5.2), di^1 is broader then di^2, while the latter is caniniform. In *C. guereza* di1 has a straight incisal margin with three mamelons; there is a lingual heel near the base of the crown. The di^1 of *T. francoisi* has a more rounded incisal margin and a concave lingual fossa. In both species, dc is a partially mineralized crown, pyramidal in form, with a stout lingual pillar. Our description of deciduous anterior teeth of *Trachipithecus* is similar to that figured for *T. cristata* by Swindler (2002; see figure 4.4 therein).

The dp^3 and dp^4 each have four prominent cusps; the mesial and distal pair are connected by transverse crests. Each has a large central basin, with smaller fovea mesial and distal to transverse ridges. The mesial fovea of dp^3 is a triangular projection, while it has a rounded mesial margin in dp^4. The transverse crests are higher in dp^4 than dp^3 in *C. guereza*. Parastyles are present in *Trachypithecus*, as observed for *T. cristata* (Swindler, 2002). Three cusps of M^1 are mineralized in *C. guereza*; the hypocone remains unmineralized. All four M^1 cusps are mineralized in *T. francoisi*. M^2 is unmineralized (and has not progressed to even the bud stage – Smith et al., 2015).

5.3.3 Mandibular Dentition

5.3.3.1 *Subfamily Cercopithecinae*

Neonates of *Allenopithecus nigroviridis* (Figure 5.9), *Macaca mulatta*, and *Papio anubis* are described. The di_1 is tall and narrow. By comparison, di_2 is more caniniform except for

Table 5.2 Number of primary cusps and degree of mineralization of maxillary deciduous postcanine teeth in primate newborns

	dp2		dp3		dp4		Replacement teeth mineralized at birth?[3]	Source
	# Primary cusps[1]	Crown mineralization[2]	#Primary cusps	Crown mineralization	# Primary cusps	Crown mineralization		
HAPLORHINES								
Allenopithecus nigroviridis	–	–	4/4	C	4/4	C	None	Paddock et al., in press
Alouatta spp.	1–2/1–2	C	1–2/1–2	C	4/4	PC	None	Swindler, 2002; Paddock et al., in press
Aotus nancymaee	2/2	PC/C	2/2	PC/C	4/4	C	None	Paddock et al., in press
Callicebus cupreus	1/1	C	2/2	C	4/4	C	(I1)	Paddock et al., in press
Callithrix spp.	1/1	PC	2/2	PC	3/3	PC	None	Paddock et al., in press
Cebuella pygmaea	1/1	C	2/2	PC	3/3	PC	(I1, C)	Paddock et al., in press
Cercopithecus spp.	–	–	4/4	C	4/4	C	None	Swindler, 2002
Chlorocebus aethiops	–	–	4/4	C	4/4	C	None	Swindler, 2002
Colobus guereza	–	–	4/4	C	4/4	C	None	Paddock et al., in press

Erythrocebus patas	–	–	4/4	C	4/4	C	None	Swindler, 2002
Leontopithecus rosalia	1/1	C	2/2	PC	3/3	PC	None	Smith et al., 2011; Hershkovitz, 1977
Macaca mulatta	–	–	4/4	C	4/4	C	None	Paddock et al., in press
Macaca nemestrina	–	–	4/4	C	4/4	C	None	Paddock et al., in press
Papio anubis	–	–	4/5	C	4/4	C	None	Paddock et al., in press
Pan troglodytes	–	–	2/2	C	4/4	PC	None	Siebert and Swindler, 1991
Pithecia pithecia	1/1	C	2/2	C	4/4	C	I1	Paddock et al., in press
Pongo pygmaeus	–	–	2/2	C	4/4	C	None	Winkler et al., 1991
Saguinus spp.	1/1	PC	1/1	PC	3/3	PC	None	Paddock et al., in press
Saimiri boliviensis	2/2	PC	2/2	PC	4/4	PC	None	Paddock et al., in press
Trachypithecus spp.	–	–	4/4	C	4/4	C	None	Paddock et al., in press; Swindler, 2002

Table 5.2 *(cont.)*

	dp2		dp3		dp4		Replacement teeth mineralized at birth?[3]	Source
	# Primary cusps[1]	Crown mineralization[2]	#Primary cusps	Crown mineralization	# Primary cusps	Crown mineralization		
Tarsiiformes								
T. syrichta			1/1		1/1		I1, I2, P2 to P4	Paddock et al., in press
STREPSIRRHINES								
Cheirogaleus medius	1/1	C	1/1	PC	3/3	PC	I1, (I2), C, P2	Paddock et al., in press
Eulemur spp.	1/1	C	1/1	PC	3/3	PC	(C)	Paddock et al., in press
Galago spp.	1/1	C	1/1	PC	4/4	C	C, P2	Paddock et al., in press
Hapalemur griseus	1/1	PC	1/1	PC	3/3	C	None	Paddock et al., in press; Tattersall and Schwartz, 1974
Lemur catta	1/1	C	1/1	PC	3/3	PC	None	Paddock et al., in press
Lepilemur leucopus	1/1	C	1/1	C	3/3	C	C, P2, P3, P4	Paddock et al., in press

Microcebus murinus	1/1	C	1/1	PC	3/3	PC	I1, C	Paddock et al., in press
Nycticebus pygmaeus	1/1	C	1/1	PC	4/4	PC	I1, I2, C, P2	Paddock et al., in press
Otolemur spp.	1/1	C	1/1	PC	4/4	PC	C, (P2)	Paddock et al., in press
Perodicticus potto	1/1	-	1/1	-	3/3	-	?	Paddock et al., in press
Propithecus coquereli	-	-	1/1	C	3–4/3–4	C	I1, C, P3, P4	Paddock et al., in press
Varecia spp.	1/1	PC	1/1	PC	3/3	IC	None	Paddock et al., in press
Scandentia								
Tupaia belangeri	1/1	IC	1/1	IC	3/3	IC	None	Paddock et al., in press

Notes: 1, number of primary cusps: expressed as number present at birth/number present on completed crown; 2, crown mineralization at birth: IC, isolated cusps; PC, partial crown (e.g., crown base poorly mineralized or isolated hypocone); C, crown complete or nearly so (e.g., basin only incomplete); 3, replacement teeth mineralized at birth: parentheses indicate variably mineralized.

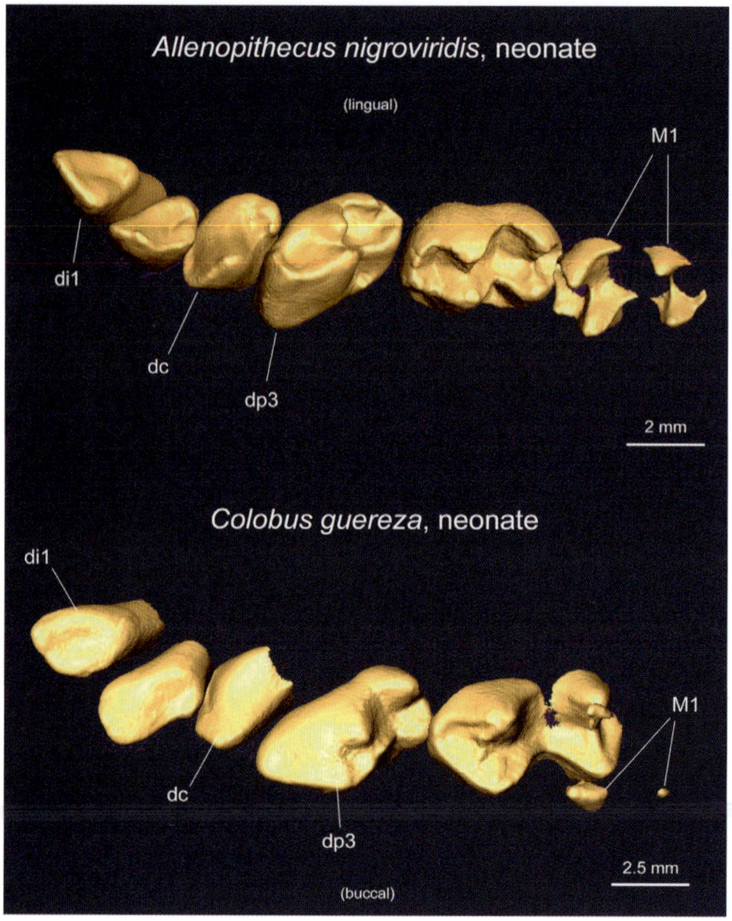

Figure 5.9 Mandibular dentition of a newborn cercopithecine (top) and colobine (bottom).

the great buccolingual flattening of the cusp. Other cercopithecines have been described similarly (Swindler, 2002). The dc is broad at the base and slightly rounded on its lingual surface. As Swindler (2002) studied older subadults, the base of the cusp of dc was further developed. He noted a raised distal heel or ledge on the crown base of dc in numerous cercopithecines. In *Cercopithecus* spp., there is a longitudinal ridge.

The dp_3 is compressed and ovoid in shape; dp_4 is nearly quadrate. Each has four primary cusps and a mesial marginal ridge that at least partly outlines the trigonid basin. Protocristids, if present, are low (present on dp_3 of *Allenopithicus*, dp_4 of *Papio*, and both dp's in *Macaca*). In dp_3, the mesial margin is raised to a tip, in what resembles a paraconid; this has been observed before in catarrhines, but the homology remains uncertain (see discussion by Swindler, 2002). In dp_4, the marginal ridge is raised on both lingual and buccal sides. In dp_3, the mesial cusp pair is more closely apposed than the distal cusp pair. No hypoconulids are observed. There are no major distinctions with cercopithecines described by Swindler (2002).

5.3.3.2 *Subfamily Colobinae*

In newborn *Colobus guereza* (Figure 5.9) and *Trachypithecus francoisi* (Box 5.2), di_1 has a straight incisal margin; it inclines distally downward in the former. The incisal margin of di_2 reaches a rounded central peak in *Colobus* and the tooth is caniniform in *Trachypithecus.* Swindler (2002) describes di_2 as caniniform in other colobines. The dc has a more rounded profile in occlusal view and appears partially unmineralized at the crown base.

The dp_3 has four prominent cusps and an extended, raised mesial ledge with an extended trigonid basin. From dp_3 to M_1, teeth become increasingly quadrate in the occlusal view. *Colobus* lacks transverse ridges connecting the mesial and distal cusp pairs, while these ridges are prominent in *Trachypithecus.* The mesial shelf bordered by the paracristid is more elevated in deciduous premolars than in M_1. There is also a mesial cusplet on dp_4 in *Trachypithecus.* Swindler (2002) states a paraconid is present 80% of the time on dp_3 of colobines. Three cusps of M_1 are mineralized in *Colobus* (entoconid remains unmineralized), whereas four cusp tips of M_1 are mineralized in *Trachypithecus*, as well as a mineralized crest between the mesial cusp pair.

5.4 Morphology of the Dentition in Newborn Platyrrhine Monkeys

5.4.1 General Considerations

There is arguably more ontogenetic dental variation among New World monkeys than among Old World monkeys. There is also much variation in the status of crown maturation in newborn platyrrhines. Callitrichines have, perhaps, the slowest perinatal rate of postcanine tooth mineralization, as evidenced by their incomplete dp4 crowns. *Alouatta*, *Aotus*, and many other platyrrhines are far more advanced in cusp mineralization.

5.4.2 Maxillary Dentition

5.4.2.1 *Family Atelidae*

In *Alouatta*, di^1 has a rounded incisal margin. The di^2, although buccolingually compressed like di^1, possesses a caniniform cusp. The dc is very broad and flattened buccolingually near the cusp tip; the base is broad with a pronounced lingual cingulum.

The dp^2 and dp^3 have pronounced paracones and a pronounced lingual cingulum with two or more small cusps. Swindler (2002) noted two cusps on dp^3 of *A. palliata.* The dp^4 has a quadrate shape and four cusps. A distinct but low crista oblique is present, which traverses to the base of the metacone. Three styles are present on the buccal side of dp^4 (para-, meso-, metastyles); only the parastyle is evident on dp^3. M^1 has three isolated mineralized cusps. M^2 is not present, even as an identifiable tooth germ (Smith et al., 2015).

Histological observations reveal that the permanent incisors are at the cap or bud stage, and other replacement teeth are not even that advanced at birth (Smith et al., 2015).

5.4.2.2 *Family Cebidae*

In most cebids di^1 has a rounded incisal margin. Some callitrichines present exceptions: in *Leontopithecus rosalia* and *Cebuella pygamaea* the tooth may be caniniform. This is variable in the latter species, and probably in *Leontopithecus* as well (Hershkovitz, 1977, described the cusp apex as rounded). The di^2 is more typically caniniform except in *Saimiri* and a P14 *Cebus apella*, in which the incisal margin is rounded. The dc is similar in form to di^2, but with a broader base; the base of dc may be deficient compared to the incisors (i.e., it is slower to mineralize), except in *Cebuella*, which has the best mineralized deciduous teeth of all callitrichines. The presence of para- and distostyles on the anterior teeth of some callitrichines was noted by Hershkovitz (1977). Among newborns and infants in which the crown is mineralized enough to detect them, distostyles are more common than parastyles, and are seen in some (not all) individuals of *Cebuella*, *Callithrix*, and *Saguinus* spp. When present, they are most prominent in the dc.

In *Saimiri* (Figure 5.10), features of the dp's resemble the illustration provided by Hershkovitz (1977, figure V.28) for *S. sciureus*. A P14 *Cebus* is highly similar in dp morphology. The dp^2 has a large paracone, a smaller protocone, and para- and distostyles. With a more expanded cingulum, dp^3 and dp^4 each have a more quadrate appearance. The dp^3 has two primary cusps (paracone, protocone), para- and distostyles, and a metaconule is seen in *Saimiri*. The dp^4 has four cusps, the paracone, protocone, metacone, and hypocone; para- and metastyles are also present in *Saimiri*. A crista oblique is pronounced, traversing to the base of the metacone. The distolingual side of the deciduous premolars appears least mineralized, with the lingual cingulum of dp^3 and dp^4 incompletely mineralized in all newborns. Smith et al. (2011) demonstrated that the posterior-most maxillary teeth, especially dp^4, are large in *Saimiri* relative to their facial size. This may partially explain the absence of a maxillary sinus in the genus (Rossie, 2006).

Callitrichines deserve a separate discussion. The dp^2 is nearly caniniform, although it possesses distinct mesial and distal ridges. The dp^3 also has a small protocone and a parastyle but the lingual portion of the tooth is poorly mineralized at birth in *Callithrix* and *Saguinus* spp. The dp^4 has three cusps, with the paracone the most pronounced. The trigon basin is incompletely mineralized in all callitrichines at birth. In addition, lingual cingula are observable in older infants (e.g., *Cebuella*) but are not yet mineralized in newborns. Hershkovitz (1977, figure V.28 therein) indicated stylar cusps on dp^4 of marmosets. Fewer are observed here, but it is possible that the small size of the teeth in the species may not allow the detection of such small cusplets even at our highest resolution. Hershkovitz wrote that the molarization of deciduous premolars was "more extensive" in *Leontopithecus* compared to the same teeth in other callitrichines, but in all callitrichines, dp^4 is distinctly more similar to M^1 than more mesial teeth.

The paracone and protocone of M^1 are mineralized in *S. boliviensis*. M^1 is rarely mineralized in most callitrichines, with the exception of *L. rosalia* (rarely beyond early bell stage) and *C. pygamea*, in which some cusp tips may be mineralized in newborns. These observed were confirmed using histology by Smith et al. (2015).

No replacement teeth are mineralized at birth except for the tip of I^1 in some specimens of *Cebuella*. Histological observations reveal that the I^1 is at the cap or early bell stage at birth in most cebids (Smith et al., 2015).

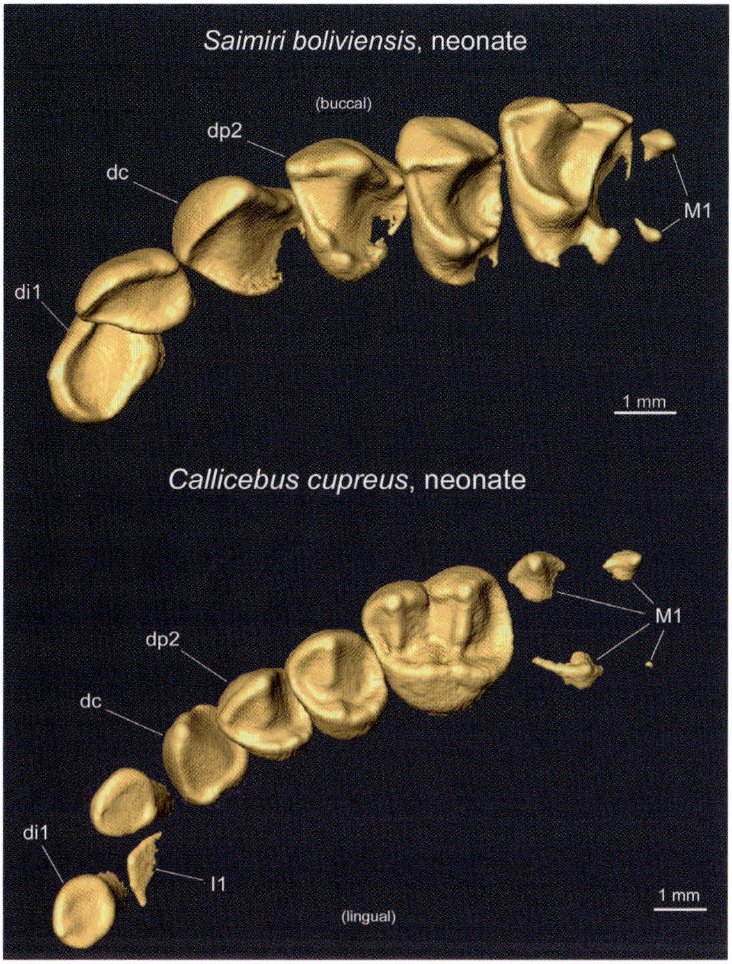

Figure 5.10 Maxillary dentition of a newborn cebid (top) and pithecid (bottom).

5.4.2.3 *Family Pitheciidae*

In newborn *Aotus nancymaae*, *Callicebus cupreus* (Figure 5.10), and *Pithecia pithecia*, di^1 and di^2 are broad at the incisal margin. The di roots are beginning to mineralize in *C. cupreus* and *P. pithecia*. The two incisors are nearly equal in width in *Callicebus*, but di^1 is far wider than di^2 in the other species. The di^2 has a pointed apex at the center of the incisal margin. The dc is robust; the base of the crown remains poorly mineralized.

The dp^2 is slightly elongated buccolingually, with a raised paracone. In *Pithecia* and *Aotus* there are para- and distostyles. Lingually is a second cusp, a distinct protocone in *Aotus* and *Callicebus*, but a minute cusplet in *Pithecia* (the para- and distostyles are raised relative to the cingulum). The dp^3 and especially dp^4 are more quadrate in form. A small tubercle on the postcingulum resembles the postprotostyle described on replacement premolars in *Pithecia* (Swindler, 2002). In dp^4 the same features are present except a distinct metacone is seen, and a more prominent cusp is

Table 5.3 Number of primary cusps and degree of mineralization of mandibular deciduous postcanine teeth at birth

	dp2		dp3		dp4				
Mandibular teeth	# Primary cusps[1]	Crown mineralization[2]	# Primary cusps[3]	Crown mineralization	# Primary cusps[4]	Paraconid?	Crown mineralization	Replacement teeth at birth?[5]	Source
HAPLORHINES									
Allenopithecus nigroviridis	–	–	5/5	C	5/5	+	C	None	Paddock et al., in press
Alouatta spp.	1/1	C	4/4	C	4/4	–	C	None	Swindler, 2002
Aotus nancymaee	1–2/1–2	PC	2/2	C	4/4	–	C	None	Paddock et al., in press
Callicebus cupreus	1/1	C	2/2	C	4/4	–	C	(I1, I2)	Paddock et al., in press
Callithrix spp.	1/1	PC/C	1/1	PC/C	4/4	–	C		Paddock et al., in press
Cebuella pygmaea	1/1	PC/C	1/1	PC/C	4/4	+	C	(I1, I2)	Paddock et al., in press
Cebus apella	2	?	2	?	4	–	?	None	Paddock et al., in press
Cercopithecus spp.	–	–	4/4	C	4/4	–	C	None	Swindler, 2002
Chlorocebus aethiops	–	–	4/4	C	4/4	–	C	None	Swindler, 2002
Colobus guereza	–	–	4/4	C	4/4	–	C	None	Paddock et al., in press
Erythrocebus patas	–	–	4/4	C	–	-	C	None	Swindler, 2002

Leontopithecus rosalia	1	–	2	–	4	–	–	–	Paddock et al., in press; Hershkovitz, 1977
Macaca mulatta	–	–	4/4	C	4/4	–	C	None	Paddock et al., in press
Macaca nemistrina	–	–	4/4	C	4/4	–	C	None	Paddock et al., in press
Pan troglodytes	–	–	4/4	C	5/5	–	C	None	Siebert and Swindler, 1991
Papio anubis	–	–	4/4	C	4/4	–	C		Paddock et al., in press
Pithecia pithecia	1/1	C	2/2	C	4/4	–	C	None	Paddock et al., in press
Pongo pygmaeus	–	–	4/4	C	5/5	–	C	None	Winkler et al., 1991
Saguinus spp.	1/1	PC	1–2/2	PC	4/4	–	PC	None	Paddock et al., in press
Saimiri boliviensis	1/1	C	2	C	4/4	–	C	None	Paddock et al., in press
Trachypithecus spp.	–	–	4/4	C	5/5	+	C	None	Paddock et al., in press; Swindler, 2002
Tarsiiformes									
T. syrichta	–	–	1	C	1	–	C	I1, C, P2–P4	Paddock et al., in press

Table 5.3 *(cont.)*

Mandibular teeth	dp2 # Primary cusps[1]	dp2 Crown mineralization[2]	dp3 # Primary cusps[3]	dp3 Crown mineralization	dp4 # Primary cusps[4]	dp4 Paraconid?	dp4 Crown mineralization	Replacement teeth at birth?[5]	Source
STREPSIRRHINES									
Cheirogaleus medius	1/1	C	1/1	PC	4/4	–	C	I1, I2, C, P2	Paddock et al., in press
Eulemur spp.	1/1	C	1/1	PC	3–4/3–4	+	C	(I1, I2, C)	Paddock et al., in press
Galago spp.	1/1	C	1/2	PC	4/4	+	C	I1, I2, C, P2	Paddock et al., in press
Hapalemur griseus	1/1	C	1/1	PC	4/4	+	C	None	Paddock et al., in press; Tattersall and Schwartz, 1974
Lemur catta	1/1	C	1/1	PC	4/4	+	C	None	Paddock et al., in press
Lepilemur leucopus	1/1	C	1/1	C	3/3	+	C	I1, I2, C, P2, P3, P4	Paddock et al., in press
Microcebus murinus	1/1	C	1/1	PC	4/4	–	C	I1, I2, C, P2	Paddock et al., in press
Nycticebus pygmaues	1/1	C	1/?	PC	4/4	–	PC	I1, I2, C, P2	Paddock et al., in press
Otolemur spp.	1/1	C	1/2	PC	4/4	+	C	I1, I2, C, P2	Paddock et al., in press

Perodicticus potto	1	?	1	?	4?	?	?	?	Paddock et al., in press
Propithecus coquereli	1/1	C	–	PC	4/4	–	C	I1, I2, P2, P4	Paddock et al., in press
Varecia spp.	1/1	PC	1/1	PC	3/3	+	PC	None	Paddock et al., in press
Scandentia									
Tupaia belangeri	1/1	IC	1/1	IC	4/4	+	IC	None	Paddock et al., in press

Notes: 1, Number of primary cusps: expressed as number present at birth/number present when crown is fully mineralized; 2, crown mineralization at birth (only described for neonates): IC, isolated cusps; PC, partial crown (e.g., crown base poorly mineralized or isolated entoconid); C, crown complete or nearly so (e.g., basin only incomplete); 3, for dp_3, this number excludes a paraconid, which may be present in some species (see text); 4, for dp_4, this number excludes the paraconid, if present, but may include the hypoconulid (e.g., *L. catta*); 5, replacement teeth mineralized at birth: parentheses indicate variably mineralized.

seen on the postcingulum, constituting a hypocone in our view. A crista obliqua is distinct near the base of the metacone and transects the common border of the trigon and talon basins, before diminishing to a low ridge near the protocone. Small para- and metasytles are visible. M^1 has four well-mineralized cusps and a mostly mineralized trigon basin. The talon basin is unmineralized.

A portion of I^1 is mineralized in *Callicebus* (Figure 5.10) and *Pithecia*.

5.4.3 Mandibular Dentition

5.4.3.1 *Family Atelidae*

Our descriptions of *Alouatta seniculus* concur with those of *A. palliata* by Swindler (2002), except as noted. The incisal margin of di_1 in is nearly straight whereas it comes to a rounded tip in di_2, similar to. The dc is broad-based; the crown is incompletely mineralized.

The dp_2 is nearly oval in occlusal profile, while dp_3 and dp_4 have a more quadrate shape. The dp_2 has one cusp connected to a raised mesial ridge (the protocristid), which descends toward the lingual margin, where a small tubercle might be regarded as a rudimentary metaconid; mesial and distal marginal ridges create distinct foveae on either side of this ridge. In *A. palliatta*, dp_2 is described as caniniform compared to the other dp's (Swindler, 2002). In dp_3 and dp_4, four cusps are present. The trigonid basin is larger than the talonid basin in dp_3, while this is reversed in dp_4. Swindler describes dp_3 of *A. palliatta* as relatively simple, being bicuspid and lacking the cristid obliqua. In *A. seniculus*, dp_3 and dp_4 both have cristid obliqua connecting to the protocristid near the protoconid. On the distal marginal ridge of dp4 there is a very small elevation that could be considered a rudimentary hypoconulid, which is rarely present in *Alouatta* according to Swindler (2002). M_1 has three mineralized cusps, each more raised than in dp_4. M_2 is not mineralized.

5.4.3.2 *Family Cebidae*

The anterior teeth are all buccolingually compressed. The incisal border of di_1 reaches a central point in marmosets, but has a rounded margin in most cebids. The di_2 is caniniform in callitrichines, and slopes gently downward on the distal end in *Saimiri* and *Cebus*. The crown base of the dc of *Saimiri* and *Cebus* is far broader than the incisors. The dc has some similarity to di_2 in callitrichines, with a more projecting cusp and more prominent distolingual heel. In callitrichines, anterior teeth are distinctly more mineralized than the postcanine teeth.

The dp_2 and dp_3 each have a highly projecting protoconid that is connected to mesial and distal marginal ridges. Hershkovitz (1977) indicated distinct distolingual heels for dp_2 and dp_3 in callitrichines. In all cebids, what is a distolingual heel in dc is present and larger in dp_2, and in dp_3, the same portion of the tooth is a small talonid basin in *Saimiri* (Figure 5.11) and *Cebus*. In callitrichines, there is a gradient of progressive enlargement of the heel from dc to dp_3, and then an abrupt morphological shift to a highly quadrate and molarized dp_4. In all newborn callitrichines, mineralization of the distolingual part of dp_3 and dp_4 is deficient; in some specimens, the entoconid and part of the talonid basin may remain unmineralized. Hershkovitz noted the conulids of dp's were well-developed in the single juvenile *Leontopithecus*

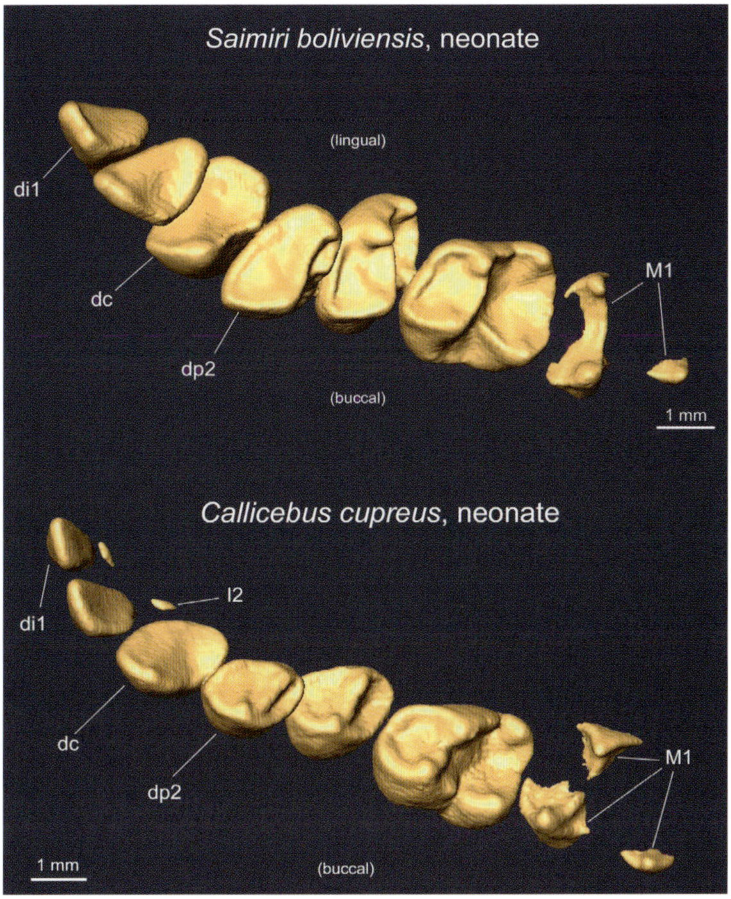

Figure 5.11 Mandibular dentition of a newborn cebid (top) and pithecid (bottom).

he examined. In our sample, these small cusps may escape detection due to the limits of resolution for very small teeth.

In *Saguinus* spp., M_1 is usually unmineralized at birth (the protoconid is mineralized in a single specimen of *S. geoffroyi*). In other cebids, only the protoconid of M_1 is mineralized in most neonates; the metaconid tip is also mineralized in one *C. jacchus* and one *Cebuella* neonate. Replacement incisors are mineralized at the cusp tip in *Cebuella*.

5.4.3.3 *Family Pitheciidae*

In newborn *Pithecia* and *Callicebus* (Figure 5.11), di_1 is narrow with a straight incisal margin; di_2 is similarly narrow. In *Aotus*, di_1 and di_2 widen toward the apex where the incisal margin is flattened and complicated with mamelons. The dc is robust in *Aotus* and *Pithecia*, but is buccolingually flattened in *Callicebus*. The dc's are poorly mineralized at the crown base.

In all three species, dp_2 and dp_3 are bicuspid with protocrostids. The metaconids of dp_2 are less pronounced in *Callicebus* and *Pithecia* compared to *Aotus*. The dp_3 of

Pithecia is more robust than in the other species with larger trigonid and talnid basins; distally there is a miniscule pair of cusps (presumptive reduced hypoconid, entoconid). The dp_4 has four primary cusps, a prominent protocristid, a small trigonid basin, and a broader talonid basin; the trigonid basin is especially small in *Aotus*. Three to four cusps of M_1 are present, with the distal pair less mineralized than the mesial pair. Part of the talonid basin is unmineralized in all three species. The protoconid of M_2 is mineralized in the *Pithecia* newborn.

The edges of the incisal margins of I_1 and I_2 are mineralized in *Callicebus* (Figure 5.11).

5.5 Morphology of the Dentition in Newborn Tarsiers

5.5.1 General Considerations

Mineralization of permanent teeth is on an advanced schedule in tarsiers compared to nearly all other primates. Cusps are a particular focus of mineralization as opposed to crests (Box 5.3). Tarsiers have an advanced mineralization schedule of replacement teeth. However, M1 is more advanced in mineralization than any deciduous or replacement tooth.

5.5.2 Maxillary Dentition

In newborn *Tarsius syrichta* (Box 5.3), di^1 has a minute peg-shaped cusp and full root; di^2 is absent in all newborns. The dc has a conical cusp and a slight cingulum distolingually, with a partially resorbed root.

The dp^2 is absent, as reported for *T. bancanus* by Luckett and Maier (1982). The dp^3 has one cusp with a pronounced cingulum on all sides. A parastyle is present and roots are nearly completely formed. The dp^4 is triangular in occlusal profile, with a large paracone and smaller metacone and protocone; a parastyle is present. A sharp crest connects the buccal cusps. Roots are nearly completely formed. M^1 is well-mineralized at birth, although the cingulum is incompletely mineralized. M^2 has three mineralized cusps; the basin is unmineralized. The paracone of M^3 is mineralized.

I^1, I^2, P^2, P^3, and P^4 crowns are all partially mineralized at birth. Luckett and Maier (1982) note only I^1, I^2, C, and P^2 in a newborn *T. bancanus.*

5.5.3 Mandibular Dentition

The di_1 has a small rounded cusp and full root; di_2 is absent or represented by a small vestigial fragment. The dc has a conical cusp, a lingual heel, and a nearly complete root.

The dp_2 is absent. Both dp_3 and dp_4 have highly projected protoconids. The dp_3 is a single-rooted, caniniform tooth. The dp_4 has two small distal cusps, but no observable metaconid, and two stout roots. M_1 is fully mineralized at birth except for parts of the cingulum. Part of the buccal cingulum is visible in a P6 specimen. M_2 is mineralized except for the middle of the talonid basin and the cingulum. The protoconid and metaconid of M_3 may be mineralized at birth.

I_1, C, P_2, P_3, and P_4 are partially mineralized at birth. A portion of the roots of I_1 and P_2 is mineralized.

5.6 Morphology of the Dentition in Newborn Strepsirrhines

5.6.1 General Considerations

Lemurids stand apart from other strepsirrhines in their relatively slow mineralization of M1 compared to other families, although there are variations among lemurids in the degree of mineralization of M1 and the replacement teeth at birth. Cheirogaleids and galagids have advanced mineralization of certain replacement teeth at birth. The single available lorisid has the most well-mineralized replacement teeth of any primates aside from indriids and tarsiers.

In dp_4, strepsirrhines, like some cercopithecoids, possess a cusp on the leading edge of the paracristid – a possible paraconid. In no species is it as large as most of the other primary cusps.

5.6.2 Maxillary Dentition

5.6.2.1 Family Cheirogaleidae

In *Cheirogaleus medius* (Figure 5.12) and *Microcebus murinus*, di^1 and di^2 are both small, with peg-like cusps. The dc is taller, more robust and conical except for a slight flattening of the lingual surface. In all anterior teeth the roots have begun to mineralize.

The dp^2 is caniniform but smaller than dc. The dp^3 is broader than dp^2 with a pyramidal cusp; the crown base is incompletely mineralized. The dp^4 is triangular-shaped with three primary cusps; the trigon basin is incompletely mineralized. The cingulum, if present, is unmineralized. Tattersall and Schwartz (1974) noted that *C. major* and *Phaner* have three-cusped dp^4's. M^1 is present with three cusps. A single cusp of M^2 is mineralized in *Microcebus*, *Cheirogaleus*, and *Mirza coquereli.*

Histological information provides definitive status of the replacement teeth in cheirogaleids (Smith et al., 2015). The tip of the I^1 cusp is mineralized in *Cheirogaleus* and *Microcebus*, whereas it is only advanced to the early bell stage at best in newborn *Mirza*. I^2 is distinctly less developed than I^1, showing no mineralization as yet in most newborns, but the cusp tip is mineralized in *Cheirogaleus*. The crown tip of P^2 is mineralized in newborn *Cheirogaleus*.

5.6.2.2 Family Lemuridae

In newborn lemurids, di^1 and di^2 are small and rounded buccally. The dc is mesiodistally broad and buccolingually compressed, with a cusp that curves distally. The roots of the di's have initiated mineralization in some newborns.

The dp^2 is small, caniniform, and buccolingually compressed, although not to the degree of dc. The dp^3 is rounded buccally, and lingually there is a distinct vertical ridge in *Eulemur* spp. and *Hapalemur griseus*; distal to the lingual ridge is a concave depression in both species. In *Lemur catta* dp^3 has a simpler pyramidal shape, lacking the distolingual concavity. The crowns of dp^2 and dp^3 are only partially mineralized in all lemurids at birth; other morphologies are apparent in older infants (see below). Among lemurs, dp^2 and dp^3 are most robust in *Hapalemur* (Box 5.4). The dp^4 is triangular in profile with three primary cusps. There is some variation in the smaller cusps of dp^4, which is difficult to attribute as species-specific based on the small samples. In *E. collaris*, a parastyle is present, as is a small paraconule (the latter is present in only two of the three neonates and also the older infant). The crista oblique is incomplete. Lingual to the protocone are two isolated cusplets, the incipient protostyle and hypocone. In *E. mongoz*, a parastyle and a smaller metastyle are present. The crista oblique is low and connects to the base of the metacone and bears a small metaconule. In a single newborn *E. rubriventer*, a pronounced parastyle is present, as is a small paraconule. The crista obliqua is distinct (far more so than in *E. collaris*) and bears two metaconules (see Figure 5.4). Distal to the protocone is a triangular shelf; because it is distinct from the partially mineralized lingual cingulum, this can be considered a pseudohypocone (Swindler, 2002), and this specimen is distinct from the other *Eulemur* spp. in this respect. There are also two small cusps on an incipient lingual cingulum, the protostyle and a small true hypocone. The trigon basin is poorly mineralized. *L. catta* has three primary cusps, a parastyle and a metastyle, but the lingual cingulum is as yet

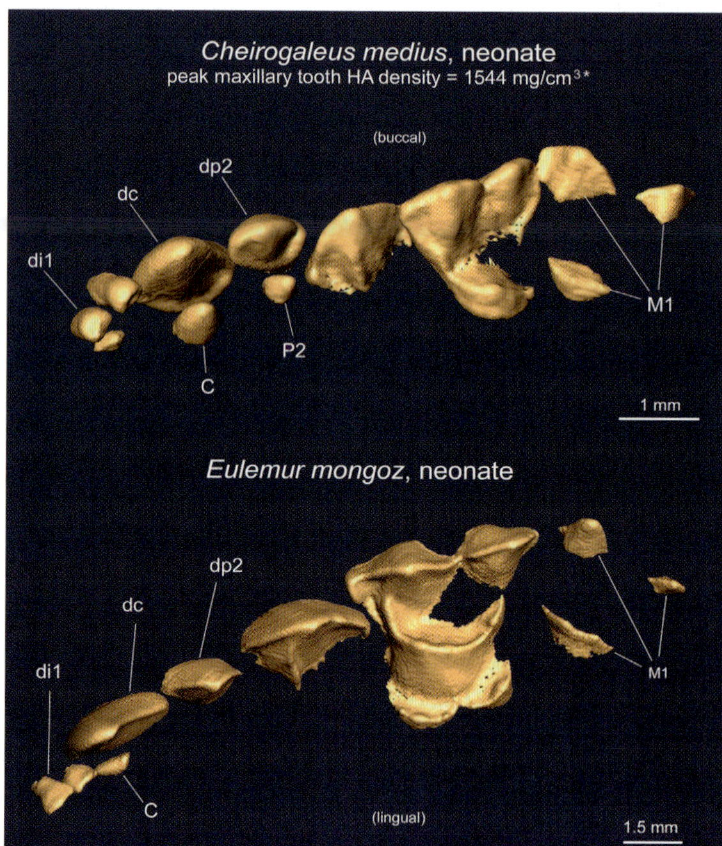

Figure 5.12 Maxillary dentition of a newborn cheirogaleid (top) and lemurid (bottom).

unmineralized at birth. In *Eulemur* and *Lemur* the trigon basin is unmineralized in the center. *Hapalemur* has the best mineralized dp4, with a complete basin and lingual cingulum; there are three buccal stylar cusps (Box 5.4). At birth, *Varecia* species have the least mineralized teeth of any primate that we have examined. The dp^4, for example, has three isolated mineralized cusps.

M^1 is present as three mineralized cusp tips in most newborn lemurids. In *Varecia* spp. only the paracone of M^1 is mineralized, while in *Hapalemur* M^1 mineralization is relatively advanced: the mesial margin is clear and the basin is beginning to mineralize (Box 5.4).

No portions of I^1, I^2, or replacement premolars are mineralized in lemurids at birth. The cusp tip of C is mineralized in one newborn *Eulemur* specimen.

5.6.2.3 *Family Lepilemuridae*

In newborn *Lepilemur leucopus*, only one deciduous incisor is present. It is nearly horizontal in position, and possibly in the process of being shed.

All deciduous teeth have fully mineralized crowns and well-formed roots. The dp^2 has one cusp and is buccolingually compressed. In dp^3 there is one prominent cusp which leads to a mesial and distal ridge; a parastyle is present. In dp^4 three primary cusps are prominent; there is a parastyle and metastyle and a distinct lingual heel. All permanent molars are well-mineralized, including basins and cingula.

The cusp apices of C and the three replacement premolars are present.

5.6.2.4 *Family Indriidae*

In newborn *Propithecus coquereli*, di^1 and di^2 are both small and buccolingually flattened. The dc is mesiodistally broad with a slightly concave lingual surface. However, dc seems proportionally far smaller than in other lemuroids.

The dp^3 is broad with one cusp that connects to mesial and distal marginal ridges; it is concave lingually. The dp^4 is broad with a prominent paracone and metacone, and a distinct but smaller protocone. Some specimens possess a hypocone, such that the tooth may be triangular or trapezoidal in profile. As described in M^1 and M^2 (Tattersall and Schwartz, 1974), dp^4 possesses a well-developed buccal cingulum with one to three styles; if a single style is present it is the parastyle. The contours of the crowns of M^1 and M^2 are nearly completely formed at birth. The paracone and protocone of M^3 are mineralized.

I^1 is very well mineralized, already revealing its spatulate shape, in newborns. I^2 is unmineralized at birth. The cusp tip of C is present lingual to dc. The paracone of P^4 is more advanced than P^3, already assuming a pyramidal shape.

5.6.2.5 *Family Galagidae*

In *Otolemur* (Figure 5.13) and *Galago* spp., di^1 and di^2 are both peg-like and slightly concave ventrally. The dc is much larger and broader with sharp mesial and distal marginal ridges. Lingually there is a broad, rounded longitudinal ridge with concave depressions on either side. Roots of the anterior teeth are partially mineralized.

The dp^2 is smaller than dc, caniniform, and has prominent para- and distostyles in *Otolemur* spp. The dp^3 is triangular in occlusal profile with a small parastyle visible in one *Otolemur* specimen. Part of the crown is unmineralized in all specimens, so

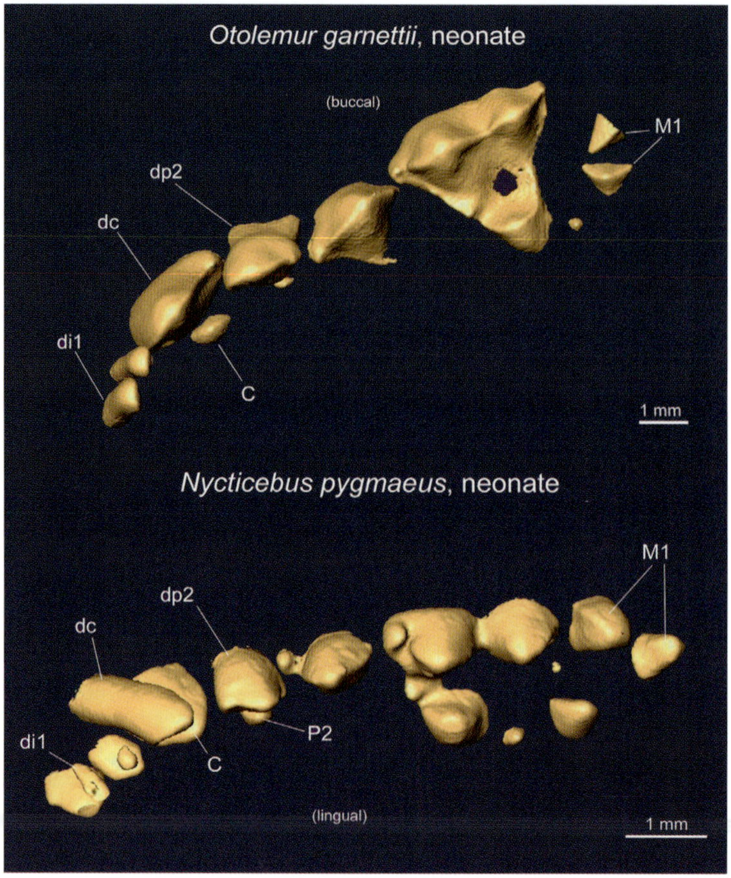

Figure 5.13 Maxillary dentition of a newborn galagid (top) and lorisid (bottom).

parastyles may be more common. The dp^4 is very large and has three primary cusps and an incompletely mineralized basin in most specimens. A parastyle and metastyle are evident in all specimens. A paraconule is visible in our single specimen of *O. garnettii*. The tip of the hypocone is mineralized in some specimens. M^1 has three mineralized cusps. Only the tip of the paracone of M^2 is present if the tooth has initiated mineralization.

The apical cusp of C is mineralized at birth. The tip of the paracone of P^2 is also present in a newborn *O. garnettii*. In *Galago* spp., a smaller mineralized portion of P^2 is also present at birth.

5.6.2.6 Family Lorisidae

In newborn *Nycticebus pygmaeus* (Figure 5.13), di^1 is small and broad. The di^2 is slightly smaller than di^1. The dc is narrow and cylindrical except for a tapering, triangular tip. A skeletonized *Perodicticus potto* skull in the collection of the Carnegie Museum (#CM69182), obtained from the Pittsburgh Zoo, was examined for deciduous tooth morphology. Dentally, it is considerably more mature than the P0 newborn lorisoids described above, instead comparable with a P26 infant *Otolemur crassicaudatus* in our sample. The crowns of di^1 and di^2 in *Perodicticus* are lingually flat, with rounded buccal surfaces; incisal margins are rounded. The dc resembles that of other lorisoids, possessing a buccolingually flattened single cusp.

The dp^2 of *Nycticebus* and *Perodicticus* has a single cusp with short mesial and distal ridges emanating away from the tip. There are distinct para- and distostyles in dp^2 of the latter; they may be present in *Nycticebus* as well, but may be too small to detect at our highest scan resolution. In both lorisids, dp^3 has a conical paracone with an adjacent parastyle; the distal crown is as yet unmineralized in *Nycticebus*, but a distostyle is visible in *Perodicticus*. In *Nycticebus*, dp^4 is a triangular-shaped mass with three primary cusps, all merged. In one specimen from a litter of three (Figure 5.13), the trigon basin is unmineralized in the center and the hypocone is an isolated cusp. In a sibling the basin is nearly fully mineralized. A parastyle is present, as well as a prominent paraconule. A crista oblique is not apparent, as yet, if there is one. In *Perodicticus*, dp^4 is a triangular-shaped mass with three primary cusps; there is a small cingulum flaring distolingually, with a slightly raised hypocone. In *Nycticebus*, M^1 is represented by three cusps and the tip of the paraconule. Neither the parastyle nor the hypocone has started to mineralize. M^1 is fully mineralized in the infant *Perodicticus*, with four primary cusps.

In *Nycticebus*, four replacement teeth are mineralized, I^1, I^2, C^1, and P^2. Only the tip of the cusp of P^2 is seen. The same teeth are presumed to have mineralized in *Perodicticus* based on large alveolar crypts. A small crypt for P^3 is also visible.

5.6.3 Mandibular Dentition

5.6.3.1 Family Cheirogaleidae

In newborn *Cheirogaleus* (Figure 5.14), *Mirza*, and *Microcebus*, the deciduous anterior tooth resembles the adult tooth comb, as described by Tattersall and Schwartz (1974).

The dp_2 is caniniform, but with a more pronounced lingual heel. The dp_3 is narrower near its apex compared to dp_2, and has a more pronounced longitudinal ridge in *Cheirogaleus* and *M. murinus* (the tooth is not mineralized enough to detect such a ridge in *Mirza*). A lingual cingulum, described in *C. major* by Tattersall and Schwartz (1974), is not yet mineralized if present. The dp_4 is broad and has a prominent protoconid, hypoconid, and entoconid; the metaconid is present, but only barely elevated above the level of the protocristid. A cristid obliqua, projecting mesially from the hypoconid, is well-mineralized and contacts the protocristid at the base of the protoconid. The mesial margin of the tooth projects mesial to the metaconid. It lacks a distinct cusp (i.e., a paraconid), but it is more pronounced than the same region of M_1. The trigonid basin is shallower, elevated, and better mineralized than the talonid basin. Smaller cusplets (e.g., metastylid) may be present based on slight swellings, but this is uncertain (perhaps due to limits of scan resolution in view of small tooth size). M_1 is represented by three to four cusps; the entoconid, if present, is an isolated cusp tip. One to two of the mesial cusps of M_2 are present in *Microcebus* and *Cheirogaleus*; none are detected in *Mirza*.

In *Cheirogaleus*, the cusp tips of I_1, I_2, C, and P_2 are mineralized. In *Microcebus*, the cusp tips of I_1, I_2, and C are mineralized.

5.6.3.2 Family Lemuridae

In all newborn lemurids, di_1 to dc resemble the adult tooth comb (e.g., Figure 5.14). In all except *Varecia* spp. (Box 5.5), most newborns have initiated mineralization of the roots for the anterior teeth.

The dp_2 is caniniform. It is roughly triangular in occlusal profile in *Varecia* and *Eulemur* spp. due to a pronounced lingual longitudinal ridge. This ridge is present and distinct in *Hapalemur*, but the lingual surface is concave overall. A lingual ridge is not distinct in *Lemur*. The dp_3 also has a single primary cusp, but with a broader triangular base in *Varecia* and *Eulemur* spp. In *Hapalemur*, this tooth has a pronounced concavity distal to a vertical lingual ridge. In *Lemur*, dp_3 has a slight concavity lingually, like dp_2. In all lemurids, dp_3 is poorly mineralized distally at birth, and thus the crown morphology is as yet unclear (older specimens reveal additional cusps, see below). The dp_4 is the most molariform deciduous tooth, with multiple cusps, and some intraspecific variation. In all lemurids, the buccal primary cusps are more prominent than the lingual cusps. Only *Hapalemur* possesses a prominent entoconid. In *Eulemur* spp. the entoconid is variable. It is prominent in *E. mongoz* but small in *E. rubriventer* and *E. collaris*. In *L. catta* there is no entoconid. At birth, *Varecia* spp. have no mineralization distolingually,

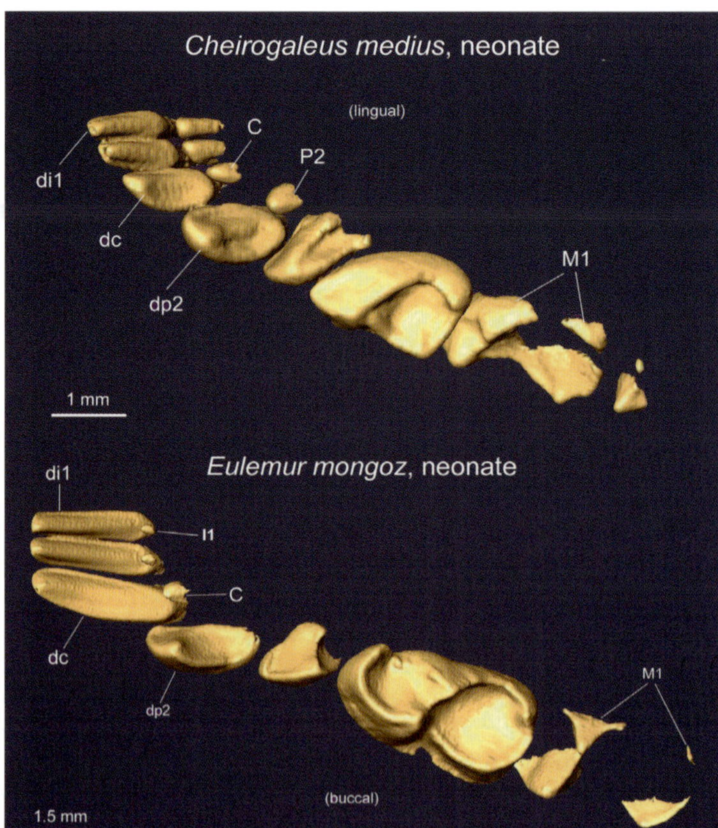

Figure 5.14 Mandibular dentition of a newborn cheirogaleid (top) and lemurid (bottom).

and in older infants there is no entoconid. *Lemur* possesses a cusp on the distal limit of the talonid basin, positioned identically to what Swindler (2002) interprets as a hypoconulid on M_1. In all lemurids, the paracristid is elevated; the similar portion of M_1 slopes downward. In all specimens the paracristid leads to a projecting keel, but in all specimens this leading edge bears a small cusp, variable in prominence, which resembles a paraconid. Tattersall and Schwartz (1974) also described this cusp in infant *Lemur.* A cristid obliqua intersects the protocristid centrally between the mesial cusp pair in *Lemur* and *Eulemur* and connects to the base of the protoconid in *Hapalemur.* This crest is unmineralized at birth in *Varecia* spp. (Box 5.5). A variable number of small secondary cusps are found lingual to the talonid (except in *Varecia*, in which the region is unmineralized). These range from zero to two, with intraspecific variation found in species represented by more than two. A metastylid is usually present in newborn lemurids.

M1 possesses four mineralized cusps in newborn *Eulemur* spp. and *Lemur*; the mesial pair are connected. In *Varecia* spp. only the protocristid is mineralized. M1 is most mineralized in *Hapalemur*: only a small portion of the talonid basin and its lingual margin are deficient.

In *Hapalemur*, *E. collaris*, and *E. mongoz*, the cusp tips of I_1, I_2, and C_2 are mineralized at birth. In the newborn *E. rubriventer*, only the cusp tip of C is mineralized. No replacement teeth are mineralized in newborn *Lemur* or *Varecia*.

5.6.3.3 *Family Lepilemuridae*

In newborn *Lepilemur leucopus*, the deciduous anterior teeth resemble an adult tooth comb; the roots have initiated mineralization

The dp_2 has one cusp and it is buccolingually compressed. In dp_3 there is one prominent cusp which leads to mesial and distal ridges that end in small cusps. In dp_4 three primary cusps are prominent; there is no entoconid. The cristid obliqua is pronounced and intersects the protocristid at its center.

M_1 and M_2 are both well-mineralized, including the basins, with pronounced cusps. M_3 is also well-mineralized, except for the distolingual border. For instance, the entoconid is not yet apparent at birth.

I_1 and I_2 and C are partially mineralized. The permanent premolars have initiated mineralization; P_2 and P_4 are more mineralized than P_3.

5.6.3.4 *Family Indriidae*

In newborn *Propithecus coquereli*, di_1 and di_2 are narrow with a lingual longitudinal ridge along the upper half of the crown. The anterior teeth do not resemble a tooth comb because the dc is far smaller in size.

The dp_2 has one cusp with mesial and distal marginal ridges; it is buccolingually compressed. The dp_3 is smaller and rounded buccally with a lingual depression. The dp_4 has four primary cusps, a long shallow trigonid basin, and a deeper talonid basin. The cristid obliqua is pronounced and intersects the protocristid at its center. M_1 and M_2 are both well-mineralized, including the basins, with pronounced cusps. Three cusps of M_3 are mineralized; the entoconid is not yet apparent at birth.

I_1 and I_2 are partially mineralized. P_2 and P_4 are mineralized, although only the cusp tip.

5.6.3.5 *Family Galagidae*

In *Galago* and *Otolemur* (Figure 5.15) spp., the deciduous anterior teeth resemble the adult tooth comb.

The dp_2 is long and caniniform, but more robust than dc. The dp_3 has a maximum of three cusps at birth. A large cusp mesial to the protoconid resembles a paraconid. However, dp_3 is less mineralized than adjacent premolars, especially distolingually; its complete morphology is discussed below based on older infant specimens. The dp_4 has four primary cusps and a variable number of smaller cusps. One or more of these occur on the paracristid. On the leading edge of the paracristid on dp_4, all galagids possess a small cusp, the presumptive paraconid. The talonid basin is deep and poorly mineralized in the center in most specimens; the trigonid is shallow and elevated by comparison.

M_1 resembles dp_4 closely, except that in the former there is no cusp at the leading edge of the paracristid. M_2 has one to two mineralized cusps.

I_1, I_2, and C are all small and located lingually to di_1, di_2, and c, respectively. The cusp tip of P_2 is mineralized in some newborns.

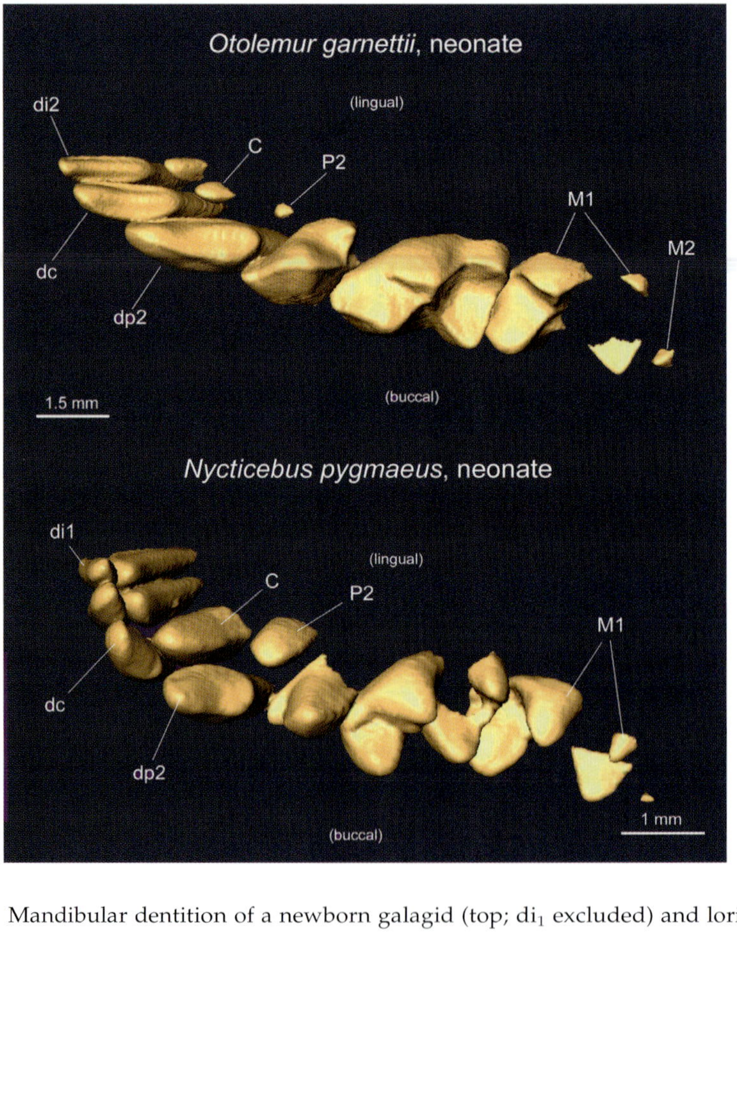

Figure 5.15 Mandibular dentition of a newborn galagid (top; di_1 excluded) and lorisid (bottom).

5.6.3.6 *Family Lorisidae*

In lorisids, the newborn/older infant deciduous anterior teeth resemble the adult tooth comb (e.g., *Nycticebus pygmaeus*: Figure 5.15). The deciduous incisors each have a lingual longitudinal ridge and flanges flaring mesially and distally.

In newborn *Nycticebus*, dp_2 is caniniform and broader than dc with a slight lingual heel. The dp_3 has one mineralized primary cusp, the protoconid, and a smaller mesial cusp that could be a paraconid. Both dp_2 and dp_3 crowns are incompletely mineralized. The dp_4 has four prominent primary cusps, and a small cusplet between the protoconid and metaconid. The talonid basin is incompletely mineralized in the smaller of the triplets examined, but nearly fully mineralized in a larger sibling. Deciduous premolar morphology is similar in *Perodicticus*. However, certain features are clearer, perhaps because the crowns are more fully mineralized in this older infant. The dp_3 has a prominent protoconid, like *Nycticebus*, but a large heel projects distolingually from the base of the crown. The dp_4 of *Perodicticus* has four primary cusps; the paracristid is obscured by the more fully erupted dp_3. A cristid obliqua intersects the protocristid close to the metaconid.

In *Nyticebus*, M_1 has two mesial cusps that are connected while the hypoconid and entoconid are isolated cusps. The protoconid of M_2 is mineralized. The crown of M_1 appears fully mineralized in the infant *Perodicticus*.

In *Nycticebus*, most of the crowns of I_1, I_2, C, and P_2 are mineralized at birth. C and P_2 are partially mineralized in *Perodicticus*, but I_1 and I_2 are not visible.

5.7 Comparative Dental Development in Primates in the Perinatal and Early Postnatal Period

5.7.1 Dental Tissue Maturation

5.7.1.1 *Patterns of Crown Mineralization in the Perinatal Period*

The process of dental crown formation has been well studied in humans (e.g., Kraus and Jordan, 1965), and several anthropoid species have been studied in some depth (Swindler and McCoy, 1964; Swindler et al., 1968; Tarrant and Swindler, 1973). A limited sample of fetal and older specimens allows us to provide a slightly broader context. Because M_1 develops earlier than M^1, we focus on lower jaw dentition in this section.

In deciduous premolars of fetal *Lemur* (Figure 5.16) and *Alouatta*, the protoconid is best developed, while the entoconid is smallest. M_1 is best developed in the fetal *Propithecus*, and barely formed (protoconid only) in the fetal *Lemur*. In fetal *Alouatta* and *Otolemur*, the M_1 entoconid is the sole unmineralized cusp. In M_1, the mesial cusp pair is present in the absence of the distal cusp pair in older specimens of the above species; mesial cusp pairs are better connected to each other (by the protocristid) than are the distal pair. Finally, the trigonid basin may be mineralized when the talonid basin is not, but the reverse is never the case.

With the above information in mind, it may be surmised that mineralization initiates buccomesially at the protoconid, proceeds lingually to the metaconid, then extends to the hypoconid; finally the distolingual cusp (entoconid) and adjacent parts of the crown mineralize. The crown basin and marginal ridges mineralize first mesially and then distally. Evidence from this small sample of strepsirrhines and *Alouatta* is

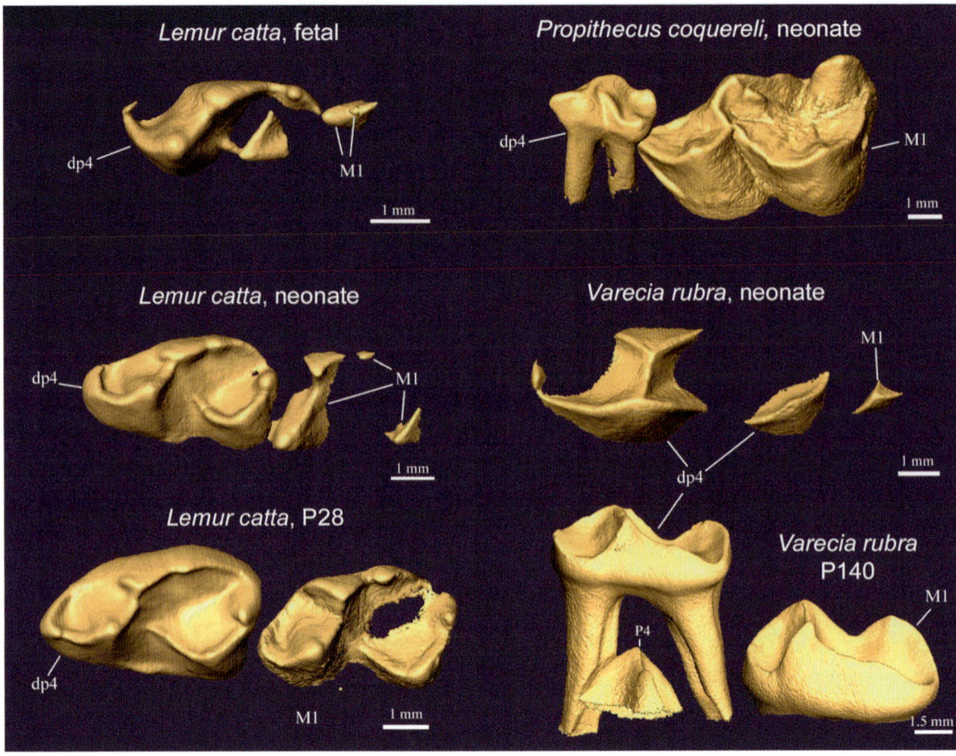

Figure 5.16 Crown maturation of dp_4 and M_1 in three strepsirrhines. At left, progression of three ages in *Lemur catta*, showing that formation of cusps largely conforms to descriptions of anthropoids (see text). At top right, note that even at birth, the M_1 crown of *Propithecus coquereli* is better mineralized than in older infant *L. catta*. At right middle and bottom, *Varecia rubra* is shown. At birth, *Varecia* spp. have less-developed postcanine teeth than any other known strepsirrhine. Note the older subadult, nearly at weaning age, still retains dp_4, and lacks roots for M_1.

entirely consistent with previous results on other anthropoids (Swindler and McCoy, 1964; Swindler et al., 1968; Tarrant and Swindler, 1973); this suggests Swindler's (2002) observation that there is a common pattern of molar mineralization in anthropoids may be extended to primates generally (Figure 5.16).

5.7.1.2 *Phylogenetic Patterns of Crown Mineralization*

In Table 5.4 the state of crown mineralization in newborn non-human primates is summarized for dp_4, M_1, and M_2. Although many species are described based on limited samples, numerous trends are apparent at the family/subfamily level. Among anthropoids, callitrichines stand out as having the least-developed dp_4 and permanent molars. Monson et al. (2019) suggest callitrichines have inhibition of M3 development, and this could relate to slow development of postcanine primary teeth. All other anthropoids have a well-mineralized dp_4. In most platyrrhines, M_1 is in early stages of mineralization at birth. *Alouatta*, *Pithecia*, and *Saimiri* are exceptions, with three or four cusps at least partially mineralized. The single newborn *Pithecia* is particularly advanced, being the only anthropoid in which M_2 has initiated mineralization in our collection. In most Old

Table 5.4 State of selected postcanine mandibular crown mineralization in newborn primates

	dp4		M1		M2	
Species[1]	**Cusps**[2]	**Basins**[3]	**Cusps**	**Basins**	**Cusps**	**Basins**
HAPLORHINES						
Allenopithecus nigroviridis	+++	+++	+++	–	NA	NA
Alouatta seniculus	+++	+++	++	–	NA	NA
Aotus nancymae	+++	+++	+	–	NA	NA
Callicebus cupreus	+++	+++	+	–	NA	NA
Callithrix spp.	+++	++	+	–	NA	NA
Cebuella pygmaea	+++/++	+++/++	+/–	–	NA	NA
Colobus guereza	+++	+++	+	–	NA	NA
Leontopithecus rosalia	+++	++	+	–	NA	NA
Macaca mulatta	+++	+++	++	–	NA	NA
Papio anubis	+++	+++	++	–	NA	NA
Pan troglodytes	+++	+++	++	–	NA	NA
Pithecia pithecia	+++	+++	++	–	+	–
Pongo pygmaeus	+++?	+++	++	–	NA	NA
Saguinus spp.	+++	++	+/–	–	NA	NA
Saimiri boliviensis	+++	+++	++	–	NA	NA
Trachypithecus francoisi	+++	+++	++	–	NA	NA
Tarsius syrichta	+++	+++	+++	+++	+++	++
STREPSIRRHINES						
Cheirogaleus medius	+++	+++	+++/++	++	+	–
Eulemur spp.	+++	+++	++	–	NA	NA
Galago spp.	+++	+++	+++	++/+	+	–
Hapalemur griseus	+++	+++			NA	NA
Lemur catta	+++	+++	++	–	NA	NA
Lepilemur leucopus	+++	+++	+++	+++	+++	+++
Nycticebus pygmaeus	+++	++	+++	–	+	–
Otolemur spp.	+++	+++	+++	++/+	+	–
Propithecus coquereli	+++	+++	+++	+++	+++	+++
Varecia spp.	++	++	+	–	NA	NA

Notes: [1]all species newly described here except *Pan* (Seibert and Swindler, 1991) and *Pongo* (Winkler et al., 1991); [2]cusp: +++ all cusps mostly mineralized, ++ some cusps unmineralized or only cusp tips, + most cusps unmineralized, – no mineralized cusps; [3]basins: +++ both basins mostly or completely mineralized, ++ one basin (e.g., talonid) mostly or completely unmineralized, + both basins mostly unmineralized, – no mineralization of basins; NA, not applicable (tooth germ not yet formed).

World monkeys the four cusps of M_1 have initiated mineralization, far outpacing hominoids at birth.

Tarsius syrichta is similar to *T. bancanus* (Luckett and Maier, 1982) at birth. The first two permanent molars are well advanced in mineralization; M_3 already has two mineralized cusps.

Newborn *Propithecus* (Figure 5.16) conforms to previous descriptions of indriids (Godfrey et al., 2004), having the best mineralized crowns of all strepsirrhines except *Lepilemur leucopus*. The entire crowns of deciduous teeth are fully mineralized and the roots are well-formed. While the deciduous teeth are better formed than other primates, hydroxyapatite density levels in the M1s are highest among all teeth at birth (unpublished data); the reverse is true of all other primates except *Tarsius*. Galagids are second only to indriids and *Lepilemur* in the maturational state of M_1 at birth; even the crown basins have begun to mineralize. The dp_4 and M_1 of lemurids are far less advanced by comparison, but *Varecia* spp. stand out as the least mineralized. Even dp_4 is poorly mineralized. In the lemurid M_1, the basins typically have not mineralized as yet (Box 5.5). In *Propithecus*, by contrast, the trigonid and even the talonid basins are mineralized at birth in both M_1 and M_2 (Figure 5.16).

5.7.1.3 *Rate of Postnatal Crown Mineralization*

A broad comparison of newborns suggests a variable timing of mineralization among lemurids (Box 5.5). By far, *Varecia* spp. are the least developed of the lemurids in terms of mineralization at birth. The distolingual margin and the entire talonid basin of dp_4 remain unmineralized. The protoconid of M_1 is the sole mineralized part. The bases of all crown teeth in *Varecia* are also unmineralized. In newborn *Lemur*, all cusps and both basins of dp_4 are mineralized at birth; M_1 is as yet unmineralized distolingually and in the talonid basin. *Eulemur* spp. have initiated more mineralization of the entoconid; in addition, several *Eulemur* neonates have initiated mineralization of deciduous incisors, and even the deciduous canine in one case (Box 5.5). Thus, overall, the degree of postnatal cusp mineralization among these lemurids is *Varecia* spp. < *Lemur catta* < *Eulemur* spp. at birth. In specimens of each species that are approximately 1 month of age, *Varecia variegata* remains the least mineralized at M_1. However, permanent anterior teeth and the cusp tip of P_2 have initiated mineralization. In this aspect, *Varecia* rivals similarly aged *Lemur* and *Eulemur*, at least in the small available sample. This suggests the rate of crown mineralization differs among lemurids.

Among the strepsirrhines studied to date, galagids possess the most highly mineralized crowns and roots at birth aside from indriids (Figure 5.15; Box 5.5). An age comparison of the mandibular tooth row in *Galago moholi* indicates early postnatal crown maturation is rapid. By 1 month of age, infant *G. moholi* have completed mineralization of the M_1 crown, nearly completed mineralization of the M_2 crown, and initiated mineralization of M_3 (mesial cusp pair only). In addition, the replacement teeth I_1 through P_2 are extensively mineralized.

Among anthropoids in our collection, dental maturation is fastest in *Cebuella*. In newborns, M_1 is more mineralized than in other callitrichines, and the cusp tips of two replacement teeth (I_1, I_2) are present. In an infant specimen (which resembles a specimen recorded at 1 month of age), P_2 and P_4 have initiated mineralization. *Saguinus oedipus* provides a useful contrast, because at 1 month of age, M_1 is still

Table 5.5 Mandibular tooth calcification: first permanent molar (M_1) and incisors (I_1)

Species	M_1: age at crown completion (years)	I_1: Onset of mineralization (years)	Ref.	Weaning age (years)	Ref.
Pan troglodytes	~2	0.5	1	4.6–5	2
Macaca nemestrina	0.89–0.97	2.11–2.28	3	0.66–1	2
Saguinus oedipus	> 0.083, but < 0.32	~ 0.083	4	0.14	2
Cebuella	~ 0.083	perinatal*	4	0.25	2
Lemur catta	> 0.083	> 0.055, but < 0.083	4	0.50	5
Varecia variegata	> 0.083	> 0.033, but < 0.068	4	0.30	5
Propithecus coquereli	~Birth	Prenatal	4	0.39	5
Galago moholi	< 0.083	Prenatal	4	0.25	5

References: 1, Anemone et al., 1991; 2, Kappeler and Pereira, 2003; 3, Sirianni and Swindler, 1985; 4, new data; 5, Zehr et al., 2014; *, tooth cusp/crown is calcified in some but not all newborns.

incompletely mineralized (lacks an entoconid), and only the very tip of I1 is mineralized. In a specimen 117 days old, even though more replacement teeth are visible, only the cusp tips of the anterior teeth are mineralized.

The novel information above on similar chronological ages hints that there are subtle differences in the postnatal rate of mineralization within primate suborders (Table 5.5). As expected based on observations of maxillary cusp mineralization in newborns (Smith et al., 2015), cheirogaleids, indriids, and galagids have more rapid mineralization of the first permanent molar and the first permanent incisor (the first replacement tooth to mineralize) compared to all other primates except tarsiers. Lemurids are delayed by comparison, but still may well reach these two dental milestones before weaning. Among anthropoids, previously studied catarrhines delay mineralization of these teeth considerably, but at least some platyrrhines mature more rapidly (Table 5.5).

5.7.2 Gingival Emergence

Gingival emergence, when a cusp of a tooth crown has pierced the gingiva, is a discrete and potentially easily observed characteristic of the newborn primate. Our comparative knowledge of neonatal status of gingival eruption is fairly broad, thanks to an ambitious effort by B. Holly Smith and colleagues, who compiled data from decades of studies on captive primates (Smith, 1989; Smith et al., 1994). They established the singularly slow gingival emergence of deciduous teeth in the extant catarrhines. This is a broader characteristic of anthropoids generally, with some exceptions (e.g., callitrichines and *Saimiri boliviensis*: Smith et al., 2015). *Saguinus* spp. are generally less advanced than other callitrichines. *Cebuella* has the fastest rate of emergence of any anthropoid. Great apes, in contrast, delay emergence of any teeth for months (e.g., Robinow et al., 1942; Fooden and Izor, 1983; Kuykendall et al., 1992).

In *Tarsius* spp., the rate of gingival emergence is advanced, and some teeth are already shed at birth (Luckett and Maier, 1982; Smith et al., 2015). Luckett and Maier describe evidence suggesting fragmentation of some shed teeth ("odontoclasts" adjacent to eosinophilic fragments). However, they also observed certain deciduous teeth (di^2) loosely connected to gingiva. Thus early resorption followed by shedding of larger portions of deciduous teeth seems probable. Loss of dp2 and the deciduous incisors is rapid, occurring within the first postnatal week (Luckett and Maier, 1982; Smith et al., 2015). Also remarkable is the advanced state of P2, which is described as "nearly erupted into the oral cavity" by Luckett and Maier (1982, p. 21). P2 is also far advanced in its tissue maturation, with far less stellate reticulum than other replacement teeth (Smith et al., 2015).

Among strepsirrhines, most small-bodied strepsirrhines have more emergent teeth at birth than large-bodied species. Nearly all larger-bodied lemurs have no emergent teeth at birth (Smith et al., 1989; Smith et al., 2015). This includes the folivorous *Hapalemur*, which resembles all other lemurids. *Propithecus coquereli*, and presumably all other indriids, have far faster emergence of deciduous teeth compared to like-sized lemuroids, occurring shortly after birth (Eaglen, 1985) or even by birth (Smith et al., 2015). Cheirogaleids are variable, with newborns reported to have between zero and two emergent teeth within the maxilla and possibly more in the lower jaw (Smith et al., 1994; Smith et al., 2015). Lorisoids are, as a group, relatively advanced in gingival emergence. In five galagids and one lorisid studied histologically by Smith et al. (2015), all had at least two emergent maxillary teeth.

5.7.3 Dental Eruption

Dental eruption refers to the extended process by which teeth achieve a state of full occlusion with the opposing tooth row. Unlike gingival emergence, it may be quite difficult to observe erupted teeth in the newborn primate. The initial phase emphasizes the very definition for studies that have compared eruption schedules: a tooth is erupting once a portion of it (typically a cusp tip) protrudes beyond the confines of the alveolar crypt (Godfrey et al., 2004).

A greater pool of data exists on dental eruption compared to observations on gingival emergence. Schultz (1935) undertook a study of the eruption of permanent teeth based on large numbers of catarrhine skulls, and added a limited amount of data on other primates (Schultz, 1935, 1960). These observations were generally confirmed by subsequent studies. Among anthropoids, the pace of deciduous tooth eruption is relatively more rapid in monkeys, beginning within days or weeks of birth (e.g., Chase and Cooper, 1969; Johnston et al., 1970; Glassman, 1983), compared to slower rates of eruption in apes and humans (Enlow, 1990; Mooney et al., 1991; Anemone et al., 1996). Tarsiers vastly outpace other haplorhines in their mineralization and eruption schedules (Luckett and Maier, 1982; Box 5.6). With only limited information on non-anthropoids, a "fast" and "slow" growing concept emerged, later called "Schultz's rule" (Smith, 2000). Smith (2000, p. 213) defined it as "the tendency for replacing teeth to come in relatively early in slow-growing, longer-lived species."

In 1985, Eaglen provided the first detailed information on strepsirrhines. In a sample of six species, he noted marked variation, and the especially rapid development in *Propithecus*. This raised the likelihood of non-phylogenetic influences on the eruption of deciduous dentition, such as dietary specialization, body size, or phylogeny. Using larger samples, Godfrey and colleagues (2001) and Henderson (2007) analyzed alveolar eruption based on skeletonized skulls. This recent work has revealed clear exceptions to Schultz's rule, with certain folivorous species having especially rapid dental development (see Chapter 10) that would not be predicted based on other somatic growth patterns (Godfrey at al., 2001, 2003, 2004, 2005). Dental development can be targeted by selection independently of other somatic growth patterns.

A commonality of all primates examined to date, except tarsiers and indriids, is that postcanine teeth are as yet unerupted at birth. In other respects, postcanine teeth vary. Relatively larger postcanine teeth (perhaps especially dp4) correlate with relatively longer midfaces (Smith et al., 2017a). In strepsirrhines, it is also noteworthy that relative midface length is *negatively* correlated with permanent molar volume (Smith et al., 2017a). Thus, deciduous teeth likely have the biggest impact on perinatal jaw bone modeling patterns, with replacement teeth as added factors in some species.

Recently, much attention has focused on life-history correlates of dental growth and eruption. Brain size (rather than overall body size) is strongly correlated with the pace of dental development (Smith, 1989; Godfrey et al., 2001). Brain development, in turn, strongly correlates with certain life-history variables, such as age at weaning. Similarly, dental development is correlated (negatively) with age at weaning: earlier weaning is linked with a more precocious dental arcade (Godfrey et al., 2001). Thus, life-history factors may be at work on the brain and teeth simultaneously, perhaps having metabolic influences on both.

5.7.3.1 Comparative Observations at Similar Time Points: Chronological and Dental Age

In the present work, most specimens are of known actual or chronological age. We are thus able to bracket the newborn dental status of dental eruption by comparing to earlier and later stages. In several lemuroid and callitrichine species we have neonates and older infant specimens for comparison (specimens considered older infants had recorded ages of P26 to P40). To appreciate how perinatal dental eruption rates vary, we will adopt criteria which are typically used to describe postnatal development of primates with slow dental maturation and apply them to strepsirrhines and small-bodied monkeys. The following are modified from Schultz (1935, 1960):

- **1:** prior to eruption of any deciduous premolars,
- **2:** dp2 erupting,
- **3:** dp3 erupting,
- **4:** full deciduous eruption, prior to M1 erupting in both jaws,
- **5:** M1 not fully in place, but has initiated eruption in both jaws.

Across this early postnatal window, lemurids may be contrasted to indriids and galagids (Figure 5.17; Box 5.5). *Lemur*, *Eulemur*, and *Varecia* are all born with incomplete eruption of premolars (stages 1 to 3), while newborn *Propithecus* has fully erupted deciduous dentition (stage 4) and newborn galagids (*Otolemur* and *Galago*) have

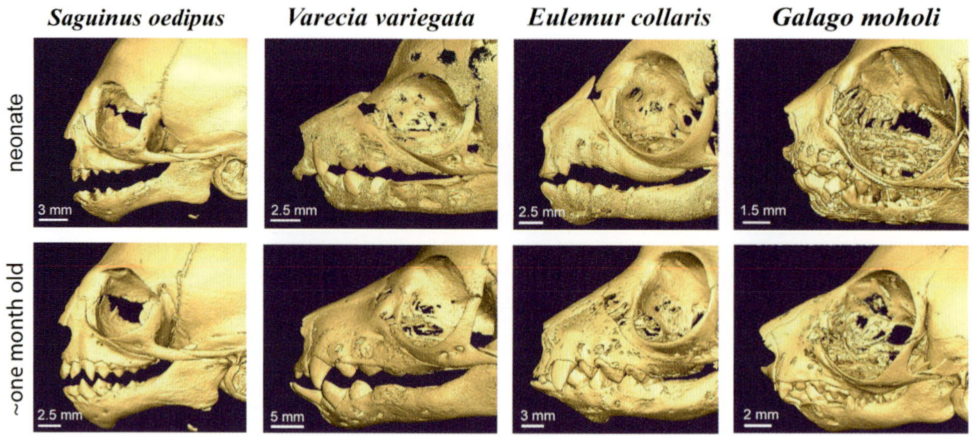

Figure 5.17 Facial profiles of a tamarin (*Saguinus*), ruffed lemur (*Varecia*), collared lemur (*Eulemur*), and bushbaby (*Galago*) at birth and at 1 month of age.

initiated M1 eruption (stage 5). The approximately month-old specimens are not at different stages compared to the newborns, although the anterior teeth and individual premolars are notably more erupted (Figure 5.17).

Age series of anthropoids reveal a contrast to all strepsirrhines. *Cebuella* is most advanced at birth, born with dp3 erupted (stage 3); *Saguinus* spp. may be stage 2 or 3 (Figure 5.17). In both species eruption may be rapid in early postnatal development: 1-month-old *Cebuella* and *Saguinus* are at stage 4.

These observations are in keeping with previous observations on older subadults (Godfrey et al., 2001, 2004). Indriids and galagids have rapid dental eruption schedules while lemurids are much slower. Small-bodied platyrrhines have relatively rapid eruption schedules; the pace of eruption may be slower in some larger-bodied anthropoids (Godfrey et al., 2001).

A striking qualitative change between newborn and approximately month-old specimens is in degree of snout projection, especially in lemurids and *Galago*. Outgrowth of the face may not proceed in tandem with eruption rates. Indeed, spatial constraint within the viscerocranium (see Vinyard and Hanna, 2005) could explain early eruption of deciduous teeth in indriids. However, the role of the dentition in facial outgrowth may relate more directly to tooth germ growth and shifting tooth position than to eruption schedules.

Box 5.1 Mineralized Tissues of the Tooth Crown

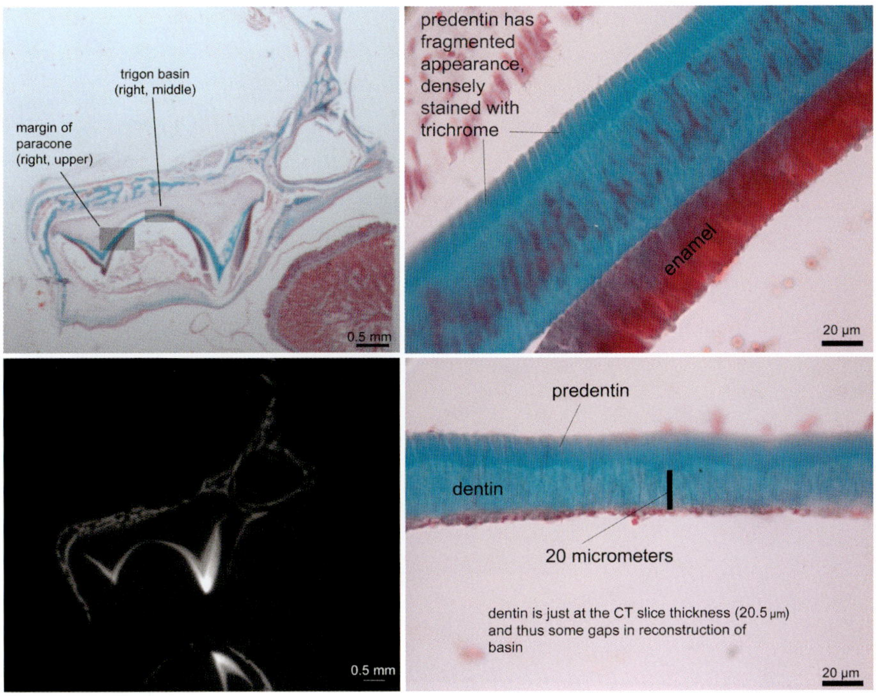

Above, the first permanent molar in a newborn tarsier, shown in histological section (top left) and a CT slice from a reconstruction of the same specimen, after it has been virtually reconstructed, rotated and "resliced" so that CT slices are in the same plane as histology. The mineralized tissues of teeth, dentin and enamel, are visible using both methods, although the most mineralized enamel (at the cusp tips) becomes fragmented when decalcified for histology (tip left). Mature enamel is more densely radio-opaque than dentin (see lingual cusp tip, bottom left). The dentin of developing teeth is unmineralized (predentin) at the side facing the pulp cavity (top right, position indicated in larger box, top left). In contrast, enamel is less mineralized, at first, on the crown surface (more mineralized enamel stains lightly in decalcified sections). The CT reconstructions reliably detect either tissue when its thickness in greater than the scan resolution (bottom right, location indicated by smaller rectangle in top left).

Box 5.2 Tooth Rows of Newborn Langur

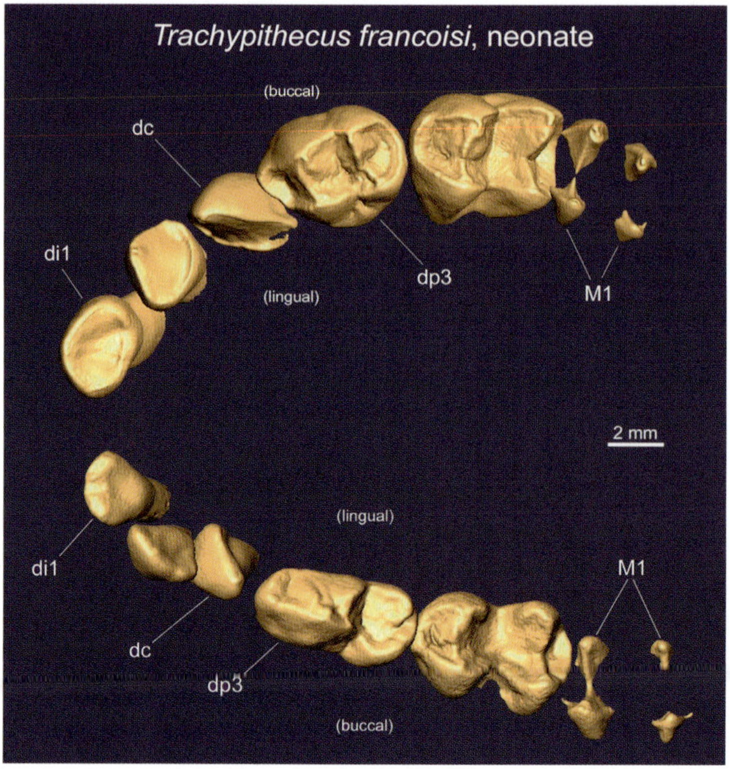

Above, occlusal views of maxillary (top) and mandibular (bottom) tooth rows in *Hapalemur griseus* at birth.

Box 5.3 Tooth Rows of Newborn Tarsier

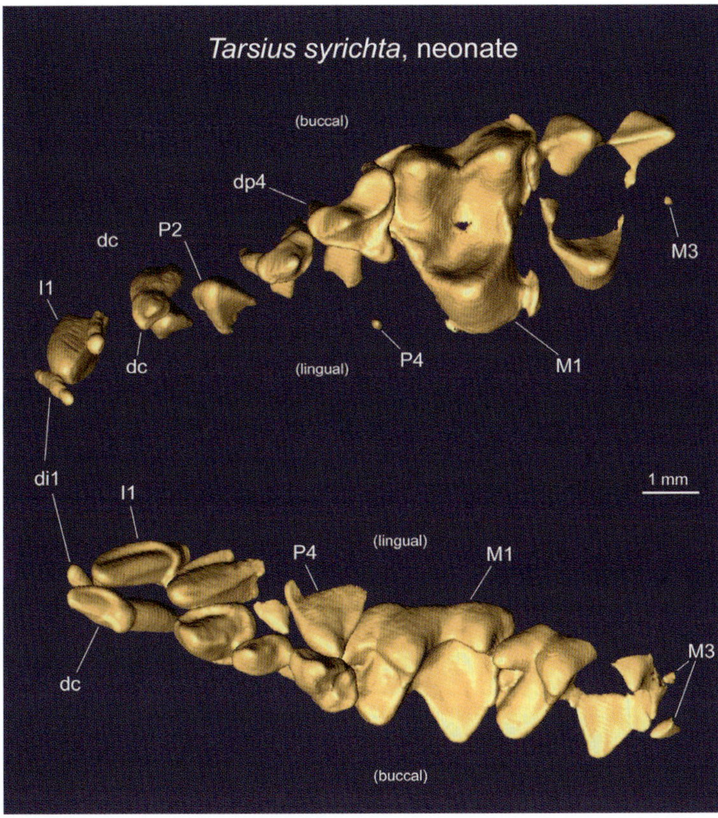

Above, occlusal views of maxillary (top) and mandibular (bottom) tooth rows in *Tarsius syrichta* at birth. Note: the right jaws were used in this reconstruction, but the image was reversed for ease of comparison to other plates.

Box 5.4 Maxillary Teeth of Newborn Bamboo Lemur

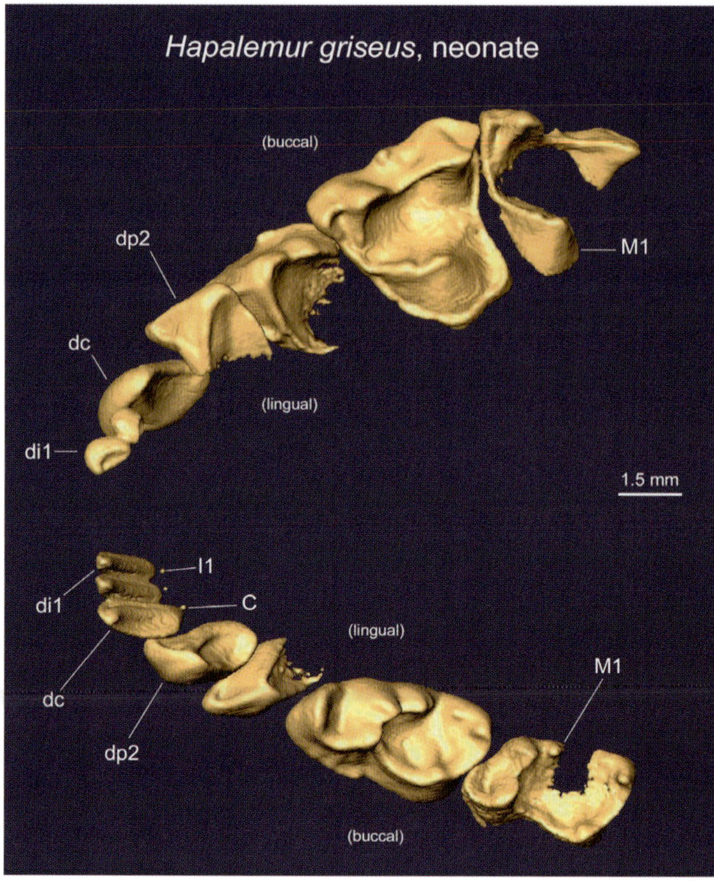

Above, occlusal views of maxillary (top) and mandibular (bottom) tooth rows in *Hapalemur griseus* at birth.

Box 5.5 Ontogeny of Tooth Crowns in Lemurs and Galagos

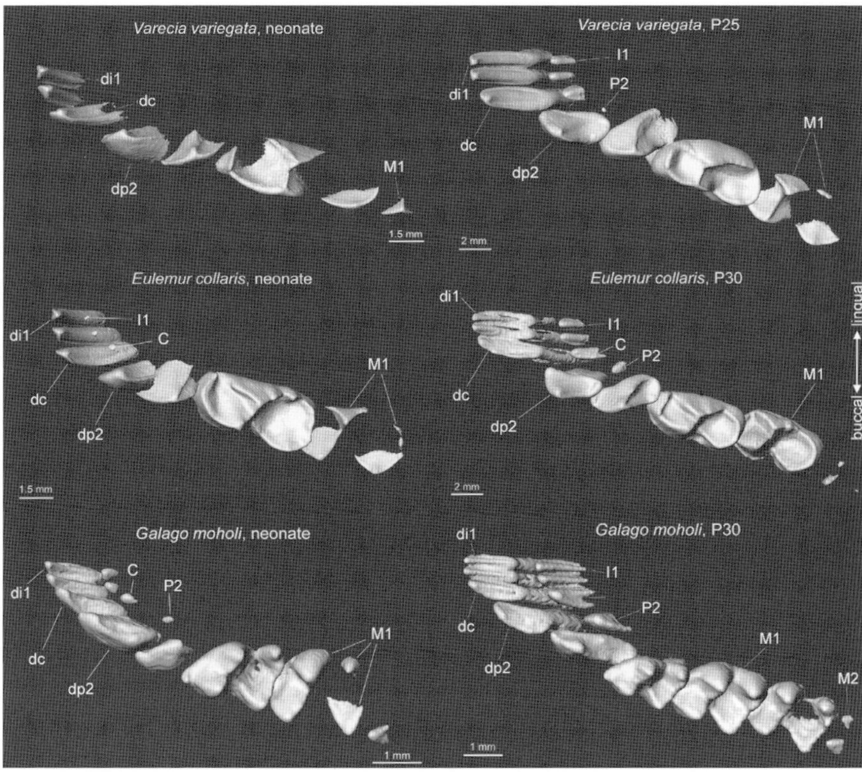

Box 5.6 Dental Mineralization and Eruption in *Tarsius syrichta*

Below, crania of a fetal (top) and 1-day-old (bottom) tarsier. In the fetal tarsier (at 80% of the cranial length of newborn specimens) mineralization of postcanine cusps is progressing rapidly, with the long cusps as the initial foci of mineralization (a). Three-dimensional reconstructions from histology sections of the same specimen reveal that the tooth germs are larger still, occupying much of the maxilla (max). At birth (c), M1 and M2 crowns are complete and some deciduous teeth are shed or nearly shed.

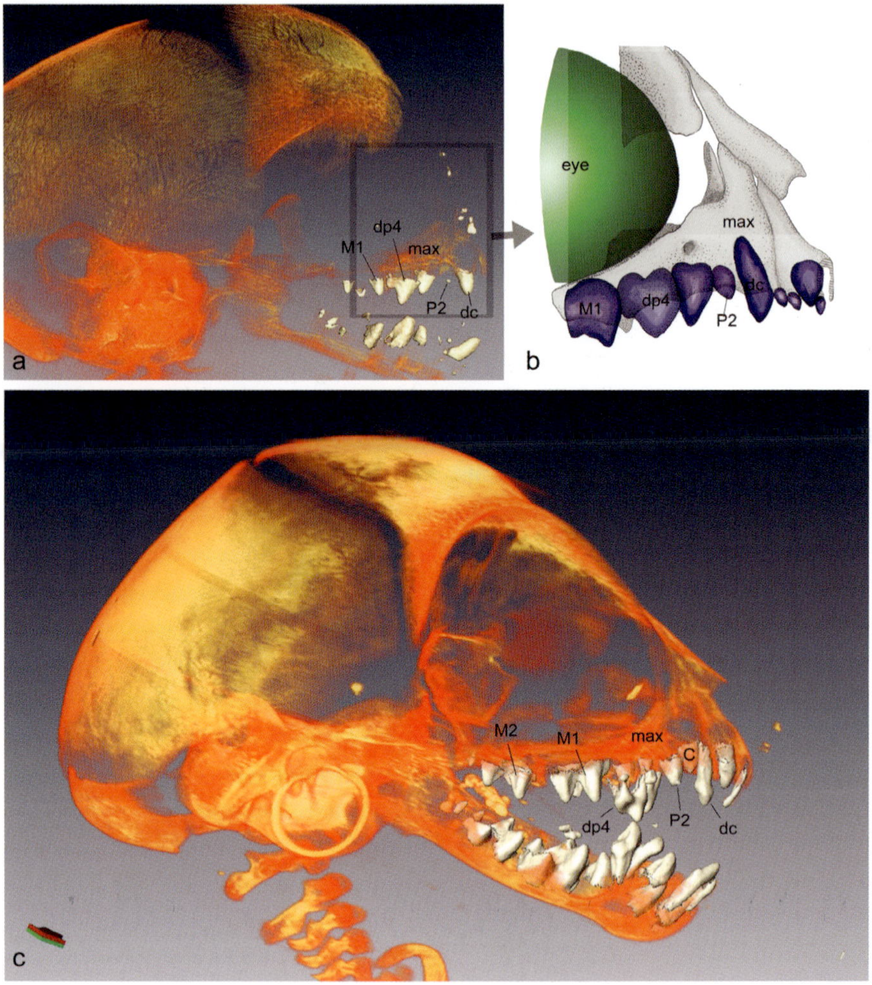

6 The Postcranial Axial Skeleton

6.1 Introduction

The axial skeleton includes the skull, vertebrae, ribs, and sternum. The vertebral column is divided regionally into cervical, thoracic, lumbar, sacral, and caudal (coccygeal) regions (Ankel-Simons, 2007). All primates have seven cervical elements. Thoracic vertebrae and associated ribs are typically between 12 and 13 in number (but 15–16 in lorises). Lumbar vertebrae range between four and eight. The sacrum is a composite structure of fused vertebrae numbering between three and seven elements. The caudal (tail) region varies the most, numbering between 25 and 31 vertebrae in most species, but fewer in large lemurs and lorises (7–12), and reduced to a vestige (although critical to support provided by the muscular pelvic floor) in hominoids. The sternum is primitively segmented, like the vertebrae.

The axial skeleton undergoes mesenchymal condensation, chondrification, and ossification prior to any elements of the appendicular skeleton (Figure 6.1). Typical vertebrae have three pairs of chondrification centers, the centra, neural arches, and costal elements. The latter are found in pairs at *every* vertebral level as derivatives of a laterally positioned chondrification center, but not all costal elements become ribs.

By late fetal stages, a typical mammalian vertebra has three ossified elements: the centrum and the right and left halves of the neural arch. Exceptions to this pattern include the first and second cervical vertebrae (C1, C2; Figure 6.2). In most mammals, C1 lacks a centrum because this element (i.e., the dens, or odontoid process) has been fused with the body of C2. Although it lacks a centrum, C1 is notable for the presence of a ventral arch. The neural arches and centra have differing patterns of ossification during prenatal development. The neural arches commence ossification first in the lower cervical region (possibly simultaneous with upper thoracic region) and this progresses in cranial and caudal directions. In contrast, the centra of the lower thoracic and upper lumbar region are the first to commence ossification, then progressing cranially and caudally. This is a sequence common to most mammals with only few exceptions (Hautier et al., 2010). Humans follow the prenatal sequence of ossification

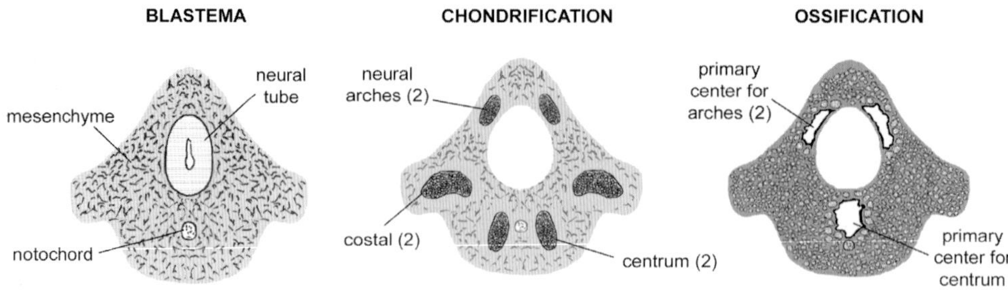

Figure 6.1 Embryogenesis of vertebrae.

for most mammals, as described above. Ossification commences relatively early prenatally (late embryonic stage), during the second month of human development.

The centrum and neural arches are joined at synchondroses to one another and/or to other elements, depending on vertebral level (Scheuer and Black, 2004; Jinkins, 2000). Because the costal elements become separate bones at thoracic levels, joined by synovial and syndesmotic joints, the thoracic vertebrae have the simplest synchondroses. Dorsally, the neural arches are joined at the intraneural synchondrosis (or the posterior synchondrosis, an alternative term for human anatomy). Ventrally, the right and left neural arches are joined to the centrum by the neurocentral synchondroses. In the cervical region, the union of the costal chondrification center with the lateral mass (transverse process) is an added complexity (Figure 6.2). Here, the ventral bar (or stem) of the lateral mass is considered a derivative of the costal chondrification center (Scheuer and Black, 2004). An additional synchondrosis at C2 connects the dens to the centrum (odontocentral). For C1, the allocentral synchondrosis connects the dorsal arch to the ventral arch (or anterior arch, of human anatomy). Finally, the dens itself ossifies from bilateral centers, connected by a longitudinal intraodontoid synchondrosis. However, this is fused at birth and evidenced by a thin line. All of these synchondroses are growth centers; the cartilage most closely adjacent to bone bears resemblance to an epiphyseal growth plate (Box 6.1).

Ribs derive from the costal chondrification centers, but separate from the remainder of the thoracic vertebrae prenatally. Most articulate with the sternum, at least indirectly, via costal cartilages (Ankel-Simons, 2007; Gebo, 2014). The more cranially situated pairs of ribs articulate dorsally with the vertebral body or between bodies and also with the transverse processes; ventrally, these same ribs articulate directly with the sternum via costal cartilages. These are referred to as true ribs. Caudally, additional ribs articulate indirectly with the sternum in that their costal cartilages merge with that of the last true rib. These are the arch ribs, as they veer cranially toward the ventral midline, creating the costal arches (the apex of the arch on each side is adjacent to the xiphoid process). Most caudally, two to three pairs of ribs do not have associated costal cartilages. These are called free or floating ribs, which are simple in form in the adult. They possess no tubercle and correspondingly do not articulate with the transverse processes.

During chrondrification and ossification, joints are at first undifferentiated pools of mesenchyme known as "interzones." These later go on to some degree of cavitation (resulting from some localized apoptosis to form joint cavities) and persisting

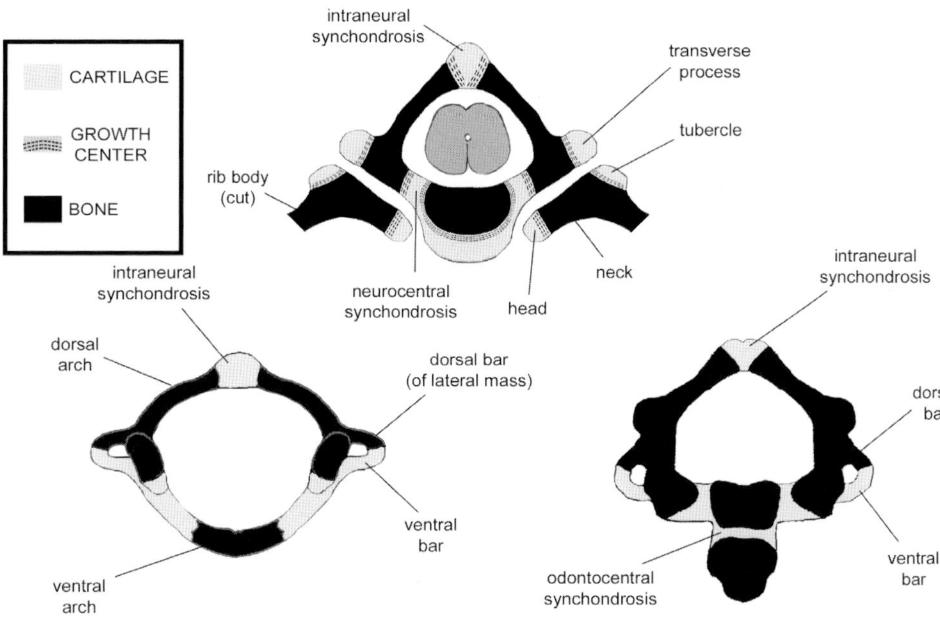

Figure 6.2 Vertebral synchondroses.

mesenchyme differentiates into synovial joint structures such as joint capsules, or symphyseal joint elements, such as fibrocartilage.

Some terms are used to reflect the embryological origin of adult structures; some are stage-specific. The term centrum, for example, refers to the ossified element in the region of the vertebral body. However, the centrum at birth has not yet formed epiphyses on cranial and caudal ends. Therefore, the term is not precisely equivalent to vertebral body. Other terms are descriptive but obscure the underlying homology. For instance, the costal elements in the cervical region ossify as part of the lateral mass, or transverse process. (The dorsal bar is considered serially homologous with the thoracic transverse process, while the remainder of the lateral mass is formed from costal centers.) In the lumbar region, the costal component projects from the neural arch and *is* the adult transverse process. Thus, not all transverse processes of vertebrae in adults are serially homologous.

6.2 The Axial Skeleton of Hominoids

The axial skeleton of hominoids is notable in its shape and its abbreviated length. In shape the thoracic region is dorsoventrally compressed, in contrast to the more cylindrical thorax of all other primates. Hominoids lack a tail, and have reduced caudal elements, the coccygeal vertebrae (between three and four elements) which form important attachment sites for ligaments and muscles of the perineum.

At birth, the state of vertebral ossification in hominoids is exemplified by the chimpanzee. Ossification has initiated throughout all cervical to sacral elements; the extent of ossification of coccygeal vertebrae may vary. They remain cartilaginous in newborn humans, but they were described as ossified in late fetal and infant gorillas by Symington (1889). Based on descriptions of humans, chimpanzees, and orangutans, there is limited or no fusion across the intraneural synchondrosis in the newborn. Examination of a 5-day-old *Pan troglodytes* in the KUPRI collection (#9905) shows that in the cervical and upper thoracic regions the neural arches are widely patent dorsally (Figures 6.3 and 6.4). This synchondrosis is widest (craniocaudally) at C1, and then becomes progressively less wide caudally. The neural arches of T8–T11 are closely approximated and become directly apposed in our reconstruction (Figure 6.4); it is impossible to know whether this is an artifact of the CT reconstruction or whether initial fusion is proceeding. There appears to be dorsal fusion of the neural arches at S1, but more caudal sacral elements are still patent (Figure 6.5). All neurocentral synchondroses are patent.

6.2.1 Cervical Region

The cervical vertebrae, numbering seven, possess two to four separate ossified elements in all hominoids, including the centrum, neural arches (two), and dens (C2 only). All hominoids have at least two ossified elements of the atlas at birth, comprising right and left sides of the neural arch. Each possesses only a ventral notch at the location of the transverse foramen (Figure 6.3). In humans the ventral (anterior) arch ossifies postnatally, in the first or second year (Castellana and Kosa, 1999). Schultz describes the ventral arch in infant samples of chimpanzees and orangutans (1940, 1941), but the age range of the group is relatively broad (as defined by Schultz, the stage ends with M1 eruption). In particular, Schultz does not specifically state whether the arch is present in the specimens designated as newborns. In the KUPRI newborn *Pan* specimen, an ossified ventral arch is lacking. This suggests that ossification of the ventral arch is delayed in hominoids, although more data on the other apes are needed.

The vertebral artery rests in a deep groove along the dorsal side of the facet. The lateral mass, as well as the transverse foramen that it surrounds, are incomplete at birth in the C1 of hominoids. The dorsal bar of the transverse process is well-ossified, with a blunt end oriented ventrolaterally (Figure 6.3). Because the foramen is partially surrounded by cartilage, the site of the foramen is indicated by a notch in the osseous C1. Caudally, a smaller and flatter articular facet (the first postzygapophysis) is present on the neural arch. Nutrient foramina are frequently visible just dorsal to the facet, the same position as in newborn humans.

C2 includes four separate ossified elements in the newborn (dens, centrum, and two neural arches). In humans, fusion of the dens to the centrum occurs postnatally; this is also described in the chimpanzee and orangutan (Schultz, 1940, 1941). Postnatally, hominoids develop a secondary ossification within the cuneiform cartilage, which extends cranially from the dens (Jinkins, 2000); after ossification, it is connected to the dens by the intraodontoid synchondrosis. The cuneiform cartilage is unossified at birth in humans (and it has not yet been described in newborn apes).

The centrum is roughly quadrangular in shape when viewed from the caudal direction. It typically has dorsally situated nutrient foramina. The dens is a separate, rounded nodule. The neural arches each take an angulated turn from the lateral mass (Figure 6.3). The prezygapophysis is positioned toward the more dorsal half of the

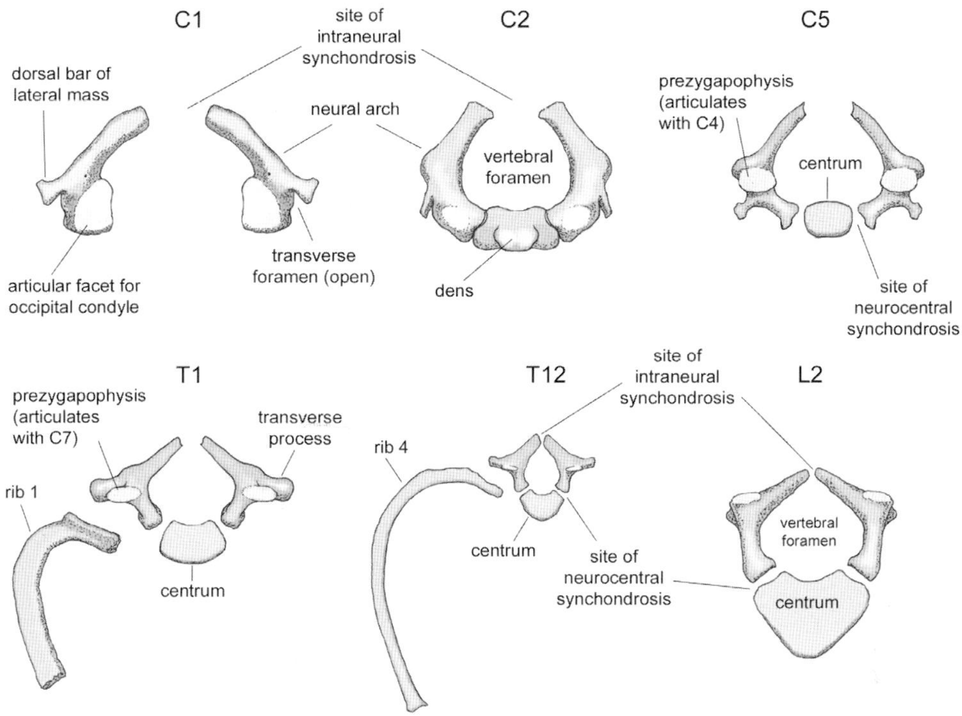

Figure 6.3 Separate vertebrae (cranial view) in 5-day-old *Pan troglodytes.*

cranial surface. Dorsally, the right and left neural arches end at blunt edges where they meet the intraneural synchondrosis.

There are three separate elements for each vertebra from C3 through C7: the centrum and right and left neural arches. There is a greater approximation of the neural arches at the intraneural synchondrosis in lower cervical vertebrae compared to C1–C2. Cranially, all neural arches have large concave prezygapophyses (the superior articular facets of humans). Each is smooth dorsally, where it is covered by developing articular cartilage. The perimeter of the facet is flattened ventrally (Figure 6.3), presumably where the prezygapophysis remains cartilaginous and is likely still enlarging.

6.2.2 Thoracic Region

In thoracic vertebrae the three ossified elements are separate in most newborns. Schultz (1940, 1941, 1944) describes intraneural synchondroses as ossifying "shortly after" birth. Based on a 5-day-old chimpanzee (Figure 6.4), mid-thoracic levels fuse first. Thoracic vertebrae possess projecting, although still incomplete, transverse processes which are spatially separated from the ribs (Figures 6.3 and 6.4). In the first one to two thoracic vertebrae, the centrum is more quadrate, as in C2–C7. More caudally, the centra become progressively more rounded ventrally. There are a total of 12–13 ribs in hominoids. Schultz (1940) detailed how anomalous ribs may form in the lower

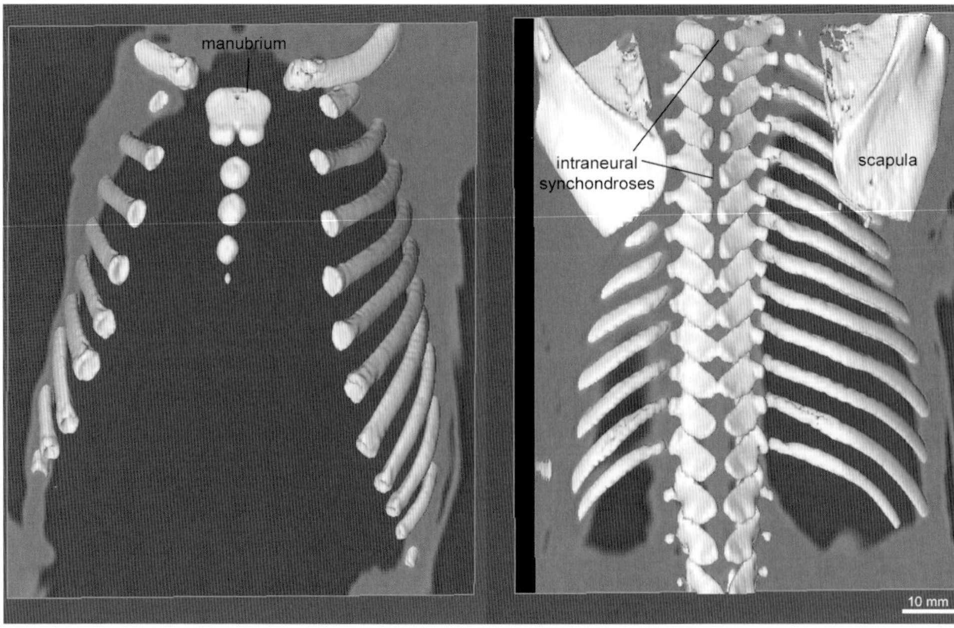

Figure 6.4 Thoracic skeleton of a newborn (5-day-old) *Pan troglodytes*.

cervical levels, or more rarely at L1 in chimpanzees. True and arch ribs have a clear, rounded neck and angle that inclines toward a flattened rib body. In the skeletonized newborn, all ribs lack a well-defined tubercle or an enlarged head because each of these parts is still mostly cartilage. The free ribs are simpler in form. They have no angle, and do not develop a tubercle. Functional correlates of the shape of the thoracic rib cage are discussed in Chapter 9.

Hominoids are notable, relative to other primates, in possessing a large sternal body which is much longer than the more cranial manubrium or the caudal xiphoid process (Ankel-Simons, 2007). The sternum is primitively even more highly segmented: multiple segments called sternebrae contribute to the adult form. The first sternebra forms the manubrium, and the body forms from multiple sternebral segments (usually four in humans and chimpanzees, perhaps a more variable number in orangutans), which fuse together in a rostrocaudal direction. The xiphoid process is unossified at birth, although Schultz (1941) notes it may ossify "early" postnatally in the orangutan. Note that these segments, present at birth, do not correspond to the full number of ossification centers. Most sternebrae coalesce from multiple ossification centers. Evidently these may be paired, as reflected in the longitudinally divided sternebrae in many orangutans of different ages (Schultz, 1941). Schultz (1940) suggested this may not be the case in chimpanzees, because divided sternebrae were not observed in his samples of chimpanzees. However, the KUPRI neonate possesses a manubrial sternebra with a caudal cleft, suggesting the merging of two centers. In a later publication, Schultz (1969) illustrated an "infant" *Gorilla* with a large single ossified manubrial element, and then two more rows of paired sternebrae. Randall (1944) illustrated a series of single sternebrae in *Gorilla* infants and older subadults with paired sternebrae caudal to the manubrium. In considering the vastly more studied human sternum, Scheuer and Black (2004) indicated that each sternebra may or may not form from multiple ossification centers. The manubrium alone may form from one to three, and these authors note there is typically one "major" and numerous "minor" centers that coalesce. In sum, data on hominoids indicate that development of the sternum is variable, and it may coalesce from a series of single or paired sternebrae.

6.2.3 Lumbar Region

In lumbar vertebrae the three ossified elements (centrum, bilateral neural arches) are separate in newborns. Lumbar vertebrae have poorly projected transverse processes in *Pan* (Figure 6.3). Although not specifically described by Schultz, we presume the transverse processes are not yet fully ossified at birth, as is the case in humans (Scheuer and Black, 2004).

6.2.4 Sacral and Caudal Regions

In the adult, the hominoid sacrum is a fused bone comprising five to six vertebral elements (Ankel-Simons, 2007). The last lumbar vertebra may partially or completely fuse with S1 in rare cases (Schultz, 1941). At birth, the segments are separate in humans. In Schultz's illustrations (1940) they are separated in *Pan* as well. A CT reconstruction of the KUPRI newborn appears to suggest a slight merging of the centra, but this is likely artifactual (e.g., insufficient resolution). Each of the first two sacral levels is represented by five separate ossified elements: neural arches (two), centrum, and lateral elements (two). The lateral elements (or ala for S1) are the costal components. Schultz (1940, 1941) indicated the same five elements are present in *Pan* and *Pongo*, referring to costal elements that fuse to the centrum during the "infantile" period, implying they are initially separate. Reconstructions of the KUPRI newborn suggest lateral elements are present for the first three sacral levels (Figure 6.5).

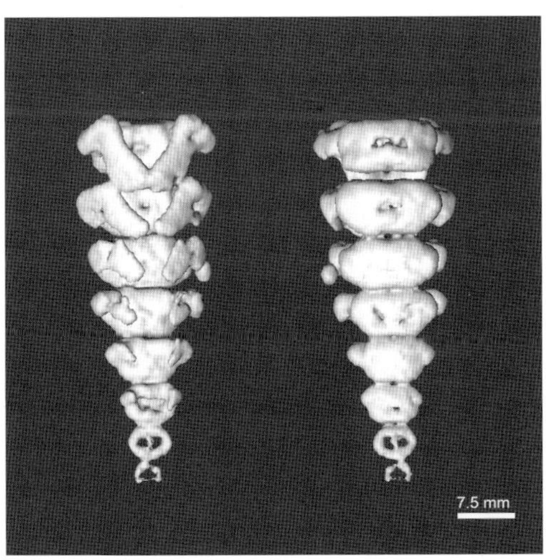

Figure 6.5 Sacrum and caudal vertebrae one and two (of three) in 5-day-old *Pan troglodytes*.

The caudal (coccygeal) vertebrae number from three to four in adult hominoids. At birth, the status of ossification in the region may vary. In humans, they remain cartilaginous. In the KUPRI newborn, the first caudal vertebra appears well-ossified, the second partially so, and the third is not detectable (and presumed cartilaginous). In *Gorilla*, ossification of these vertebrae may be more advanced, perhaps initiating prenatally (Symington, 1889).

6.3 Postcranial Axial Skeleton in Newborn Old and New World Monkeys

6.3.1 Cervical Region

As in hominoids, in all other anthropoids the major components of the cervical vertebrae are present as separate ossified masses at birth. The intraneural synchondrosis is relatively widest at C1, and then becomes progressively less wide caudally (Table 6.1). This synchondrosis has some bony bridging at lower cervical levels in *Cebuella*. The centrum and neural arches are connected by wide neurocentral synchondroses in all species except *Cebuella*, which exhibits bridging in cervical, mid- to lower thoracic, lumbar, and sacral levels (Table 6.2). The extent of ossification of the ventral bar of the transverse process varies.

6.3.1.1 C1 (Atlas)

As in hominoids, the lateral mass is incomplete at birth in C1 of most monkeys (Figure 6.6). In C1 and any other cervical levels in which the transverse foramen is unenclosed, the blunt end of the dorsal bar is an interface with a cartilaginous growth center (Box 6.1). Because the foramen is partially surrounded by cartilage, the site of the foramen is indicated by a notch in the osseous C1. In most monkeys this notch is widely open (Figure 6.6). The notch remains barely open in one newborn *Callicebus*, suggesting bony bridging is imminent. The greatest degree of ossification is observed in reconstructions of newborn cercopithecoids (e.g., *Colobus*, *Trachypithecus*), in which bone encircles the transverse foramen. The ventral bar may be thin, suggesting some cartilage remains.

In contrast to hominoids, all cercopithecoids and nearly all newborn platyrrhines examined here have an ossification center already apparent in the ventral arch of C1. In perinatal *Callicebus cupreus* and *Aotus nancymaae*, the ventral arch remains unossified at birth. Examination of older *Aotus* infants reveals that the species has a slow postnatal trajectory toward ventral arch ossification (see below).

The intraneural synchondrosis is located in the midline between the dorsal ends of the neural arches. Synchondroses also separate the ventral margin of the neural arches from the ventral arch, regardless of whether ossification has commenced in the latter (Box 6.1).

Nutrient foramina are frequently visible just dorsal to the postzygapophysis, the same position as in newborn humans.

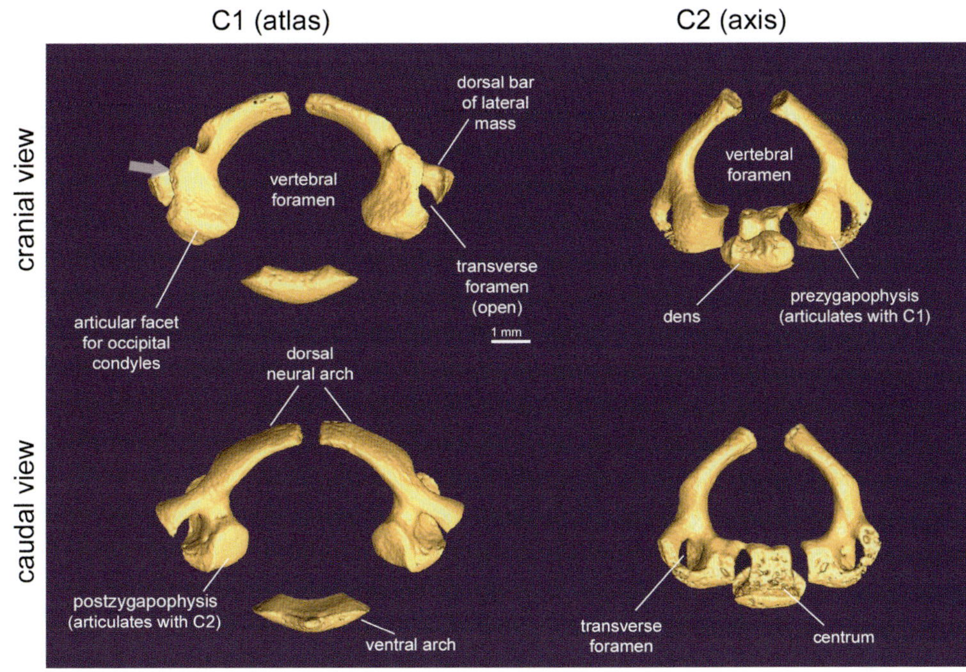

Figure 6.6 Atlas and axis of newborn *Saguinus oedipus.*

6.3.1.2 C2 (Axis)

In C2, the cartilaginous points of union (e.g., intraneural, dentocentral, and neurocentral synchondroses) are patent in all monkeys at birth. Histological examination indicates growth centers within each of the synchondroses (Box 6.1).

Details on the transverse foramen at C2 level in newborn cercopithecoids are not available in our collection. In platyrrhines, the transverse foramen is more frequently encircled by bone than "open" (in contrast to C1, where it is enclosed in part by cartilage). Among the platyrrhines examined here, bridging had commenced on at least one side in all cebids (excepting *Alouatta*, in which C2 is unavailable). The *Pithecia* neonate has an enclosed transverse foramen. However, in several samples of *Aotus* and two samples of *Callicebus*, the site of the foramen is represented by a notch. In one *Callicebus* and one *Aotus*, there are initial signs of small bridges of bone between the dorsal and ventral bars.

6.3.1.3 C3–C7

In platyrrhines, C3 through C7 have the same features as described for hominoids, although ossification may be more advanced in some monkeys. The extent of ossification of cartilage surrounding the transverse foramen differs among species from C3 to C6, but even where some lower-level vertebrae have well-ossified lateral masses, C7 has less complete ossification than C6. In the platyrrhines, especially small-bodied species, C3–C6 have very advanced ossification of the lateral mass, and thus a transverse foramen that is enclosed or nearly so. In the largest platyrrhine examined,

Pithecia pithecia, closure of the transverse foramen is less advanced by comparison to smaller species. Our postcranial material of cercopithecoids is limited, and thus we rely mainly on the literature, which mainly concerns *Macaca* spp. (Van Wagenen and Asling, 1964). *Macaca* is similar to most platyrrhines regarding the intraneural and neurocentral synchondroses. There is a greater approximation of the neural arches in lower cervical vertebrae compared to C1–C2. *Cebuella* shows bridging at C4–C7 (progressively more in a caudal direction) across the intraneural synchondrosis. No monkeys had fusion of C3–C7 neurocentral synchondroses at birth. As with the intraneural joints, the neural arches become progressively less widely separated from the centrum at lower cervical levels.

6.3.2 Thoracic Region

As in hominoids, thoracic vertebrae have three ossified elements at birth (two arches, centrum), while the costal element is a separate rib. Costotransverse synovial joints are incompletely formed at birth in at least some species (Box 6.2). The centra undergo a transition to more rounded ventral surfaces in the caudal direction (Figure 6.7). The rounded surface owes its shape to the radial path of the growth zone adjacent to the centrum. The transverse processes are more elongated at cranial levels than at more intermediate levels. Histology of selected specimens indicates that correspondingly more of the transverse process is cartilaginous as the series progresses from upper to middle thoracic levels. However, the most caudal thoracic vertebrae exhibit a more lumbar-like morphology (as in all adult primates), with a shorter transverse process (as yet, unossified – Figure 6.8).

The neurocentral junction remains cartilaginous at birth in all platyrrhines. Insufficient detail was offered by Van Wagenen and Asling (1964) to assess whether the same is true of *Macaca*. In some platyrrhines, fusion across the intraneural synchondrosis may be commencing in at least some more caudal thoracic vertebrae. Both of these joints contain growth centers. In each case, the growth centers are bipolar (growth occurs in two directions). Growth in the intraneural synchondrosis is occurring both transversely and dorsally, as the proliferating chondrocytes are in a somewhat arc-like arrangement from the ossified right and left neural arches (Box 6.2 – see b,e,i). The neurocentral synchondroses project dorsally (and obliquely lateralward) from the dorsal aspect of the centrum. As bone is forming at two poles, these are simultaneously adding bone to the dorsal side of the centrum and forming the pedicle as a ventral projection from the neural arch. The ossified centrum itself is associated with a growth center at birth, which adds bone on its ventral surface (Box 6.3). When bony bridging across the intraneural synchondrosis is seen in newborn monkeys, it occurs at lower levels (between T9 and the last thoracic vertebra) in all species except in *Cebuella*. *Cebuella* has at least some degree of ossification across all thoracic intraneural joints. *Aotus* presents another extreme, with no bridging across this joint at any thoracic level (Tables 6.1 and 6.2).

In all monkeys, individual ribs resemble those of hominoids. The head is unossified at birth, as is most of the tubercle. However, the gap at the costotransverse joints appears smaller than in other primates, perhaps reflecting a more advanced degree of rib ossification in *Cebuella* and *Callithrix*.

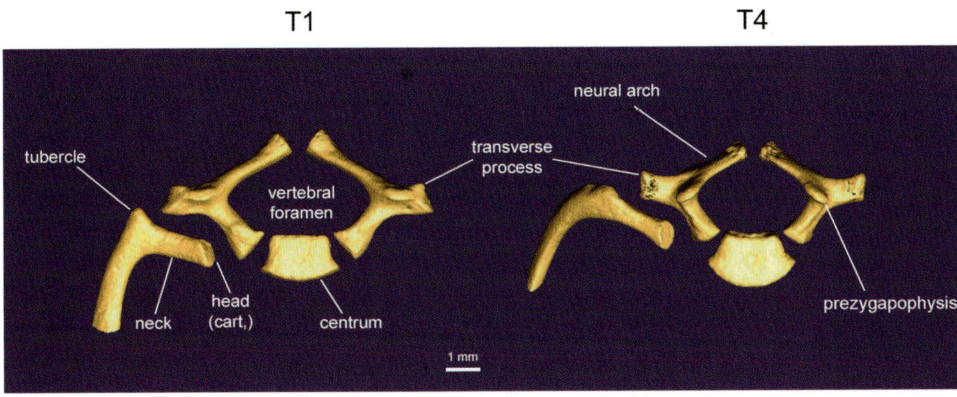

Figure 6.7 Thoracic vertebrae of a newborn *Aotus nancymaae*, cranial view, showing serial differences in shape of centra and size of intraneural synchondrosis.

In newborn monkeys, the sternum possesses four to six sternebrae segments. The segments have a graded decrease in size caudal to the manubrium.

6.3.3 Lumbar Region

As in hominoids, lumbar vertebrae typically have unfused ossified elements in newborn monkeys at cranial lumbar levels, separated by synchondroses. In tamarins, histology verifies patency of these same joints (Box 6.3). In most monkeys, some degree of bridging across the intraneural synchondrosis is seen in at least some lumbar vertebrae at birth. However, compared to hominoids the degree of bridging is greater in monkeys, particularly at more caudal lumbar levels. In *Cebuella* and *Callithrix*, for example, the degree of bridging is advanced to complete fusion across the joint, although a dorsal depression on each fused neural arch indicates the site of an as yet unossified spinous process. Histological examination reveals a growth center within the cartilage of the spinous process (Box 6.3).

The transverse processes become increasingly longer in caudal sequence. At the L2 level (Figure 6.8), where the transverse processes are relatively short, there may be no bony projection at birth – they remain mostly cartilaginous (Box 6.3). On the more caudal lumbar vertebrae, these processes project as small stems from the inferolateral margins of the neural arches. Ultimately, this portion of the neural arch merges with the centrum, and thus the transverse process merges with the vertebral body, as in adults (Ankel-Simons, 2007). The centrum is convex ventrally. Dorsally it is flattened in profile, but has a central fossa with nutrient foramina.

6.3.4 Sacral and Caudal Regions

The centra of sacral vertebrae of newborn monkeys (numbering three in all monkeys) are unfused at birth, as in hominoids. This is described in *Macaca mulatta* (Van Wagenen and Asling, 1964) and one might speculate it is true of all anthropoids, based on the lack of sacral fusion in newborn hominoids (see above) and all platyrrhines

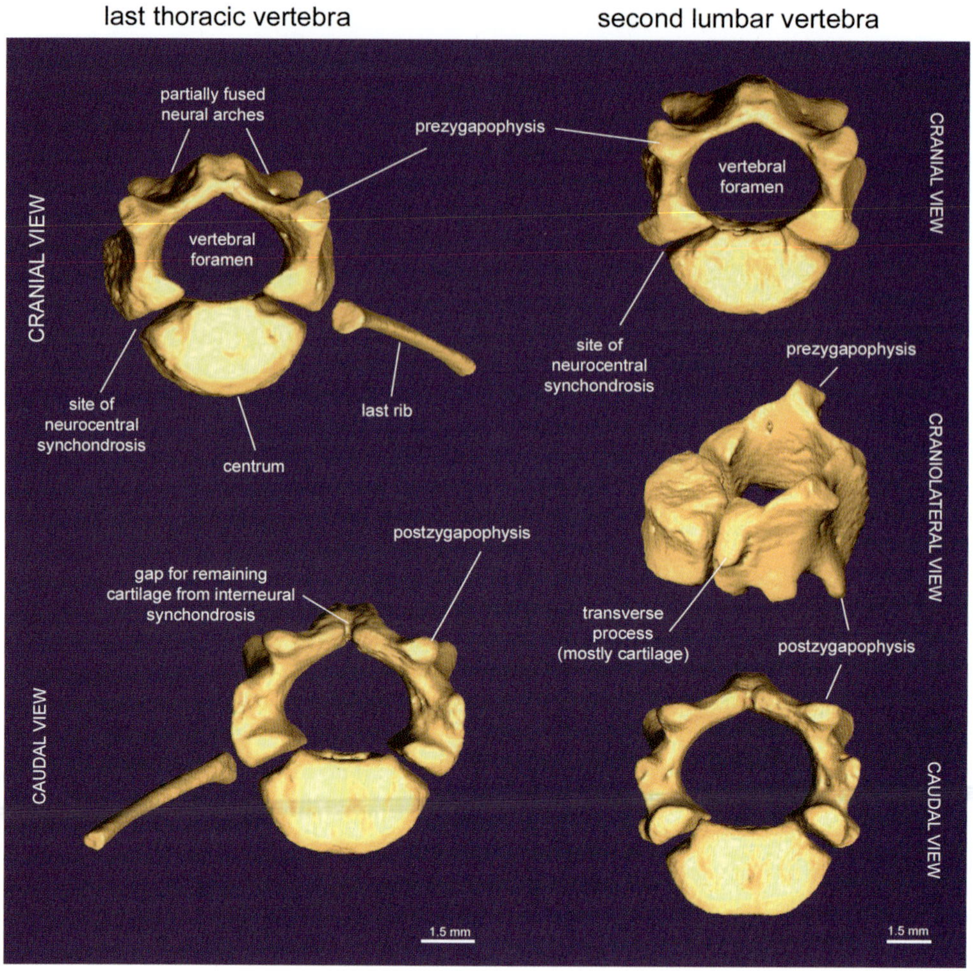

Figure 6.8 Last thoracic and second lumbar vertebrae in *Pithecia pithecia*.

studied here. Within each segment, the degree of coalescence of the centrum and two neural arches varies. A lateral element (precursor to the ala at S1) is also present as a separate, small, rounded, ossified element at birth in macaques (Van Wagenen and Asling, 1964) and platyrrhines.

The neural arches show some degree of fusion in all platyrrhines. Most species have at least partial bridging, but in *Saguinus* the two sides may remain unfused at birth. *Cebuella* and *Callithrix* had the most complete fusion across the intraneural joints. Bony bridging across the sacral neurocentral synchondroses is observed in *Cebuella*, *Callithrix*, and *Aotus*.

At birth, a true caudal vertebra has an established cylindrical diaphysis with cartilaginous growth zones at cranial and caudal ends (Box 6.4). The most proximal caudal vertebrae have neural arches; these cover the last projecting nerves of the spinal cord. Similar to lumbar vertebrae, these arches have no bony attachment to the

Table 6.1 Intraneural synchondrosis

Species	C1	C2	C3–6	C7	T1–T4	T5–T8	T9–last T	L1–L3	L4–last L	S1–S3	Source
Platyrrhines											
Alouatta seniculus	U	U	NA	NA	NA	NA	NA	NA	NA	NA	Present sample
Aotus nancymaae	U	U	U	U	U	U	U	B	B	B	Present sample
Callicebus cupreus	U	U	U	U	U	U	U→B	?	B	B	Present sample
Callithrix penicillata	U	U	U	U	U	U	U→B→F	F	F	B→F	Present sample
Cebuella pygmaea	U	U	U→B	B	F	F→B	B	B	B	B→F	Present sample
Pithecia pithecia	U	U	U	U	U	U	B	B	B	B	Present sample
Saguinus midas	U	U	U	U	U	U	U or B	U or B	U or B	U or B	Present sample
Saguinus oedipus	U	U	U	U	U	U	U	U→B or B→F	B→U	U	Present sample
Cercopithecoids											
Macaca spp.	U	U	U	NA	NA	NA	NA	NA	NA	NA	Present sample; Van Wagenen and Asling, 1964
Trachypithecus francoisi	U	NA	NA	NA	NA	NA	NA	NA	NA	NA	Present sample

Abbreviations: B, bone bridging seen across synchondrosis; F, synchondrosis is fused, completely bridged by bone; U, synchondrosis is unfused, remains a cartilaginous growth site; NA, material unavailable; →, transition in the series.

Table 6.2 Neurocentral synchondrosis

Species	C1	C2	C3–6	C7	T1–T4	T5–T8	T9–last T	L1–L3	L4–last L	S1–S3	Source
Platyrrhines											
Alouatta seniculus	–	U									Present sample
Aotus nancymaae	–	U	U	U	U	U	U	U	U	U→B	Present sample
Callicebus cupreus	–	U	U	U	U	U	U	U	U	U	Present sample
Callithrix penicillata	–	U	U	U	U	U	U	U	U	U→B	Present sample
Cebuella pygmaea	–	U	U	U	U→B	B	B	B	B	B	Present sample
Pithecia pithecia	–	U	U	U	U	U	U	U	U→B	U→B→F	Present sample
Saguinus midas	–	U	U	U	U	U	U	U	U	U	Present sample
Saguinus oedipus	–	U	U	U	U	U	U	U	U	U	Present sample
Cercopithecoids											
Macaca spp.	–	U	U	NA	NA	NA	NA	NA	NA	NA	Present sample; Van Wagenen and Asling, 1964
Trachypithecus francoisi	–	U	NA	NA	NA	NA	NA	NA	NA	NA	Present sample

Abbreviations: B, bone bridging seen across synchondrosis; F, synchondrosis is fused, completely bridged by bone; U, synchondrosis is unfused, remains a cartilaginous growth site; NA, material unavailable; →, transition in series.

vertebral bodies (unfused neurocentral synchondroses). Hemal processes (i.e., chevron bones) are found on the ventrodistal margin of the more proximal vertebrae in the series. These are partially ossified at birth, close to their articulation site. Chevron bones embrace a caudal artery, which passes ventrally along the base of the tail (Box 6.4).

6.4 Postcranial Axial Skeleton in Tarsiers at Birth

6.4.1 Cervical Region

General features and joints resemble anthropoids at birth in most respects, especially platyrrhines.

6.4.1.1 C1 (Atlas)

The dorsal arch is slightly curved. The lateral mass flares widely. The dorsal bar extends far ventrally; only a small portion of the ventral bar is unossified. Thus, the transverse foramen is nearly encircled by bone. A ventral arch is ossified with a bilobed appearance in the ventral view. Observed from the cranial perspective, the ossified part of the ventral arch takes the form of two conjoined cup-shaped and thin-walled islands of bone. All synchondroses are widely patent.

6.4.1.2 C2 (Axis)

The axis is represented by four separate ossified elements in the newborn primate, two neural arches, the centrum, and the dens. The cartilaginous points of union (e.g., intraneural, dentocentral, and neurocentral synchondroses) are widely patent. The neural arches are less curved compared to the dorsal arch of C1, and with broad, rounded apices facing the intraneural synchondrosis. The transverse foramen is an open notch, bracketed by projecting dorsal and ventral bars.

6.4.1.3 C3–C7

The remaining cervical vertebrae possess three ossified elements at birth, separated by synchondroses (Figure 6.8). The centra are slightly elongated ventrodorsally except for C7, which is nearly square. Each has a dorsally situated fossa with paired nutrient foramina. The neural arches are slightly rounded dorsally. In contrast to anthropoids, intraneural synchondroses become slightly wider from C3 to C7. All lateral masses have a well-ossified dorsal bar. C7 is the only level not to exhibit an ossified ventral bar. However, ossification of the ventral bar is incomplete throughout the series, and C3–C6 have open, notched transverse foramina.

6.4.2 Thoracic Region

General features of the thoracic vertebrae resemble those of all other primates at birth (Figure 6.9). All synchondroses remain unfused at upper and mid-thoracic levels. In the lower thoracic levels, the last one or two intraneural synchondroses have bridging in two of the newborn tarsiers. The neurocentral junction remains cartilaginous throughout the entire thoracic series.

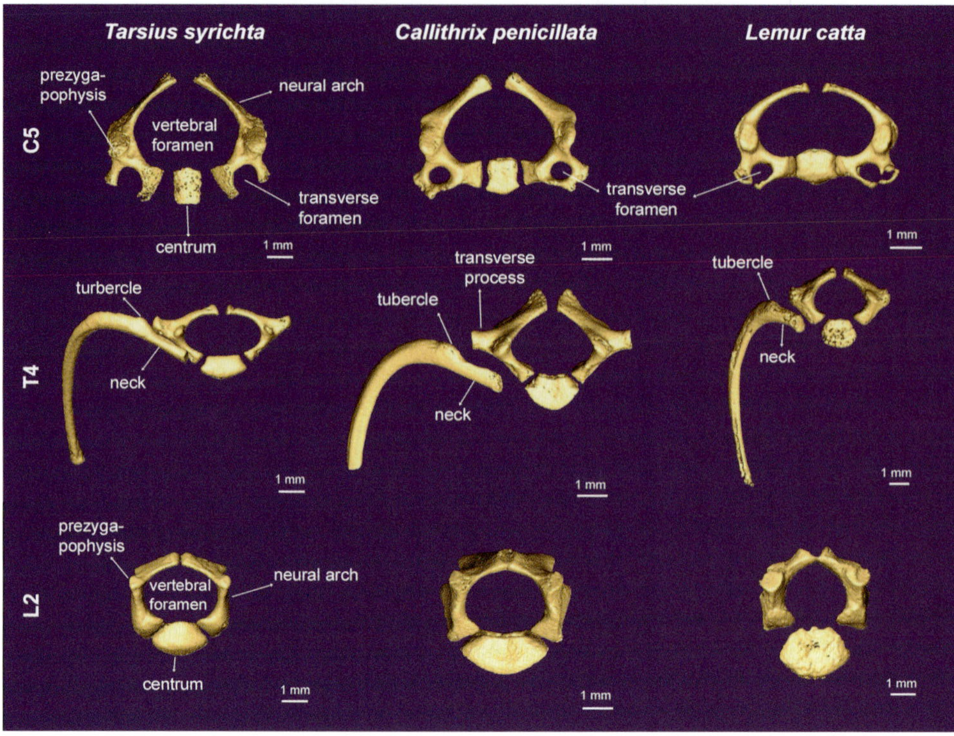

Figure 6.9 C5, T4, and L2 levels in three primates at birth; shown in cranial view. Scale = 1 mm.

The centra are similarly shaped compared to all other primates at birth. At upper levels they are nearly quadrate. As the series progresses caudally the ventral side becomes progressively convex and rounded (Figure 6.9). In transverse profile, the vertebral canal is somewhat dorsoventrally compressed compared to many other species, although this may also be said of some small strepsirrhines (Figure 6.10). Neural arches are extremely well-defined laterally. The transverse process projects to a proportionally greater degree compared to other primates. Likewise, the articular tubercles of the ribs are better-defined than in other newborn primates. As a result, the costovertebral joint space is extremely narrow, even compared to callitrichines (Figure 6.9).

6.4.3 Lumbar Region

The state of ossification of the lumbar series closely resembles that of *Saguinus* or *Calllithrix* at birth. At upper levels, lamina are craniocaudally broad and intraneural synchondroses are narrow. The synchondrosis is fully bridged by L3. The transverse processes are barely ossified at L1; the osseous stem of transverse processes becomes progressively elongated in more caudal lumbar levels. Neurocentral synchondroses remain unfused at all levels.

6.4.4 Sacral and Caudal Regions

The neural arches of sacral vertebrae are lumbar-like. Fusion across the intraneural synchondroses is less extensive than the last lumbar vertebrae, showing some porosity in reconstructions. Neurocentral synchondroses show an initial state of fusion, with partial bridging. The sacral alae of S1 are present as rounded, separate ossification centers.

Synchondroses of proximal caudal vertebrae initially resemble the sacral levels: partial bridging only. By the fourth caudal level, the elements are a fused mass with a central canal.

6.5 Postcranial Axial Skeleton in Strepsirrhines at Birth

6.5.1 Cervical Region

6.5.1.1 General Considerations

General features and joints resemble haplorhines overall, except that the ventral arch of C1 remains unossified at birth in all strepsirrhines studied to date. No species are known to have bridging across neurocentral or intraneural synchondroses (Tables 6.3 and 6.4).

6.5.1.2 C1 (Atlas)

The ventral arch is unossified at birth; the dorsal arches may be curved towards the intraneural joint, or straight, as in *Lemur* (Figure 6.9). Large nutrient foramina are occasionally visible cranially (Figure 6.9). The dorsal bar of the lateral mass is well-ossified in all strepsirrhines, oriented ventrolaterally; no species had any ossification in the more ventral or lateral part of the lateral mass.

On the cranial surface of C1, the articular facet for the articular condyle is rough and supports growing hyaline cartilage. The ventral side of the condyle is flattened in superior profile (Figure 6.9). Histology reveals cartilage on all surfaces of the condyle, and the ventral edge has long columns of proliferating chondrocytes, indicating ventromedial growth of the neural arch and articular facet at this margin.

6.5.1.3 C2 (Axis)

The axis is represented by four separate ossified elements, widely separated by synchondroses. The centrum is quadrate or trapezoidal and smaller than the dens in most species. In larger species (*Propithecus, Lemur*; Figure 6.10) the two may be more nearly equal in size. The dens is rounded and rests on a broader caudal base. The latter is connected via the odontocentral synchondrosis to the centrum of C2. The rounded appearance of the dens is in contrast to the "lobed" appearance in all anthropoids. The dorsal arch either makes an angular (e.g., *Galago, Lemur, Propithecus*) or rounded (e.g., *Cheirogaleus*) turn toward the midline.

6.5.1.4 C3–C7

The remaining cervical vertebrae possess three ossified elements at birth. The intraneural synchondroses do not show a caudal gradient in diminishing width, as

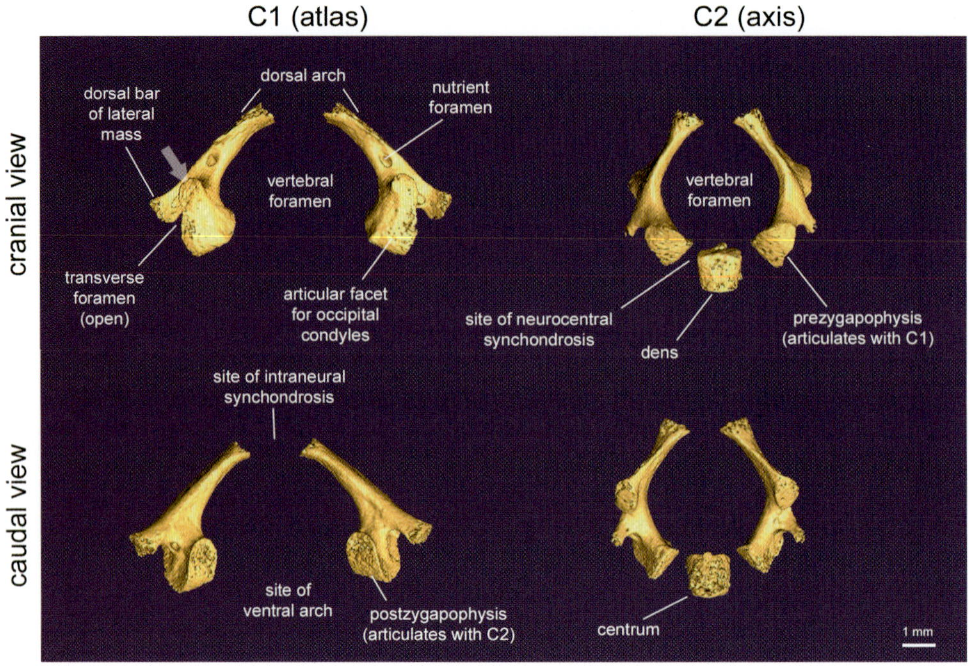

Figure 6.10 Atlas and axis of newborn *Lemur catta*.

observed in monkeys. At lower cervical levels the synchondrosis may actually widen. The transverse foramen is well surrounded by bone in *Cheirogaleus medius*, but not in some larger lemuroids (e.g., *Lemur*, *Eulemur*). As with C2, the dorsal bar is always better ossified than the ventral bar in C3 through C7. In C3–C6, the ventral bar is partially ossified, but at C7 it remains unossified. The neurocentral synchondroses are unfused at all cervical levels.

6.5.2 Thoracic Region

In thoracic vertebrae there are three ossified elements present in newborns (two neural arches and centrum) with varied degrees of synchondroseal fusion among strepsirrhines at birth (Tables 6.3 and 6.4). The centrum goes through the same craniocaudal shape gradient described above for other primates. The neural arches have poorly developed transverse processes at birth in most strepsirrhines, with the dorsal surface being flat, and the ventral surface becoming more rounded and convex. In *Galago* they are beginning to ossify and in *Cheirogaleus* they are especially well ossified (Figure 6.11).

In larger lemuroids, the arches and centra are widely separated at synchondroses of upper thoracic levels. *Lemur* and *Eulemur* have some bridging across the intraneural synchondroses at lower thoracic levels, but no bridging of any neurocentral synchondroses. *Propithecus* and *Varecia* have unfused neurocentral and intraneural synchondroses throughout the entire thoracic series.

The heads of ribs remain unossified at birth in all strepsirrhines and the articular tubercle is poorly formed in most. *Cheirogaleus* is an exception, with the articular

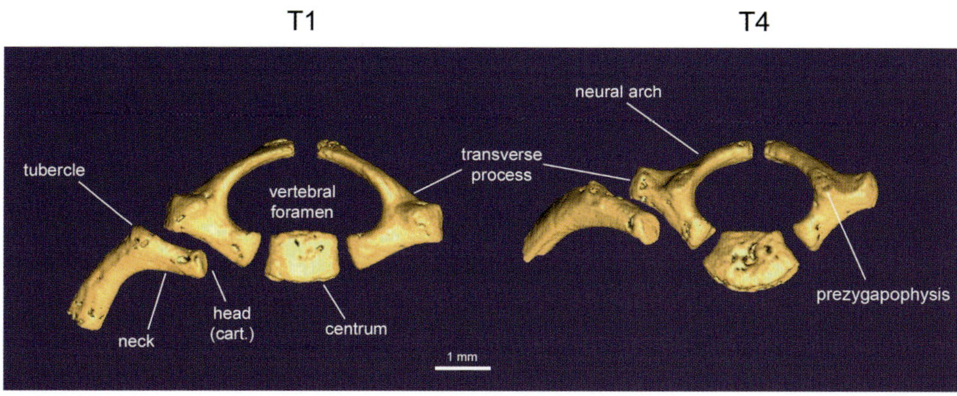

Figure 6.11 Thoracic vertebrae of a newborn *Cheirogaleus medius*, cranial view, showing serial differences in shape of centra and size of intraneural synchondrosis.

tubercle beginning to project and a relatively narrow costotransverse joint space at birth (Figure 6.11).

Newborn strepsirrhines possess four to seven sternebrae.

6.5.3 Lumbar Region

In lumbar vertebrae there are three ossified elements present in newborn strepsirrhines (two neural arches and centrum). These elements appear notably more porous in reconstructions compared to thoracic and cervical levels, particularly in the centrum. This is especially so in *Varecia* (Figure 6.12), and suggests relatively poor mineralization compared to more cranial regions at this age. Lumbar transverse processes are poorly ossified in all strepsirrhines at birth.

Neurocentral synchondroses are unfused at all lumbar levels in all strepsirrhine species under review here, and intraneural junctions remain cartilaginous at all lumbar levels in *Propithecus* and *Varecia*. At least some bridging across the intraneural joints is seen in *Lemur* and *Eulemur*, and fusion is even more advanced in *Cheirogaleus* and *Galago*.

6.5.4 Sacral and Caudal Regions

Each sacral vertebrae of strepsirrhine newborns includes three ossified elements at birth: the centrum and two neural arches. The neural arches may have initiated fusion at the intraneural and neurocentral synchondroses, as seen in *Galago* and *Cheirogaleus*; fusion is more advanced in the intraneural synchondrosis. In other strepsirrhines the right and left arches and the neurocentral junctions remain widely patent at birth (e.g., *Lemur*; Figure 6.13). Wide gaps between consecutive sacral vertebrae exist in all strepsirrhines at birth.

The most proximal caudal vertebrae have neural arches; these cover the last projecting nerves of the spinal cord. Hemal processes (i.e., chevron bones) embrace a caudal artery, which passes ventrally along the base of the tail. The caudal vertebrae of *Nycticebus* lack the growth centers seen in those of other primates (Box 6.4).

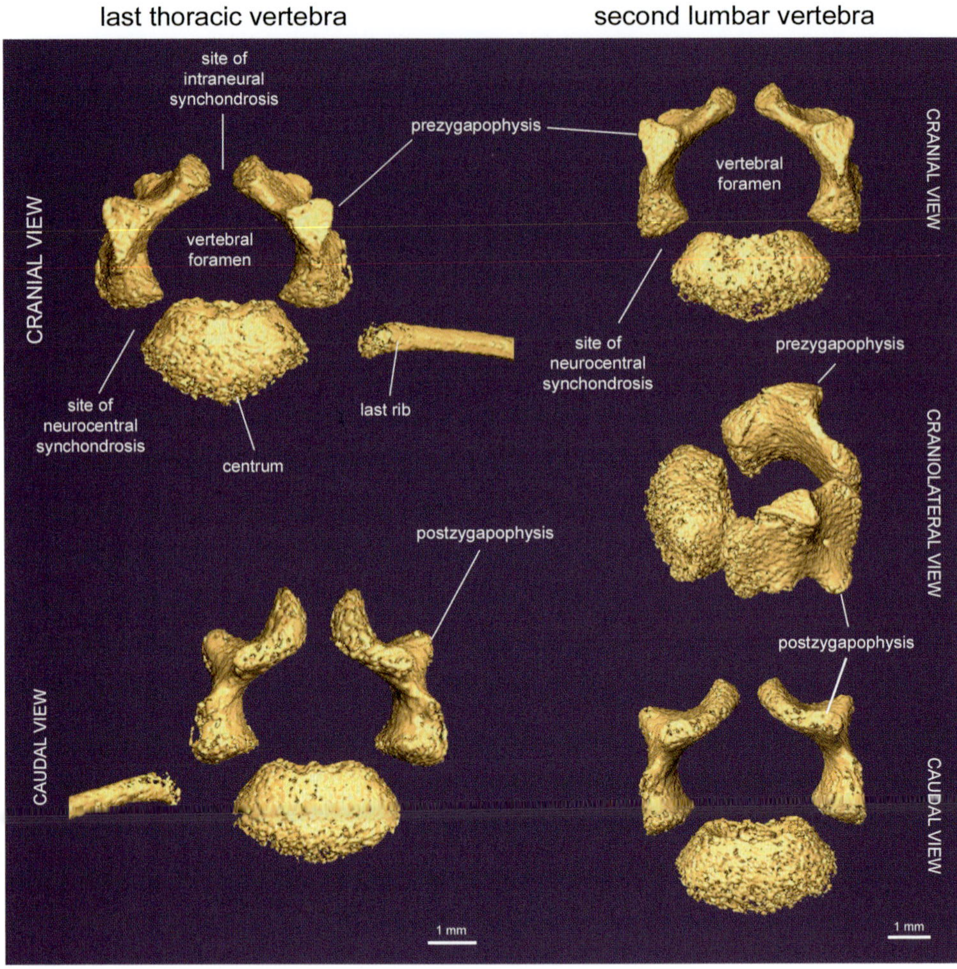

Figure 6.12 The last thoracic vertebra and second lumbar vertebra in newborn *Varecia variegata*.

6.6 Patterns of Ossification of the Postcranial Axial Skeleton in the Early Postnatal Period

6.6.1 Unossified Components

Ossification of secondary centers of ossification (i.e., epiphyses) has not yet commenced in the axial skeleton at birth. As in humans, such sites probably ossify well after the primary synchondroses fuse. In at least some other primates, these secondary centers appear more rapidly, especially in primates that mature quickly. *Galago* has ossified cranial and caudal epiphyses (or annular rings) of vertebral centra visible at 1 month of age, appearing first in thoracic vertebrae.

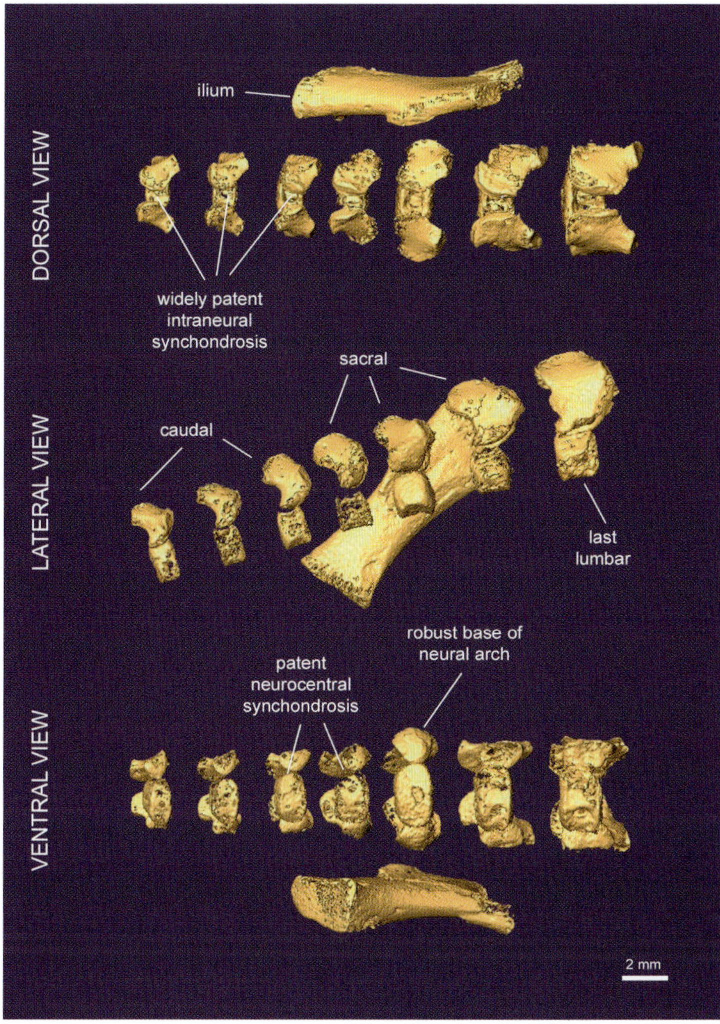

Figure 6.13 Series of vertebrae extending from last lumbar to third caudal vertebrae in newborn *Lemur catta*.

6.6.2 Early Postnatal Transformations

Certain patterns described above in newborns, such as patency or fusion of intraneural synchondroses, reflect a segmental gradient of ossification rates across the vertebral column. This is reflected postnatally as well. A comparison of the lumbar region in newborn and 1-month-old *Saguinus oedipus* (Figure 6.14) shows that vertebrae with greater fusion of the intraneural synchondrosis also have more ossification of the spinous process at older ages, a reflection of a persisting growth center (Box 6.3). Transverse processes also become more ossified craniocaudally through the series (Figure 6.14), also reflecting a craniocaudal gradient in the degree of growth center activity.

Table 6.3 Intraneural synchondroses

Species	C1	C2	C3–6	C7	T1–T4	T5–T8	T9–last T	L1–L3	L4–last L	S1–S3	Source
Cheirogaleus medius	U	U	U	U	U	U	B	B	F	F	Present sample
Eulemur collaris	U	U	U	U	U	U	B → F		F	U	Present sample
Galago moholi	U	U	U	U	U→B→F	B→F	B→F	F	F	F	Present sample
Lemur catta	U	U	U	U	U	U	U	U→B	U	U	Present sample
Propithecus coquereli	U	U	U	U	U	U	U	U	U	U	Present sample
Varecia spp.	U	U	U	U	U	U	U	U	U	U	Present sample

Abbreviations: B, bone bridging seen across synchondrosis; F, synchondrosis is fused, completely bridged by bone; U, synchondrosis is unfused, remains a cartilaginous growth site; →, transition in caudal series.

Table 6.4 Neurocentral synchondroses

	C1	C2	C3–6	C7	T1–T4	T5–T8	T9–last T	L1–L3	L4–last L	S1–S3	Source
Cheirogaleus medius	U	U	U	U	U	U	U	U	U	U	Present sample
Eulemur collaris	–	U	U	U	U	U	U	U	U	B	Present sample
Galago moholi	–	U	U	U	U	U	U	U	U	U	Present sample
Lemur catta	–	U	U	U	U	U	U	U	U	B	Present sample
Propithecus coquereli	–	U	U	U	U	U	U	U	U	U	Present sample
Varecia spp.	–	U	U	U	U	U	U	U	U	U	Present sample

Abbreviations: B, bone bridging seen across synchondrosis; F, synchondrosis is fused, completely bridged by bone; U, synchondrosis is unfused, remains a cartilaginous growth site; →, transition in caudal series.

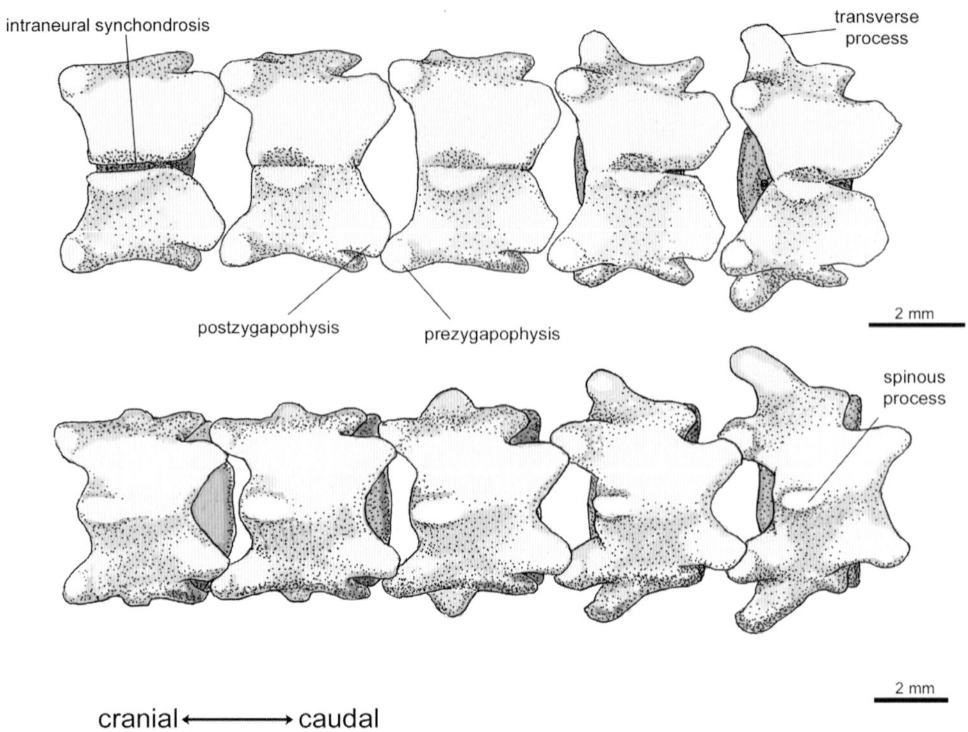

Figure 6.14 Lumbar vertebrae in *Saguinus oedipus* at birth (top) and 1 month of age (bottom).

Our limited comparative sample of older infants indicates that postnatal ossification rates of vertebrae vary among primates. For example, the ventral arch of C1 is slow to ossify in lemurids. No ossification is seen by approximately 1 month in *Lemur*. In *Eulemur collaris*, ossification of the anterior arch has commenced by approximately 1 month, but only a small midline mass of bone is present. In galagids, ossification likely commences soon after birth. In 17- and 20-day-old *Galago moholi* the ventral arch is ossified, with a lobed appearance suggesting the union of two ossification centers (Figure 6.15). In terms of absolute age, anthropoids also vary significantly. The ventral (anterior) arch delays ossification until 1 or 2 years in humans, whereas it is ossified at birth in most monkeys. However, in a 14-day-old *Aotus nancymaae*, the ventral arch has commenced ossification, although only as a thin plate on the presumptive ventral face of the arch (Figure 6.15). This indicates that ossification of the ventral arch is slower in *Aotus* than in most other monkeys. Other maturational aspects of primary vertebral ossification centers are also likely to vary, but are beyond the scope of this book.

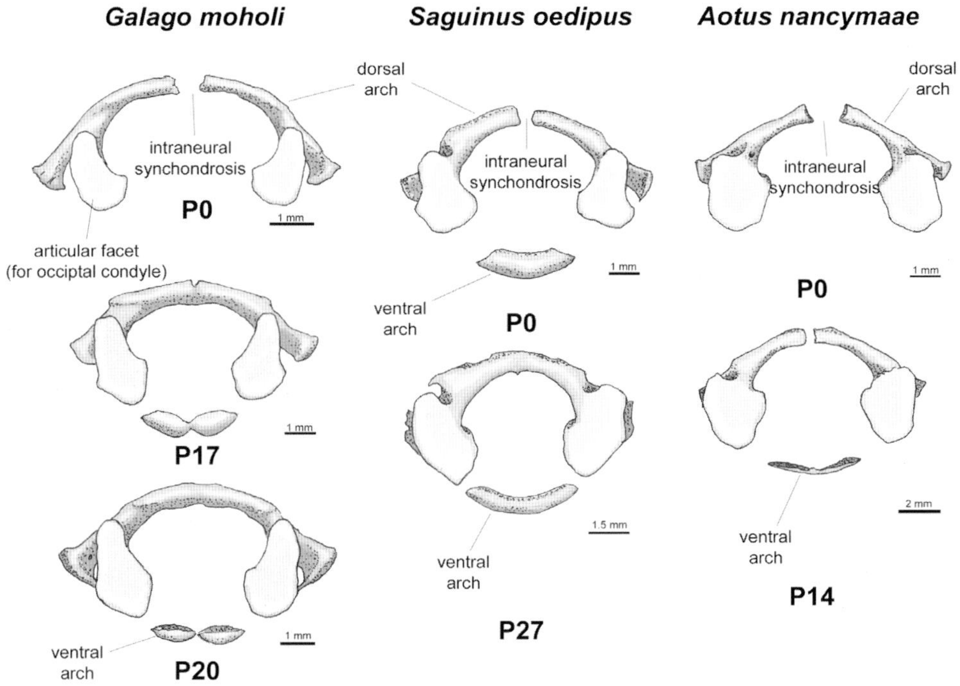

Figure 6.15 Early transformation of C1 in three primate species.

Box 6.1 Cervical Vertebrae: Ossified Elements and Synchondroses at Birth

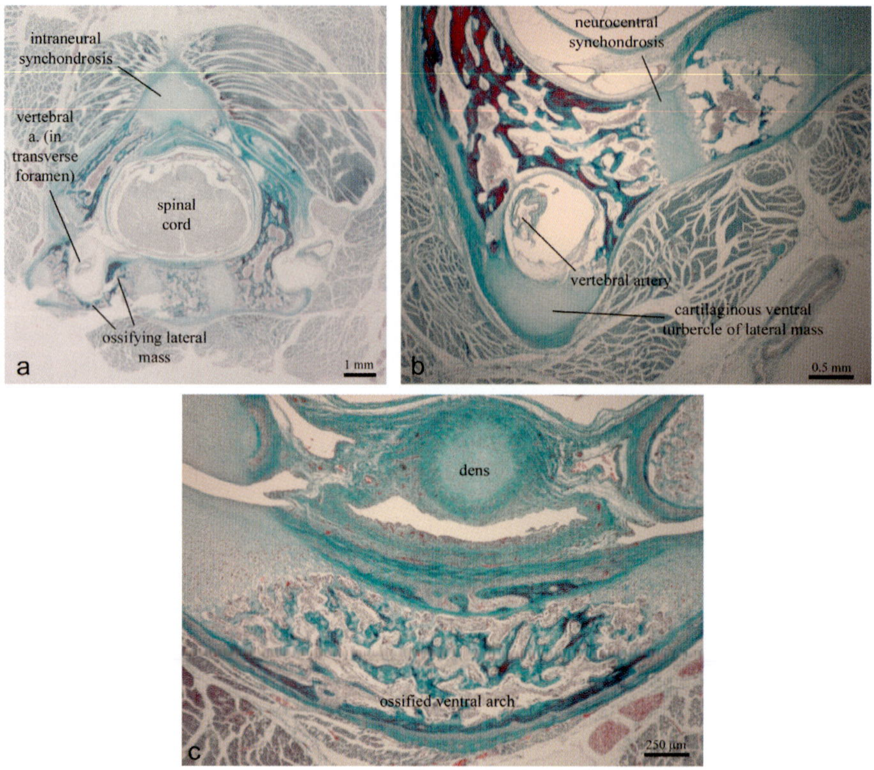

Above: Lower-level cervical vertebra in (a) *Saguinus oedipus* and (b) *Eulemur collaris*. In most newborn primates the intraneural and neurocentral synchondroses are unossified and active growth centers. Ossification of the transverse foramen that surrounds the vertebral artery varies. If the lateral mass is in an advanced stated of ossification (a), an enclosed transverse foramen is present at birth. In all newborn primates, at least some vertebrae have partially unossified lateral masses, resulting in an "open" transverse foramen, which is indicated by a notch in skeletonized specimens. (c) C1 of many primates lacks an ossified ventral arch at birth. In most monkeys (c, *Saguinus*) the ventral arch is ossified and has bilateral cartilaginous joints (allocentral synchondroses) with the dorsal arch. On the dorsal side of the ventral arch, C1 has a synovial joint with the dens of C2. This joint is precociously formed in *Saguinus*, with a synovial membrane and joint space (*) already visible.

Below: Isolated exoccipital and dorsal arch (left side only) of *Galagoides demidovii*. Two views of the bones are shown at the top. Sectional planes with histology are shown at the bottom. Note that the cranial side of C1 is still largely cartilaginous at birth. This includes some growth centers that are actively expanding the ventral dimensions of the cranial articular facet.

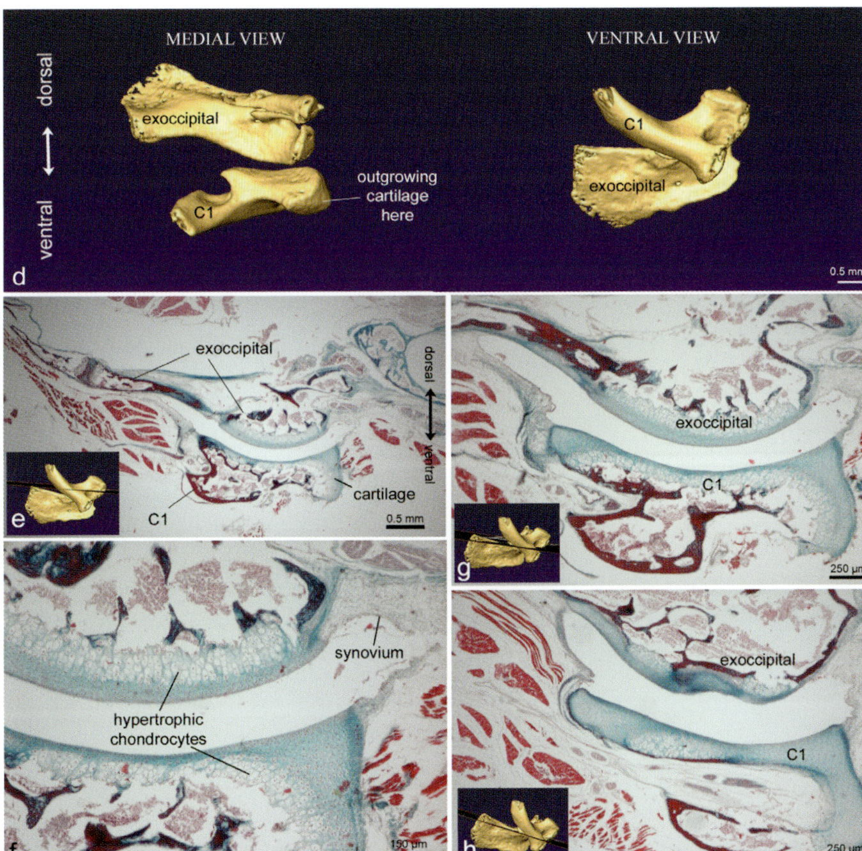

Box 6.2 Thoracic Vertebrae: Ossified Elements and Synchondroses at Birth

Below: Upper-level thoracic vertebrae of newborn *Saguinus oedipus*, showing the ossified elements and synchondroses. (a) Separate ossified elements in the thoracic region include the centrum, neural arches (two), and ribs (two). (b) The centrum and neural arches are joined by cartilage. (c–e) Enlarged views of cartilage (location of these insets indicated in the top row). Note that hyaline cartilage (light blue–green tissue) is seen at joints (synchondroses) at the tips of processes or tubercles, as well as at the peripheral margins of the centrum. The ribs are forming joint cavities at birth. Joint cavities are beginning to form between the ribs and transverse processes. Note that all cartilage, including that located on tubercles (f), the periphery of the centrum (g), and synchondroses (h,i) have growth characteristics. Specifically, columns of hypertrophic and proliferating chondrocytes indicate unidirectional or even radial expansion of cartilage (f–i: blue arrows).

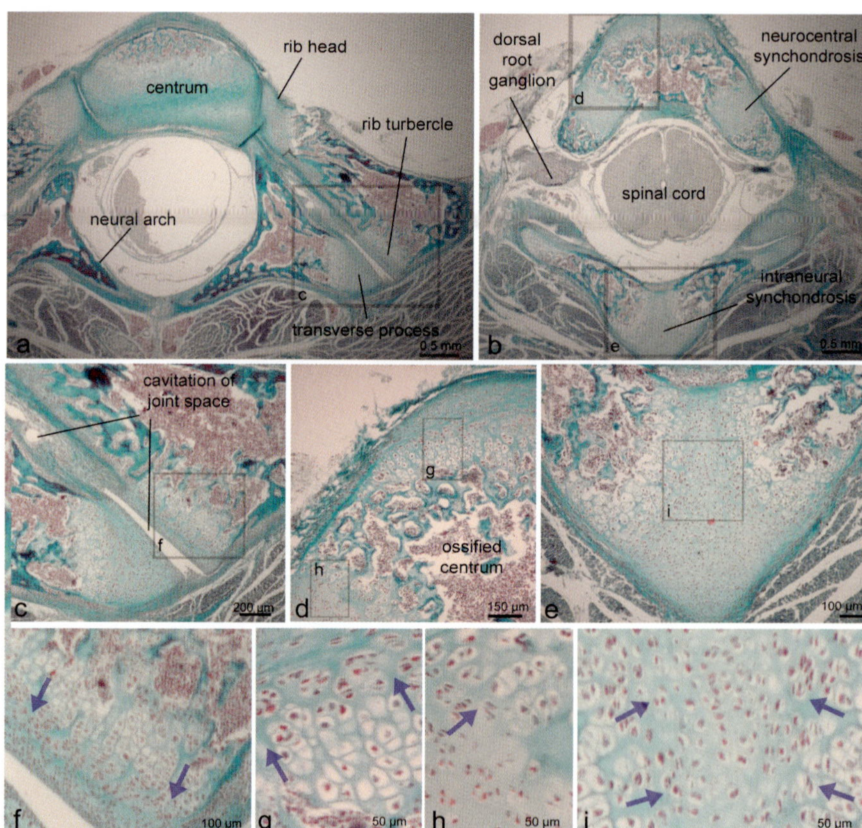

Box 6.3 Lumbar Vertebrae: Ossified Elements and Synchondroses at Birth

Below: Mid-level lumbar vertebrae of newborn *Saguinus oedipus*, showing the ossified elements and synchondroses. (a) Separate ossified elements in the mid and lower lumbar region include the centrum and fused neural arches. These ossifications are separated by neurocentral synchondroses on each side. In cleaned skeletal specimens, a small depression on the dorsal side of the neural arches indicates the location of the spinous process, still cartilaginous. (b) Magnified view of the spinous process showing a growth center with proliferating and hypertrophic chondrocytes (blue arrows indicate the direction of growth). (c) Section through lumbar vertebra cranial to the spinous process. The transverse process is a small blunt projection tipped with cartilage. (d) An enlarged view of the transverse process, showing growth at the tip (arrows).

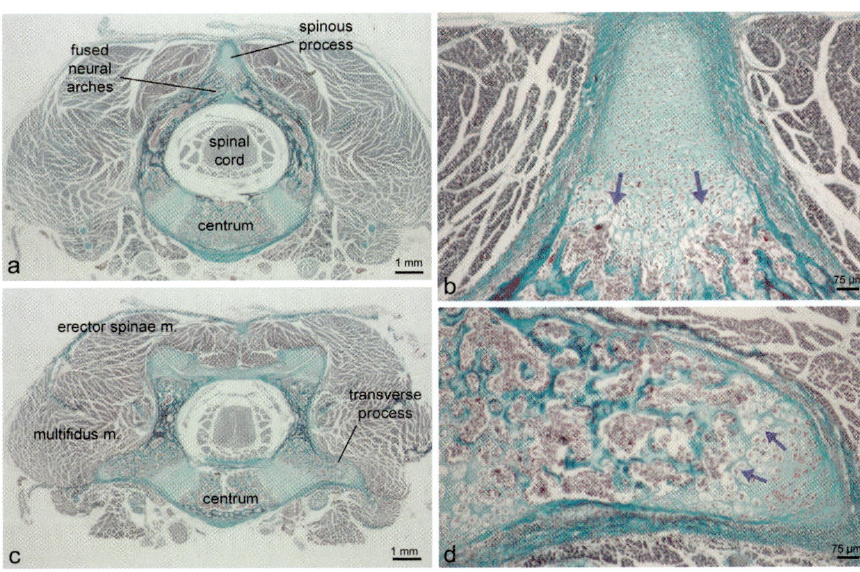

Box 6.4 Sacral and Caudal Vertebrae: Ossified Elements and Synchondroses at Birth

Below: Histological sections of newborn primates showing the sacral region of *Lemur catta* (a), and caudal elements of *Saguinus oedipus* (b) and *Nycticebus pygmaeus* (c). (a) Section from sacral region deviates from the transverse plane of the bone, showing the ossified sacral centra (S1–S3) and the merged sacral elements. At the top, the plane is tilted further dorsally, showing the first neural arch. (b) Caudal vertebrae of primates with long tails have elongated centra in the proximal tail, with epiphysis-like cartilaginous caps on either side (each is an active growth center). Ventrally are partially ossified chevron bones that partially surround the caudal artery. (c) *Nycticebus* has a vestigial tail in which caudal elements are small, round ossification centers at birth, which lack the disk-shaped cranial and caudal cartilaginous ends. The joint spaces (*) are less well-formed at birth compared to species with tails.

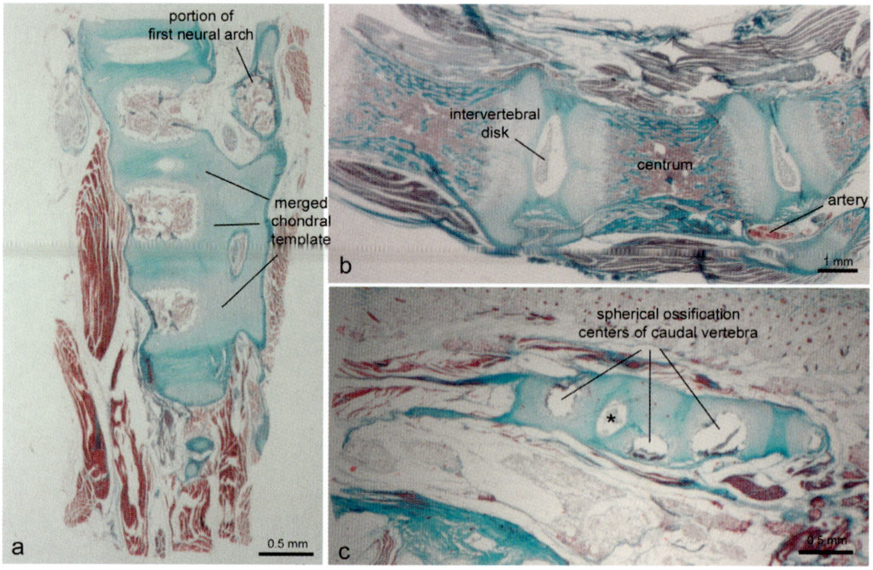

7 The Pectoral Girdle and Forelimb Skeleton

7.1 Introduction

The primate pectoral girdle includes the scapula and the clavicle. The clavicle articulates with the manubrium medially, and with the acromion process of the scapula laterally. The glenoid fossa of the scapula articulates with the humerus, the bony element of the brachium, or arm. Adult primate scapulae vary in craniocaudal to proximodistal breadth ratios (Ankel-Simons, 2007; Gebo, 2014), but at birth these differences are less salient. The shape of the thorax has a direct bearing on the elements of the pectoral girdle, in both position and proportions (see below).

The antebrachium, or forearm, contains the radius and ulna. The wrist and hand include carpals, metacarpals, and phalanges. In most primates, there are nine carpals. The proximal row includes the scaphoid, centrale, lunate, triquetral, and pisiform; the distal row includes the trapezium, trapezoid, capitate, and hamate. In some species there are only eight carpals because the scaphoid is equivalent to a fused scaphoid–centrale. The distal forelimb may be characterized by digital reduction (digit 2 of lorises; digit 1 of brachiating primates) or even loss (e.g., digit 1 of *Ateles*).

Development of the limbs in non-human primates has been best studied in anthropoids, with thorough descriptions of staged embryos in the baboon (Hendrickx, 1971) and the common marmoset (Phillips, 1976). Among strepsirrhine species, Butler (1983) studied the lesser bushbaby. In early limb development, there is a broad similarity among primates in the superficial form of limbs (Figure 7.1) and the sequence of morphogenesis (Box 7.1). In primates, as all mammals, the forelimb is visible as a bud prior to the hindlimb (e.g., Hendrickx, 1971; Hughes and Hall, 1998; Nolte et al., 2009; Keyte and Smith, 2010). Subsequently, at each successive embryonic stage, the forelimb assumes an advanced morphological state relative to the hindlimb. The forelimb and hindlimb have a developmental similarity in that there are three regions of each. Proximally, the stylopod is supported by one bone element (humerus/femur). The middle region is supported by two elements, the radius/tibia and the ulna/fibula (the zeugopod). The distal region is supported by carpals/tarsals (wrist/ankle bones) and the bones of the hand/foot (the autopod). With this framework in mind, serially

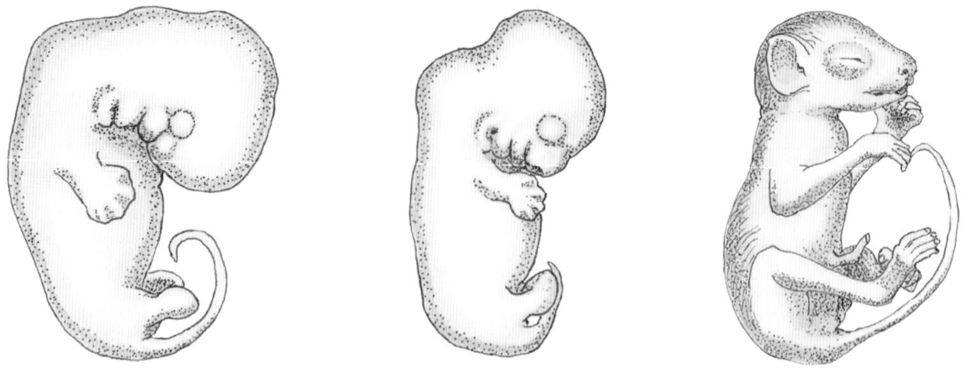

Figure 7.1 Embryos of two non-human primates, compared to a fetal stage. Illustrations after photographs of embryos and fetuses from Hendrickx (1971) and from archival images in the Bluntschli collection (American Museum of Natural History).

homologous parts of limbs may be identified in embryos by observing fore- and hindlimbs at Carnegie stages 17–20, with identical features across species (Figure 7.1). At these stages, fore- and hindlimbs have a similar orientation: they project ventrally and the palms of the hands/soles of the feet face the opposing side. The digits or digital rays are clearly defined, and the thumb and great toe both face in the cranial direction; likewise, the radius and tibia are on the cranial side of their respective limbs. This primitive orientation is altered by embryonic limb rotation, a process that spans late embryonic and early fetal stages, and results in opposing directionality of elbow/wrist versus knee/ankle flexion (see Carlson, 2014, for further details).

The forelimb skeleton begins mesenchymal condensation after this process is initiated in the axial skeleton. In humans, the major embryonic events may be summarized as follows. Limb buds derive from lateral plate mesoderm at a time (P28 in human embryos) that the paired parasagittally located somites are already parceled into different regions, including the sclerotomes, which provide mesenchymal condensations for the vertebrae, ribs, and part of the cranial base. The radius and ulna are apparent as mesenchymal condensations at 33 days fertilization age in humans (Scheuer and Black, 2004) and the digital bones condense shortly after. Starting in week 5, the upper limb bones chondrify, completing the process by the end of week 6. At this point chondrification is overlapping in the axial and appendicular skeleton. By the end of week 7, the primary ossification centers are radiographically visible in the upper limb skeleton, as is the case for the thoracic axial skeleton to which it articulates.

At the end of the embryonic period, the form of the forelimb is superficially clear, with completely free digits and joint contours at the elbow and wrist. However, some key features that typify primates at the family and even species level are not yet established. Familiar examples include distinctive characteristics of the first digit in humans compared to other primates (the relatively long thumb; the relatively greatly adducted great toe), which become gradually shaped as digits and joints develop during the early fetal stages. In addition, the rapid growth during the fetal period may establish a degree of adult-like limb proportions (Figure 7.1).

Ossification of long bones was briefly described in Chapter 2, and microscopy pertaining to primates is discussed in Box 7.1. In early fetal development, most long bones of the fore-/upper limb begin a prolonged phase of rapid expansion in bone length from both ends via interstitial cartilage growth. Exceptions include the distal phalanges (which possess only a proximal epiphysis), some carpals/tarsals, and the clavicle. The clavicle is a bone of dual modes of ossification in humans (Scheuer and Black, 2004), although development in other primates is poorly studied. The sternal end forms via endochondral ossification (Boxes 7.1 and 7.2), while in the lateral half a second primary center of ossification forms within the membrane that connects the developing scapula with the manubrium. There is much variation in mammals, with some having solely intramembranous ossification, and others having endochondral centers at both acromial and sternal ends (Großmann et al., 2002; and see Hall, 2015). In mammals lacking a clavicle, a collagenous membrane exists in its place, serving as an attachment site for the same muscles that attach to the clavicle.

The scapula commences ossification in the "neck" region, or the side closest to the glenohumeral joint, at the base of the spine. Endochondral ossification progresses directionally along the vertebral–glenoid axis (Box 7.2). In humans, the more superior and inferior (cranial–caudal in other primates) parts are laid down intramembranously, expanding from endochondral parts (Ogden and Phillips, 1983). Thus, part of the flat, blade-like part of the body is a thin but dense intramembranous plate.

Long bones grow rapidly during fetal development. The most important centers of growth are within the metaphyseal region (the zones between diaphysis and epiphyses), and the adjacent portion of the epiphysis. The zones within the cartilaginous epiphysis and metaphysis resemble a rather expanded epiphyseal growth plate (Box 2.2). Schultz (1926) illustrated that the basic locomotor differences in hominoids could be detected in newborns: newborn apes already possess a much higher intermembral index (forelimbs relatively longer than hindlimbs) than humans. Terrestrial or arboreal quadrupeds exhibit relatively longer hindlimbs; this can even be recognized at late fetal stages (Figure 7.1). However, fully "adult-like" limb proportions are not established at birth in humans. While the newborn human is clearly proportionally distinct from brachiating hominoids (Schultz, 1926), much of the disproportionate increase in growth of the lower limb is delayed until the postnatal period. In numerous somatic and neural dimensions, humans are especially slow to develop (Leigh and Park, 1998). The length of the long bones of the upper and lower limbs in newborns provides a measuring stick for the contribution of fetal growth toward establishing adult limb proportions (see Chapter 11).

7.2 The Pectoral Girdle and Forelimb Skeleton of Hominoids

7.2.1 The Pectoral Girdle

7.2.1.1 Scapula

In adults, the hominoid scapula is dorsally situated on a dorsoventrally compressed thoracic rib cage; in other primates the scapula is located on the sides of more cylindrical rib cages. Put another way, the flattened surfaces (supra- and infraspinous fossae) of the hominoid scapula are most closely oriented within a coronal plane, while that of non-hominoids is most closely oriented within a parasagittal plane. Compared to most other primates, hominoids have either craniocaudally wide scapulae (apes), or relatively equal proportions of glenoid–vertebral border length to craniocaudal length (humans). To some extent, these two scapular dimensions are less dichotomized in neonates, although subtle differences are described below. The adult clavicle is relatively long in hominoids compared to other primates.

In the scapula of newborn hominoids, most features are recognizable, including the spine, the infraspinous and supraspinous fossae, and the angles (Figure 7.2). The spine ends abruptly at its lateral limit in skeletonized material of newborns because the acromion process is cartilaginous in all newborn hominoids. Moreover, no trace of the coracoid process is visible on the body, except for a flat surface where a synchondrosis exists at birth. The presence of ossified secondary centers is rare in newborn humans. The body has a rather fan-shaped profile in newborn humans, with cartilage forming the acromion and coracoid processes and the glenoid region and also forming extensions all along the vertebral border and inferior angle (Schwartz, 2007). In *Pan*, the scapula possesses the same truncated form at birth, and subsequent growth greatly extends the caudal angle, yielding a scapula greatly elongated in craniocaudal relative to mediolateral dimensions (Figure 7.2). In humans, multiple nutrient foramina are visible, but Scheuer and Black (2004) assert that the primary artery during development penetrates the base of the spinous process, implying a primary vascular role for the suprascapular artery. Multiple foramina are similarly present in the late fetal/neonatal chimpanzees observed here. A prominent nutrient foramen is visible on the axillary margin, close to the lateral angle, and along the path of the scapular circumflex artery (Figure 7.2).

Pan and humans are later to ossify the coracoid center(s) than *Pongo* and *Hylobates*. In support of this, there is no detectable coracoid ossification in the KUPRI *Pan* neonate. In addition, Torre and Giacobini (1978) observed a coracoid ossification in a prematurely born *Pongo*. The situation in *Gorilla* is unknown, to our knowledge.

As in humans, the acromion process of apes remains mostly cartilaginous until adult age, with separate ossification centers still visible in some dental adults. The secondary centers along the vertebral margin and inferior angle likewise have prolonged presence, and the inferior angle is seen to remain cartilaginous in some adult specimens.

7.2.1.2 *Clavicle*

At birth, the sternal (proximal) end of the clavicle retains a cartilaginous epiphysis, which ossifies and fuses late relative to other epiphyses in humans, *Pan*, and *Pongo*. Neither medial nor lateral epiphyses, which are irregular bony caps in juveniles, are ossified at birth in any hominoids. The shaft has a pronounced curvature at birth. The curvature is also visible in radiographs of fetal stages, revealing precocial morphogenesis to a greater degree than other long bones. However, the sternal side lacks curvature in neonatal *Pan* (Figure 7.2), suggesting the degree of curvature becomes enhanced postnatally. In contrast, Scheuer and Black (2004, p. 252) say the clavicular diaphysis has "essentially adult" morphology at birth. Recent work by Barros (2013) provides some clarification, indicating that although the curvature of the sternal end of the clavicle is less plastic than that of the acromial (distal) end, the clavicular curvatures are not fully established until about the time the shaft stops growing. This suggests primary curvature may be established earlier in some primates than others, but a broad comparative study including non-hominoids is lacking.

At birth, the tubercles and ridges of the clavicle are not yet prominent in humans or *Pan*, and were not described in other apes. The groove for the subclavius muscle, visible in juveniles, is not apparent in newborn hominoids (Figure 7.2), giving the shaft a more cylindrical appearance overall compared to juveniles and adults. In a comparison of one prematurely born *Pongo* to one human neonate, Torre and Giacobini (1978) suggested the clavicle of *Pongo* was more heavily mineralized, based on radio-opacity.

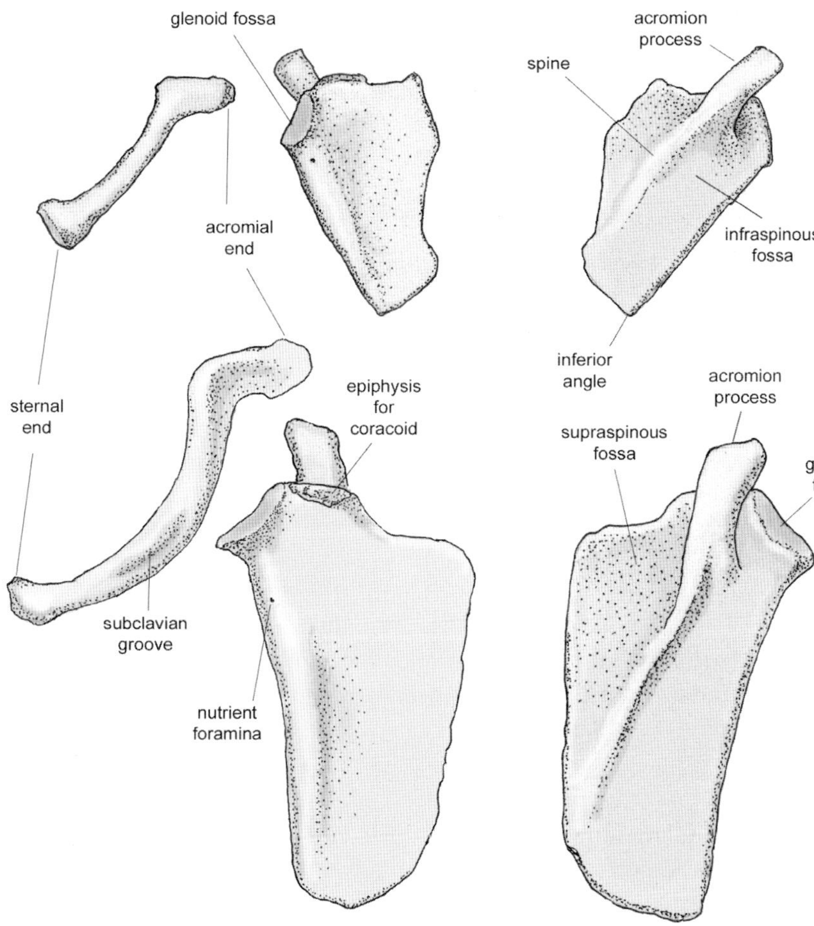

Figure 7.2 Pectoral girdle of perinatal and older subadult chimpanzee. Perinatal specimen, FMNH 122719; juvenile, FMNH 44390. Note an ossification nucleus for the coracoid is not yet present in the newborn. It is not illustrated for the juvenile, but likely had initiated ossification in this individual.

7.2.2 The Forelimb Skeleton

7.2.2.1 Humerus

The humerus of hominoids exhibits between one and two ossified elements at birth (Figure 7.3; Table 7.1). In humans, all secondary centers remain cartilaginous at birth, while in *Pan* (Figure 7.3) and *Gorilla* a secondary ossification center has commenced mineralization in the head of the humerus and perhaps distally in *Pan* (Table 7.1). More broadly, humans and *Pongo* are the least progressed (no appearance of secondary centers at birth) while *Hylobates* has progressed the farthest, with the center for the greater tubercle mineralized, in addition to that of the head. Schultz (1944) does describe an older infant *Hylobates* that is less progressed in ossification of secondary centers (while they are relatively dentally advanced as newborns), suggesting some variability.

The diaphysis of newborn hominoids exhibits an intertubercular groove proximally, bordered by lateral and medial lips (Figure 7.3). Otherwise, the surface is relatively smooth. In cross-section, the diaphysis is nearly round proximally, except ventrally where there is a slight groove for the long head of the biceps brachii. It is more narrow and rounded at the midshaft, while distally the diaphysis is "pinched" in the center, where the proximal part of the coronoid fossa is formed ventrally and the olecranon fossa dorsally.

7.2.2.2 Radius

In hominoids, the radius has a well-ossified diaphysis at birth. One secondary center is observed in the distal epiphysis in newborn *Pan*, *Gorilla*, and *Hylobates* (Figures 7.3 and 7.4). Schultz recorded no distal secondary ossification in *Pongo*, although Winkler et al. (1991) recorded it to be ossified in one newborn. The distal epiphysis ossifies postnatally in human newborns. The number of perinatal apes examined is still relatively few. In sum, differing sources agree that *Gorilla* and *Pan* are most likely to have initiated ossification of the distal radial epiphysis (Schultz, 1941; Winkler et al., 1991; Diogo et al., 2015). In contrast, most descriptions of *Pongo* denote its lack of ossification (Schultz, 1941; Winkler et al., 1991).

Scheuer and Black (2004) remarked on the lack of curvature of the radial diaphysis in newborn humans, aside from a slight lateral diversion of the proximal end. In contrast, the diaphysis has a pronounced, laterally convex, curvature in newborn greater apes (Figure 7.3, and see Torre and Giacobini, 1978; Diogo et al., 2015). Proximally, a rounded or oval plateau indicates the position of the radial tuberosity, the projection itself still cartilaginous.

7.2.2.3 Ulna

The perinatal ulna of hominoids is quite adult-like in form, with a well-defined trochlear notch (described as shallow at birth in humans) bracketed by the coronoid and olecranon processes. The trochlear notch bears a longitudinal central ridge, reciprocal in form with the trochlea of the humerus, which is as yet cartilaginous. The olecranon process is partially cartilaginous. A skeletonized perinatal ulna appears deficient on the proximal side of the olecranon, where two ossification centers form during late childhood in humans (8–10 years) and similarly remain unossified in early juvenile apes.

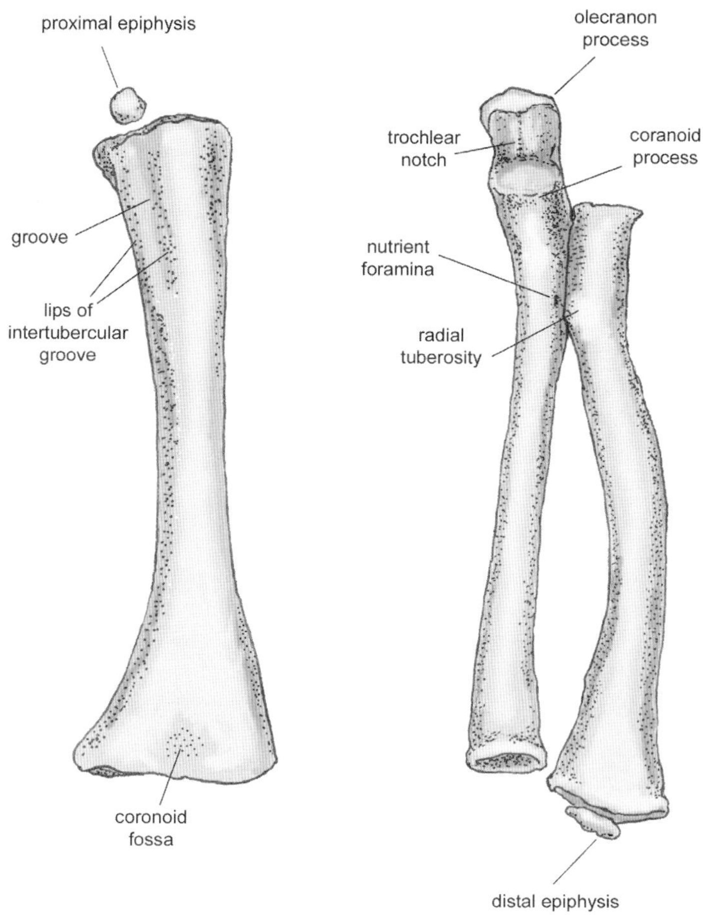

Figure 7.3 Humerus (left) and forearm bones of a perinatal chimpanzee, shown in ventral view.

7.2.2.4 *Wrist and Hand*

Between eight and nine wrist bones, or carpals, are present in the hominoid hand. Developmental distinctions separate humans and African apes from all Asian apes. In the latter, a ninth carpal bone, the os centrale, is typically present in subadults. In *Pan*, *Gorilla*, and humans, the os centrale is fused with the scaphoid. Kivell and Begun (2007) emphasize that this fusion is an early developmental event in these hominoids. The two bones *may* fuse in *Pongo* and hylobatids (Schultz, 1941; Kivell and Begun, 2007). However, the fusion occurs later in life, and Kivell and Begun (2007) suggest a distinct mechanism, such as fusion via ossification of syndesmotic connections, for *Pongo*.

None of the carpals are ossified at birth in humans. In the apes, between two and three distal carpals initiate ossification perinatally (Schultz, 1969; Kivell, 2016). Schultz's often reprinted figure of neonatal forelimbs includes five hominoids (1969, figure 60) and suggests hominoids are less precocious than at least some other catarrhines. For the neonate, the most frequent condition for apes should be considered uncertain as yet, due to the small number of known-age specimens examined, and the inherent challenges of ruling out premature specimens (see Chapter 1). However, the figure reveals a notable distinction between *Pongo* and other great apes, in that two carpals are present in most apes, the hamate and capitate. In *Pongo* a third ossified carpal is illustrated, the scaphoid; this is confirmed by Winkler et al. (1991), and reflects a basic difference in the sequence of ossification among hominoids (Kivell, 2016). The initiation with capitate–hamate appears to be a hominoid commonality, with humans and African apes being similar in that the triquetral is the first proximal carpal to ossify, while it appears to be among the last in Asian apes.

The diaphyses of all metacarpals and phalanges are well-ossified at birth. In great apes and humans, there are no ossified epiphyses at birth, while in *Hylobates* the distal epiphyses of the metacarpals are ossified (Table 7.1).

7.3 Pectoral Girdle and Forelimb Skeleton in Newborn Old and New World Monkeys

7.3.1 The Pectoral Girdle

In all non-hominoid anthropoid newborns, the scapula is positioned laterally, that is, on the sides of the thorax. As in adults, this correlates with dorsoventral elongation of the thorax (relative to hominoids). However, the thorax is proportionally small in the newborn, such that the scapulae may overlap the cervical region more than in adults. Thus, the "scapulothoracic joint" may move in a slightly different body plane in neonates than adults.

The acromioclavicular joint is not apparent in CT reconstructions because the acromion process remains cartilaginous at birth, as does the medial clavicular epiphysis. The clavicle varies in degree of curvature; the acromial end invariably tilts slightly in the dorsal direction toward the acromion.

7.3.1.1 Scapula

Generally, the body of the scapula is broadly triangular in newborn monkeys compared to adults; in adults the scapula is proportionally more elongated from glenoid to vertebral border (Figure 7.4). In part, this reflects unossified secondary centers along the vertebral margin. However, the degree to which the curvature along the axillary margin increases from birth to adult age suggests differential growth along the craniocaudal length, or compensatory remodeling after the secondary centers have ossified. In either case, the scapular body caudal to the spine is smoother in the newborn compared to adults. For example, flanges for attachment of the teres major muscle are not yet apparent (Figure 7.4).

Most newborn monkeys have a relatively small supraspinous fossa. The vertebral border of the scapula in newborn monkeys has a gently rounded appearance, even in species in which it is flattened in adults. The glenoid fossa and acromion processes are not yet completely ossified, as may be inferred from newborn to adult comparisons (Figure 7.4).

A coracoid ossification center is visible in most of the species that are newly described here, with the exception of one *Callicebus* specimen (records and somatic measurements indicate prematurity) and a single *Pithecia* specimen.

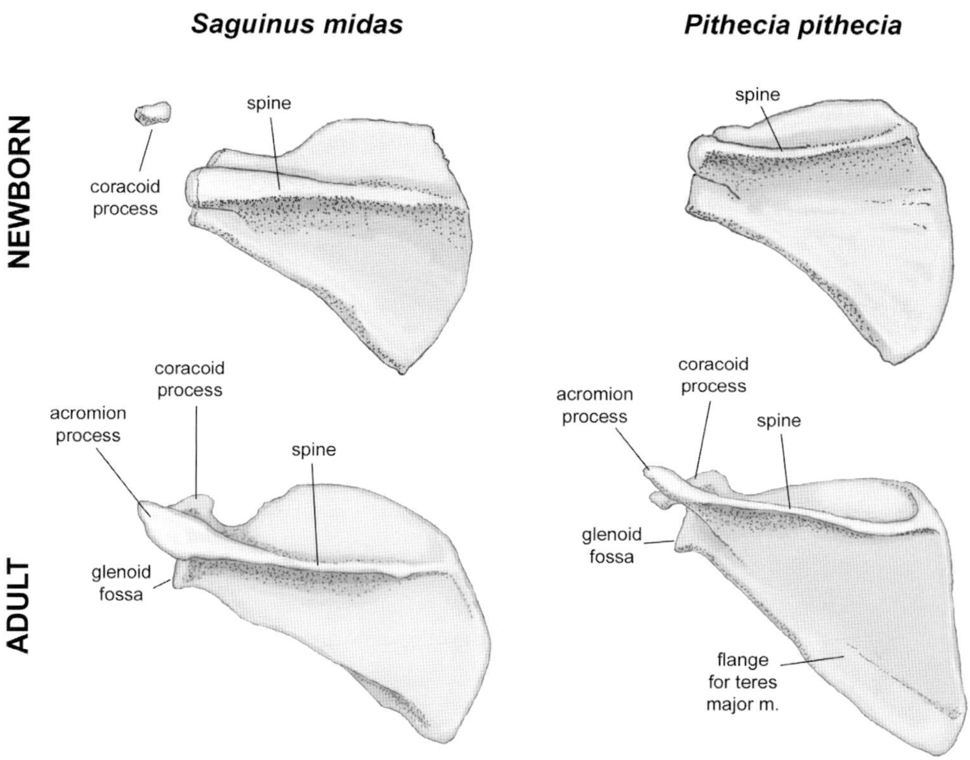

Figure 7.4 Scapula, lateral view, of two newborn anthropoids compared to those of adults, showing proportional differences.

In most newborns, we observe multiple nutrient foramina, as is the case in humans. Very prominent nutrient foramina are present along the axillary margin near the glenoid region in many specimens (e.g., in at least one specimen *of Aotus*, *Callicebus*, *Callithrix*, and *Cebuella*), as observed in *Pan* (see above). In *Leontopithecus* and *Pithecia*, in contrast, the nutrient foramen is large in the suprapinous fossa (and lacking along axillary margin in the former). Thus, it appears more than one vessel may provide the primary supply to initiate ossification of the scapula.

7.3.1.2 *Clavicle*

There is a double curve to the clavicle in monkeys, including all the species newly examined here. The curvature matches that seen in humans in that it is ventrally convex on the medial end and concave laterally. The curvature is somewhat more subtle than in humans, especially in some species (*Pithecia*). The sternal end is rounded (e.g., callitrichines) or may have more "squared" surfaces, and in some cases is also flattened (*Callicebus*, *Aotus*). The acromial end may be blunt or round-ended and flattened. The nutrient foramen is typically found on the caudal side; i.e., matching the inferior position seen in humans.

7.3.2 The Forelimb

7.3.2.1 Humerus

The humerus of Old World monkeys examined to date is far in advance of the ossification for hominoids and also for platyrrhines (Table 7.1; Figures 7.5 and 7.6). At least three secondary centers are ossified at birth in *Macaca* spp., *Papio* spp., *Cercopithecus nictitans*, and *Theropithecus gelada* (Glaser, 1970; Schultz, 1969). In addition to the diaphysis, the secondary center for the humeral head is ossified in all known Old World monkeys at birth. Schultz's observations indicate additional ossification centers may be present at birth, such as the greater tubercle and one or two distal centers (capitulum and/or trochlea). It seems likely that we do not yet know the degree of variability, as the more proximal skeleton has received little study. However, they appear to far outpace ossification in all other known anthropoids. Few newborn New World monkeys have initiated ossification of secondary centers, but one *Aotus nancymaae* examined here exhibits a very minute ossification center for the proximal humeral head, and *Cebuella* has an ossified distal epiphysis.

Cross-sectional contours of the humeral diaphyses of newborn monkeys are very similar to those described above for hominoids. The proximal end of the diaphysis slopes backward at the dorsal margin. This reflects the posterior position of the humeral head, which helps to distinguish the humerus in anthropoid monkeys from

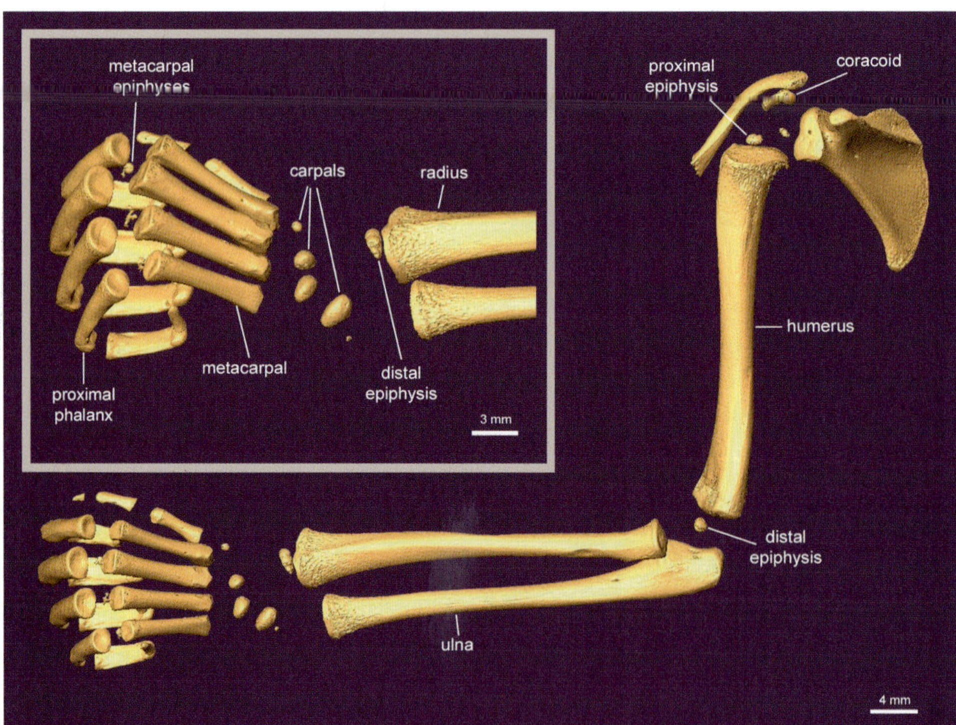

Figure 7.5 Forelimb and shoulder girdle of newborn *Trachypithecus francoisi.*

hominoids. In newborn platyrrhines the humeral diaphysis has a prominent ridge on the proximal ventral side, the future lateral lip of the intertubercular groove.

In degree of curvature, the humeral diaphysis of the newborn monkey varies. It is nearly straight in *Aotus* and *Callicebus*, but curves slightly dorsally in many species (e.g., *Cebuella* and *Leontopithecus*).

The position of the nutrient foramina is highly variable among species, and we lack the samples to establish the extent of intraspecific variation. A common location among the monkeys examined here is dorsal (either at midshaft or more proximally). Within the same individual, foramina may be located both medially at the midshaft, and dorsolaterally at the proximal one-third (as is the case in two *Callicebus*). In *Leontopithecus*, it is observed dorsally (midshaft) and medially (distal one-third).

The entepicondylar foramen, a greatly variable trait among platyrrhines (Garbino and de Aquino, 2016), is already evident as a notch or foramen at birth in *Saguinus*, *Aotus*, and *Pithecia*.

7.3.2.2 *Radius*

The radius is more ossified in newborn Old World monkeys than in New World monkeys; at least the distal epiphysis has commenced ossification (Table 7.1). While Glaser (1970) detected no epiphyses in the radii of platyrrhines at birth, ossified distal foci occur in some newborn *Aotus*, *Callithrix*, and *Cebus* (Table 7.1). Findings tabulated by Hofmann et al. (2007) indicate at least some newborn *Callimico* possess one ossified epiphysis in the forearm or arm, but which is unstated. In our collection, we observed ossification of the distal epiphysis in newborn *Cebuella*.

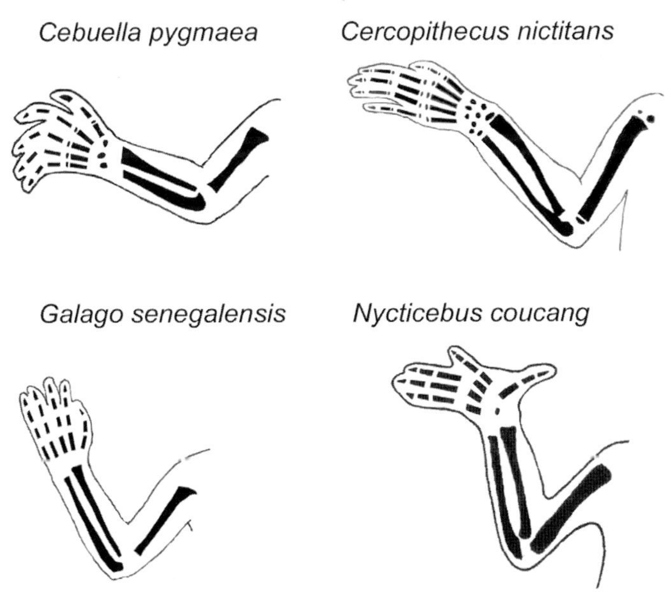

Figure 7.6 Schematic view of the newborn forelimb, redrawn after Glaser (1970), showing the primary ossification centers of long bones, carpals (in some), and the secondary centers of ossification of long bones (in *Cercopithecus* only).

The radius is minimally curved in most of the monkeys newly examined here. In all cases there is at least a lateralward flare of the proximal radius. The radius of one *Leontopithecus* specimen had a more pronounced primary curvature, laterally. The brachial tuberosity is usually visible at least as a slightly raised flattened surface. The ends of the diaphysis may exhibit a raised flange that nearly or completely encircles the bone (Figure 7.7, e.g., *Pithecia*).

The nutrient foramina of the radius vary little, occurring medially within the proximal one-third of the bone.

7.3.2.3 Ulna

Ossification is most advanced in newborn cercopithecoids. Glaser notes that the distal epiphysis is ossified in *Macaca* and *Cercopithecus* specimens (but also states that in one *Cercopithecus* specimen, both proximal and distal epiphyses are still devoid of bone as yet). No platyrrhines have commenced ossification at this site.

The curvature of the ulna is a dorsal convexity; typically subtle in newborns. No interosseous crest is present; the supinator crest is typically subtle, if detected. The trochlear fossa is shallow due to the largely cartilaginous olecranon and coronoid. The brachial tuberosity is not yet prominent. The radial notch is apparent laterally on the proximal end, and distally there may be a ring-like flange encircling the shaft.

A nutrient foramen may be visible laterally or ventrally. In either case it occurs within the proximal third.

7.3.2.4 Wrist and Hand

All monkeys possess nine carpals, and their state of ossification is quite varied at birth. Two studies provide a very meaningful perspective on the state of ossification of the wrist bones. Newell-Morris et al. (1980) had the benefit of a very large sample size of macaques (127 fetuses and infants). They concluded that, based on observed variation in the cross-sectional age sample, "the presence of a particular center, or even of several centers, cannot be assigned to a specific gestational age, and the timing of ossification can be described only in very general terms" (p. 428). Intraspecific variation may be significant, particularly for the carpal bones. Indeed, a wide range has been reported for some species (Table 7.1). A second issue may influence the wide range for some species – the state of ossification varies at birth simply because ossification has just commenced in some species. As discussed earlier (Chapter 1), histological methods identify bone as a tissue type better than CT or other radiographic methods. Regarding carpals, Phillips (1976) detected more carpals using Alizarin red staining than using radiographs.

With the caveats of intraspecific variation and degree of ossification in mind, the previous findings on anthropoids may be summarized as follows. All cercopithecoids studied to date have more ossified carpal bones at birth compared to any hominoids and most platyrrhines; they also exhibit a narrower number range of ossified carpals at birth. This indicates the carpals of cercopithecoids ossify more quickly, and quite possible earlier (fetally) than other anthropoids, yielding more consistent data for newborns. Existing data on platyrrhines suggest great variation with no indication that body size or phylogeny has a significant influence (Figure 7.4). We suggest that of all anthropoids, the platyrrhine carpus may be ossifying most specifically within the "newborn" window (P0 to P7) as conceived in this book. As such, both intraspecific variation and degree of ossification may contribute to the relatively high variability of reports on the platyrrhine carpus.

Ossification of the hand skeleton follows a similar trend, with cercopithecoids far exceeding the ossification pace of most other anthropoids. *Hylobates* and some callitrichines, with ossified metacarpal epiphyses at birth, may most closely approach this. No other newborn anthropoids are reported to have ossified epiphyses in the metacarpal or phalanges.

The wrist and hand skeleton is one of the better-studied postcranial regions in *Macaca nemistrina*. An interesting ontogenetic trend is noted regarding the sexes: females mature more quickly prenatally (Newell-Morris and Tarrant, 1978), and in a small subsample of longitudinally studied macaques, males had more unossified wrist and hand bones at birth, and delayed ossification postnatally of certain hand/wrist bones (Newell-Morris et al., 1980). Other species have not been sufficiently studied to assess for sex differences.

7.4 The Pectoral Girdle and Forelimb Skeleton in Tarsiers at Birth

7.4.1 Pectoral Girdle

Niemitz and Sprankel (1974) previously described the relatively advanced state of ossification of the postcranial skeleton in one newborn and one 10-day-old *Tarsius bancanus*.

7.4.1.1 Scapula

The profile is acutely triangular with a straight vertebral border. The spine is thin and leans slightly cranially. The coracoid ossification has commenced in all newborn *T. syrichta*. Niemitz and Sprankel did not specifically address its presence, but in one newborn *T. bancanus* newly examined here the center is likewise present (Table 7.1).

7.4.1.2 Clavicle

The shaft has a primary curvature at midshaft, which is ventrally convex. The sternal end is round; the flat acromial leans caudally toward the acromion.

7.4.2 The Forelimb

7.4.2.1 Humerus

No secondary ossification centers have commenced ossification in the newborn tarsier, based on individuals ranging from birth to 10 days (Table 7.1).

The diaphysis has a slight curvature, a dorsal convexity. Proximally the metaphyseal region notably lacks the dorsally projecting "shelf" that supports the humeral head in monkeys. Distally there is an entepicondylar foramen, as evidenced by a deep notch, which is distally enclosed by the cartilaginous epiphysis.

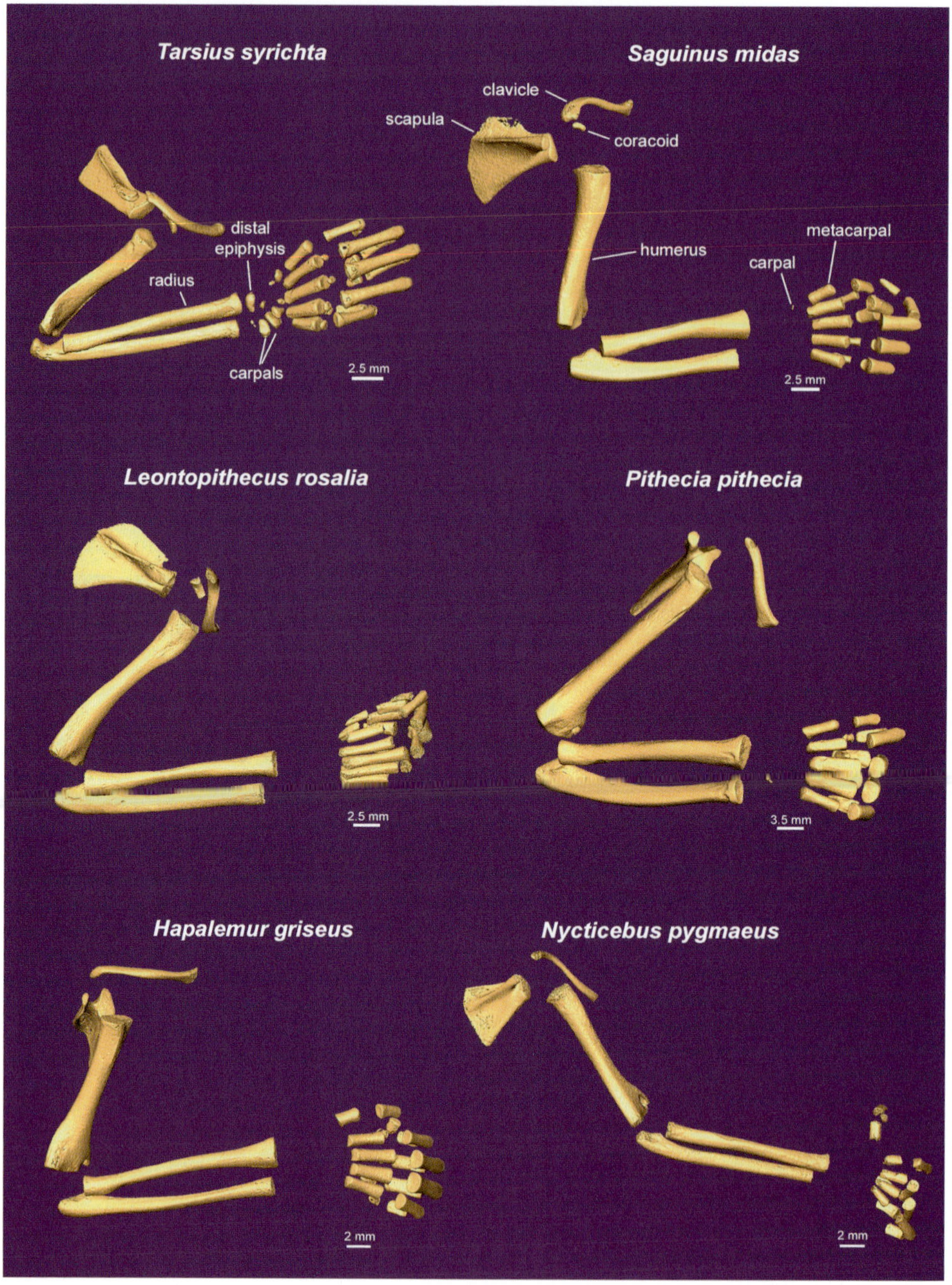

Figure 7.7 Comparison of the pectoral girdle and forelimb skeleton in a 6-day-old tarsier compared to selected anthropoids (upper right, middle row) and strepsirrhines (bottom row).

7.4.2.2 Radius

No secondary ossification centers have commenced ossification proximally, but distally the secondary ossification has commenced at birth, at least unilaterally (Table 7.1). In 6- and 10-day-old individuals, both distal epiphyses are osseous.

The diaphysis is relatively uncurved at birth and is relatively smooth in most individuals, while the distal end has a slightly raised lip encircling it.

7.4.2.3 Ulna

No secondary ossification centers have commenced ossification proximally or distally.

The diaphysis is relatively uncurved at birth. The trochlear notch is relatively deep compared to newborn monkeys, with correspondingly better-developed olecranon process.

7.4.2.4 Wrist and Hand

All individuals examined here and by Niemitz and Sprankel suggest the ossification of the carpus rivals cercopithecoids. During the first postnatal week, up to 6–7 carpals are radiographically detected. Tarsiers are also more advanced in hand ossification than most other haplorhines (excepting cercopithecoids) in that the distal metacarpal heads (II–V) ossify within the first week.

7.5 Comparative Anatomy of the Postcranial Axial Skeleton in Strepsirrhines at Birth

7.5.1 Pectoral Girdle

7.5.1.1 Scapula

The scapula in strepsirrhines is acutely triangular in shape, being narrow craniocaudally (Figures 7.7 and 7.8). The vertebral margin is greatly simplified in form compared to the adult, bearing a flattened rather than rounded contour, and lacking the caudal flare (where the teres major originates) that is seen in adults (Figure 7.8). The vertebral margin as well as the glenoid region is flattened in many specimens; histology demonstrates these flat edges are sites of articulation with secondary centers (see below). The spinous process likewise ends abruptly as the acromion remains cartilaginous in newborns. The primary center of ossification for the coracoid process is mineralized at birth in at least one specimen of all newly described species, and is connected to the body of the scapula by a synchondrosis.

The scapular spine is rather straight, even in species which go on to develop some curvature. However, it "scrolls" caudally, such that it is concave facing the infraspinous fossa; this is especially pronounced in newborn bushbabies. However, as it approaches the acromion the spine begins to bend in the opposite direction.

The osseous proportions of the infraspinous and supraspinous fossae change markedly from birth (Figure 7.8). Some notable differences from adults are largely due to large expanses of the scapular blade, glenoid region, and vertebral margin that remain cartilaginous at birth (Box 7.2). Thus, skeletonized specimens are substantially reduced in glenovertebral length compared to their actual dimension. Scapular muscles of the rotator cuff are attaching partially to the cartilaginous

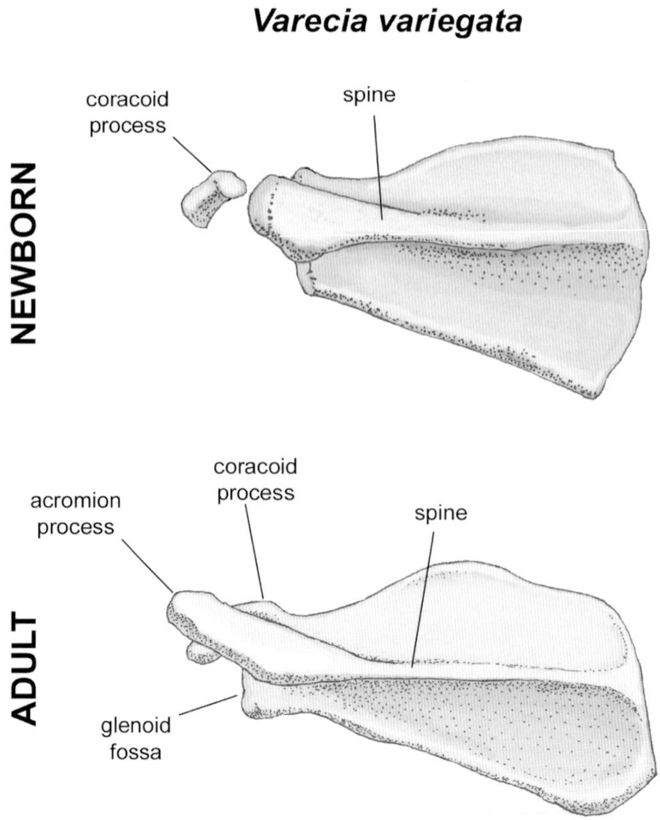

Figure 7.8 Scapula of newborn *Varecia variegata* compared to the adult, showing proportional differences.

parts of the blade, especially near the vertebral border (Box 7.2). A cartilaginous growth center is intrinsic to each of the cartilaginous regions, as well as in the apex of the scapular spine. At each pole, these growth zones increase glenovertebral length, while the spine grows to propel the acromion outward (Box 7.2). The expanse of cartilage along the vertebral margin should not be assumed to grow synchronously along its craniocaudal extent. Differential growth along the axis may produce expanded caudal edges (Figure 7.8); of interest is how much of these expansions occurs via primary growth as opposed to postnatal biomechanical demands.

Multiple nutrient foramina are typically visible; they may be near the base of the spinous process, on the axillary margin, and within the fossa.

7.5.1.2 Clavicle

The sternal head is rounded, while laterally the acromial end is flatter. Curvature varies significantly at birth. The acromial end is curved in most strepsirrhines, such that at least the lateral 20% of the length of the ossified clavicle tilts toward the cartilaginous acromion process of the scapula (Figure 7.7). This bend is at least slightly

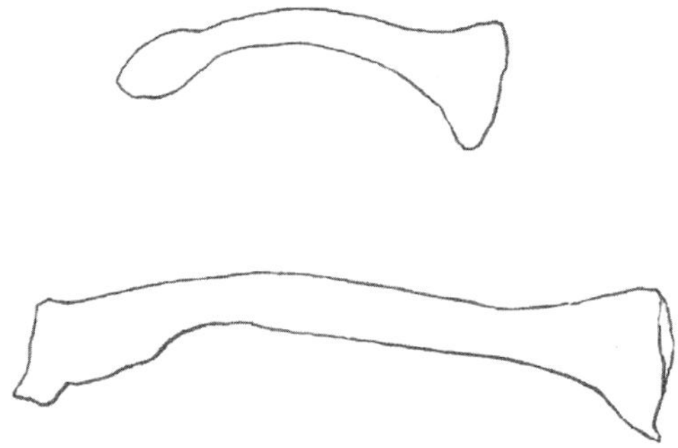

Figure 7.9 Cranial view of clavicle of *Varecia variegata* in newborn (top) and adult (bottom), showing proportional differences. Left side = acromial end; top = ventral.

in the caudal direction in all species, and most often it also bends dorsally. However, in *Nycticebus* (Figure 7.7) and a late fetal *Loris*, the acromial end curves ventrally, as it does in hominoids. However, there is no curvature at the sternal end in newborn lorises. In many newborn strepsirrhine species, there is a subtle curve throughout the length of the bone, with a ventral convexity (e.g., *Cheirogaleus*, *Hapalemur*, *Lemur*, *Varecia*, *Galago*).

A comparison of neonatal to adult clavicles suggests that curvature may diminish with age in larger strepsirrhines (Figure 7.9), except for the caudal and dorsalward bending of the acromial end (this may be the "primary" curvature of the newborn clavicle in strepsirrhines). Thus, the main thrust of diaphyseal growth may be more linear in these species. However, a thorough ontogenetic study of the clavicle remains needed in strepsirrhines.

Although the clavicle of many mammals (when present) develops partially from a medial endochondral center while the remainder is intramembranous, in some mammals there is cartilage at both lateral and medial sides (Großmann et al., 2002). This appears to be the case in at least some bushbabies (Box 7.2), unlike in humans. Further work on ontogenetic diversity of primate clavicles is needed.

7.5.2 The Forelimb

7.5.2.1 *Humerus*

The strepsirrhine humerus is represented only by the diaphysis at birth, at least in all of the species examined here as well as some previous observations (Table 7.1). In shape, the diaphysis is generally similar to that of other primates at birth, although proportions vary (see Chapter 11). The shape of the newborn diaphysis appears more "robustly" proportioned (i.e., relatively more broad across the diaphysis) than that of adults, especially in larger-bodied species (Figure 7.10).

Distally, the humeral shaft flares widely, with a smooth ventral surface. In lemuroids, the medial side exhibits a deep notch at the distal margin, reflective of the incomplete osseous boundary of the entepicondylar foramen. In some newborns the foramen may be almost enclosed distally by bony bridging (e.g., *Propithecus*), and in a 7-day-old *Varecia* it is completely enclosed. Thus, closure may be occurring perinatally in lemuroids. On the lateral side, there may be a slight supracondylar ridge present, but in no newborn does it flare broadly into a brachioradialis ridge. Newborn lorisoids have osseous bridging that completes the osseous perimeter of the entepicondylar foramen.

Location of the nutrient foramen appears less variable in strepsirrhines than other primates in that it is frequently medial, although the proximodistal locality differs broadly among specimens. In our largest samples of newborns (*Propithecus* and *Lemur*), it is most frequently at the distal two-thirds of proximodistal length.

7.5.2.2 *Radius*

No secondary ossification centers have commenced ossification (Table 7.1).

The diaphysis is relatively uncurved at birth and is relatively smooth in most individuals. While the brachial tuberosity may be evident, it is barely raised.

7.5.2.3 *Ulna*

No secondary ossification centers have commenced ossification proximally or distally.

The diaphysis has a ventrally concave curvature at birth (Figure 7.7). The trochlear notch is relatively shallow.

7.5.2.4 *Wrist and Hand*

Ossification of the carpus in strepsirrhines is less advanced than in any other primates. Available data suggest ossification may be slightly more precocious in lorisoids than lemuroids (Table 7.1; Figures 7.6, 7.7, and 7.11). As in all monkeys, most strepsirrhines

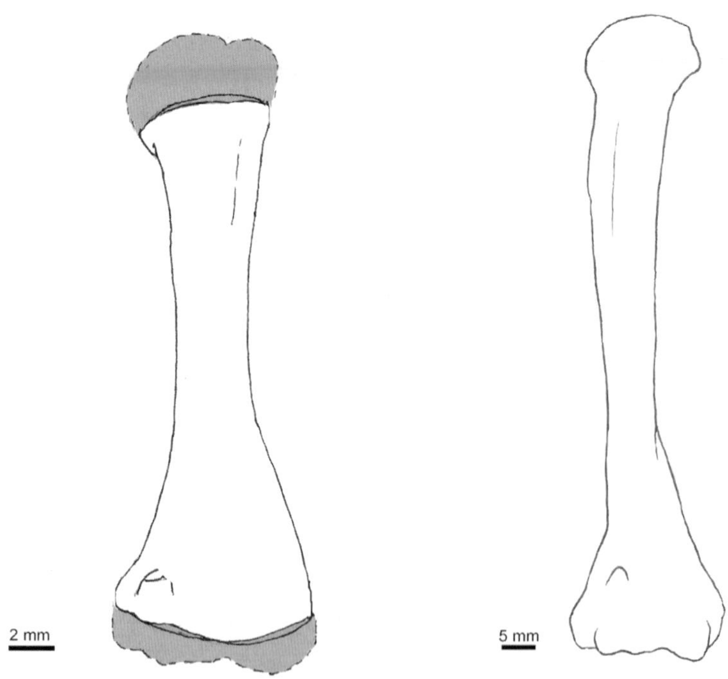

Figure 7.10 Humerus of *Varecia variegata* in newborn (left) and adult (right), showing proportional differences. Cartilaginous epiphyses are indicated in gray.

Table 7.1 Ossification of the coracoid and secondary centers of ossification of the forelimb in newborn non-human primates

Species	Coracoid	Prox. Hum.	Dist. Hum.	Prox. Rad.	Prox. Ulna	Dist. Rad.	Dist. Ulna	Dist. MC	Carpal notes	Source
HAPLORHINES										
Aotus nancymae	X	(X)	U	U	U	X	U	II–IV	1–2, LUN (CAP)	This book
Callicebus cupreus	X	U	U	U	U	U	U	U	1, LUN	This book
Callimico goeldii	?	?	U	U	U	?	?	?	5–8	Hofmann et al., 2007
Callithrix jacchus	?	?	U	U	U	(X)	U	(X)	6–8, HAM, CAP, TRZM, OC (TRI, LUN, SCAPH, TRZD)	Phillips, 1976; Hofmann et al., 2007
Callithrix penicilatta	X	U	U	U	U	X	U	I–IV	7	This book
Cebuella pygmaea	X	U	U	U	X	U	U	I–V	0–5	Glaser, 1970; this book
Cebus albifrons	?	?	?	?	?	(X)	?	?	1–6: TRI (PISI, HAM, CAP, OC, TRZM)	Watts, 1990
Cebus apella	?	U	U	U	U	(X)	U	U	1–3: TRI, SCAPH (HAM)	Glaser, 1970; Watts, 1990
Cebus capucinus	?	?	U	U	U	U	?	?	3	Glaser, 1970
Cercopithecus nictitans	?	X	X	X	U	X	X	X	7	Glaser, 1970
Gorilla gorilla	?	X	U	U	U	X	U	U	0–1	Schultz, 1969; Diogo et al., 2015
Hylobates lar	X	X	U	U	U	X	X	II–V	2: CAP, HAM	Schultz, 1944
Lagothrix jacchus	?	U	U	U	U	U	U	U	2	Glaser 1970

Table 7.1 *(cont.)*

Species	Coracoid	Prox. Hum.	Dist. Hum.	Prox. Rad.	Prox. Ulna	Dist. Rad.	Dist. Ulna	Dist. MC	Carpal notes	Source
Leontopithecus rosalia	X	U	U	U	U	U	U	U	0–2	Glaser 1970; this book
Macaca mulatta	?	X	X	U	U	X	(X)	II–V	5–8: SCAP, LUN, TRI, CAP, HAM (TRZM, TRZD, OC)	Watts, 1990; Van Wagenen and Asling, 1964
Macaca nemistrina	?	?	?	?	?	X	X	MC II–V	8 (OC ossifies postnatally)	Newell-Morris et al., 1980
Pan troglodytes	U	X	(X)	U	U	X	(X)	U (m); X (II, IV: f)	2: CAP, HAM	Watts, 1990; Nissen and Riesen 1949; Schultz, 1969
Pithecia pithecia	U	U	U	U	U	U	U	U	1: LUN	This book
Pongo pygmaeus	X	U	U	U	U	U	U	U		Schultz, 1941
Saguinis midas	X	U	U	U	U	U	U	U	0–1: TRZM	This book
Saguinus oedipus	X	U	U	U	U	U	U	U	2–3	Hofmann et al., 2007; this book
Saimiri boliviensis	?	(X)	U	U	U	X, (X)	X, (X)	X (II–IV)	TRI, PISI, HAM, CAP, OC, TRZM	Watts, 1990; Galliari, 1988 (*, this author indicates these bones ossify postnatally)
Trachypithecus francoisi	X	X	X	U	U	X	X	X	5	This book
Tarsius spp.	X	U	U	U	U	(X)	U	U	6–7	Niemitz and Sprankel, 1974; this book

Tarsius syrichta	X	U	U	U	U	X	U?	U/ P6: X (II–V)	TRZM, SCA, CAP, HAM, LUN?, TRI, TRZD?, PISI?)	This book
STREPSIRRHINES										
Cheirogaleus medius	(X)	U	U	U	U	U	U	U	0	This book
Eulemur fulvus	?	U	U	U	U	U	U	U	0	Glaser 1970
Eulemur mongoz	X	U	U	U	U	U	U	U	0	This book
Galago spp.	X	U	U	U	U	U	U	U	0	Glaser, 1970; this book
Galagoides demidovii	?	U	U	U	U	U	U	U	0	This book
Hapalemur griseus	X	U	U	U	U	U	U	U	0	This book
Lemur catta	X	U	U	U	U	U	U	U	0	Glaser, 1970
Nycticebus coucang	?	U	U	U	U	U	U	U	1	Glaser, 1970
Nycticebus pygmaeus	X	U	U	U	U	U	U	U	0	This book
Otolemur crassicaudatus	X	U	U	U					0–2	Newman and Hendrickx, 1984; this book
Propithecus coquereli	X	U	U	U	U	U	U	U	0	This book
Varecia spp.	X	U	U	U	U	U	U	U	0	This book
SCANDENTIA										
Tupaia belangeri	?	U	U	U	U	U	U	U	0	Glaser, 1970

Abbreviations: U, unossified; X, ossified; CAP, capitate; HAM, hamate; LUN, lunate; OC, os centrale (if present); PISI, pisiform; SCA, scaphoid; TRI, triquetral; TRZD, trapezoid; TRZM, trapezium. Parentheses indicate variable state of ossification at birth.

possess a ninth carpal, the os centrale. Unlike monkeys, some strepsirrhines have been described to have a fused scaphoid–os centrale as adults (see Kivell and Begun, 2007, for a review). Two of the genera in which fusion has been described, *Hapalemur* and *Propithecus*, are shown as histologically sectioned newborns in Box 7.3. In these specimens as well as a galagid, the os centrale can be seen as a distinct element; the interface of the os centrale to the scaphoid appears to include a joint space in all three species. These observations suggest that if fusion occurs, it is not a prenatal coalescence of two bones, as in humans. The joint has a notably fibrous union with the scaphoid in these three newborn primates (Box 7.3). Ossification of these ligaments may present an avenue for occasional postnatal coalescence of these bones in some strepsirrhines; this might explain some discrepant findings in adults (see Kivell and Begun, 2007).

7.6 Patterns of Ossification of the Postcranial Axial Skeleton in the Early Postnatal Period

7.6.1 Secondary Centers of Ossification

7.6.1.1 Pectoral Girdle

In apes, ossification of the acromion or coracoid processes and the fusion with the body of the scapula is quite prolonged postnatally. According to Schultz's descriptions, the

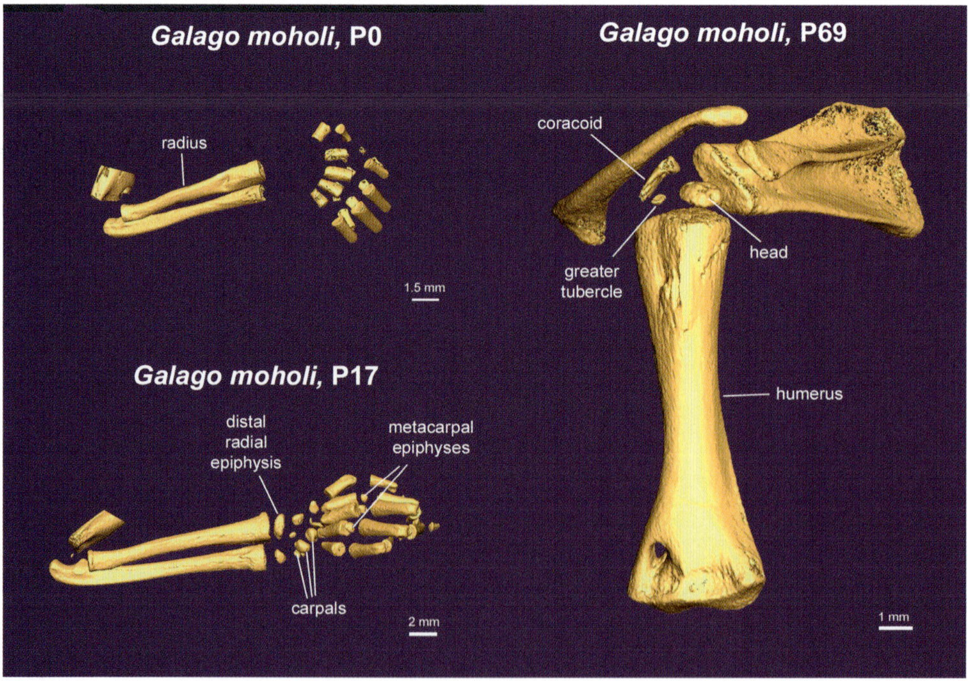

Figure 7.11 Age changes in ossification of the shoulder of *G. moholi*. Left, the wrist in a newborn and an older infant. Right, the shoulder of a juvenile.

coracoid may even remain unfused to the body in some apes with adult dentition. Similar information on strepsirrhines is currently lacking, but the coracoid remains unfused and the acromion mostly unossified in a ~ 3-month-old *Galago moholi* (Figure 7.11).

7.6.1.2 Forelimb Elements

Excepting humans, ossification status of secondary centers of ossification (i.e., epiphyses) in the forelimb is likely more advanced *prenatally* in anthropoids compared to

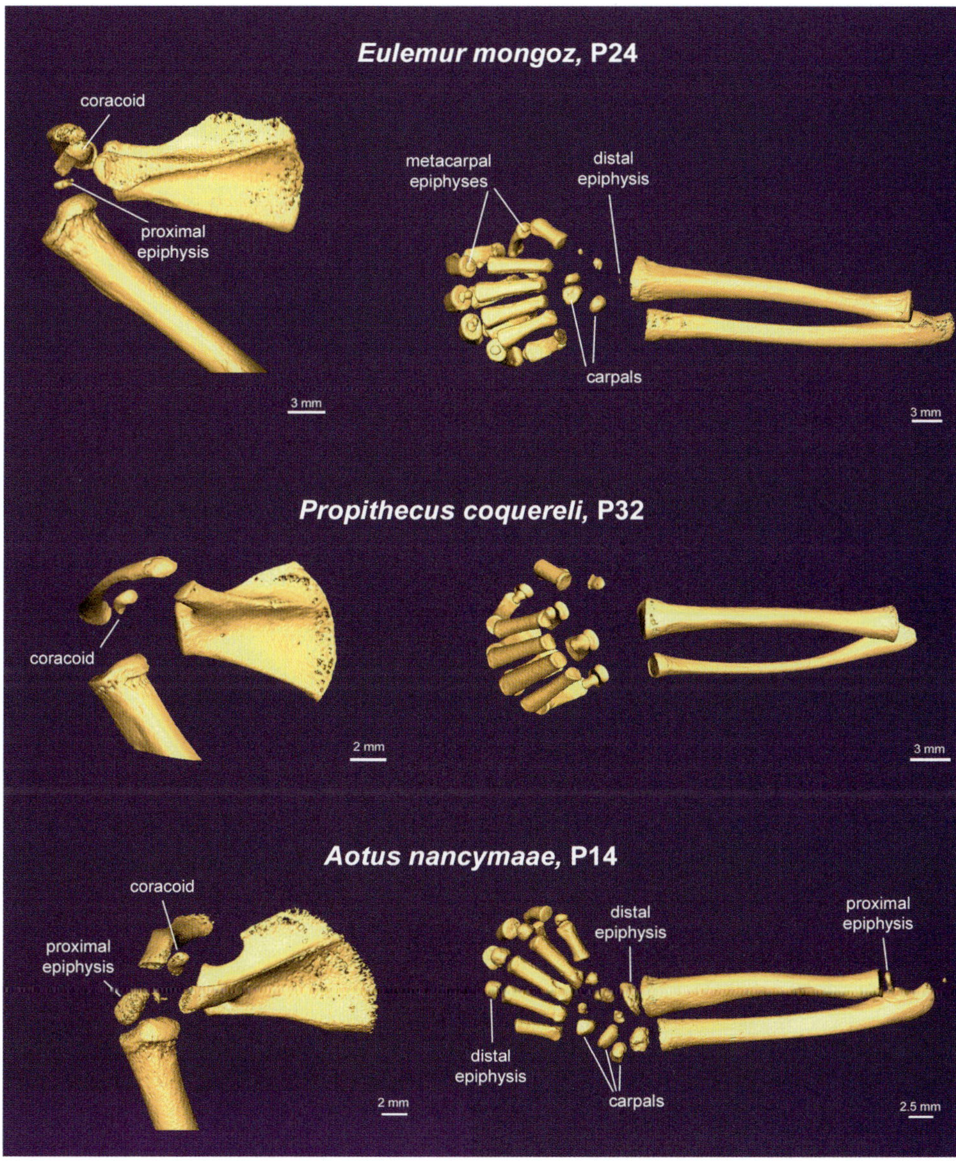

Figure 7.12 Older infant forelimb skeleton in infant strepsirrhines (top, middle) and anthropoid (bottom). Left, lateral view of shoulder; right, dorsal view of wrist.

other primates, based on the state of newborns (Table 7.1). Postnatally, pace of ossification in secondary centers likely diverges among strepsirrhines. This is most notable in Galagids, which rapidly "catch up" to the degree of ossification in the proximal humerus, wrist, and hand seen in newborn cercopithecoids (Figure 7.11).

7.6.2 Early Postnatal Transformations

Growth of the forelimb girdle is likely prolonged in many primates. Union of the coracoid with the body of the scapula is delayed until after 1 month in all species known to us, including *Galago*, *Lemur*, *Otolemur*, *Eulemur* spp., *Propithecus*, and *Saguinus*. Similarly, the wide separation at the acromioclavicular (Figures 7.11 and 7.12) and sternoclavicular joints at these ages indicates the clavicle continues to lengthen.

Our sample of older infants suggests stark differences in postnatal pace of ossification of limb epiphyses. Lemurids develop secondary centers of ossification at the shoulder, wrist, and hand within the first postnatal month, while similarly aged *Propithecus* are less developed (Figure 7.12). No large-bodied lemuroids rival galagids in how rapidly carpals and secondary ossification centers for long bones appear. Relatively large samples of *Callithrix jacchus* examined by Phillips (1976) confirm that the species' well-ossified wrist corresponds to a rapid postnatal ossification of the entire upper limb. By P13 the distal epiphyses of the proximal phalanges and distal humeral epiphysis are ossified, and by P32 all limb bone secondary centers except the proximal radial and ulnar epiphyses are ossified. The latter are ossified by P41. The state of forelimb ossification in a single available P30 tamarin (*Saguinus oedipus*) indicates a less-advanced state compared to *Callithrix* (as suggested by Hofmann et al., 2007), continuing the same trajectory from birth (Table 7.1). *Aotus nancymaae* may develop similarly to *Callithrix* in postnatal pace of forelimb ossification. In a single P14 specimen, there are already three secondary centers ossified for the humerus (proximal and distal epiphyses, plus the greater tubercle), both proximal and distal epiphyses for the radius and ulna, seven carpals and the epiphyses for metacarpals I–V. Although they are described being altricial in some respects (poorly furred and highly dependent on parents for warmth – Wolovich et al., 2008), in early skeletal development *Aotus* compares well with other platyrrhines and may even have relatively rapid limb ossification.

Box 7.1 Early Limb Development in Primates

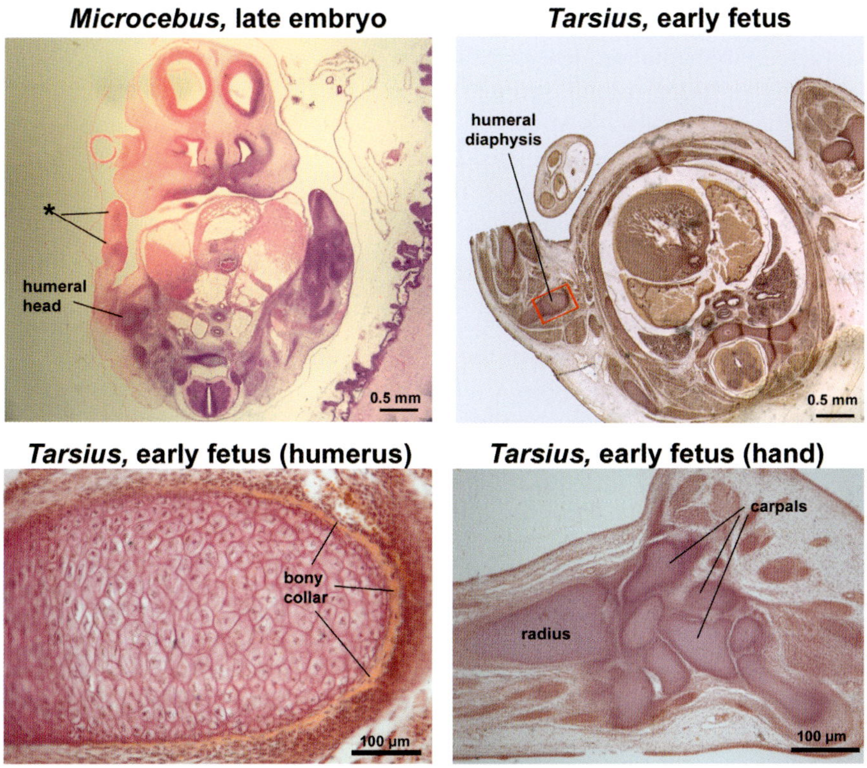

In early development (embryonic to earliest fetal stages), there is both a "cephalocaudal" sequence of limb morphogenesis (i.e., forelimbs ahead of hindlimbs – see Figure 7.1), and a proximodistal sequence. Above, transverse sections through non-human primates are shown, including an embryonic mouse lemur (Bluntschli collection) and an early fetal tarsier (Hubrecht collection). Top left: in late embryos the proximal limb has recently chondrified (e.g., humerus) before the distal limb (*, mesenchymal condensations in hand plate). Top right: in early fetal development all limb bones have chondrified (forelimb is sectioned), and are discretely identifiable within surrounding soft tissues. Bottom left: in the same specimen, the primary ossification of the humerus has just initiated. Note the bony collar surrounding the diaphysis, and the enlarged chondrocytes. Bottom right: wrist region of the same specimen. Note all bones are fully chondrified, but the radius as yet lacks a bony collar.

Box 7.2 Histology of the Pectoral Limb Girdle at Birth

Pectoral girdle microanatomy in a dwarf galago (a–e: *Galagoides demidoff*) and a cotton-top tamarin (f–h: *Saguinus oedipus*). (a–c) Cross-sectional levels, in a lateral to medial sequence, are shown through the shoulder.
(a) Acromioclavicular joint, spine of scapula; two sequential levels showing the ossified spine of the scapula (b), and endochondral ossification occurring toward the vertebral end of the blade in the supraspinous region (c). The acromion and the root of the spine remain cartilaginous. Note that the acromial end of the clavicle is cartilaginous in newborn *Galagoides* (d), unlike humans or *Saguinus* (the latter sectioned but not shown). In plate (e), the cartilaginous portion of the infraspinous fossa is shown. The green arrows indicate the directional growth imparted by proliferation of chondrocytes. The radial pattern carries the base of the spine and the acromion lateralward (d). The infraspinous fossa also has a subtly radial pattern – i.e., the blade is widening slightly along the craniocaudal axis of the vertebral border.

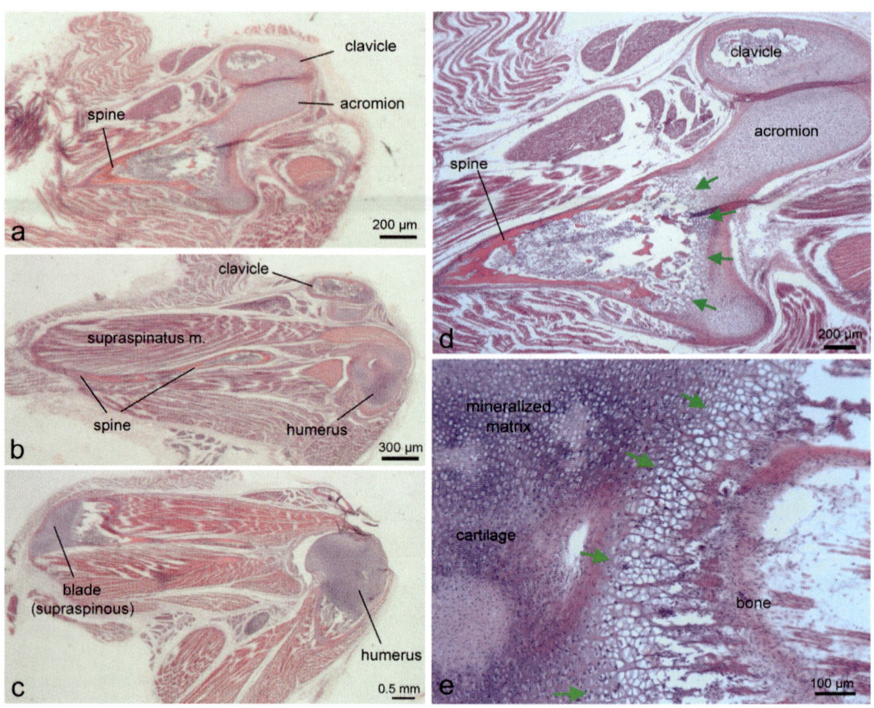

Below (f) is a histological section through the infraspinous fossa and the root of the scapular spine in *Saguinus oedipus.* The fossa is cartilaginous on the vertebral end. Thus, skeletonized scapulae appear substantially more abbreviated along the glenovertebral axis than they actually are, and muscular origins are shared by bony and cartilaginous surfaces. The glenohumeral synovial joint structure is fully formed, including the fibrocartilaginous labrum. Red arrows indicate cartilage growth vectors; there is overall dorsoventral lengthening of the scapula in progress, with active growth at the glenoid (g) and vertebral (h) borders.

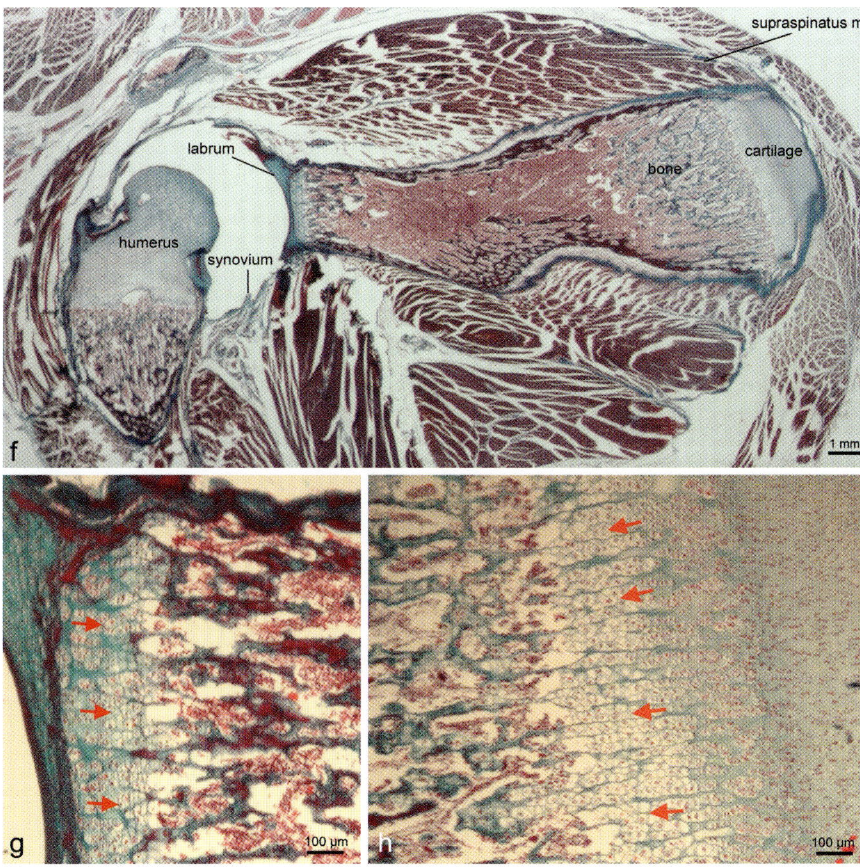

Box 7.3 Microanatomy of the Wrist

Below, histological sections through the wrist and hand of *Hapalemur griseus* (a–c), *Propithecus coquereli* (d), and *Galagoides demidoff* (e) are shown. Most strepsirrhines lack any ossification of carpals at birth, as is the case for all species below. Most primates possess nine carpals. The presence of os centrale is variable (see text). In all three strepsirrhines below, this bone is a distinct element at birth; in *Hapalemur* and *Propithecus* there is a joint cavity formed between it and the scaphoid (b). Fibrous connections to adjacent carpals are impressive, including some fibrocartilaginous ligaments in *Hapalemur* (c).

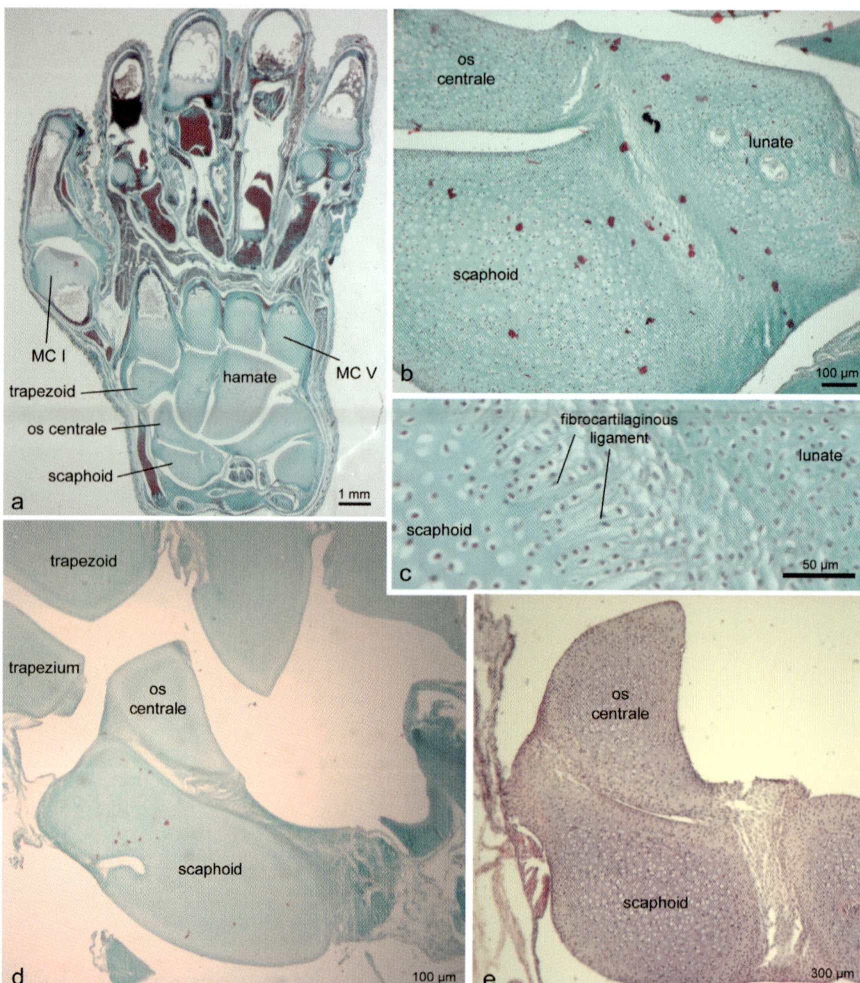

8 The Pelvic Girdle and Hindlimb Skeleton

8.1 Introduction

The primate pelvic girdle includes the right and left os coxae (innominates, pelvic bones), which are invariably present as three ossified parts, united by synchondroses. The two os coxae articulate with one another at the pubic symphysis ventrally. The os coxae articulate with the sacrum at auricular surfaces. The acetabular fossa of the os coxa (a socket-shaped joint which faces laterally) articulates with the proximal head of the femur, the bone of the thigh. The femur articulates with the tibia, the primary bone of the leg for weight support. The fibula articulates with the tibia distal to the tibiofemoral joint, and at the ankle joint. The ankle and foot contain tarsals, metatarsals, and phalanges. Adult proportions vary according to locomotor specializations (Napier and Walker, 1967; Ankel-Simons, 2007; Gebo, 2014).

There are seven tarsals. The bones of the leg articulate with the talus, which in turn is supported inferiorly by the calcaneus. The navicular articulates with the talus on its distal side; the navicular and calcaneus articulate with a distal row of four tarsals, the cuboid, the entocuneiform, mesocuneiform, and ectocuneiform (the last three also called the medial, intermediate, and lateral cuneiforms, respectively). The first metatarsal and its phalanges are divergent in non-human primates, and as such impart a grasping function to the foot that is unlike the derived foot of hominins.

Embryonic development of the hindlimbs was discussed in Chapter 7. During fetal development, typical long bones and some tarsals become elongated by cartilaginous growth centers (synchondroses) positioned at proximal and distal poles of the diaphysis. Growth centers are continuous with cartilaginous epiphyses, and resemble an expanded epiphyseal growth plate (Box 8.1). The ilium has growth centers aligned with the long axis of the body. More detail on microanatomy of developing limb bones is included within Boxes 8.1–8.3, following this chapter.

8.2 The Pelvic Girdle and Hindlimb Skeleton of Hominoids

8.2.1 The Pelvic Girdle

Each os coxa has three regions, which correspond to ossification centers: the ilium, pubis, and ischium. All three regions contribute to the acetabulum (Figure 8.1). The form of the os coxa is highly distinctive in humans in that the ilium flares widely. Historically, this has been interpreted as an "obstetric by-product" (that is, relating to the functional demands of parturition), and a trade-off which imparts a cost to female locomotion (see DeSilva and Rosenberg, 2017). However, recent work indicates that wider-set os coxae actually confer a locomotor advantage for shorter-legged individuals (Gruss et al., 2017). In all other primates, the ilium flares little, and is elongated in the trunk's long axis.

The ossified ilium, ischium, and pubis are widely separated in newborns, connected by synchondroses. This is described in all hominoid genera.

In isolated elements of the newborn, the ilium of apes is already dissimilar to that of humans in morphology (Figure 8.1). In humans the greatest width and height appear nearly equal, whereas in the newborn ape, the ilium is already a craniocaudally long bone. In the medial view, the auricular surface for the sacroiliac joint is oval and follows the trunk's long axis in apes, situated against the dorsal margin of the iliac

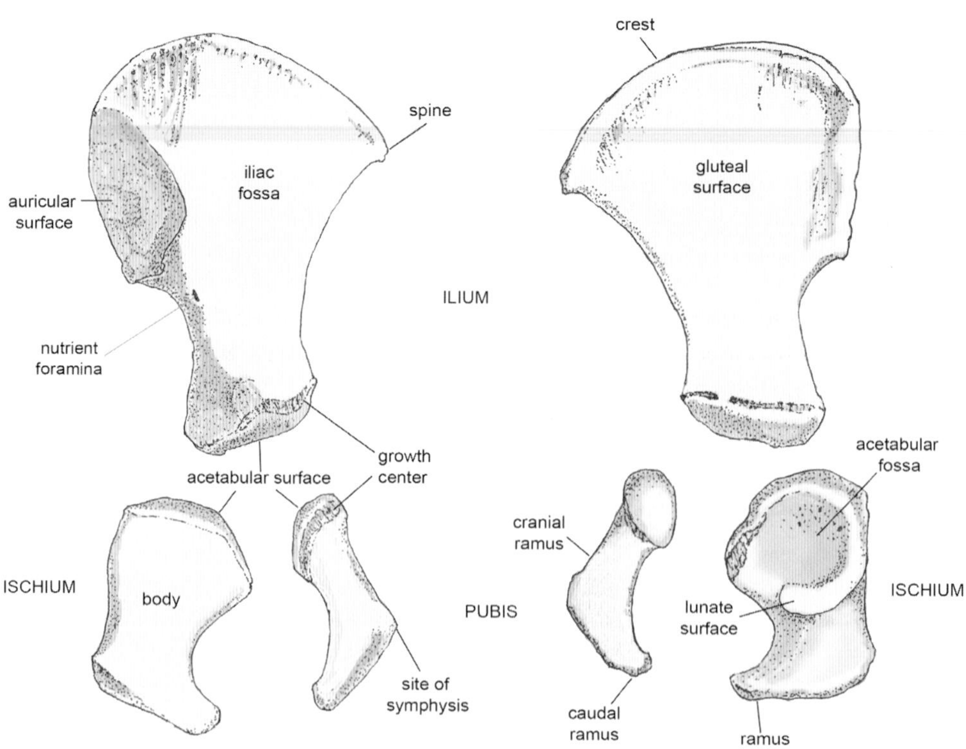

Figure 8.1 Pelvic girdle of perinatal chimpanzee, shown in medial (left) and lateral (right) views. Based on perinatal specimens, FMNH 122719 and KUPRI 9905.

blade (Figure 8.1). In humans, it is obliquely displaced from the posterior margin of the iliac blade. Also dorsally, the greater sciatic notch is already deeply curved in newborn humans, while in apes the dorsal surface adjacent to the posterior iliac spine has only a subtle curvature (Figure 8.1). The iliac blade itself is fan-shaped in the newborn ape, although not as widely flaring as in humans.

The isolated pubis is roughly similar in human and ape newborns in that the inferior ramus for articulation with the ischium forms a pronounced tip. The bone is narrow along the superior pubic ramus, and ends at the acetabular surface, which in the newborn appears like a rounded "head." The pectin of the pubis (a sharp ridge atop the superior ramus) is readily apparent in human newborns, but in apes there is no raised ridge (Figure 8.1).

The ischium is very similar in newborn humans and apes. In both, it is wide at the acetabular side, and bears the most distinct and adult-like acetabular morphology of all three primary bones. There is a deep acetabular notch on the lateral surface, while the medial surface is far smoother. The bone is roughened along the synchondroseal margin where it articulates with the other primary elements. Ventrally there is a gentle concave curvature – a boundary of the obturator foramen. The caudal end (inferior end for humans) has a boot- or apostrophe-like appearance, with the toe leaning toward the pubis to complete the ischiopubic ramus (Figure 8.1).

8.2.2 The Hindlimb Skeleton

As in the forelimb, the long bones of the stylopod and zeugopod undergo primary ossification to form long bone diaphyses prenatally, while most secondary ossification centers ossify postnatally. Thus, CT and other radiological methods will not allow a full appreciation of limb bone length, because the majority of thigh and leg bone epiphyses are mostly cartilaginous in all newborn primates.

8.2.2.1 Femur

The femur of hominoids exhibits between one and two ossified elements at birth (Figure 8.2). In newborn humans, the secondary centers of the distal epiphysis normally commence ossification shortly before term, and in apes they are also commonly present. In apes, the proximal epiphysis is typically ossified by the end of infancy. However, in the 5-day-old *Pan* the proximal epiphysis of the femoral head has formed a minute ossified nucleus (Figure 8.2), suggesting that this may variably be present very close to birth or shortly after. The femoral head begins to ossify at about year 1 in humans.

In apes and humans, the diaphysis of the femur flares both proximally and distally into wide metaphyseal zones (Figure 8.2). The proximal surface billows to a rounded top. This end leans more medially, creating a ledge supporting the femoral head which most often remains entirely cartilage at birth. Distally the bicondylar form of the epiphysis is reflected by the narrower central zone of the metaphyseal surface, which is depressed relative to the wider, rounded billowing surfaces on medial and lateral sides. The dorsal side is identifiable because on the proximal end there is a raised platform for development of the lesser trochanter (Figure 8.2).

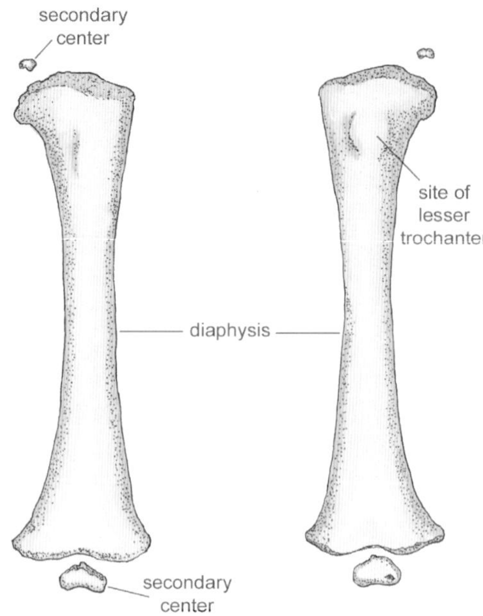

Figure 8.2 Femur of perinatal chimpanzee, shown in ventral (left) and dorsal (right) views. Based on 5-day-old specimen, KUPRI 9905.

8.2.2.2 Patella

The patella is the largest sesamoid bone of the body. It is entirely cartilaginous in all hominoids at birth. In humans it may begin to ossify as early as 18 months, but more typically within a range of 3–6 years. The timing may be similar in *Pan*; it is ossified in most infants. In *Hylobates*, however, the patella is not ossified until animals reach the first stage of mixed dentition (Schultz's Juvenile I phase).

8.2.2.3 Tibia

In newborn humans, the diaphysis and the proximal epiphysis of the tibia are usually ossified at birth (80% per Scheuer and Black, 2004). There are varied reports on apes. We observe an ossification within the epiphysis of the 5-day-old *Pan*, and Diogo et al. (2015) figured CT reconstructions of a newborn gorilla with an ossified proximal epiphysis. It is ossified in newborn *Hylobates*. This suggests that ossification of this epiphysis is relatively common at birth in most hominoids. The distal epiphysis remains unossified in newborn hominoids.

The form of the tibia shares basic similarities among hominoids. The tibia flares widely at the proximal diaphysis (Figure 8.3). It widens at least to a lesser degree distally. Proximally the metaphyseal surface may be rounded, and the perimeter is flared outward, especially on the medial side. The tuberosity is raised but not yet prominent. In *Homo* and *Pan*, a nutrient foramen is visible on the dorsal surface, within the proximal third. The shaft is smooth, rounded at the midshaft, and mostly non-curved. Proximally, the metaphyseal region is curved slightly backward and medially.

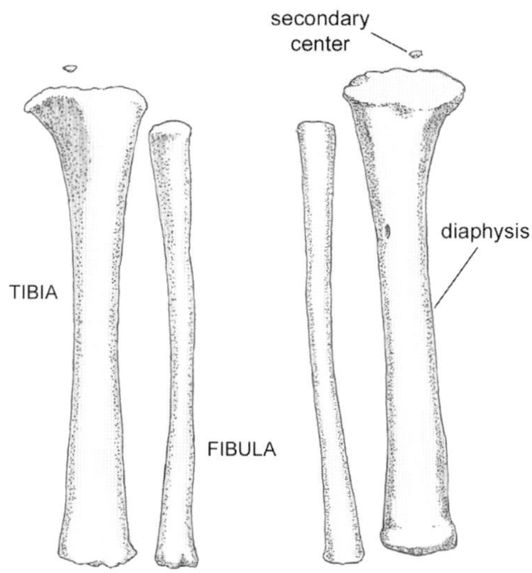

Figure 8.3 Tibia and fibula of perinatal chimpanzee, shown in ventral (left) and dorsal (right) views. Based on 5-day-old specimen, KUPRI 9905.

8.2.2.4 Fibula

Only the diaphysis of the fibula is ossified at birth. In humans the bone is described as relatively straight at birth, rounded proximally, and somewhat flattened distally in the mediolateral direction. In the newborn chimpanzee examined here, there is a subtle curvature (convex laterally). An illustration of newborn *Hylobates* also indicates a curved fibula (Schultz, 1944, figure 19).

8.2.2.5 Ankle and Foot

In humans, primary ossification centers of the calcaneus and talus are present at birth; an ossified nucleus of the cuboid is often but not always present. In apes the same three bones are commonly present as ossified centers, plus the lateral cuneiform and medial cuneiform in some individuals (Table 8.1). In both humans and great apes, it is relatively common to have primary ossification of tarsals limited to the talus and calcaneus at birth, as in the 5-day-old KUPRI *Pan* (Figure 8.4). Newborn hylobatids are understudied by comparison to *Pan*, but they may be more advanced in tarsal and metatarsal ossification (Table 8.1).

In humans the calcaneal primary ossification center is elongated in newborns and the talus is rounded or ovoid. The same descriptor appears to apply to apes (Figure 8.4). Distally and proximally there are synchondroseal interfaces.

In all metatarsals and phalanges, the diaphysis is well-ossified at birth. In all hominoids, the distal epiphyses of the metacarpals and phalangeal epiphyses remain unossified at birth (Figure 8.4). The first metatarsal is very robust in compared to II–V in the newborn *Pan* (with a greater contrast to II–V than in humans). In *Pan*, the midshaft is narrow and the metaphyseal regions flare broadly, with the proximal end blunt and a rounded distal end (Figure 8.4).

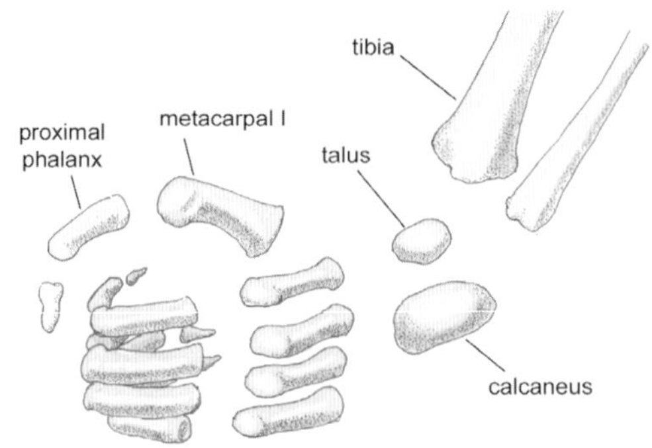

Figure 8.4 Ankle and foot of perinatal chimpanzee, shown in dorsolateral view. Based on 5-day-old specimen, KUPRI 9905.

8.3 Pelvic Girdle and Hindlimb Skeleton in Newborn Old and New World Monkeys

8.3.1 The Pelvic Girdle

8.3.1.1 Os Coxa

There is a basic similarity of the ilium among perinatal anthropoid monkeys: it is broad and slightly flared to a narrow fan shape on the cranial end, while the caudal end is more bluntly rectangular. Both margins bear smooth edges indicative of synchondroseal growth. The auricular surface has a pronounced rounded lip ventrally and caudally (Figure 8.5). Although the iliac blade is always broader than the acetabular end of the bone, the blade is less fan-shaped than in the perinatal *Pan*.

The ischium is similar to that of hominoids except more robustly proportioned. The pubis is wide, blunt-edged on the caudal end and has a narrower curved acetabular end. Both ischium and pubis have a curved edge for the obturator foramen border.

8.3.2 The Hindlimb

8.3.2.1 Femur

The femur of Old World monkeys is more advanced in ossification than in all platyrrhines and some hominoids (Figure 8.6; Table 8.1). Three secondary ossification centers are ossified at birth in *Macaca* spp., *Papio* spp., *Cercopithecus nictitans*, and *Theropithecus gelada* (Table 8.1), and at least the proximal and distal epiphyses contain an ossified nucleus in all known Old World monkeys. In contrast, we observed ossification in only the distal epiphysis in some, but not all, newborn platyrrhines (e.g., *Cebuella* and *Saimiri*).

Distally, the diaphysis is oval and mediolaterally broad. Proximally, the perimeter is irregular. Dorsomedially, a ridge projects, the base for the lesser trochanter. In all specimens examined, the metaphyseal region leans medially into a small shelf, as in *Pan* (Figure 8.2). The diaphysis itself is relatively straight in all anthropoids. The dorsal side is slightly concave. Nutrient foramina are on the dorsal side in all species examined here.

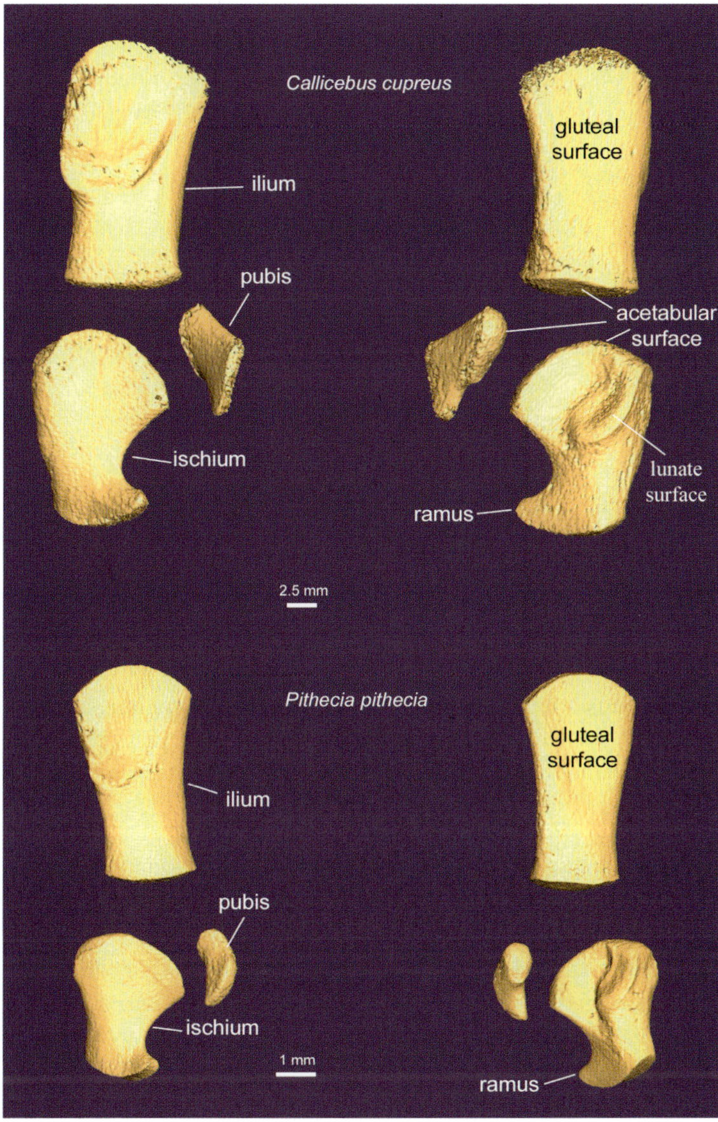

Figure 8.5 Left os coxa of two anthropoids. Left, medial view; right, lateral view.

8.3.2.2 Patella

The patella remains entirely cartilaginous in all monkeys at birth.

8.3.2.3 Tibia

Previously described cercopithecoids, with the exception of *Nasalis*, have three tibial ossification centers present at birth, including proximal and distal epiphyses (Table 8.1). Among platyrrhines, the epiphyses remain unossified at birth in most species. Occasional presence is attributed to *Saimiri* and *Aotus* (Table 8.1), and *Cebuella* (see atlas).

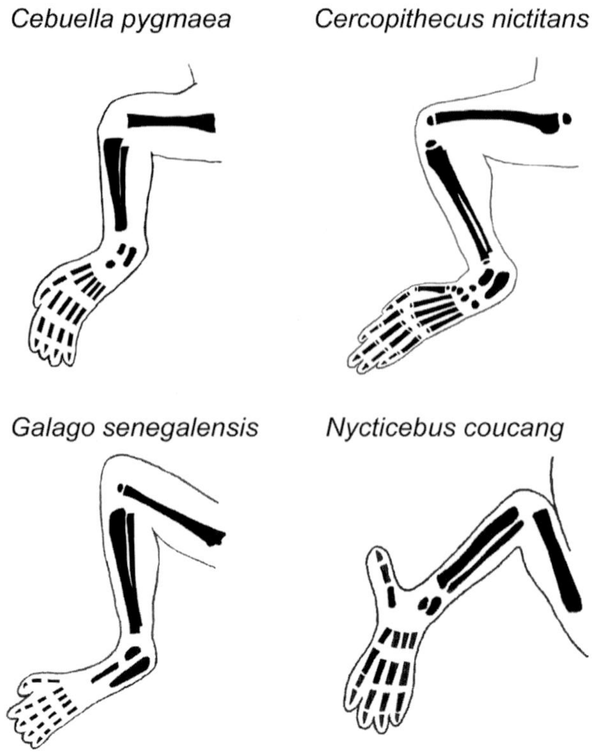

Figure 8.6 Schematic illustration showing hindlimb ossification of newborn primates (redrawn after Glaser, 1970).

In external form, the tibia resembles the description for hominoids. The diaphysis is relatively straight in the single *Pithecia* and *Leontopithecus* specimens described here, except with a slight dorsal bending proximally. In *Callithrix*, *Callicebus*, and *Saguinus*, there is more pronounced dorsoventral curvature, with a ventral convexity extending near to the midshaft. The nutrient foramen is invariably dorsal, at the proximal third of the diaphysis.

8.3.2.4 *Fibula*

Some cercopithecoids have an ossified distal epiphysis of the fibula, but none has been described in newborn platyrrhines. No ossification of the proximal epiphysis has been described in any newborn monkeys (Table 8.1).

The fibula is slightly flared in diameter proximally and distally. The diaphysis is either straight or slightly bowed laterally.

8.3.2.5 *Ankle and Foot*

The calcaneus and talus are ossified in all newborn anthropoids (and indeed in all primates – Table 8.1). The calcaneus is the first to ossify in *Callithrix*, with the talus ossifying relatively late in fetal development (Phillips, 1976). Among more distal tarsals, there is much variation. In a large sample of *Macaca nemistrina*, Newell-Morris et al. (1980) found all tarsals ossified in "most or all" newborns. In *Macaca mulatta*, the constant number was five, with the navicular and/or mesocuneiform being rarely present (these bones are not constantly present until week 2). Reports on small samples of other cercopithecoids indicate a similarly large number of tarsals are ossified at birth (Table 8.1). The only platyrrhine examined in relatively large samples is *Callithrix jacchus* (Phillips, 1976). The number of ossified tarsals reported within the first week is 2–5. However, the large sample and varied methods employed by Phillips allow us to draw certain inferences. First, Phillips detected more tarsals with Alizarin red whole-tissue preparations than with radiographic examination. This suggests that some more distal tarsals are just commencing ossification perinatally, and may be difficult to detect without methods that demonstrate immature bone tissue. In addition, more tarsals were detected radiographically in P1–P3-aged specimens (3+) than in P0 specimens (2). Taken together, these observations suggest that the first postnatal week constitutes a rapid ossification phase for tarsals in marmosets (*Callithrix* and *Cebuella*).

The following broad patterns are indicated by data in Table 8.1. Of all anthropoids, cercopithecoids possess the most well-ossified tarsus at birth, although colobines remain less studied. Among platyrrhines, marmosets (especially *Callithrix* spp.) possess a more fully ossified tarsus at birth than most other species (Figures 8.6–8.8). *Callimico* may rival marmosets (0–7 ossified in a small sample of newborns – Table 8.1). It is most common that only two are detected in newborn platyrrhines, with the exception of *Saimiri* and *Lagothrix*, each of which may possess three ossified tarsals (Table 8.1). In this light, hominoids (including humans, which often have three ossification centers at birth) are not especially slow to ossify the ankle. Another pattern that emerges from these data is that hominoids and the few well-known cercopithecoids follow a different sequence of ankle ossification than other anthropoids: the first three tarsals to ossify are the calcaneus, talus, and cuboid. In all other anthropoids studied to date, the navicular or one of the cuneiforms ossifies before the cuboid (Figure 8.8). This may well be a basic difference between catarrhines and platyrrhines, and a functional basis deserves further exploration.

Ossification of more distal elements, aside from primary ossification centers, is rare at birth. In some cercopithecoids, *Saimiri*, and marmosets, metatarsal epiphyses are ossified; in some cercopithecoids even more distal epiphyses are also ossified (Figure 8.6).

The morphology of newborn tarsal ossification centers bears very little resemblance to the adult form, excepting that the calcaneus may easily be differentiated from the talus by its greater size and relative length (Figures 8.7 and 8.8). The talus and other bones (when present) are rounded or oval. The calcaneus bears distinctive planes that correspond to synchondroses, and medially, the site of the future sustentaculum tali is visible as a small bar (Figure 8.6).

Primary ossification centers for metatarsals are highly cylindrical while proximal and intermediate phalanges are blunt proximally and rounded distally. The first metatarsal (MT) lacks the relative robusticity that is already apparent in newborn hominoids. In most of the species newly examined here, MT I is nearly the same diameter as adjacent metatarsals, although shorter. In the *Pithecia* specimen, MT I has a broad base, a slender midshaft, and a rounded distal end (Figure 8.8). The distal phalanges resemble adult morphology in that the flattened end of nail-bearing species is already clear.

8.4 Pelvic Girdle and Hindlimb Skeleton in Tarsiers at Birth

8.4.1 Pelvic Girdle

8.4.1.1 Os Coxa

The ilium is more gracile and relatively elongated compared to anthropoids. It is more similar to lorisoids in proportional length and shape, being rounded in the center, and flaring equally at both cranial and caudal ends (Figure 8.7). The ischium is similar on the acetabular side to the above descriptions, but distally it is broad and lacks a pronounced ischial ramus in osseous form. The pubis is similar to other primates.

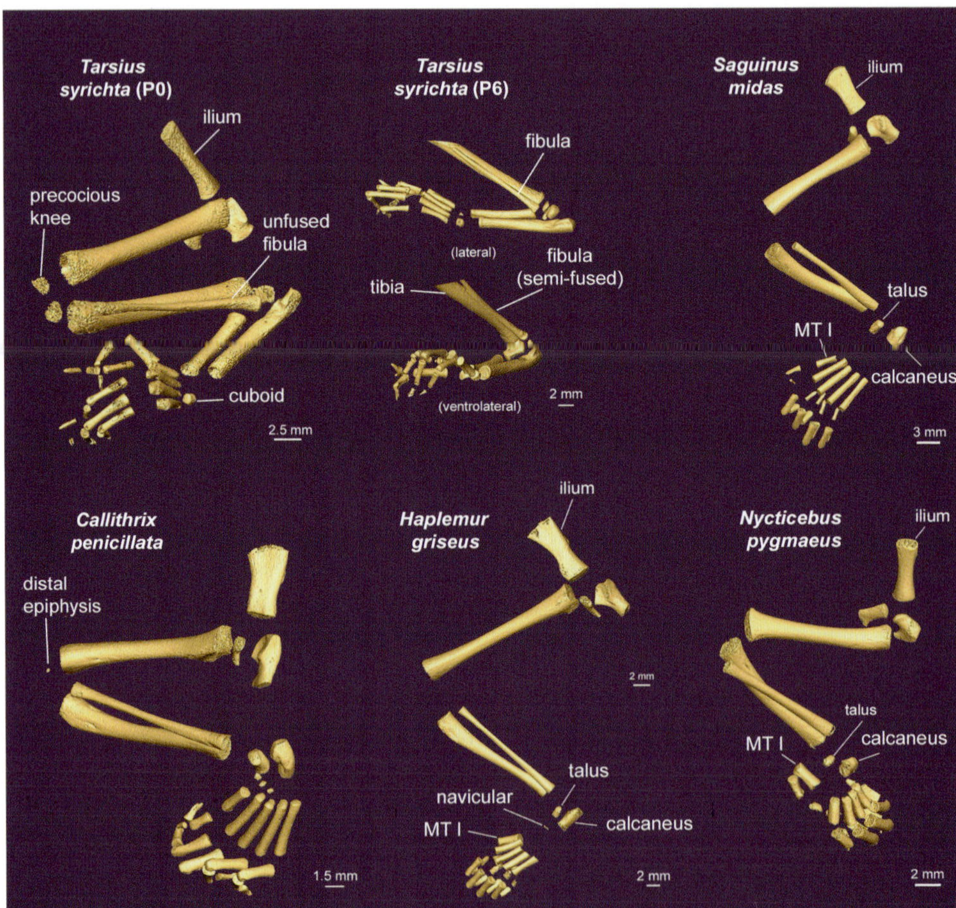

Figure 8.7 Comparison of tarsier left pelvic girdle and hindlimb to selected anthropoids and strepsirrhines.

8.4.2 The Hindlimb

8.4.2.1 Femur

Ossification of the femur in tarsiers is relatively advanced, with ossified nuclei of its distal epiphysis. The diaphysis tilts medially on the proximal end. Distally, there is also a slight medial curve, and the distal epiphysis is displaced both medially and dorsally relative to the long axis of the diaphysis.

8.4.2.2 Patella

The patella remains entirely cartilaginous in newborn tarsiers.

8.4.2.3 Tibia and Fibula

The proximal epiphysis of the tibia is ossified in *T. syrichta* and *T. bancanus*. The distal epiphyses of the tibia and fibula are ossified in both *T. syrichta* and *T. bancanus*. The distal ends of tibia and fibula of adult tarsiers are completely fused (Ankel-Simons, 2007). In P0 specimens, the tibia and fibula remain unfused, or barely fused. In the P6 *T. syrichta* fusion is underway, but has not progressed to the distal end of the two bones (Figure 8.7).

8.4.2.4 Ankle and Foot

Niemitz and Sprankel (1974) described the neonatal tarsier as both "nidifugous" (precocial), based on numerous criteria indicated by Portmann (see Chapter 2), but also "nidiculous" (altricial) in degree of perinatal ossification. With the broader available data for comparison (Table 8.1), *Tarsius* is easily among the more precocious of living primates in terms of ankle and foot ossification. *Tarsius* develops as many as five tarsal ossifications within the first postnatal week. This is more than about half of known anthropoids, and more than any known strepsirrhine (Table 8.1; Figure 8.8). Moreover, very few primates, aside from some cercopithecoids, *Saimiri*, marmosets, and *Tarsius*, possess ossified metatarsal epiphyses at birth. Thus, *Tarsius* is not particularly altricial in hindlimb development. Indeed, compared to the nest-bound, newborn *Tupaia*, all primates have a better ossified tarsus (Table 8.1). The navicular is the third tarsal to ossify, as is the case for other non-catarrhine haplorhines.

Two tarsals, the calcaneus and navicular, are already elongated within the first postnatal week. The navicular is cylindrical. The calcaneus has the most adult-like form of the tarsals, possessing a heightened posterior region, with a slight medial extension beneath the talus (Figure 8.8). The talus is rounded posteriorly and narrows anteriorly, reflecting the incipient neck. The cuboid and ectocuneiform, where present, are rounded nodules.

Metatarsals and phalanges are similar in general form to the descriptions above, but the phalanges are proportionally longer compared to newborn monkeys and strepsirrhines (Figure 8.8). The distal phalanges supporting grooming claws (II, III) are more slender than the others; I, IV, and V are flattened distally.

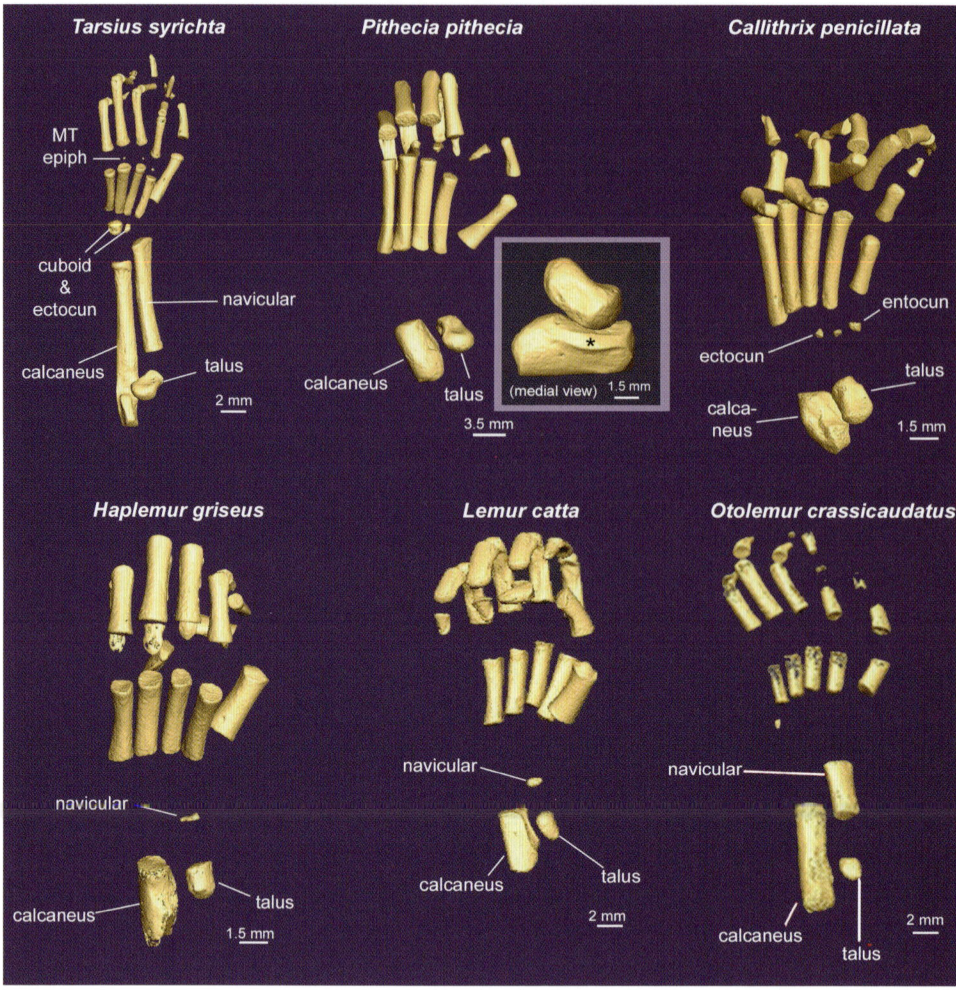

Figure 8.8 Left foot skeleton of tarsier compared to anthropoids and strepsirrhines. Ectocun = ectocuneiform. * = initial ossification of sustentaculum tali.

8.5 Pelvic Girdle and Hindlimb in Strepsirrhines at Birth

8.5.1 Pelvic Girdle

8.5.1.1 Os Coxa

Compared to anthropoids, the strepsirrhine ilium is less flared along the cranial margin at birth. This is especially true of lorisoids, in which the dorsoventral breadth of cranial and caudal ends is similar (Figure 8.7). In addition, the ilium of newborn lorisoids is a more rounded bone overall than in most lemurs. *Cheirogaleus* has a similarly rounded acetabular end, but the iliac blade is more flattened than in lorisoids. In large-bodied lemurs the cranial end flares into a broader iliac blade, although still not as broad as in anthropoids; the ventral flare is more pronounced (Figures 8.7 and 8.9). The ventral margin of the iliac blade flares out near the acetabulum in some specimens (perhaps an incipient iliac spine – e.g., Figure 8.7). In the large-bodied *Propithecus* the overall form of the ilium is more robust.

As in other primates, the cranial and caudal ends of the bone are smooth surfaces corresponding to synchondroses. The proportions of the ilium are craniocaudally abbreviated in newborns, indicative that the cranial synchondrosis expands the length of iliac blades; the age contrast may be most pronounced in large-bodied lemurids (Figure 8.9).

The ilium, pubis, and ischium are widely separated by synchondroses at birth. The ischium and pubis are similar in form to the description for tarsiers, above, but the ischial ramus remains almost entirely cartilaginous. The surfaces bordering the obturator foramen are curved but not to the same degree as in anthropoids.

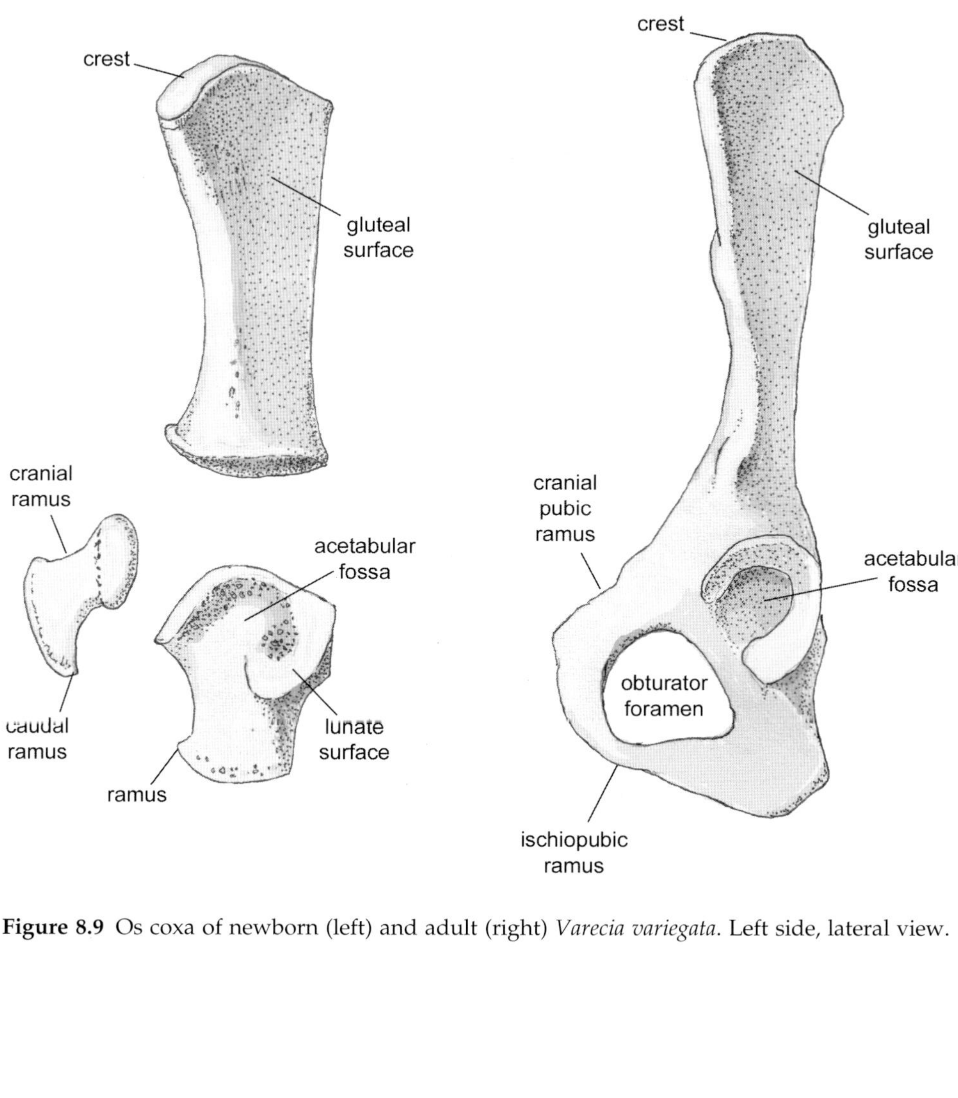

Figure 8.9 Os coxa of newborn (left) and adult (right) *Varecia variegata*. Left side, lateral view.

8.5.2 The Hindlimb

8.5.2.1 Femur

Among strepsirrhines, galagids possess the best ossified femur, with the distal epiphysis commonly ossified. The proximal epiphysis remains unossified in all known newborn strepsirrhines.

The femoral diaphysis of all strepsirrhines is distinct from anthropoids in that the proximal end has a sharply sloping synchondroseal surface on the medial side. The transition to this inclined plane is abrupt. Most of the proximal end is a flat plane, transverse to the bone's long axis, but near the medial side the plane abruptly leans distalward. This reflects the position of the femoral head, which departs at a right angle. The shaft is uncurved. In larger-bodied strepsirrhines (lemurs and *Propithecus*), the diaphysis appears relatively long, especially compared to cheirogaleids and lorisoids. However, by comparison to adults, all strepsirrhines have a proportionately stouter diaphysis at birth (e.g., *Varecia*: Figure 8.10).

The nutrient foramen is typically medially located at about the proximal one-fourth of the length of the bone. *Propithecus* alone departs from this, with the foramen appearing more dorsally, also at the proximal one-fourth point.

8.5.2.2 Patella

The patella remains entirely cartilaginous in newborn strepsirrhines.

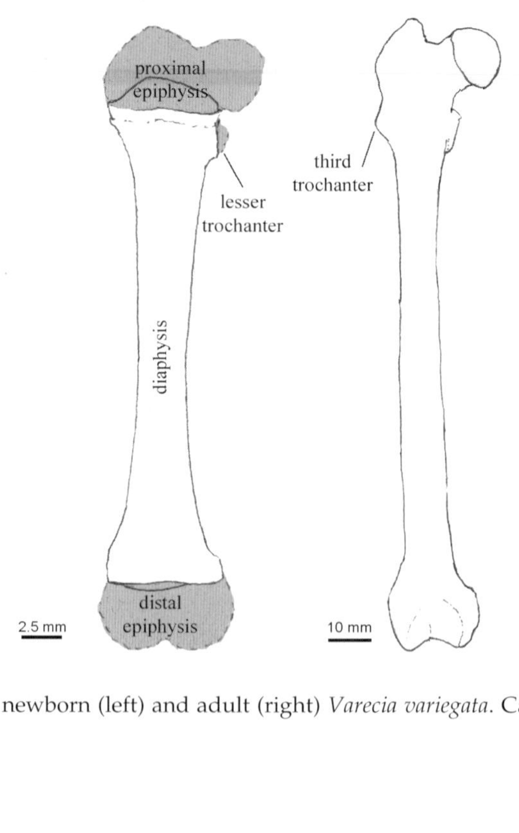

Figure 8.10 Femur of newborn (left) and adult (right) *Varecia variegata*. Cartilage shown in gray.

8.5.2.3 Tibia and Fibula

The tibia is better ossified in galagids than in other strepsirrhines, with the proximal epiphysis ossified at birth. In one newborn *G. moholi*, the distal epiphysis has a very small nucleus of ossification (Figure 8.11). No other secondary ossification centers in the tibia or fibula were observed in newborn strepsirrhines.

8.5.2.4 Ankle and Foot

Lemuroids appear to commonly possess only two ossified tarsals at birth: the calcaneus and the talus. In one *Lemur* specimen, the navicular is present as a small ossified nodule. Among lorisoids, the tarsus is better ossified in galagids. In three genera of galagids, the navicular (in addition to the calcaneus and talus) is well advanced in ossification and elongated in association with the calcaneus. No ossified epiphyses of metatarsals or phalanges are described in any strepsirrhines at birth (Table 8.1).

Of the tarsals, the earliest bone to form in strepsirrhines (as in all primates) is the calcaneus. In some newborn strepsirrhines, such as cheirogaleids and *Nycticebus*, the calcaneus has a smooth, nearly ovoid appearance. In larger-bodied lemuroids there are more distinctive features, some of which correspond to synchondroseal interfaces rather than adult morphology. Two factors create this impression. First, bones of smaller species inevitably are revealed with less detail due to limits of CT scan resolution. However, small galagids do have distinctively elongated calcanei with raised surfaces posteriorly. The different appearance of tarsals among small strepsirrhines is best understood with reference to histology of ankle joints (Box 8.3). The calcaneus of galagids is far better ossified at birth than cheirogaleids, and this contributes to the lack of distinctive features in the latter.

In larger lemurids, the calcaneus is roughly rectangular in profile, with a sloping posterior surface, flattened plantar, lateral, and anterior surfaces, and a medial surface that bears a distinctive ridge. On the surface facing the talus, there is a slight depression. *Propithecus* has similar features, but with a more abbreviated length. A comparison to histology reveals the basis for the distinctive surfaces. The sloping posterior surface corresponds to the synchondrosis to the, as yet, still cartilaginous calcaneal tuberosity. Anteriorly, the body of the calcaneus remains largely cartilaginous. Finally, the medial side of the bone remains highly cartilaginous; the medial ridge is the site of the enlarging sustentaculum tali. The posterior, anterior, and medial surfaces are all interfaces for active growth sites. The ossification nucleus is fairly centralized in the cartilaginous template of the talus in most newborns lemuroids (Box 8.3).

Although galagids have relatively advanced ossification of the calcaneus at birth, this mainly reflects rapidity of growth. Portions that remain largely cartilaginous are the same as in lemurids, and also include the bone supporting the posterior talar facet (Box 8.3). The navicular has a similarly rapid pace of growth in newborns, based on histology (Box 8.3). The talus is a more rounded, oval nodule in newborns. Histology reveals this is because these centers are, as yet, centrally within the cartilaginous bone templates, but may have extended toward the dorsum of the neck region (Box 8.3).

Metacarpal bones are represented by cylindrical primary centers, and the first metacarpal stands out as a shorter and more robust diaphysis than the others. Metatarsals of newborn galagids have relatively shorter diaphyses compared to lemurids

Table 8.1 Ossification of hindlimb elements at birth in primates and the tree shrew (*Tupaia*)

	Prox. femur	Dist. femur	Prox. tibia	Dist. tibia	Dist. fibula	Talus	Calcaneus	Navicular	Cuboid	Meso cun.	Ecto cun.	# Tarsals	MT	Source
HAPLORHINES														
Aotus nancymae	U	X	(X)	U	U	X	X	X	X	U	?	2 to 5	U	This book
Callicebus cupreus	U	U	U	U	U	X	X	U	U	U	U	2	U	This book
Callimico goeldii	U	X	U	U	U	X	X	?	?	?	?	0 to 7	?	Hofmann et al., 2007
Callithrix jacchus	U	(X)	U	U	U	X	X	U	?	?	?	2 to 5	X	Phillips, 1976; Hofmann et al., 2007
Callithrix penicillata	U	X	U	U	U	X	X	U	U	X	X	5	U	This book
Cebuella pygmaea	U	U	U	U	U	X	X	X	U	U	U	3		Glaser, 1970; this book
Cebus apella	U	U	U	U	U	X	X	U	U	U	U	2	U	Glaser, 1970
Cebus capucinus	U	U	U	U	U	X	X	U	U	U	U	2	U	Glaser, 1970
Cercopithecus nictitans	X	X	X	X	U	X	X	X	X	X	X	7	X (II–V)	Glaser, 1970
Gorilla gorilla	U	X	X	U	U	X	X	U	X	?	?	3 to 4	U	Winkler et al., 1991; Diogo et al., 2015
Hylobates lar	U	X	X	U	U	X	X	U	X	X	X	5	U	Schultz, 1944
Lagothrix lagotricha	U	U	U	U	U	X	X	?	?	?	?	3	U	Glaser, 1970
Leontopithecus rosalia	U	U	U	U	U	X	X	U	U	U	U	2	U	This book
Macaca mulatta	X	X	X	X	X	X	X	(X)	X	X	X	5 to 7	X	Van Wagenen and Asling, 1958

Macaca nemistrina	?	?	?	X	X	X	X	X	X	X	X	7	U	Newell-Morris et al., 1980
Mandrillus sphinx	?	?	X	?	?	?	?	?	?	?	?	?	?	Kirkwood and Kember, 1993
Nasalis larvatus	X	X	U	U	U	?	?	?	?	?	?	6	U	Schultz, 1941
Pan troglodytes	X	X	(X)	U*	(X)	X	X	U	(X)	(X)	(X)	2 to 5	U	Nissen and Riesen, 1949; Winkler et al., 1991
Pithecia pithecia	U	U	U	U	U	X	X	U	U	U	U	2	U	This book
Pongo pygamaeus	U	(X)	U	U	U	X	X	U	(X)	U	(X)	2 to 4		Schultz, 1941; Winkler et al., 1991
Saguinus fuscicollus	?	?	?	U	U	X	X	?	?	?	?	?		Glassman, 1983
Saguinus midas	U	U	U	U	U	X	X	U	U	U	U	2	U	This book
Saguinus oedipus	U	U	U	U	U	X	X	U	U	U	U	2	U	Hofmann et al., 2007; this book
Saimiri boliviensis	U	X	(X)	U	U	X	X	U	U	(X)	(X)	3	MT II–V	Galliari, 1988
Tarsius bancanus	X	X	X	X	?	X	X	X	U	U	U	3	U	Niemitz and Sprankel, 1974
Tarsius syrichta	U	X	X	X	X	X	X	X	X	U	(X)	5	X (II–IV)	This book
STREPSIRRHINES														
Cheirogaleus medius	U	U	U	U	U	X	X	U	U	U	U	2	U	This book
Eulemur collaris	U	U	U	U	U	X	X	U	U	U	U	2	U	This book
Eulemur mongoz	U	U	U	U	U	X	X	U	U	U	U	2	U	This book

Table 8.1 *(cont.)*

	Prox. femur	Dist. femur	Prox. tibia	Dist. tibia	Dist. fibula	Talus	Calcaneus	Navicular	Cuboid	Meso cun.	Ecto cun.	# Tarsals	MT	Source
Galago spp.	U	X	(X)	(X)	U	X	X	X	U	U	U	3	U	Glaser, 1970; this book
Galgoides demidovii	?	U	U*	U	U	X	X	X	U	U	U	3	?	This book
Hapalemur griseus	U	U	U	U	U	X	X	(X)	U	U	U	2 to 3	U	This book
Lemur catta	U	U	U	U	U	X	X	(X)	U	U	U	2 to 3	U	Glaser, 1970; this book
Nycticebus coucang	U	U	U	U	U	X	X	U	U	U	U	2	U	Glaser, 1970
Nycticebus pygmaeus	U	U	U	U	U	X	X	U	U	U	U	2	U	This book
Otolemur crassicaudatus	U	X	(X)	U	U	X	X	X	(X)	U	U	3 to 4		Newman and Hendrickx, 1984
Propithecus coquereli	U	U	U	U	U	X	X	U	U	U	U	2	U	Glaser, 1970; this book
Varecia spp.	U	U	U	U	U	X	X	U	U	U	U	2	U	This book
SCANDENTIA														
Tupaia belangeri	U	U	U	U	U	U	X	U	U	U	U	1	U	Glaser, 1970

Abbreviations: U, unossified; X, ossified; *, incipient ossification seen via histology; ectocun., ectocuneuform; mesocun., mesoconeiform. Parentheses indicate variable state of ossification at birth. Note: endocuneiform unossified in most newborn primates, excepting some cercopithecoids.

(Figure 8.8). The phalangeal bones resemble the description above for other primates, except that the distal phalanx of the second digit alone departs from the dorsoventrally flattened appearance; digital epiphyses remain unossified. The abbreviated length of the second digit of lorises is not yet obvious in the newborn.

8.6 Patterns of Ossification of the Postcranial Axial Skeleton in the Early Postnatal Period

8.6.1 Secondary Centers of Ossification

8.6.1.1 Pelvic Girdle

Secondary centers of ossification are slow to appear in humans and do not fuse until the third decade of life. Our limited sample of older infants reveals no secondary centers by 1 month of age in various strepsirrhines or *Saguinus*.

8.6.1.2 Hindlimb Elements

There are few comparative observations on timing of appearance of hindlimb epiphyses in anthropoids. Generally, catarrhines are more precocious than platyrrhines at the knee joint (Table 8.1). Postnatal ossification rates likely vary drastically among strepsirrhines. Our available samples suggest extreme differences in rate from very fast postnatal ossification of hindlimb epiphyses and tarsals (galagids – Figure 8.11) to very slow (*Propithecus*, see below). Among strepsirrhines, only

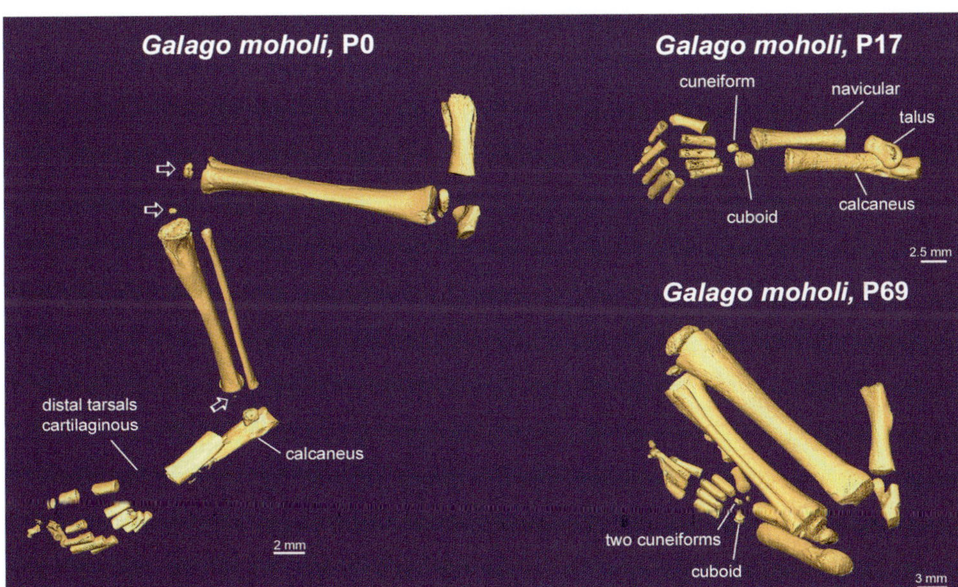

Figure 8.11 Age changes in the hindlimb of *Galago moholi*, which has precocious ossification of the tibia (open arrows = secondary ossification centers). Elongating of proportions of the navicular continues postnatally. The distal tarsal row begins ossifying within weeks, but is still incomplete in an older (P69) juvenile.

galagids have perinatal ossification at the knee, and this rapid pace of ossification continues postnatally (Figure 8.11). Study of a broad range of primates across postnatal ages is still needed.

8.6.2 Epiphyseal Growth Zones

Because most secondary centers of ossification do not form prenatally in most primates, a discrete epiphyseal plate is not present in most locations. Nonetheless, the chondrocyte zones of the growth plate exist adjacent to the metaphysis even before the epiphyses ossify (see Chapter 2). A study by Kirkwood and Kember (1993) is the only work that provides cellular measurements of the growth plate in newborn primates compared to other mammals, specifically studying tibial growth plates. Based on estimates such as the number of cells produced (by proliferation) per day, the frequency of cell division in the zone of proliferating chondrocytes, and the overall growth rate in micrometers per day, the authors found four anthropoids of diverse body size were "outperformed" in these parameters by selected rodents and artiodactyls, suggesting growth of the tibia is slow relative to other mammals. In a broad array of mammals studied (including three strepsirrhines), the depth of the zone of proliferating chondrocytes and the number of chondrocytes within it are quite variable. The size (diameter) of individual chondrocytes is not as variable, even if comparing species of vastly disparate body sizes (chondrocytes, like most other body cells, do not scale closely with body size: Morgado et al., 1990, and see Savage et al., 2007; Lui and Brown, 2011).

Kirkwood and Kember's results, although based on small samples, are not surprising in a broad sense. Primates are unified by a prolonged subadult phase, with accelerations and decelerations; most mammals have a smoother, continuous postnatal growth curve (Bogin, 1999b). In primates, growth deceleration is notable during the juvenile period (Watts and Gavin, 1982; Bogin, 1999b).

Chondrocyte proliferation, hypertrophy, and associated matrix production have been studied at great length in other mammals (e.g., Hunziker et al., 1987; Farnum et al., 2008; Rolian, 2008; Cooper et al., 2013). All of these growth plate variables play some role in the lengthening of long bones. Even though chondrocyte hypertrophy has been emphasized by many authors in its great importance to epiphyseal growth (Cooper et al., 2013; and see Burr and Organ, 2017), the region is necessarily constantly replenished in its depth until eventual fusion occurs across the synchondrosis. It is the number of chondrocytes recruited into hypertrophy, or the rate of recruitment, that matters more than the size of individual chondrocytes. Thus, variation in many parameters could potentially underlie differences in limb proportions in primates, as is indicated in other mammals (Rolian, 2008). A cursory histological comparison of the proximal tibial growth plate region in four strepsirrhines (Figure 8.12; Box 8.2) indicates that within strepsirrhines alone, there may be significant variation in the number of chondrocytes in proliferating and hypertrophic zones among species. The extent to which this variation among species may be explained by body size, life history, or locomotor behavior is completely unclear. This indicates that Kirkwood and Kember's categorization of growth plates (see above) underestimates variability among species or even higher taxonomic groups of primates.

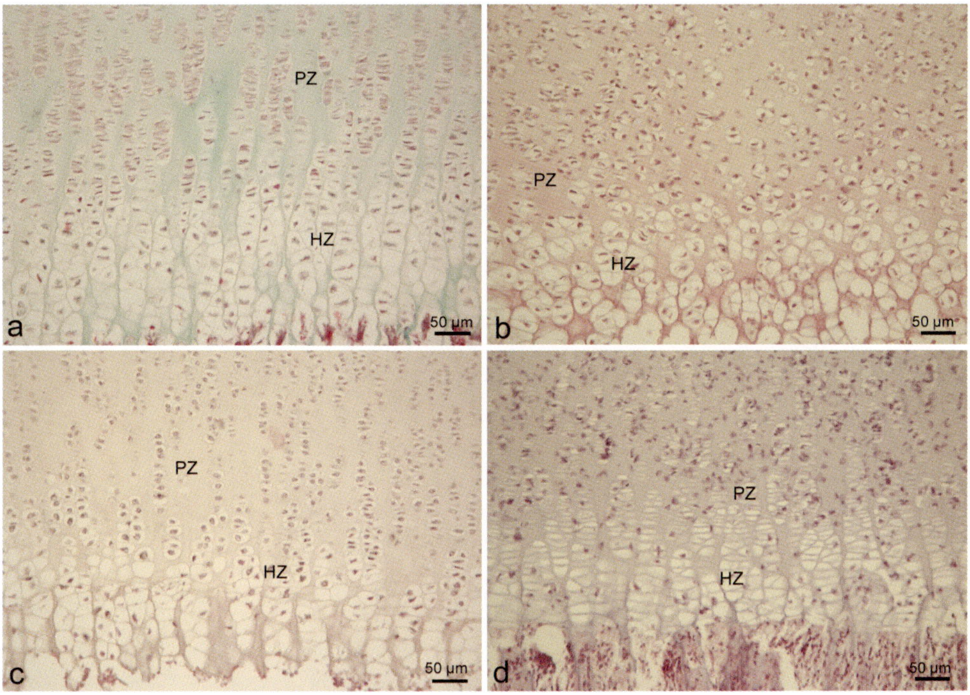

Figure 8.12 Growth plate regions of the proximal tibia in four species of strepsirrhines, *Propithecus coquereli* (a), *Cheirogaleus medius* (b), *Otolemur crassicaudatus* (c), and *Galagoides demidovii* (d). Note the great differences in the number of cells within the zones of proliferating (PZ) and hypertrophic (HZ) chondrocytes among species. Also note that cell size shows little if any difference among species, despite great differences in body size.

8.6.3 Early Postnatal Transformations

According to Schultz (1940, 1941), the first fusion of the os coxa in apes occurs at the ischiopubic ramus, e.g., between 4 and 15 months in *Pan*. This is relatively earlier than in humans (years 5–8). Complete fusion is a relatively prolonged process in apes; Schultz notes that in some adult *Hylobates* the process is still incomplete, although all primary elements are mostly fused. In humans, fusion of the primary elements is also a lengthy process, occurring between 11 and 17 years of age, and in other hominoids, fusion is incomplete until the permanent dentition is fully erupted (Schultz, 1940, 1941; Bolter and Zihlman, 2003). Fusion of the proximal femoral epiphysis is also delayed, although occurring in some late-stage juveniles. While precocious in knee development, great apes are born with far fewer ossified tarsals, and ossification is prolonged beyond infancy.

Our sample of older infant platyrrhines illustrates variation among other anthropoids. A P14 *Aotus nancymaae* is remarkably advanced in ossification, already exhibiting distal epiphyses for leg bones and metatarsals, and ossification nuclei for all tarsals (Figure 8.13). By P13, *Callithrix jacchus* reaches the same degree of ossification in the leg, ankle, and foot as the P14 *Aotus* (Phillips, 1976). However, a 1-month-old *Saguinus*

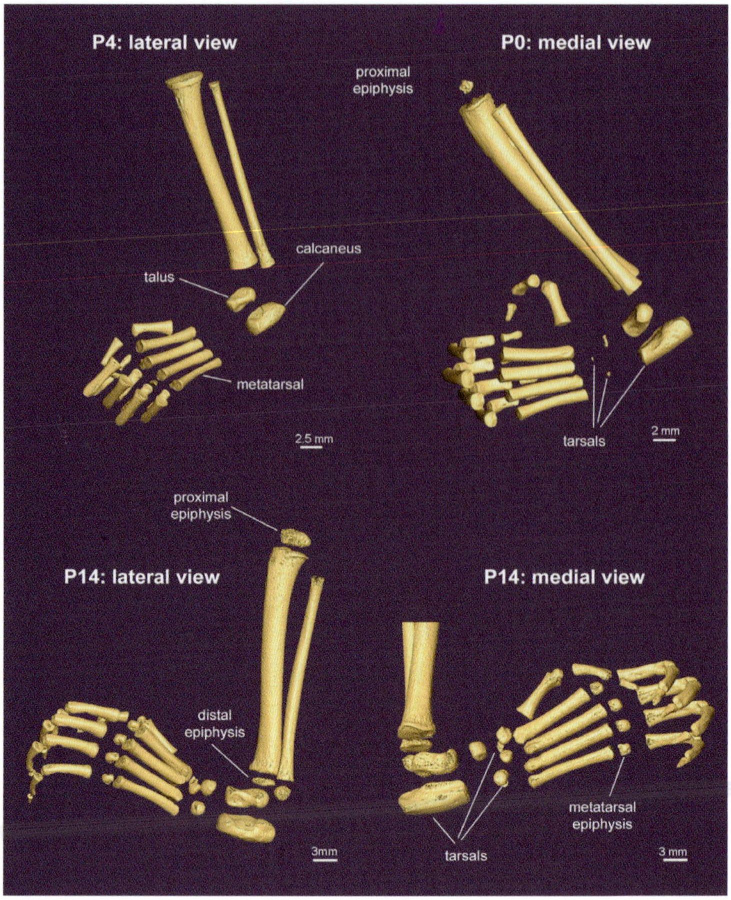

Figure 8.13 Leg, ankle, and foot bones across early postnatal ages in *Aotus nancymaae*. Perinatal specimens show variation in degree of ossification of tarsals. A 2-week-old infant is well advanced by comparison to younger animals.

oedipus is not as advanced, possessing only five tarsals that have begun ossification and lacking ossified distal leg epiphyses.

Strepsirrhines also differ at matched chronological ages. A comparison of three species at perinatal and ~1-month-old ages reveals different rates of ossification at the ankle in particular (Figure 8.14). The age series of *Otolemur* specimens reveals three to four tarsal ossification centers at birth, while at 1 month five centers are present. Thus, ankle ossification is relatively rapid prenatally and postnatally in the greater bushbaby. Two lemuroids, *Propithecus* and *Lemur*, have markedly different rates of development at the ankle. *Propithecus* possesses only the talar and calcaneal centers at both ages. In *Lemur*, it is likely that the navicular ossifies within the first postnatal week, and two additional centers are detected in a 1-month-old (Figure 8.14).

Limb bone maturation proceeds at different rates in other measures as well. *Propithecus* has significantly stronger hindlimb long bones compared to *Lemur* across this age range (Young et al., 2019). This is hypothesized to relate to the extreme

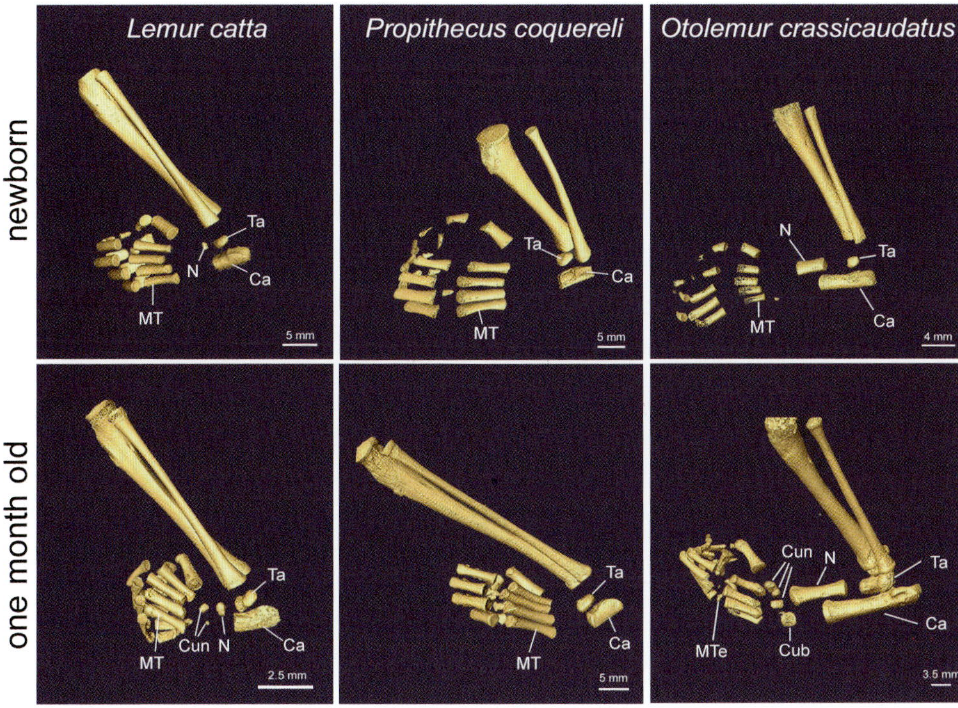

Figure 8.14 Three species of primates showing the leg and ankle skeleton at birth and at approximately 1 month of age. The skeleton of the ankle and foot is broadly similar among primates in that the calcaneus (Ca) is the first tarsal to ossify, followed by the talus (Ta), but galagids have early ossification of the navicular (N). Beyond this, the progression of ossification varies. Note that the ankle is ossifying quite slowly in *Propithecus* compared to the other species. Abbreviations: Cub, cuboid; Cun, cuneiforms; MT, metatarsals (primary ossification centers).

demands of specialized vertical clinging and leaping (VCL) locomotor behaviors of adults (i.e., the baby sifaka has a "head start" in preparing its long bones, biomechanically). Thus, while ankle/foot ossification may lag, other limb properties are advancing quite precociously in the sifaka. The slow pace of ankle ossification may be feasible because *Propithecus* does not initiate leaping for 8 weeks (Chapter 11). While *Propithecus* is initially slow to ossify their tarsals, they may be catching up with other strepsirrhines by weaning (Box 8.4).

Lesser galagids (Figure 8.11), also VCL primates, provide a useful contrast to the sifaka. Their ankle and foot ossification is notably precocious (Llera et al., 2016). Correspondingly, they initiate jumping within 10 days, although they continue to strengthen and perfect jumps in the course of the first 2 months (Doyle, 1979). However, the rapid rate of ossification also typifies non-VCL galagids, so this may simply reflect phylogeny.

Box 8.1 Microanatomy of Hindlimb Epiphyseal Cartilage: Hip Region

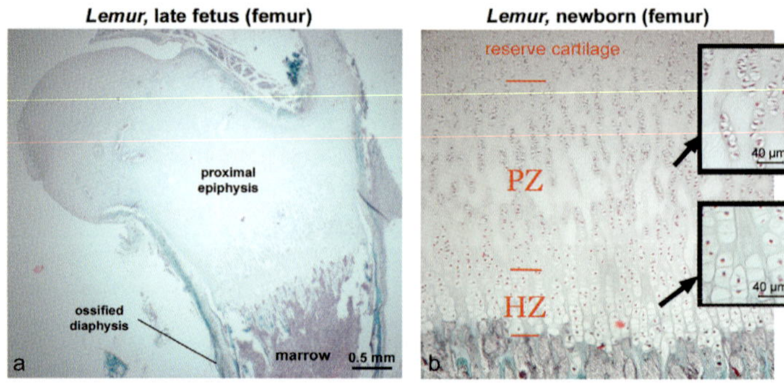

In primates, few epiphyses have initiated ossification at birth, and in many species no epiphyses are ossified. Even prior to ossification, a region of growth exists at the interface(s) of the primary and secondary ossification centers. The region is like a thickened epiphyseal plate in that the reserve, proliferating (PZ), and/or hypertrophic zones (HZ) are expanded (a,b) compared to their dimensions following secondary ossification (see Box 2.2). The thickness of this growth zone, which is nearly disk-shaped in long bones (a,c) varies. For instance, in larger species (b: *Lemur catta*), the number of cells stacked within the proliferating zone PZ may be more numerous than in smaller species (d: *Cheirogaleus medius*, newborn). Irregular bones such as the os coxae also possess growth centers, most notably near the acetabulum (e and f, enlarged from c), but with small groupings of proliferating chondrocytes.

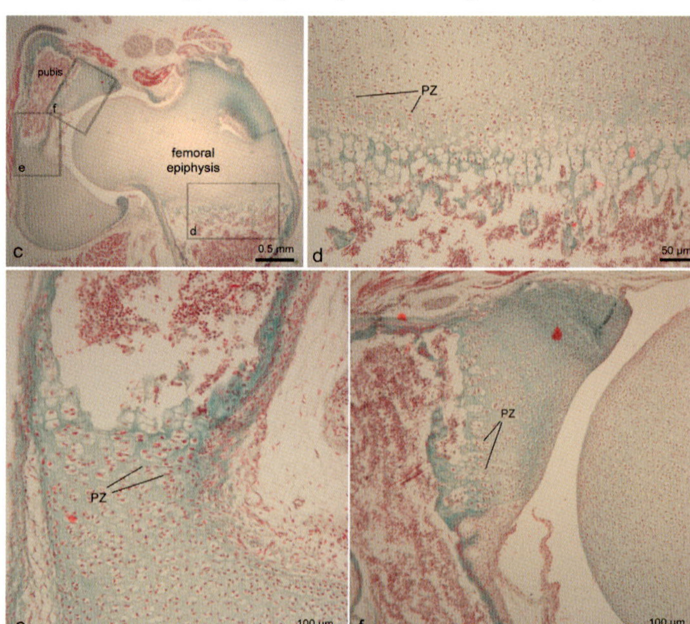

Box 8.2 Microanatomy of Hindlimb Epiphyseal Cartilage

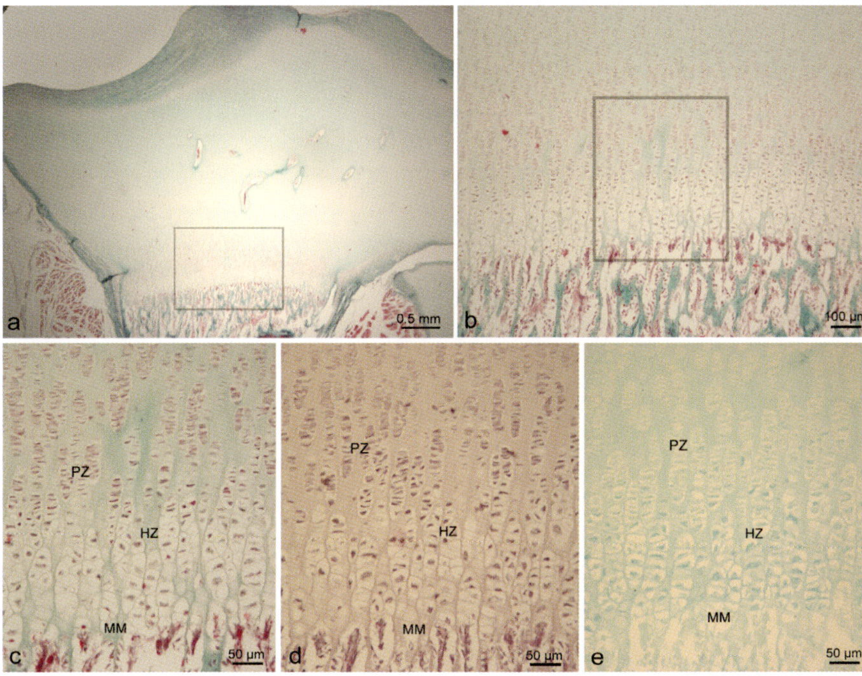

All lemuroids have entirely cartilaginous long bone epiphyses at birth, as shown in the proximal tibia of *Propithecus coquereli* (a). The apertures are cartilage canals. There is no discrete growth plate, but the organization of the cartilage adjacent to the diaphysis is distinctive (b, enlarged from box in a). The box in b is enlarged in c–e (all are closely adjacent sections stained differently). The zones of proliferation (PZ) and hypertrophy (HZ) are organized into columns. Trichrome stains (c, Gomori) reveal collagen production in the matrix (green stain). Note that there is mineralized matrix (MM) adjacent to the zone of ossification, which lengthens the diaphysis; this matrix stains darker in trichrome and hematoxylin and eosin (d). Alcian blue preparation (e) reveals sulfated glycosaminoglycans within the matrix (Frazier et al., 2008); note the reduced staining in the mineralized matrix. The matrix in the zone of hypertrophic chondrocytes (f: Gomori trichrome) is distinctive from adjacent regions in the production of type X collagen (g: adjacent section to f, prepared with immunohistochemistry using antibodies to type X collagen). *, mineralized matrix.

Postnatally, the organization of the epiphyseal zones remains the same, even once they are reduced into a growth plate that separates primary and

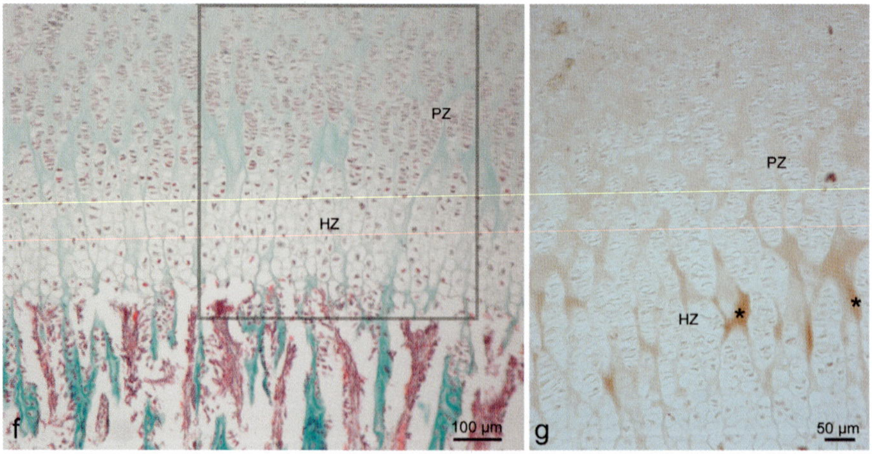

secondary ossification centers (see Chapter 2). However, a preliminary survey of perinatal and older (fetal, newborn, and ~1-month-old) infants indicates dynamic changes in cell organization and matrix composition may be occurring near the time of birth. Plate series h–j shows the proximal epiphysis in *Propithecus coquereli* specimens. Note that the number of rows of cells in the HZ appears relatively greater in newborn and older infants compared to a late fetal specimen. In addition, note that the darkened staining (using hematoxylin and eosin preparation) of mineralized matrix (*) is greatly expanded in the month-old compared to the younger specimens. This may relate to an increasingly rapid vascular invasion as the plate ossifies at the metaphyseal side. *= mineralized matrix.

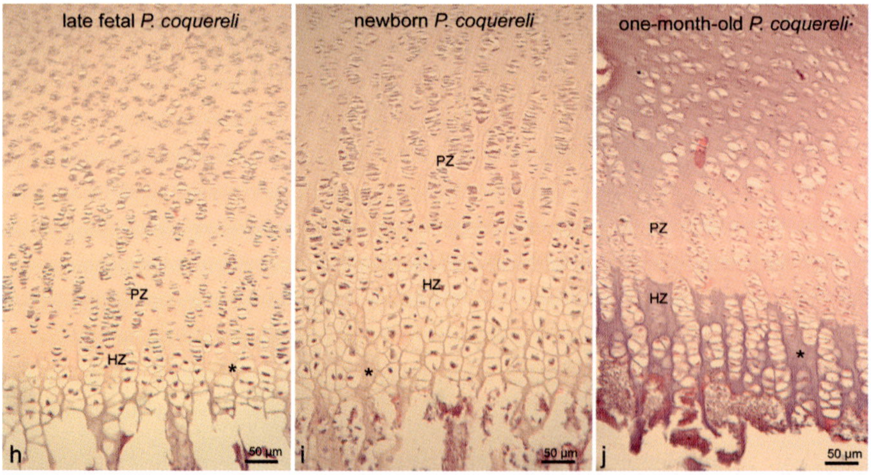

Box 8.3 Comparative Microanatomy of the Ankle: Growth Differences Among Primates

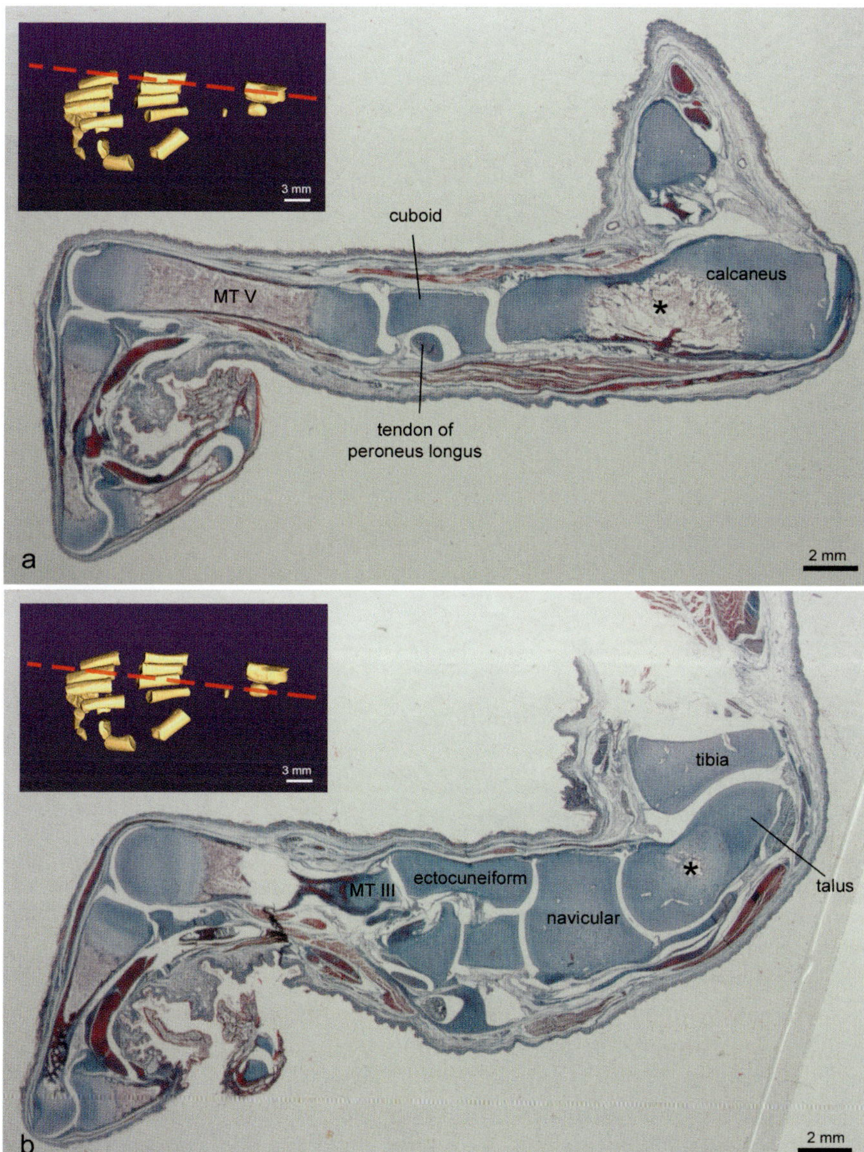

Microanatomy of the ankle/foot region in a comparative sample of primates (a,b: *Lemur catta*; c,d: *Propithecus coquereli*; e: *Cheirogaleus medius*; f: *Galagoides demidoff*; g,h: *Galago moholi*) reveals a varied state of ossification at birth.
(a) A sagittal histological section at the level of the calcaneus (plane indicated by dashed line in the inset). All primates possess an ossified nucleus (*) in the calcaneus at birth, but few have any ossification in the cuboid. (b) A parallel

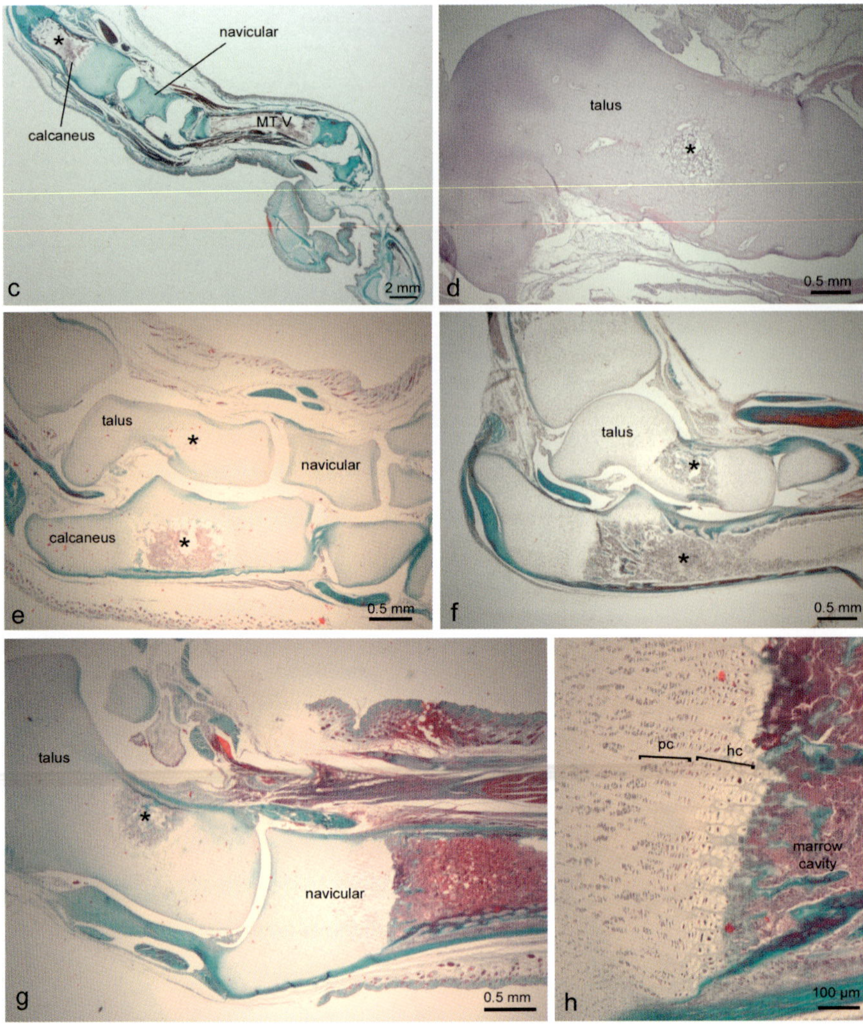

section through the talus, which has a smaller nucleus of ossification, appearing after that of the calcaneus. The navicular may or may not initiate ossification by birth. Newborn *Propithecus* (c,d) have no more than two tarsals ossified at birth, with only a small, rounded ossification nucleus for the talus. The same applies to *Cheirogaleus medius* (e).

Galagids have at least three ossified tarsals at birth. In addition to the calcaneus and talus, the navicular is ossified in all galagids. Both the calcaneus (f, *Galagoides*) and navicular (h, *Galago*) have well-ossified elongated primary ossification centers. Proximally (h) and distally are growth centers with zones of proliferating and hypertrophic chondrocytes (pc, hc).

Box 8.4 Tarsal and Foot Bones of *Propithecus coquereli* at Weaning Age

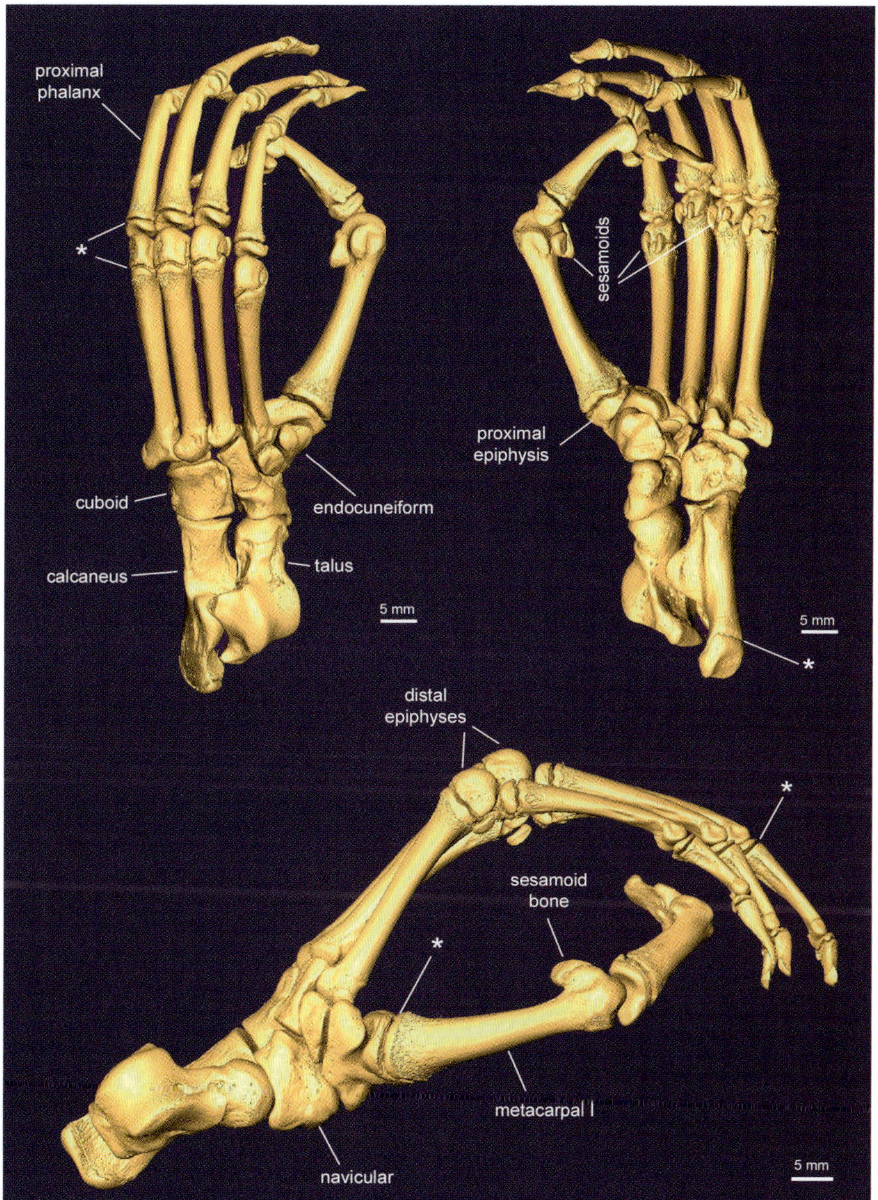

The tarsals of *Propithecus coquereli* are slow to ossify during infancy (see text). By weaning age (above, foot skeleton of a 6-month-old male), the foot skeleton is well-ossified, but still has open joints between numerous epiphyses and diaphyses (*), suggesting persistent synchondroses. Note one synchondrosis within the calcaneus distally.

9 The Newborn Primate Body Form: Phylogenetic and Life-History Influences

The research interests of Schultz (e.g., 1960) on comparative development of primates have persisted in morphological studies and have expanded to behavioral studies on subadult primates (e.g., Fairbanks, 1993; Ross, 2001; and see Bolter and Zihlman, 2007). We remain keenly interested in how long primates invest in each life stage (Chapter 2), and recognize that life-history theory pertains to the entirety of prenatal and postnatal ontogeny (Leigh and Bernstein, 2006; Leigh and Blomquist, 2007). Certain life-history characteristics of a species, such as gestational length or relative neonatal body size, are directly related to prenatal development (Schultz, 1969; Harvey et al., 1987; Martin, 1990; Ross, 2003a). In this chapter, we consider craniodental dimensions of newborn primates in the context of life-history variables. Further discussion of life history and the postcranial skeleton is found in Chapter 11.

9.1 Relative Gestation Length, Weaning Length, and Neonatal Body Size

Non-human primates vary immensely in gestation length, from 60 days (*Microcebus murinus*) to 273 days (*Gorilla gorilla*) (Kappeler and Pereira, 2003). These same species produce the smallest and largest neonates in terms of body mass (4.6 g and 2053 g, respectively). In this broad comparison, primates conform to expectations based on mammals in general; gestation is positively correlated with neonatal body size (Harvey et al., 1987). Gestation length is also positively correlated with other life-history variables. The correlations indicate that, generally, longer gestations yield larger neonates that are weaned later, become sexually mature later, and have larger brain masses, both as newborns and as adults (Harvey et al., 1987).

Harvey et al. (1987) noted that certain life-history trade-offs are inherently likely. For example, mammals that have litters tend to have small altricial young, a parceling out of gestational investment. Primates, like many mammals that typically have single young, bear well-developed and precociously mobile infants. Few primates have more than one neonate per birth. Some do have small young, that might be regarded as relatively altricial (e.g., mouse lemurs), at least for a primate. Callitrichines, however, typically have twins that are relatively large.

A notable characteristic of primates, especially among those with the most delayed reproductivity, is a relatively prolonged juvenility (Harvey et al., 1987). With this long time period in mind, the full potential cost of primate growth and development may be appreciated. Among non-human primates, the absolute time of maternal investment (from conception to weaning) may reach 5.5 years, as in *Pan troglodytes*. On the other hand, *Microcebus murinus* only invests approximately 100 days from conception to weaning. Harvey et al. also showed the close correlations between maternal body mass and many life-history variables, and some of these relationships are worth revisiting based on more recent data compilations (from Kappeler and Pereira, 2003, and Zehr et al., 2014). Maternal body size explains most of the wide range of variation in newborn body mass (correlation coefficient = 0.95; Figure 9.1), and also has strong correlations to weaning age (0.87), and moderate correlations to gestation length (0.68).

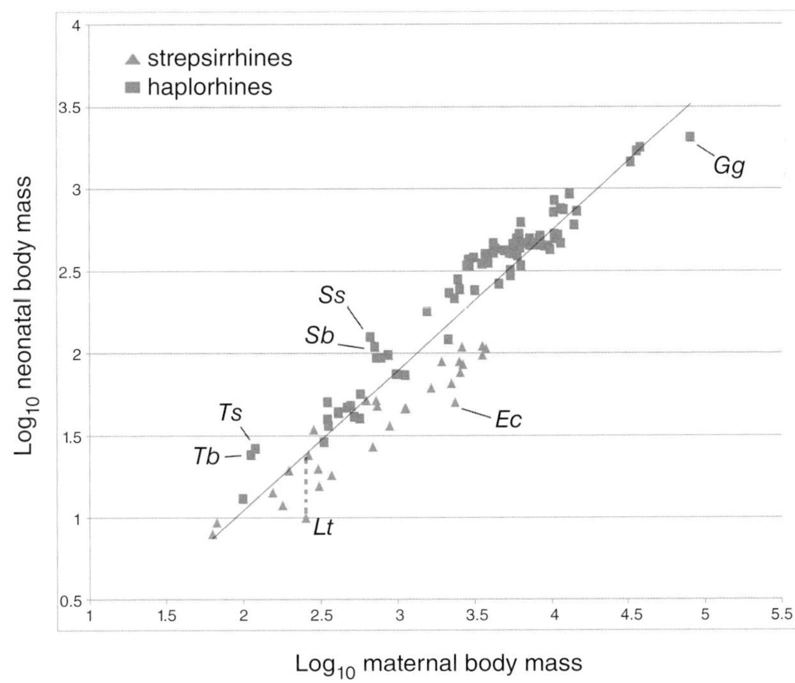

Figure 9.1 Plot of $\log_{10}$-transformed neonatal body mass versus maternal body mass of non-human primates, with a linear regression line. Broadly, most haplorhines are distributed above the predicted line, while most strepsirrhines are distributed below the line. Those distributed farthest above the line, such as the tarsiers (*Tb, Ts*), have the relatively largest newborns for a maternal body mass. Those distributed farthest below the line have the relatively smallest newborns. This distance is expressed in residual values. For example, the residual value for *Loris tardigradus* (*Lt*) is –0.388, a measure of its separation (dashed line) from the predicted line. *Ec, Eulemur collaris*; *Gg, Gorilla gorilla*; *Sb, Saimirii boliviensis*; *Ss, Saimiri sciureus*; *Tb, Tarsius bancanus*; *Ts, Tarsius syrichta.* Data from Ross (2003a).

These correlations are slightly lower than reported by Harvey et al. (1987), but with the same trends. Generally, larger-bodied mothers invest in the offspring (both prenatally and postnatally) for a longer duration. For this reason, most studies of life-history variables also reexamine data after controlling for maternal body size.

Neonatal body mass also correlates with gestation length (Table 9.1). There is greater variability in strepsirrhines (correlation coefficients: strepsirrhines = 0.36, haplorhines = 0.69); the same disparity is seen when correlating gestation length and maternal body mass (strepsirrhines = 0.35, haplorrhines= 0.71). Thus, existing data suggest strepsirrhines vary more in gestational development, especially in overall growth rate. Neonatal or maternal body mass is more strongly correlated to weaning age (Table 9.1). Interestingly, combining gestational and weaning age provides a better correlation for neonatal body mass (0.72) and maternal body mass (0.63) of strepsirrhines. This suggests strepsirrhines may invest preferentially toward either prenatal or postnatal growth.

Maternal body mass has allometric influence on all of the above life-history variables (Table 9.1). One way to control for such factors is to look at residual values. These are acquired by using a least-squares regression of the dependent variable (e.g., gestation length) compared with the independent variable (maternal body mass). The residual value for each species is a direct measure of its distance from the predicted line generated by a linear regression equation (e.g., see Figure 9.1). As a result, we have numeric indicators of which primates have *relatively* short or long gestation periods (Tables 9.2 and 9.3), or other life-history measurements. Hereafter, such life-history variables are explained as *relative* life-history values.

The addition of data acquired since the analysis by Harvey et al. (1987) yields similar results in many respects. Compared to most haplorhines, strepsirrhines have lower relative neonatal body mass compared to any anthropoid (Table 9.2; Figure 9.1). However, there is a fundamental difference in terms of maternal investment (with a caveat that alloparental provisioning occurs in many primates). Lorisoids invest relatively more in gestation and weaning periods compared to lemuroids, a finding echoing the results of Harvey et al. (1987). At the family/subfamily level, most platyrrhines have relatively large newborns, especially cebines. In Old World monkeys, cercopithecine newborns are relatively large and colobines are relatively small (Table 9.2). Ape newborns are slightly on the small side, despite having long relative gestational length and even longer weaning periods (based on later relative weaning age). Tarsiers have the largest relative newborn mass (Figure 9.1), coupled with long relative gestation length.

Table 9.1 Correlation coefficients for life-history variables of non-human primates

	Maternal body mass	Neonatal body mass
Neonatal body mass	0.95	–
Gestation length (GL)	0.68	0.71
Weaning age (WA)	0.87	0.87
GL+WA	0.86	0.87

Table 9.2 Relative life-history variables for non-human primates

	RNM	RGL	RW
HAPLORHINES			
Atelidae	–0.005	0.069	0.121
Callitrichinae	0.044	0.067	–0.082
Cebinae	0.216	0.023	0.193
Cercopithecinae	0.116	–0.015	–0.006
Colobinae	–0.026	–0.002	–0.055
Hominidae*	–0.054	0.048	0.137
Hylobatidae	0.039	0.066	0.217
Pitheciidae	0.057	–0.004	–0.143
Tarsidae	0.301	0.205	0.177
STREPSIRRHINES			
Cheirogaleidae	–0.128	–0.232	–0.036
Galagidae	–0.091	0.018	0.062
Indriidae	–0.362	–0.020	–0.160
Lemuridae	–0.289	–0.088	–0.156
Lepilemuridae	–0.325	–0.008	0.032
Lorisidae	–0.147	0.126	0.166

Abbreviations: RGL, relative gestation length; RNM, relative neonatal mass; RW, relative weaning age; *, excluding humans.

Genus-level comparisons reveal additional differences, especially among platyrrhines (Table 9.3). *Alouatta* is the main basis for the small relative size of Atelid newborns; residuals for *Ateles* are slightly on the positive side. *Callicebus*, *Callithrix*, and *Pithecia* also have negative residuals for newborn body mass against maternal body mass (Table 9.3). *Aotus*, *Cebus*, and *Callimico* have relatively large newborns. While all cercopithecines have positive residuals for neonatal body mass, this is mostly driven by the genus *Cercopithecus* (0.225), and to a lesser extent *Cercocebus* (0.157) (Figure 9.1). Counterintuitively, there is no clear relationship between relative gestation length or weaning age and relative neonatal body mass. These findings suggest that prenatal rates of growth for a given gestation length vary. Moreover, growing precociously large young is not a strategy for early independence in primates (or early reproductivity: Harvey et al., 1987).

9.2 Modular Growth and Development

The strong influence of allometry on life-history characteristics must reflect basic biological constraints on organisms in general (Harvey et al., 1987). One basis for this could be one of metabolic costs relating to development, and this has been proposed

Table 9.3 Relative life-history variables for platyrrhines

	RNM	RGL	RW
Alouatta	−0.034	0.034	0.029
Aotus	0.172	−0.023	−0.142
Ateles	0.003	0.093	0.206
Callicebus	−0.040	0.005	−0.145
Callimico	0.194	0.080	−0.133
Callithrix	−0.032	0.068	−0.071
Cebuella	0.071	0.095	0.255
Cebus	0.150	−0.005	0.190
Lagothrix	0.069	0.100	0.053
Leontopithecus	0.059	−0.019	−0.110
Mico	0.044		
Pithecia	−0.09	0.034	
Saguinus	0.028	0.089	−0.172
Saimiri	0.314	0.065	0.202

Abbreviations: RGL, relative gestation length; RNM, relative neonatal mass; RW, relative weaning age.

for the brain in particular. The brain has been proposed to have especially important links to life history, including lifespan (Sacher and Staffeldt, 1974) and age at first reproduction (Harvey et al., 1987). In addition, Smith (1989) found a strong correlation between adult (maternal) brain mass and dental development, in particular M1 eruption. This was supported later by Godfrey et al. (2001, 2003). The significance of this link has been often discussed. Vinyard and Hanna (2005) suggested the pace of dental eruption (but not crown size) may be under similar energetic constraints to the brain during development. The strong correlation between brain size and many other life-history variables indicates the need for further inquiry.

Neonatal brain mass too correlates strongly with life-history variables (Harvey et al., 1987). A strong relationship to gestation length should be expected. Primates are highly encephalized mammals, but even anthropoid brains grow prenatally at a rate similar to other eutherians (Halley, 2017). However, within primates there is significant variation in the proportion of prenatal to postnatal brain growth (Harvey et al., 1987; Leigh, 2004), and there is a wide range in the degree of infant dependence (and the rate of somatic development: Ross, 2001; Young et al., in press).

Life-history variables such as gestation length have the potential to explain independent growth rates of different parts of the body across development (Leigh and Bernstein, 2006; Halley, 2017). We thus seek to understand the modularity of growth: "different organs, or organ systems may grow over separate age spans and at very different rates" (Leigh and Blomquist, 2007, p. 402; and see Bolter and Zihlman, 2007).

9.3 Newborn Body Dimensions: Managing an Oversized Head

How do whole primate bodies grow prenatally? A relatively large sample of newborns referred to throughout this book offer a new perspective. Sitting height, overall cranial length (prosthion–inion), palatal length, cranial width (bizygomatic distance), and orbital aperture diameter (maximum height in parasagittal plane) are provided from 158 specimens judged to be newborns (see Chapter 1). These measurements have been routinely recorded on a broad array of primates (Smith et al., 2015) with added data on neonatal chimpanzees from the KUPRI collection and the literature (Table 9.4).

Table 9.4 Selected somatic[1] and dental measurements in newborn primates and *Tupaia*

Species	N	SH (mm)	CL (mm)	PL (mm)	BZW (mm)	OAD (mm)	dp4 vol. (mm^3)	M1 vol. (mm^3)
Anthropoids								
Allenopithecus	1	–	–	17.40	–	12.30	15.31	2.98
Alouatta seniculus	1	–	58.44	15.34	33.11	13.03	14.05	1.16
Alouatta sp.	1	–	58.39		–	12.70	–	–
Aotus nancymaae	4	88.95	42.01	10.71	22.90	10.31	2.12	0.85
Callicebus cupreus	1	100.29	40.65	12.76	21.83	7.78	2.99	0.50
Callithrix jacchus	16	–	27.48	7.99	15.04	6.52	0.52	0.01
Callithrix penicillata	2	80.79	29.70	7.54	15.49	6.81	0.54	0.01
Cebuella pygmaea	3	61.60	26.78	6.60	14.64	5.92	0.48	0.06
Colobus guereza	1	–	59.78	18.50	–	12.40	12.59	0.60
C. polykomos	1	–	64.12	–	–	12.37	–	–
Gorilla sp.	1	–	95.10	–	79.90	21.10	–	–
Leontopithecus rosalia	5	103.30	36.90	10.41	19.41	6.63	1.85	0.002
Macaca mulatta	2	–	62.86	19.60	44.37	14.60	15.75	1.20
Pan troglodytes	1	254.80	98.95	30.91	54.11	20.85		
Papio anubis	1	–	84.61	30.16	46.28	15.54	37.52	4.46
Pithecia pithecia	2	126.24	49.23	13.77	27.88	10.69	4.40	1.78
Pongo pygmaeus	1	–	–	28.75	–	20.18	0.52	0.22
Presbytis obscura	1	–	65.22	18.43	–	14.91	–	–
Saguinus bicolor	1	90.00	31.84	10.13	21.64	–	0.19	–
S. geoffroyi	3	–	34.86	10.46	–	6.88	0.46	0.00
S. midas	3	89.98	32.01	9.97	19.02	6.89	0.20	–
S. oedipus	12	87.19	31.65	8.62	17.07	6.69	0.26	–

Table 9.4 (*cont.*)

Species	N	SH (mm)	CL (mm)	PL (mm)	BZW (mm)	OAD (mm)	dp4 vol. (mm^3)	M1 vol. (mm^3)
Saimiri boliviensis	11	120.13	44.01	10.94	22.97	10.2	1.53	0.07
Trachypithecus francoisi	1	–	68.79	17.36	–	14.08	21.07	2.28
Tarsiiformes								
Tarsius bancanus	1	–	31.70	12.42	21.20	12.29	–	–
T. syrichta	3	57.70	28.96	9.82	16.40	9.59	0.72	2.75
Strepsirrhines								
Cheirogaleus medius	7	56.59	23.2	8.55	13.46	5.76	0.40	0.21
Eulemur collaris	3	110.84	44.37	15.82	26.08	10.27	5.45	1.05
E. coronatus	2	97.82	43.05	15.18	22.48	9.39	–	–
E. flavifrons	2	101.09	46.59	17.41	26.60	9.83	–	–
E. macaco	2	103.10	41.89	16.17	24.88	10.72	3.34	0.64
E. mongoz	2	110.10	42.53	16.44	24.47	10.41	4.75	1.16
E. rubriventer	2	113.58	47.46	17.11	26.17	11.22	6.60	1.04
Otolemur garnettii	4	83.37	35.52	12.63	19.36	9.38	2.45	1.22
O. crassicaudatus	3	90.40	37.10	14.46	17.97	8.70	2.93	1.69
Galagoides demidovii	1	51.79	21.05	7.08	–	5.90	0.48	0.53
Galago moholi	5	60.78	25.17	8.32	15.77	7.79	0.53	0.48
G. senegalensis	10	61.61	25.32	8.13	15.95	8.57	0.66	0.40
Hapalemur griseus	6	88.39	36.38	13.01	21.01	9.10	7.10	3.20
Lemur catta	5	98.61	40.26	14.56	23.50	9.86	4.01	1.23
Lepilemur leucopus	3	80.85	34.17	11.36	18.76	9.00	0.71	4.34
Mirza coquereli	2	64.86	26.17	9.00	16.34	6.94	0.34	0.07
Microcebus murinus	2	41.12	18.72	5.80	9.69	5.51	0.14	0.05
Nycticebus pygmaeus	1	53.00	28.11	9.12	16.92	7.34	0.75	0.56
Propithecus coquereli	6	114.39	46.88	17.37	28.75	11.03	1.86	19.68
Varecia variegata	2	123.50	47.83	20.46	27.02	10.88	3.62	0.16
V. rubra	6	123.10	48.28	20.21	26.55	11.30	2.73	0.15
Tupaia belangeri	2	–	22.71	11.97	10.76	4.84	0.17	0.06

Notes: [1]Somatic measurements are based on original data except for selected measurements on *Pan troglodytes* (sitting height from Schultz, 1973). SH, sitting height; BZW, bizygomatic width; CL, cranial length; OAD, orbital aperture diameter; PL, palatal length; dp4 vol., crown volume of last mandibular deciduous premolar; M1 volume, crown volume of first mandibular permanent molar; –, data unavailable.

9.3.1 Differences in Facial Projection between Strepsirrhines and Anthropoids are Detectable at Birth

Craniofacial proportions differ among adult primates at higher taxonomic levels, but with some complexity owing to allometric scaling of dental adaptations. Features such as degree of orbital convergence, frontation, and relative size are potentially diagnostic phylogenetic characteristic features at subordinal or hypordinal levels, and also have correlations with activity patterns (e.g., Ross, 1995; Kirk, 2006). In addition, adult strepsirrhines have many primitive midfacial features (see Smith et al., 2007). To what extent are these traits of adult primates manifested at birth?

Figure 9.2 is a plot of cranial length (prosthion–inion) and palatal length (prosthion–posterior mid-palatal point) against sitting height in newborn primates. The former includes the neurocranium and the latter provides a reasonable proxy for snout length. Sitting height is strongly correlated with cranial or palatal length, in all primates, anthropoids only, or strepsirrhine only (0.89–0.99). Based on this sample, $\log_{10}$ cranial and palatal lengths scale similarly in strepsirrhines and anthropoids, with a slope indicating negative allometry. Of course, the head is included within sitting height dimensions, so a close relationship is not very surprising. However, the *y* intercept for palatal length indicates that newborn strepsirrhines have longer palates relative to sitting height. This reveals that basic differences in viscerocranial profiles of adults; relatively less-pronounced midfaces in anthropoids compared to strepsirrhines are already detectable at birth. *Tarsius* falls within the range for strepsirrhines, consistent with the skull morphology (Chapter 4).

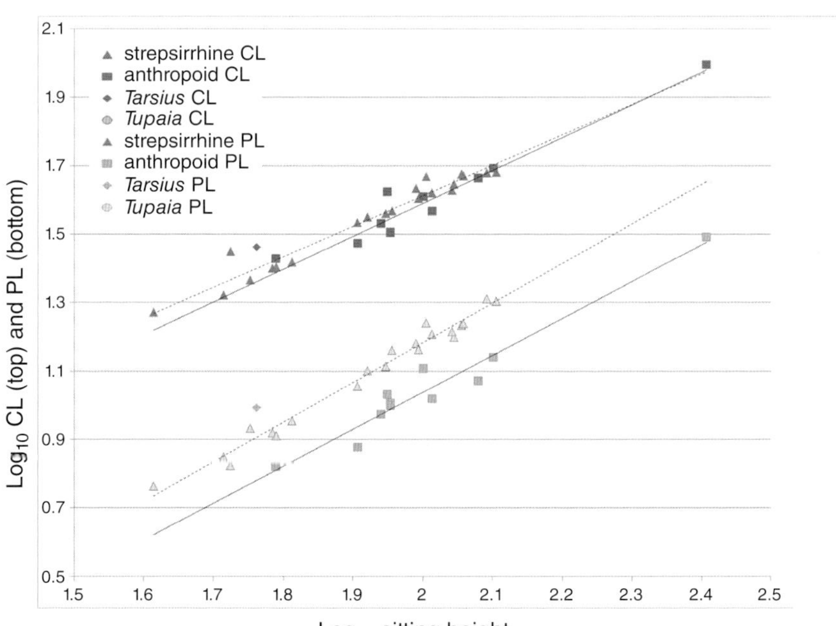

Figure 9.2 Cranial length (CL, dark gray symbols) and palatal length (PL, light gray symbols) plotted against sitting height. Regression lines: solid, anthropoids; dashed, strepsirrhines.

9.3.2 How Integrated is Prenatal Craniodental Development?

The developing head is a self-contained region in which any of the soft-tissue parts (functional matrices) may have profound influences on the skull as a whole. Thus, strong correlations among linear dimensions among regions in the newborn are expected (Table 9.5). Even though scaling differences in palatal length are clearly apparent in the newborn (Figures 9.2 and 9.3), "snout" length is uniformly diminutive in newborn compared to adult primates, even in species with pronounced snouts as adults. Just as the facial profile of the chimpanzee exhibits a pronounced vertical profile of the frontal bone ("forehead," Chapter 4), the frontal bone of larger lemurs appears more upright at birth due to a proportionally small midface and mandible (Figures 3.2 and 4.8). By contrast, in some smaller strepsirrhines and all lorisoids the midface and lower jaw may appear greatly enlarged at birth relative to adults (Figure 4.8), probably relating to delayed orbital growth combined with precocious dental growth (Cummings et al., 2012; Smith et al., 2017a).

A broader view of allometric differences among newborn primate faces is provided by Figure 9.3, which shows facial dimensions plotted against cranial length. Palatal length scales similarly in strepsirrhines and anthropoids, but with a gradistic difference similar to that seen in Figure 9.2: strepsirrhines have a greater palatal length for any cranial length. Once again, tarsiers fall within the strepsirrhine distribution at birth. The tree shrew (*Tupaia*) has a relatively long palate compared to any primate for its cranial length. Relationships of orbital and bizygomatic dimensions scale differently to cranial length in newborn anthropoids compared to strepsirrhines, as evidenced by lower slopes of regression lines in the latter (Figure 9.3). At smaller cranial lengths, strepsirrhines have relatively larger orbits and biorbital breadths than anthropoids.

Three-dimensional morphometric analyses are required to more fully understand what region(s) primarily influence the dichotomous pattern in facial projection of anthropoids and strepsirrhine. It is tempting to speculate that reduction of the nasal capsule results in anthropoid midfacial attenuation at birth. Yet, then it must be explained why tarsiers, also with reduced nasal capsules, scale with strepsirrhines. Because the palate is at the interface of two regions, this leads us to a discussion of the influence of dentition.

9.3.3 Tooth Growth and Mineralization is Frequently Disassociated from Cranial Growth

Napier and Napier (1967) suggested the snouts of anthropoids with more projecting faces were driven by the presence of well-developed teeth ("dental snouts") while facial projection in strepsirrhines relates more to the primitive architecture of the nose. Their term "olfactory snout" is since understood to be a misnomer, because outgrowth of the snout in strepsirrhines primarily increases respiratory rather than olfactory surface area (Smith et al., 2007). Is perinatal facial projection influenced by dental growth in all primates? Certainly not equally so, because the dentition varies greatly in size and eruption status in newborns (Godfrey et al., 2001, 2004; Smith et al., 2015, 2017a). A more basic question remains: does dental growth drive jaw growth? Boughner and Dean (2004) found that developing baboon mandibles possess more jaw space

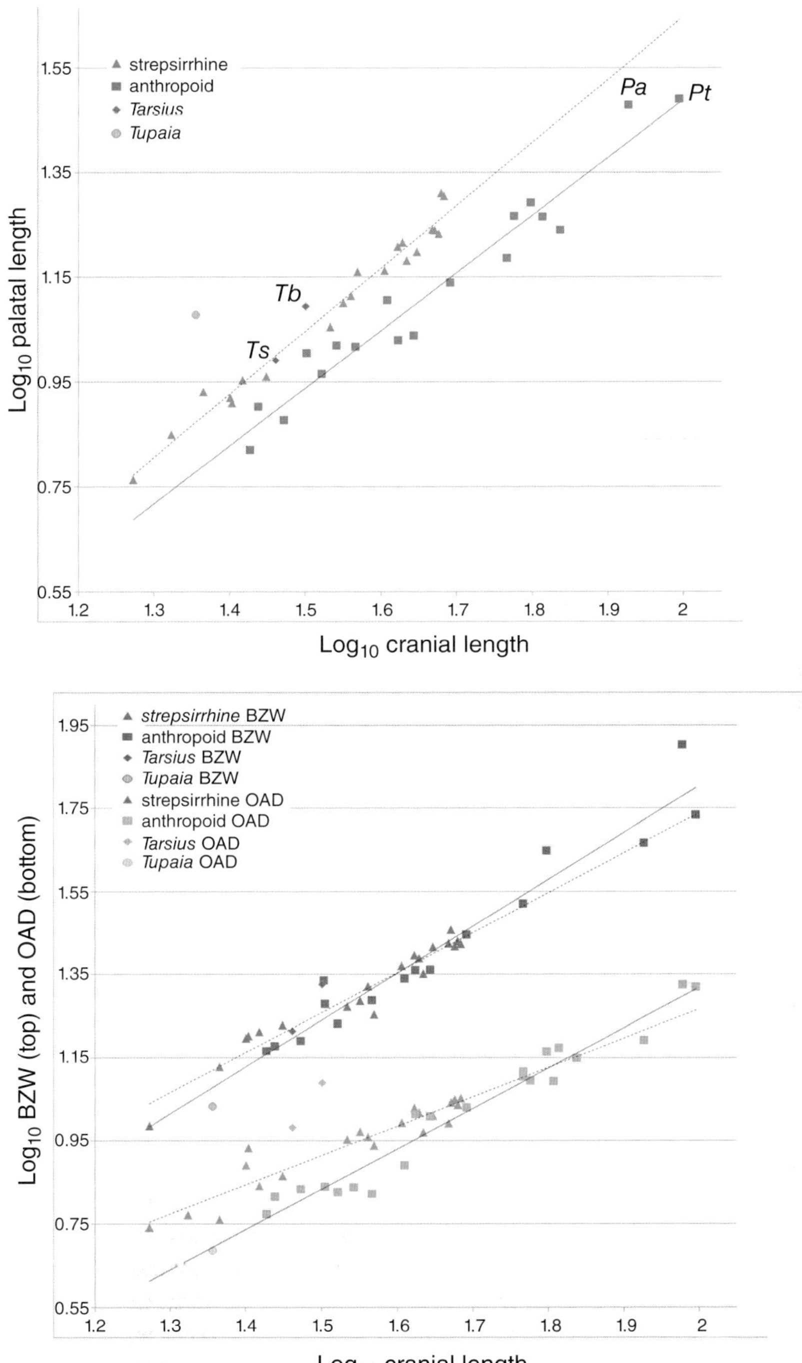

Figure 9.3 Log-transformed data of palatal length (top) and orbital dimensions (bottom) plotted against cranial length. Regression lines: solid, anthropoids; dashed, strepsirrhines. *Pa, Papio anubis; Pt, Pan troglodytes; Tb, Tarsius bancanus; Ts, Tarsius syrichta*. BZW, bizygomatic width; OAD, orbital aperture diameter.

than is easily explained by tooth crypt size. However, their observations included few or no neonates, as the infants included were of unknown age, many with advanced M1 development. It may be the case that deciduous and permanent teeth contribute quite differently to facial form. Correlations of newborn palatal length are far higher with tooth germ volume of deciduous premolars than with those of permanent molars (Smith et al., 2017a). This is also true of crown volumes, although strepsirrhines appear to drive this scaling difference (Table 9.5).

Even tooth germs that are little advanced into the late bell stage may influence jaw length (DeLeon and Smith, 2018). This may be why newborn *Varecia*, which have poorly mineralized teeth compared to other lemurids at birth, nonetheless have projecting snouts. "Snout" length also differs among newborn anthropoids (Figure 4.5), and the extent to which this is driven by growth differences in tooth germs deserves exploration.

Table 9.5 Pearson correlation coefficents of $\log_{10}$-transformed craniodental dimensions in the neonate

	CL	PL	BZW	OAD	CR dp_4 vol.
All primates					
PL	0.926				
BZW	0.969	0.921			
OAD	0.915	0.901	0.931		
CR dp_4 vol.	0.888	0.863	0.847	0.891	
CR M_1 vol.	0.389	0.502	0.479	0.614	0.527
Anthropoids					
PL	0.974				
BZW	0.970	0.973			
OAD	0.977	0.940	0.972		
CR dp_4 vol.	0.942	0.884	0.873	0.927	
CR M_1 vol.	0.783	0.729	0.793	0.843	0.836
Strepsirrhines					
PL	0.988				
BZW	0.971	0.953			
OAD	0.946	0.921	0.945		
CR dp_4 vol.	0.891	0.877	0.856	0.854	
CR M_1 vol.	0.459	0.385	0.485	0.472	0.487

Abbreviations: BZW, bizygomatic width; CL, cranial length; OAD, orbital aperture diameter; PL, palatal length; CR dp_4 vol., cube root of mandibular dp_4 crown volume; CR M_1 volume, cube root of maandibular M_1 crown volume; all correlations significant at $p < 0.05$.

9.3.4 Life History, Functional Matrices, and Modular Growth

Recently, Halley (2017) examined fetal development in 10 eutherians and two metatherians in order to assess prenatal growth rates of the brain and other body organs. Neural growth rate is notably similar among the eutherians he studied. Moreover, acceleration of overall somatic growth rate predicts growth of organs (e.g., heart, lungs) but not brain growth. This is consistent with the observation that mammals vary more in body than brain growth (Sacher and Staffeldt, 1974; Passingham, 1985). Pagel and Harvey (1988) argued that life-history variables such as gestation length or litter size have a strong influence on neonatal brain size. Other authors consider neural tissue itself to develop under a constraint because it is metabolically costly to develop and maintain (Isler and van Shaik, 2009); for this reason, Martin (1981) proposed brain mass in the newborn is linked to maternal metabolism. Halley (2017) asserted that higher encephalization at birth reflects reduced body growth rates, rather than enhanced neural growth rate. In this light, the large-headedness of the newborn primate may reflect a conservation of metabolic input toward gestational development of the brain, at the expense of other body regions.

We should await observations on prenatal strepsirrhines to better understand brain growth in primates generally, because they appear to defer more brain growth to postnatal development. Based on data from Harvey et al. (1987), the brain of newborn anthropoids (excluding humans) is 31.2–62.3% of adult brain mass, whereas the brain of newborn strepsirrhines is 29.9–47.9% of the adult mass.

It may be the case that all primates are more advanced in neural than somatic maturity at birth (Portmann, 1939; Grand, 1992; Nicolson, 1984; Derrickson, 1992). If so, this may reflect a preferential allocation of metabolic resources toward prenatal brain maturation. This is certainly behaviorally true in that immediately following birth, no primates are described to locomote in an adult-like manner, and no primates immediately ingest an adult diet (Atzeva et al., 2007; Young et al., in press). The pace at which they become independent varies immensely. While numerous studies have documented the behavioral and morphological transition from subadult to independent stages (e.g., Eaglan and Boskoff, 1978; Ravosa et al., 1993; Young and Heard-Booth, 2016; Young et al., in press), the neonatal stage in particular represents a notable transitional mark in time. At parturition, we assume selection begins to act more directly on the individual: despite maternal and/or alloparental care, infants that cannot nurse or cling to adults are highly vulnerable. Many primates can cling to mothers immediately, while a few species are nest-bound for weeks (Fietz and Dausmann, 2003; Ross, 2003a).

In humans, neonatal head size is often related to a sexually dimorphic pelvis, and specifically to the dimensions of the pelvic outlet. Comparative evidence on primates suggests complex factors affect pelvic dimorphism (Schultz, 1949; Tague, 1991; Rosenberg, 1992). While there may be limits in brain size imposed by pelvic dimensions, the flexible sutures of the neurocranium allow transient deformation during parturition, so the limitation is not strictly dimensional. The peak velocity of brain growth occurs prenatally, at least in anthropoids (Halley, 2017), so growth of the brain gradually wanes over postnatal time as other modules gain momentum.

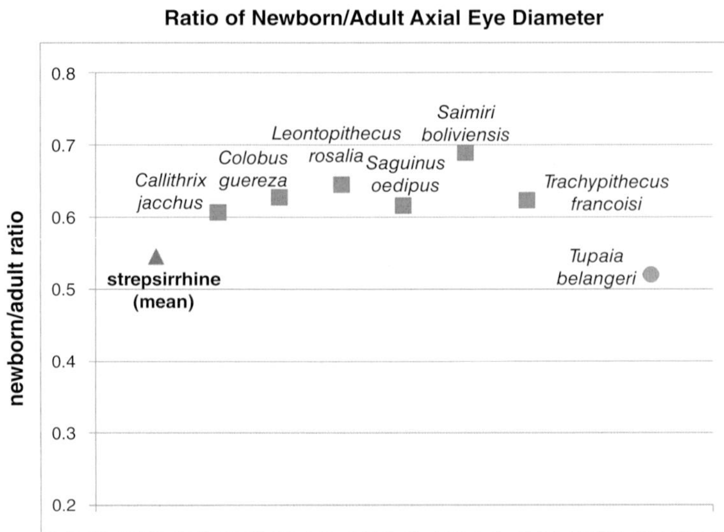

Figure 9.4 Newborn/adult eye size ratios in primates and *Tupaia*, comparing selected anthropoids (unpublished data) to average axial eye diameter in strepsirrhines (Cummings et al., 2012).

The eyes of primates are precociously functional at birth in most species, although cheirogaleids are born with closed eyes. Roughly speaking, there is a parallel to existing data on neonatal brain size. Strepsirrhines tend to have less well-grown brain *and* eyes at birth compared to anthropoids, although few anthropoids have been studied to date. Figure 9.4 shows the ratio of newborn to adult eye diameter in a small sample of anthropoids, compared to the mean for strepsirrhines. In strepsirrhines this ratio is close to that for *Tupaia*. Anthropoids uniformly cluster higher.

Tarsiers have notably smaller eyes at birth compared to other primates (relative to their adult eye size), suggesting a delayed postnatal growth spurt for the eye (Cummings et al., 2012). This contrasts with the plots on orbital aperture diameter in Figure 9.3, in which newborn tarsiers scale higher than anthropoids. This may indicate the eyes are undersized for orbit diameter. However, it is noteworthy that Figure 9.3 includes average data values from tarsiers ranging from P0 to P6, and data for *T. syrichta* have a wide range (6.4–11.5 mm). Smith et al. (2013) and DeLeon et al. (2017) interpreted perinatal changes in osteology to indicate tarsiers have a perinatal rapid "pulse" of eye growth, and speculated that eye size may be under an obstetric constraint. Intriguingly, the "partial post-orbital plates" are incomplete in newborns, with the zygomaticofrontal suture site existing as a broader separation than in other primates, connected by a broad collagenous ligament (Figure 9.5). Smith et al. (2013) did not interpret this as a homologue of the postorbital ligament seen in some other mammals, but rather as a flexible joint, which may function like a fontanelle during parturition, to protect the eyes. The eyes may actually be more recessed (and protected) in newborn tarsiers than adults.

Strepsirrhines and tarsiers have a lesser "packaging dilemma" in that their eyes become partially ectopic during development. Because of this, there may be more

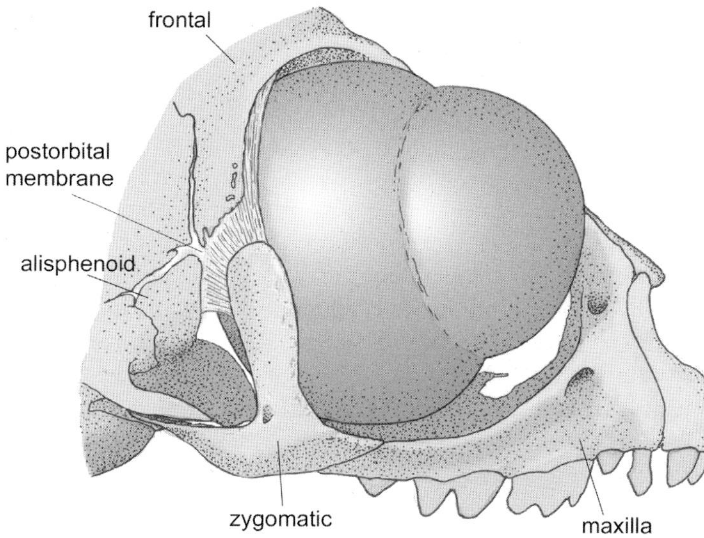

Figure 9.5 The orbit of a newborn *Tarsius bancanus*. Note lateral, expansive ligamentous support and partially recessed, partially ectopic eye.

potential for delayed growth of the eyes. This may be beneficial because tarsiers and some strepsirrhines must accommodate relatively precocious teeth (Figure 9.5).

Like the brain, the dentition has been hypothesized to be metabolically "expensive" to grow, in light of its heavily mineralized end-point (Godfrey et al., 2004). If so, primate mothers may invest in prenatal crown development, or defer this until postnatal time. There are strong phylogenetic patterns, more completely covered in Chapter 5. Indriids uniformly have a more advanced status of dental eruption at birth than other large lemurs (Smith et al., 1994; Godfrey et al., 2005). Although this is an adaptation to a highly folivorous diet, not all folivores are so advanced in development of permanent teeth. Instead, folivorous strepsirrhines are unified in their later pace of dental eruption and overall occlusal surface area (Godfrey et al., 2001). Cheirogaleids and lorisoids have extremely well-mineralized deciduous teeth and precocious development of many permanent teeth in newborns (Smith et al., 2015).

Residual analyses revealed interesting correlations of eye diameter and dental germ volumes to life-history variables, while controlling for body size and phylogeny. In analyses of strepsirrhines, relative gestation length is negatively correlated with relative deciduous tooth germ volume but positively correlated with relative permanent molar (summed) and relative eye size (Cummings et al., 2012; Smith et al., 2017a). Cranial modules thus appear to be quite independent targets of selection, and some are transiently influential on facial morphology. Generally, species with the shortest gestation rapidly prepare large teeth for mastication of solid food, but are relatively slower to grow their eyes and brains. Species with longer gestations have smaller deciduous tooth germs, but larger permanent tooth germs and eyes at birth. This may indicate maternal prenatal investment has a stronger influence on deciduous dentition than permanent dentition.

The link to weaning age, as a strategy, is unclear. However, a trend in strepsirrhines (but without significant correlation) is that strepsirrhines with a longer duration until weaning also have larger permanent molar volume at birth (Smith et al., 2017a). The implication may be that strepsirrhines invest more in prenatal growth of deciduous teeth compared to permanent teeth. A caveat is that the trajectory to ultimate tooth size is certainly unlinked from crown mineralization rate – most large-bodied lemurs have large postcanine deciduous tooth germs with large unmineralized cusp parts (Smith et al., 2015, 2017a).

As is the case for adult primates, there are some strong phylogenetic patterns that confound our ability to glean life-history influences on eye or tooth size (Cummings et al., 2012; Smith et al., 2017a). Among the factors discussed above, gestational length has the most significant influence on facial functional matrices, with a positive correlation with permanent tooth and eye size, and a negative correlation with deciduous tooth germ size, even after controlling for phylogeny (Cummings et al., 2012; Smith et al., 2017a). This suggests selection on gestation has a major influence on cranial morphology through effects on the growth of viscerocranial structures.

10 Ontogeny of Feeding

10.1 Introduction

Primates undergo three basic stages of feeding development. In utero, nutrition is gained maternally via placentation without active involvement of the feeding apparatus. Suckling commences with birth and represents the first feeding behavior that can be linked to the feeding apparatus (German et al., 1992; German and Crompton, 2000). Following suckling is a transition to solid foods. When weaning occurs and how long it takes to complete vary among primates (Lee, 1996). Similarly, weaning and post-weaning changes in diet appear to vary considerably across primates and likely represent competing influences of numerous ecological, sociological, and morphological factors. Documenting the timing and relative significance of these relationships has proven challenging.

We have two main goals of this chapter. First, we briefly discuss the essential pattern of feeding ontogeny in primates. Detailed descriptions for species are beyond our scope and are unlikely to provide summary insights into primate feeding ontogeny. Second, we examine weaning and feeding ontogeny as they relate to diet and feeding apparatus ontogeny. We do not attempt to explore the numerous ideas related to life-history explanations underlying weaning and the ontogeny of feeding (e.g., Lee, 1996, 1999; Ross, 2003b; Leigh and Blomquist, 2007). For our purposes, it is sufficient to note that numerous factors likely play a role in the feeding ontogeny of primates. We concentrate on describing and exploring how those relationships relate to the development and growth of the primate feeding apparatus.

10.2 A Brief Overview of Primate Feeding Ontogeny

The landmark events in feeding ontogeny are the advent of suckling at birth followed by weaning as they mark major changes in life history with numerous social, nutritional, and ecological sequelae. From an oral processing perspective, suckling begins the process of liquid ingestion and active transport through the oral cavity to the pharynx, while the transition to solid foods adds ingestion of solids and mastication along with active transport (Box 10.1).

For primates (and more generally therian mammals), the ability to suckle effectively has to happen at birth and is therefore critical to survival. The capacity to use the feeding apparatus to latch-on effectively and create a seal, as well as the oscillating sucks needed to acquire milk (see Elad et al., 2014) appear, based solely on observation, to be characteristic of all primates. All primates suckle at birth. We do not know how variation in lip, jaw, and/or tongue morphology, their movements, or their coordination impacts the relative abilities or efficiency of suckling across primates. We do know that inefficiency or inability to suckle can be detrimental and even fatal to infants (e.g., Mackeprang and Hay, 1972). Given the lack of comparative data related to suckling, there are numerous interesting and unexplored questions for future research. For example, do folivorous primates that wean relatively early during ontogeny have a feeding apparatus that offers them improved capacity and efficiency in suckling? Developing measures of suckling efficiency, morphological correlates of the feeding and pharyngeal apparatuses, and approaches for quantifying these parameters appears to be a wide-open area for future research.

Compared to suckling, weaning and the acquisition of an adult diet has garnered far more attention in primates (see Lee, 1996 for review). Weaning is not a single event but a process, making it deceptive and difficult to quantify accurately. Definitions vary in their focus on infants and/or mothers and can range in time from the first bite of solid food to the last suckling bout (Lee, 1996). The time infants and juveniles spend in between those events varies considerably among species and is impacted by several potential somatic, social, and ecological processes (Leigh, 1994; Lee, 1996; Rapaport and Brown, 2008). Across primates it is clear that weaning is a key life-history event as it demonstrates significant associations with adult body mass, neonatal mass, brain mass, gestation length, interbirth interval, and locomotor independence (Leigh and Blomquist, 2007; Young and Shapiro, 2018). To date, most of the research on weaning has focused on the behavioral and social aspects of these events and how juvenile primates learn to become adept foragers (e.g., Rapaport and Brown, 2008; Bray et al., 2018).

The process of weaning and acquisition of an adult diet is characterized by a progression through several milestones. Initially, there are early bouts of experimentation with solid foods while maintaining lactation. This is followed by increased feeding on solid foods and active weaning. Subsequently, juveniles tend to take on the feeding profiles of their mother. Finally, feeding becomes similar to other group members as a juvenile adopts a more adult-like feeding pattern (Nicolson, 1984; Watts, 1985; Boinski and Fragaszy, 1989; Lee, 1996; Weins and Zitzmann, 2003; Tarnaud, 2004; Schiel and Huber, 2006; Raguet-Schoefield, 2010). Across primates, the time course of this general progression varies in length and hence so does weaning age.

Once juvenile primates have transitioned to a primarily solid-food diet, the ontogeny of adult-like feeding behavior is not well documented across primates. In some cases, juveniles are thought to adopt an adult-like pattern rather quickly (e.g., Stone, 2006; Nowell and Fletcher, 2008; Bray et al., 2018), while other species are noted for having an extended period of juvenile (or more accurately, non-adult) feeding patterns (e.g., Johnson and Bock, 2004; Gunst et al., 2010; Eadie, 2015; Chalk-Wilayto et al., 2016).

It is clear that the smaller size and reduced capabilities of juveniles (either ingestive, masticatory, and/or digestive) likely constrain their potential diet (e.g., Janson and

Van Schaik, 1993; Leigh, 1994; Gunst et al., 2010; Eadie, 2015). More often than not, explanations of reduced capabilities are inferred rather than demonstrated, given the difficulties of quantifying physiological abilities in free-ranging primates. More subtle differences are also noted, such as observations that juvenile squirrel monkeys feed similarly to adults but demonstrate less skill in manual capture and processing. Consequently, less-acute visual processing can lead to reduced avoidance of less-palatable insects (Boinski and Fragaszy, 1989).

Recent work examining the food material properties (FMP) of diet during growth in several primate species provides another important perspective on the potential factors underlying the acquisition of an adult diet. One of the best case studies focuses on capuchins because of their relatively challenging diet (Fragaszy et al., 2004; Norconk et al., 2009). Capuchins as extractive foragers are generally viewed to have a difficult-to-acquire diet and a subsequent extended period of dietary maturation (Fragaszy et al., 2004; Gunst et al., 2010; Chalk et al., 2016). Recent work by Eadie (2015) demonstrated that juveniles exhibited longer rates of acquisition than adults for more challenging foods. However, Chalk et al. (2016) observed no significant differences in dietary toughness or elastic modulus between juvenile and adult capuchins. Similar overlap in FMPs was observed between juvenile and adult saki monkeys (Robl, 2008), orangutans (van Schaik et al., 2009), howlers (Raguet-Schofield, 2010), sooty mangabeys (McGraw et al., 2011), and leaf monkeys (Chalk-Wilayto et al., 2016). Alternatively, juvenile diets appeared significantly less tough compared to adults in uakaris (Bowler and Bodmer, 2011), Japanese macaques (Taniguchi, 2015), and gelada baboons (Venkataraman et al., 2014). Based on these varying signals, it is clear that numerous factors impact diet choice including FMPs, food geometry, animal size and strength, as well as learning. Complicating these relationships is the ability to modulate feeding behaviorally, such as altering bite sizes (e.g., Reed and Ross, 2010) with food properties. Determining how these and other factors impact feeding efficiency speaks to the need for additional data and a more holistic consideration of feeding throughout growth (Chalk-Wilayto et al., 2016). Collectively, it is clear that primate dietary development is a complex and varying process without singular explanations.

10.3 Feeding Ontogeny, Diet, and the Feeding Apparatus

The ontogeny of the feeding apparatus, along with the rest of the body, is tied to a multitude of influencing factors (Harvey and Clutton-Brock, 1985; Janson and Van Schaik, 1993; Smith et al., 1994; Godfrey et al., 2001). It is beyond our scope to review all of these and we alternatively focus on how feeding ontogeny relates to diet and the feeding apparatus. Interested readers can consult various reviews for a more inclusive discussion of the life-history correlates of growth in primates (e.g., Janson and Van Schaik, 1993; Smith et al., 1994; Lee, 1996; Ross and Jones, 1999; Godfrey et al., 2001; Leigh, 2004).

10.3.1 Diet

Diet is considered a significant factor influencing growth, and by extension feeding apparatus ontogeny, in multiple life-history models of primate growth. Others, such as the design constraint or brain pleiotropy hypotheses, focus on somatic or neural growth and incorporate diet only secondarily if at all (e.g., Western, 1979; Ross and Jones, 1999). The foraging independence hypothesis (Ross and Jones, 1999) links to diet through variation in the amount of learning needed to forage effectively. It predicts that folivores should develop their feeding apparatus more quickly given the ease of foraging for leaves compared to diets rich in fruits. Godfrey et al. (2001) found support for the foraging independence hypothesis after adjusting for body and brain size, with the potential exception of folivorous lemurs. The risk aversion hypothesis (Janson and Van Schaik, 1993) predicts that animals will reduce growth rates to avoid starvation in risky environments, such as those with highly varying dietary resources. Leigh (1994) identified a general trend for anthropoid folivores to exhibit more rapid growth that ends earlier than frugivores in support of the hypothesis that a more continuous, stable diet (i.e., leaves) reduces starvation risks and facilitates a more rapid growth rate. The protein richness hypothesis also predicts rapid growth in folivores, along with insectivores and seed-eating primates, given these diets' relatively greater amounts of dietary protein (Glander, 1980). After controlling for body size, there is little support for this hypothesis in primates (Godfrey et al., 2001). A final hypothesis, the food-processing hypothesis, specifically predicts that the feeding apparatus must be sufficiently developed to meet the requirements of ingestion and mastication at weaning (Eaglen, 1985). This hypothesis predicts that folivores should exhibit more advanced feeding apparatus development at weaning than frugivores, holding other variables constant. In general, Godfrey et al. (2001) found that folivores did exhibit relatively advanced dental development at weaning compared to frugivores (see Godfrey et al., 2001, figure 1).

Despite the apparently rapid pace of dental eruption and/or mineralization in the most highly folivorous primates, there are important caveats to consider. First, most of the comparative work examining the predictions of these hypotheses has focused exclusively on the teeth (e.g., Smith, 1989; Smith et al., 1994; Godfrey et al., 2001, 2004, 2005). Second, recent work showing that more frugivorous macaques do not differ markedly from folivorous colobines in several life-history variables challenges these results based on underlying issues with the original data, such as the combination of wild versus captive data, as well as theoretical challenges to the concept of a fast–slow continuum of life-history variation (Borries et al., 2011). While it is clear there is some relationship(s) between diet and feeding apparatus growth, these caveats stand as a stark reminder of the complexity of primate growth and the fact that individual factors likely have varying influences across different primate species.

10.3.2 Dental Maturation, Eruption, and Diet

Dental maturation and eruption across primates are covered in detail in Chapter 5. Our goal here is to briefly consider some of the potential relationships between dental ontogeny and diet that may help to explain variation in dental ontogeny across primates. As seen from the previous section, most of the research on feeding apparatus ontogeny has focused on the teeth.

In general, mineralization rates at birth vary considerably across primates. Among anthropoids, callitrichines appear less mineralized than other platyrrhines, with other species exhibiting some level of M_1 mineralization at birth. *Cebuella* exhibits more rapid mineralization compared to other callitrichids after birth. Only *Pithecia* appears to have begun M_2 mineralization by birth. Strepsirrhines appear on average to exhibit greater mineralization at birth than anthropoids. Among strepsirrhines, indriids exhibit rapid mineralization, followed by galagos, while *Varecia* exhibits the least amount of dp_4 and M_1 mineralization at birth. This trend persists postnatally.

Emergence of postcanine teeth generally follows patterns seen for mineralization in that strepsirrhines tend to exhibit gingival emergence before anthropoids. Among anthropoids, catarrhines have later emergence than platyrrhines, with hominoids further delayed as initial emergence can occur months after birth (Smith, 1989). A similar size-related trend is apparent in strepsirrhines, with smaller taxa demonstrating more rapid emergence than larger taxa.

Dental eruption, as marked by some level of functional occlusion, offers the most complete data set across primates. Monkeys generally erupt teeth faster than hominoids as predicted by "Schultz's Rule" (Smith, 2000). Callitrichids exhibit relatively faster eruption than other platyrrhines. Strepsirrhines also follow this general pattern, with the exception of indriids, which exhibit relatively rapid eruption. As with mineralization and emergence, indriids erupt their teeth more rapidly than galagos, who precede ruffed lemurs.

Multiple factors likely contribute to the variation in dental maturation and eruption seen across primates. Some of the differences, such as the relative delay in mineralization and emergence of permanent teeth (i.e., M_2 and M_3) in anthropoids relative to strepsirrhines (Smith et al., 1994; Smith et al., 2015, 2017a) along with the former's emphasis on deciduous dentition, remain somewhat unexplained. These basic suborder differences may simply reflect alternative strategies for maintaining functional dental batteries throughout life. Setting aside explanations, significant associations of dental development with brain size, body size, and facial size have been observed or proposed across various primate groups (Smith 1989; Godfrey et al., 2001; Vinyard and Hanna, 2005; Scott, 2012). The interactions among these potential variables, such as interactions between brain and dental development, may ultimately help explain variation in dental ontogeny across primates. For example, relative increases in encephalization in anthropoids versus strepsirrhines may play a role in their different strategies for emphasizing deciduous versus permanent dentition in ontogeny.

The rapid mineralization and eruption of the dentition in indriids is one of the best described relationships between diet and dental ontogeny in primates (Godfrey et al., 2001, 2004, 2005; Smith et al., 2015; Box 10.2). Indriids rapidly develop their postcanine dentition to provide a sufficient, functional dental battery early in life. This rapid dental maturation allows them to break down their mechanically challenging, folivorous diet (Godfrey et al., 2001). An important part of this observation is that dental ontogeny is decoupled from the otherwise slower somatic development of indriids (Godfrey et al., 2003, 2004). This decoupling demonstrates that teeth can evolve independently of the rest of the body and that diet can impact life history in different ways across primates (Godfrey et al., 2004; Smith et al., 2017a). The overall slow growth of the rest of the body fails to support the risk-aversion hypothesis (Godfrey et al., 2001, 2004). Moreover, their dental ontogeny does not follow "Schultz's rule" of dental development (Godfrey et al., 2005), while alternatively supporting a clear relationship between dental ontogeny and a folivorous diet (Godfrey et al., 2001, 2004).

The convincing case of the indriids may provide comparative insight into other groups. *Hapalemur* does not have accelerated dental development of its permanent dentition (Godfrey et al., 2003), but does present the largest relative deciduous tooth volume of any primate (Smith et al., 2017a). Thus, strepsirrhines present (at least) two approaches to dealing with the challenges of folivory: (1) rapid maturation and eruption of permanent teeth (i.e., indriids and *Lepilemur* – see Chapter 5) and (2) enlarging deciduous teeth for prolonged utilization (*Hapalemur*). Given that the functional necessity of breaking down fibrous sheets of material (i.e., leaves and grasses) relies on sufficient occlusal surface area for generating occlusal work, either approach theoretically suffices. These differing growth patterns for establishing a functional dental battery may point to different initial conditions and evolutionary pathways available to various folivorous primate groups. For example, Smith et al. (2017a) note a negative correlation between relative volume of the permanent teeth and facial length, suggesting that facial development may impact dental patterns. This is certainly a question deserving further consideration.

Interestingly, among anthropoids *Trachypithecus* and *Alouatta* also show advanced mineralization of dp_4 and M_1 at birth, suggesting that rapid mineralization may be a characteristic of many primate folivores (Smith et al., 2015). It is clear at this point that the strongest associations between diet and dental ontogeny highlight various strategies used by folivores to acquire a competent dental battery by weaning. Exceptions to the relationship between dental precocity and folivory, such as the rapid dental development of *Cebuella*, point to the multiplicity of factors driving somatic and life-history evolution in primates.

10.3.3 Musculoskeletal Ontogeny of the Feeding Apparatus and Diet

While there is a considerable body of work on dental ontogeny in primates, research into how the rest of the feeding apparatus grows is far more limited. Most of this research focuses on pairwise comparisons or evaluation of specific taxonomic groups. Moreover, there have been relatively few studies tracking age-specific changes in musculoskeletal form of the feeding apparatus. In part, the relatively low number of ontogenetic studies of musculoskeletal form of the feeding apparatus reflects the difficulty of acquiring appropriate samples for analysis.

The largest group of ontogenetic studies of the primate feeding apparatus compares growth allometries of skeletal elements among closely related species. Many of these species differ in diet. Among African hominids, Taylor (2002) identified a dietary signal linked to mandibular morphology between more folivorous gorillas compared to more frugivorous chimpanzees and bonobos. In these ontogenetic comparisons, mountain gorillas exhibited relatively wider mandibular corpora and symphyses, both of which provide increased load resistance abilities. However, gorillas did not exhibit relatively deeper corpora. Deeper corpora are frequently associated with more challenging diets, raising the possibility of alternative explanations such as wider corpora being needed to house the large postcanine teeth of gorillas (see e.g., Plavcan and Daegling, 2006; Boughner, 2011) or symphyseal morphology reflecting size-related differences in symphyseal curvature (Taylor, 2002). Mountain gorillas possessed relatively larger and taller rami compared to chimps and bonobos, suggesting they have larger masseter attachment areas (Taylor, 2002) and larger masseter muscles (Taylor and Vinyard, 2013). The mixed dietary signal in this analysis further emphasizes the multiple influences impacting the growth of the feeding apparatus.

Among Old World monkeys, Vinyard and Ravosa (1998) compared growth trajectories of the mandibular symphyses in macaques and baboons to interspecific allometric patterns of symphyseal morphology across Cercopithecinae. Prior to comparing allometric trajectories, they used metric estimators to calculate and compare relative stress estimates for lateral transverse bending, or wishboning, in cercopithecine symphyses. These results demonstrated few growth-related or interspecific differences in stress estimates. The lack of significant variation in relative stress values provided evidence for functional equivalence in symphyseal load resisting abilities throughout growth and across species of cercopithecines. Importantly, functional equivalence grounded the interpretation of allometric patterning as different morphologies presented equivalent stress-resisting abilities. The interspecific allometric patterns in symphyseal form were realized through a series of ontogenetic transpositions during growth, specifically related to symphyseal width and curvature. Because width and curvature are inversely related to each other in generating symphyseal stresses, transpositions of these two measures compensated for each other in maintaining functional equivalence in symphyseal stress estimates. Equally important, these results demonstrate that the interspecific pattern of symphyseal morphology is likely an epiphenomenon of growth-related processes (Vinyard and Ravosa, 1998). Thus, adult morphology may in some cases reflect selective processes occurring during ontogeny (see also Chapter 3). Along similar lines, Boughner (2011) demonstrated that the mandibles of baboons and chimpanzees grow differently, while specifically addressing how osseous growth accommodates the different dentition and eruption patterns of the two species.

Two studies of growth of the feeding apparatus between closely related species have been conducted in platyrrhines. Cole (1992) explored the ontogenetic allometry of the feeding apparatus in capuchins. The hard-object feeding *Cebus* (*Sapajus*) *apella* was compared to *C. albifrons* with its less-resistant diet. Tufted capuchins (*C. apella*) demonstrated significantly increased load-resistance abilities in the mandibular corpus and symphysis early in ontogeny compared to *C. albifrons*. Alternatively, estimates of bite force generation did not differ markedly between species, although both species demonstrated age-related increases in bite forces based on skeletal estimates (Cole, 1992). Cole (1992) hypothesized that the early establishment of increased mandibular robusticity in *C. apella* provides evidence for prenatal selection on masticatory apparatus form related to their more challenging diet. Interestingly, the establishment of increased load-resistance abilities early in life in tufted capuchins corresponds to the ecological data demonstrating few differences in food material properties with age in *C. apella* (Chalk et al., 2016). Alternatively, the age-related increase in bite force in capuchins may help explain why adult capuchins can acquire challenging foods more quickly than juveniles (Eadie, 2015). Ravosa and Ross (1994) explored growth allometries in howlers, comparing *Alouatta seniculus* to the smaller *A. palliata*. As predicted based on their overall similarities, they found a general pattern of ontogenetic scaling among red and mantled howlers. They did observe some scaling differences during ontogeny in the mandible which they attributed to subtle differences in diet and masticatory function between the two species (Ravosa and Ross, 1994).

Among strepsirrhines, Ravosa (1992) compared growth allometries in sifakas and bamboo lemurs, while Ravosa (2007) considered ontogenetic allometries in pottos (*Perodicticus*) and angwantibos (*Arctocebus*). The two sifaka species (*Propithecus verreauxi* and *P. diadema*) differ in adult size, but do not differ qualitatively in diet. *P. diadema* attains its larger skull dimensions largely through ontogenetic scaling by growing at a faster rate over a similar time period (Ravosa, 1992). The larger greater bamboo lemur [*Hapalemur* (*Prolemur*) *simus*] appears ontogenetically scaled in most skull dimensions compared to the smaller *Hapalemur griseus*. The exceptions primarily relate to the mandible, where *H. simus* possesses relatively larger mandibular corporal and symphyseal dimensions compared to the ontogenetic trajectory of *H. griseus*. Given that *H. simus* consumes a mechanically more challenging diet of bamboo culm pith than *H. griseus* (Yamashita et al., 2009), these differences in mandibular form likely offer improved load-resistance abilities related to this more challenging diet. Comparison of pottos and angwantibos showed that pottos were largely ontogenetically scaled, with the exception of several features related to the masticatory apparatus (Ravosa, 2007). These differences are potentially linked to the more challenging diet of pottos compared to angwantibos (Ravosa, 2007).

The timing of symphyseal fusion across primates and the functional interpretations of different patterns of fusion represents another area of research at the interface of growth and feeding apparatus functional morphology. Not unlike dental eruption, ontogenetic events in the symphysis have been linked to various feeding behaviors and mechanical factors driving symphyseal fusion (Ravosa and Simons, 1994; Ravosa, 1999). Anthropoids fuse their symphyses prior to weaning (Ravosa, 1999). This early-fusing pattern has been linked to the lateral transverse bending, or wishboning, loading regime characteristic of the suborder (Hylander et al., 1998, 2000). Unfused primate symphyses are predicted to lack sufficient morphological capacity to effectively resist wishboning. Thus, early fusion of the symphysis in anthropoids allows members of this suborder to resist symphyseal wishboning (Ravosa, 1999).

Most strepsirrhines do not fuse their symphyses during growth and maintain patent symphyseal joints throughout adulthood (Beecher, 1977). Sifakas are an exception to this pattern as they develop partial fusion associated with their convergent acquisition of the wishboning jaw-muscle recruitment pattern (Hylander et al., 2011). Several subfossil lemurs also fuse their symphyses following weaning (Ravosa and Simons, 1994). *Archaeolemur* represents the best-studied ontogenetic series of subfossil symphyses (Ravosa and Simons, 1994; Ravosa, 2007). During (likely) postweaning ontogeny, *Archaeolemur* begins fusing along the inferolabial surface of the symphysis. This continues into the subadult period when the symphysis is fused except along its inferolingual-most extent. This pattern of fusion is temporally contrasted with the early fusion of anthropoid primates. The late-fusing pattern is hypothesized to be related to dorsoventral shearing of the symphysis in these lemurs. The contrast with the early fusing pattern reflects the hypothesis that rather than preceding loading as in wishboning, the late-fusing pattern is directly in response to repeated loading throughout growth (Ravosa, 1999).

More subtle differences in the timing of fusion exist within anthropoids. For example, marmosets (e.g., *Cebuella pygmaea* and *Callithrix jacchus*) tend to fuse their symphyses earlier than other platyrrhines (Box 10.3). Unfortunately, explanations of these differences in timing remain elusive. In the case of marmosets, early fusion may reflect an overall increase in the pace of development in the feeding apparatus. This hypothesis is consistent with the rapid dental eruption observed in these species compared to most anthropoids (Smith, 1994). Alternatively, their derived tree-gouging behavior may provide a functional explanation for the rapid fusion of the symphysis. Behavioral evidence does show gouging behavior occurs in juvenile callitrichines. The overall extent of variation in timing and pattern of early fusion within anthropoids remains largely unexplored and an area for future research.

Previous studies of skull growth have also considered how the feeding apparatus, and more broadly skull form, responds to masticatory stresses experienced during ontogeny. Historically, studies of squirrel monkeys (Corruccini and Beecher, 1982) and baboons (Corruccini and Beecher, 1984) both suggested that experimentally manipulated soft diets (i.e., mechanically reduced below what animals would likely experience in the wild) result in dental misalignment, elevated craniofacial variation, reduced association among elements, and failure to develop normal occlusion. Similarly, Oyen et al. (1979) maintained that masticatory stresses played an important role in normal craniofacial development in baboons. Bouvier and Hylander (1981) provided added mechanistic evidence supporting these macroscopic observations by demonstrating that macaques fed an experimentally soft diet had lower levels of bone remodeling in their mandibles. Additionally, higher amounts of unremodeled bone and more resorption spaces were observed in animals fed soft diets (Bouvier and Hylander, 1981). Microscopic differences were correlated with gross morphological changes such as deeper mandibular corpora in macaques fed hard diets. Bouvier and Hylander (1982) extended these analyses to demonstrate that macaques fed a soft diet throughout growth also had significantly thinner mandibular condylar cartilages and less-dense subchondral bone in presumed load-bearing regions of their temporomandibular joints compared to macaques consuming hard diets during growth.

More recent work has refined and challenged certain aspects of these historical studies. Bouvier and Hylander (1996) examined secondary bone formation in presumed high- and low-strain areas of the macaque face to assess whether secondary bone remodeling relates primarily to mechanical demands of stress-related microdamage or metabolic needs (e.g., maintaining the body's calcium levels). In adults, mechanical and metabolic factors appeared to contribute equally to remodeling, while growing animals experienced higher rates of remodeling in high-strain regions of the skull. This result suggests a stronger link between masticatory stress and bone remodeling during ontogeny. In the last 10 years, Wang et al. (2010) and Daegling et al. (2013) documented spatial and growth-related variation in bone stiffness in the baboon and macaque mandibles, respectively. Daegling et al. (2013) further demonstrated that patterning of variation did not vary with postweaning age (as defined by unerupted M_2), suggesting that load history throughout growth is not a strong explanation for spatial variation in mandibular bone stiffness. While these more recent findings complicate some of the picture regarding how masticatory loading impacts adult form, they do reinforce the significance of growth and point to new areas for future work in understanding the translation of mechanical signals to bone growth in the primate feeding apparatus.

The jaw-closing muscles affect skeletal growth. This effect has been experimentally demonstrated through muscle translocation studies in macaques (Hohl, 1983). Moreover, by generating occlusal forces, the jaw muscles play a fundamental role in creating loads experienced throughout the skull during feeding. Cachel (1984) used muscle weights to examine ratios of jaw-muscle growth, finding relative increases in the posterior over anterior temporalis and masseter over medial pterygoid with age in anthropoid primates. Recently, Dickinson et al. (2018) considered how jaw-muscle fiber architecture changed with growth in crab-eating macaques (*M. fascicularis*). They found positive allometry of fiber lengths and physiological cross-sectional area (PCSA) during growth in this species. Given that fiber length is related to excursion while PCSA is linked to force production, they hypothesized that growth-related changes in muscle architecture contribute to growth (and size) related changes in both maximum gape and bite force (Dickinson et al., 2018).

The relative increase in jaw-muscle force production (PCSA) in *M. fascicularis* is consistent with earlier observations of positive ontogenetic allometry in bite force in rhesus macaques (*M. mulatta*) (Dechow and Carlson, 1990). Overall, bite forces increased with positive allometry relative to body weight, but isometrically relative to the growth of the mandible. Throughout growth, mechanical advantage for molar biting increased with age in these monkeys. This change in mechanical advantage followed a growth-related shift in muscle position, particularly the masseter, which facilitated bite force at the molars (Dechow and Carlson, 1990). Collectively, these studies demonstrate that jaw-muscle growth is not necessarily isometric and can have significant impact on age-related performance of the feeding apparatus.

10.4 Conclusion: Improving Our Understanding of Feeding Apparatus Ontogeny

Furthering our understanding of the ontogeny of the feeding apparatus and its underlying influencing factors is challenging. The questions we would like to answer about feeding apparatus growth, performance, and evolution sit at the nexus of growth and development, life history, feeding ecology, and biomechanics. The multiplicity of potential interacting influences, that likely vary with species and even individual, make this a difficult area to research and empirically answer questions.

Available samples have been an important logistical challenge historically. The relative difference in dental research versus musculoskeletal research in part reflects differences in available samples and the previous ability to incorporate radiology into dental maturation research. Along these lines, CT scanning technology can help increase samples for studies of musculoskeletal growth of the feeding apparatus. This resource has not been fully utilized to date. The other logistical challenge is the extended growth period of many primates. Addressing this issue longitudinally will require data collection by institutions over individuals. These databases including scans of primate growth have the potential to revolutionize our understanding of how growth impacts adult form and primate evolution.

A final consideration relates to our mindset in how we approach problems in craniofacial studies of primates. More frequently than not, primate craniofacial research considers adults as the epistemological essence of form. As an example, type specimens are rarely not adults. From a processual viewpoint, adulthood is more realistically viewed as a manifestation of a number of previous influences, including both evolutionary history and growth. While we emphasize the significance of evolution in shaping the primate feeding apparatus, growth processes have been relatively neglected (but certainly not ignored). Bringing our thinking around to acknowledge the primacy of growth in determining adult form will help us to garner the data and attention needed to better understand the functional morphology and evolutionary history of the primate feeding apparatus.

Box 10.1 Feeding Processes in Primates

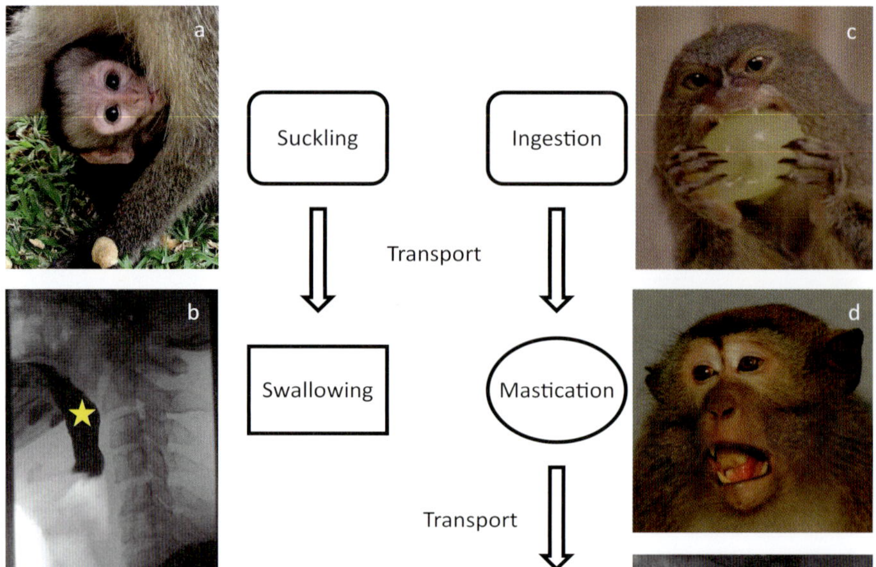

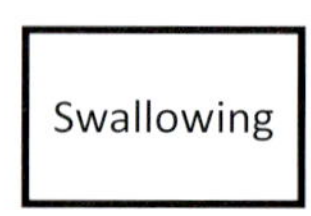

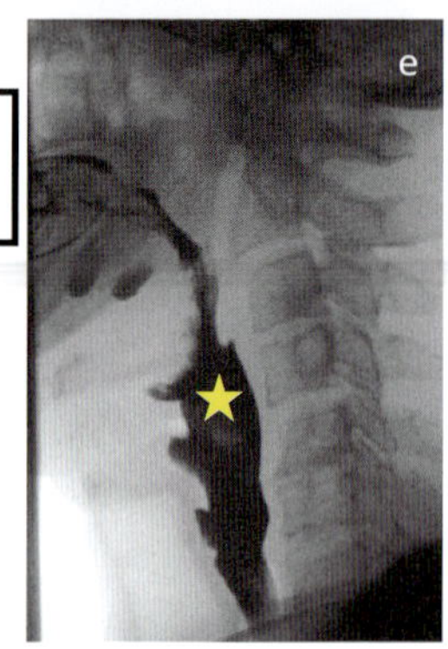

(a) Infant monkey suckling (CC), (b,e) two consecutive fluoroscopy frames of a human swallowing a barium-filled bolus. The left frame is earlier than the right frame. Yellow stars indicate bolus. Person is facing left (CC; Normaler Schluck). (c) Ingestion of grape by pygmy marmoset (CC; Mark Frauenfelder); (d) macaque masticating an apple (photo credit: Jeanette Killius).

Above: Feeding involves manipulative processes including acquisition and transport of liquids along with reduction of solid foods. Following the food-processing model (Hiiemae, 2000), we highlight suckling of liquids (left) and solid food processing (right) relating to infant through subadult primates. Infant suckling (a) involves rhythmic tongue movements to pump milk into the oral cavity (German and Crompton, 2000). A liquid bolus is transported through the oral cavity into the pharynx for swallowing (b). While suckling is characteristic of infants, the ingestion, transport, and swallowing of liquids during drinking is similar in adults. Solid food processing begins with an animal ingesting a bite of food (c) using the dentition if necessary. The bite is transported posteriorly for mastication where foods are rhythmically reduced by the postcanine dentition (d). Once a bolus is formed, it is transported through the posterior oral cavity into the pharynx for swallowing (e).

Box 10.2 Dietary Signal in Newborn Strepsirrhine Dentition

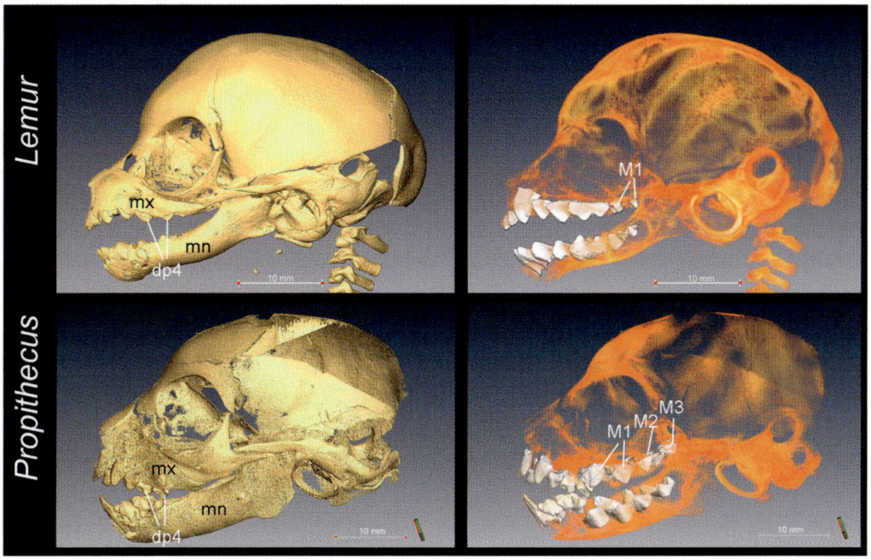

Above: Two lemuroids with differing dental adaptations to diet, at perinatal ages. *Lemur catta* (top) consumes much fruit, but also much foliage. *Propithecus coquereli* (bottom) also consumes a mixed diet, but mainly leaves. The adult dentition reveals pronounced adaptations to diet (Godfrey et al., 2001, 2004). *Propithecus* have enlarged permanent molars, which erupt rapidly. The comparison above shows that both species have alveolar eruption of deciduous anterior teeth and premolars at birth. However, when mineralized portions of teeth are shown against ghosted skulls (right side) it is clear that the permanent molars (M1–M3) are more mineralized in *Propithecus*, even though this particular specimen is at an earlier stage (late fetal) compared to *Lemur*. mn, mandible; mx, maxilla.

Box 10.3 Microanatomy of the Mandibular Symphysis in Newborn Callitrichines

Below: histological sections through the mandibular symphysis of *Cebuella pygmaea* (a,b; with inset) and *Leontopithecus rosalia* (c,d). The top row is nearer the anterior side of the symphysis and the bottom row is more posterior. In *Cebuella*, the right and left sides of the mandible are fused (open arrows) and there is only partial patency dorsally. The inset shows the fused symphysis at higher magnification, with complete fusion ventrally, a cartilage remnant, and patency dorsally. In *Leontopithecus*, the symphysis is completely unfused throughout its anteroposterior extent, the common condition for newborn platyrrhines.

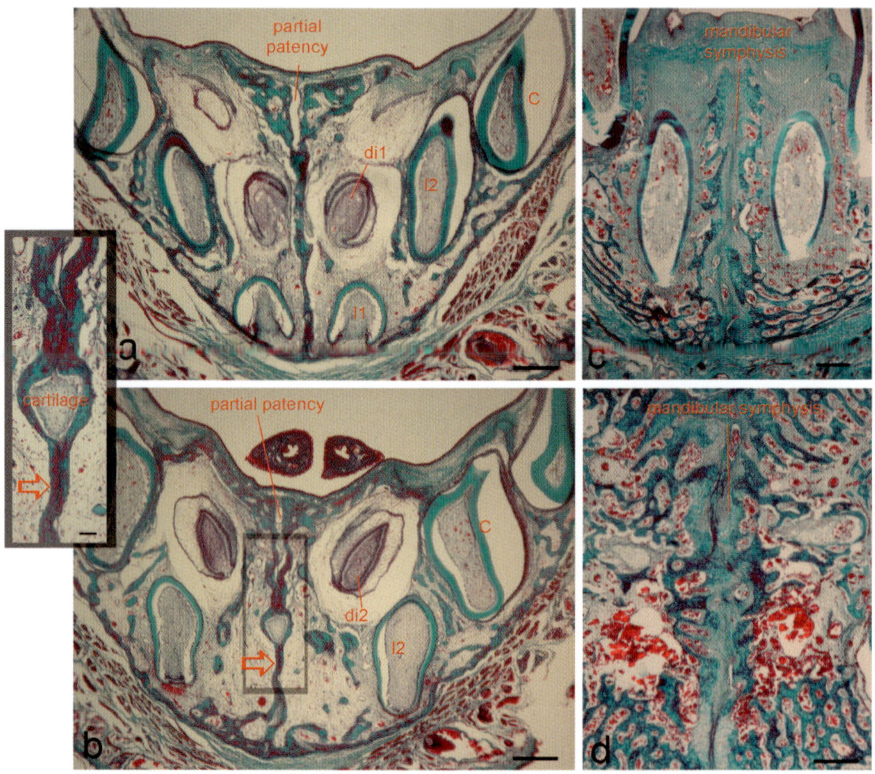

11 Ontogeny of Locomotion

11.1 Introduction

Primates, like all mammals, are characterized by determinant growth trajectories, eventually reaching a mature anatomical configuration characteristic of their species. The same general pattern of ontogenetic transition to mature stability is characteristic of locomotor behavior as well. Although many primates go through a protracted period of locomotor development (Figure 11.1), all eventually express a relatively stable set of positional behaviors that define the adult locomotor repertoire. However, most primates are also characterized by an extended period of infant and juvenile development, making them "behaviorally altricial" despite being "neurally precocial" (Grand, 1992). During this long period of immaturity, locomotor behavior and competence undergo many changes, often in concert with functional changes in postcranial anatomy as well as age-graded variation in ecological pressures (e.g., predation risk).

This chapter is organized into two main sections. In the first, we summarize available data on the time course of locomotor development in primates. In our literature review, we were able to identify qualitative or quantitative information on locomotor development in ~50 species of primates (Table 11.1), encompassing all of the major taxonomic groups of extant primates, including strepsirrhines, tarsiers, platyrrhines, cercopithecoids, and hominoids. We follow these detailed summaries with a broad review of trends characteristic of locomotor development across primates. Our review is necessarily not exhaustive. We focus primarily on early locomotor development. More information on primate locomotor development, including information on the locomotion of juvenile primates and the transition to locomotor maturity, can be found in Bezanson and Morbeck (2013), Bezanson (2017), and Young and Shapiro (2018).

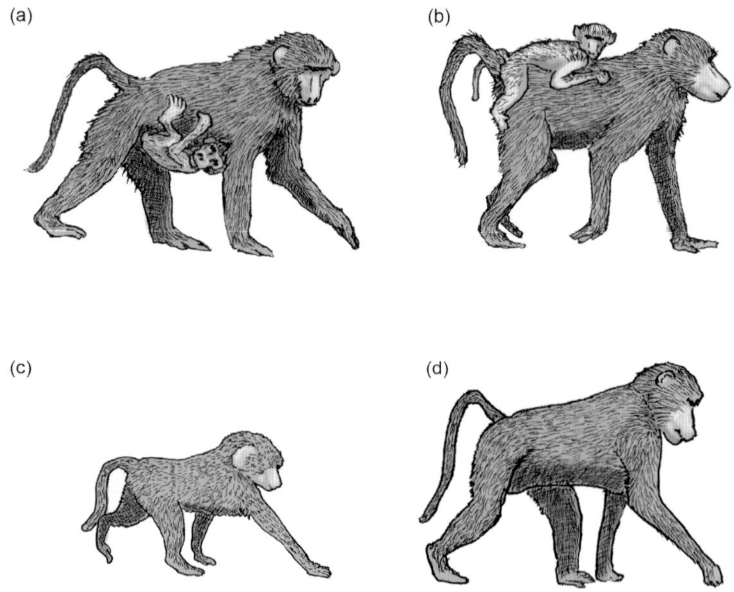

Figure 11.1 Stages of primate locomotor development, illustrated using *Papio cynocephalus* as a model (Altmann, 1980). (a) Most newborn primates are typically transported by clinging ventrally to their mother or another groupmate, although some smaller taxa (particularly strepsirrhines) will "park" their infants in nests or tree-holes during active foraging (Kappeler, 1998; Ross, 2001; Tecot et al., 2012). (b) As neuromuscular development advances and postural stability improves, older infants often transition to dorsal carriage. (c) Once immature primates are able to locomote independently, the biomechanical details of their locomotion are often different from the adult condition (for instance, note that this young baboon is using a lateral sequence gait, whereas almost all adult primates use a diagonal sequence gait; see Section 11.3). (d) Locomotor independence, and an adult-like locomotor profile, is achieved at different ages in different species of primates (see Young and Shapiro, 2018 for review).

11.2 Descriptive Accounts of Primate Locomotor Development

Most developmental studies of primate locomotion have been descriptive in nature (Table 11.1). This section summarizes these studies, paying particular attention to the order in which different skills appear. Animals are grouped using common names (e.g., "capuchin monkeys"), as some of our descriptions cover multiple independent species.

11.2.1 Hominoids

11.2.1.1 Gibbons (e.g., Hylobates, Nomascus*)*

Gibbons' first locomotor attempts are climbs made on their mothers. By 6–8 weeks, infants are able to pull themselves upright and sit up on their own. Standing and climbing are first seen between 8 and 10 weeks. By 4 months of age they are skillful climbers and are beginning to brachiate, with brachiation already being the most prevalent locomotor mode in independent infant gibbons in the wild. Infant and juvenile gibbons also choose to move on smaller diameter locomotor substrates significantly more often than larger-bodied adults, a pattern also seen in other apes. Relative to monkeys, overall locomotor development is slow, and infants are frequently carried by their mothers throughout the first 2 years. Full locomotor independence is not achieved until the animals are 7 or 8 years old.

11.2.1.2 Chimpanzees (e.g., Pan*)*

As noted above with regard to gibbons, locomotor development in free-ranging common chimpanzees (*Pan troglodytes*) proceeds at a much slower pace than observed among monkeys (Doran, 1992). At birth, chimpanzees are unable to cling or support themselves and consequently have to be cradled by their mothers during travel and resting for the first 3 months. Locomotor attempts are first seen at 5 months of age and involve climbing on lianas or the mother. Crawling and walking emerge around this time as well, although the amount of quadrupedal locomotion prior to 6 months of age is quite low overall. Most locomotion on the ground is bipedal with postural support coming from the mother or the environment. From 6 months to 2 years, infants primarily engage in suspensory locomotor behaviors. Bipedal locomotion is still used on the ground, but the frequency of quadrupedal locomotion steadily rises during this period. Early quadrupedal locomotion is primarily palmigrade. Knuckle-walking is first attempted at 12–15 months, although infants do not master this form of progression until they are nearly 2.5 years old (see also Inouye, 1994 for ontogenetic data on chimpanzee knuckle-walking behaviors). During this time, infants also frequently use the "crutch-walk," a form of quadrupedalism in which the trunk and hindlimbs are vaulted through the extended forelimbs, similar to a human using crutches (Hunt et al., 1996). From 2 years on, infants increasingly engage in more terrestrial and quadrupedal behaviors, although the frequency of arboreal suspensory locomotion is still high relative to adults. By the time infants are weaned at 5 years of age, locomotor patterns are, for the most part, identical to that of adults. Overall, young chimpanzees engage in more arboreal and suspensory positional behaviors than their adult counterparts, for whom terrestrial quadrupedalism is the most common form of locomotion.

Detailed observations on *P. troglogytes* at Ngogo, Kibale National Park, Uganda generally corroborate the findings above, for instance showing that infant locomotion commences at approximately 5 months of age, and that long-term locomotor development is marked by a gradual transition from the arboreal climbing and orthograde suspensory behavior in infants and juveniles to primarily quadrupedal walking in adults (i.e., quadrupedalism accounted for 18% of locomotion in infant chimpanzees, versus 77% in adults). Immature chimpanzees are also generally more active, and engage in a wider variety of positional behaviors, than adults. Finally, using a fine-grained analysis of locomotor behaviors, Sarringhaus et al. (2014, 2016) found that the adult-like locomotor profile is not seen until chimpanzees reach adolescence (10–13 years), indicating a more protracted developmental timetable than suggested by Doran's (1992) data.

Table 11.1 Developmental studies of primate locomotor behavior

Species	Environment	Data	References
Gibbons *Hylobates lar, Nomascus nasatus*	C, N	D	Kirkwood and Stathatos (1992); Fan et al. (2012); Fei et al. (2015)
Chimpanzees *Pan troglodytes, P. paniscus*	C, N	D, DQ, K, B	Kimura (1987); Doran (1992); Kirkwood and Stathatos (1992); Inouye (1994); Schwandt (2001, 2002); Schwandt and Marzke (2003); Sarringhaus et al. (2014, 2016)
Gorillas *Gorilla gorilla*	N	DQ	Inouye (1994); Doran (1997)
Orangutans *Pongo pygmaeus, P. abelii*	N	D	Sugardjito and van Hooff (1986); Wich et al. (2004); Thorpe and Crompton (2005; 2006); van Noordwijk and van Schaik (2005); Manduell et al. (2011); Myatt and Thorpe (2011); van Noordwijk et al. (2018)
Vervet monkeys *Chlorocebus aethiops*	C, SN, N	L, K	Lancaster (1971); Hurov (1982); Vilensky and Gankiewicz (1986, 1989); Vilensky et al. (1989); Kirkwood and Stathatos (1992)
Macaques *Macaca fuscata, M. mulatta, M. radiata*	C	L, K, B	Hildebrand (1967); Rawlins (1976); Negayama et al. (1983); Kirkwood and Stathatos (1992); Maestripieri (1995); Nakano (1996); Dunbar and Badam (1998); Kimura (2000); Wells and Turnquist (2001)
Baboons *Papio anubis, P. cynocephalus*	C	L, K	Rose (1977); Altmann (1980); Chalmers (1980); Altmann and Samuels, 1992; Kirkwood and Stathatos (1992); Shapiro and Raichlen (2005, 2006); Druelle et al. (2016, 2017b, 2018)

Langurs *Pygathrix nemaeus, Semnopithecus entellus, Trachypithecus delacouri, T. hatinhensis*	SN, N	D	Dolhinow and Murphy (1982); Dunbar and Badam (1998); Workman and Covert (2005)
Colobus monkeys *Colobus polykomos*	C, N	D	Kirkwood and Stathatos (1992); Dunham (2015)
Squirrel monkeys *Saimiri boliviensis, S. oerstedi, S. sciureus*	C	D, DQ, B	Rosenblum (1968); Schusterman and Sjoberg (1968); Castell and Sackett (1972); Elias (1977); Boinski and Fragaszy (1989); Kirkwood and Stathatos (1992); Young (2009, 2012a)
Owl monkeys *Aotus trivirgatus*	C, N	D	Kirkwood and Stathatos (1992)
Capuchin monkeys *Cebus albifrons, C. capucinus, C. olivaceus, C. apella*	C	D	Elias (1977); Escobar-Páramo (1989); Fragaszy (1989, 1990); Fragaszy et al. (2004); Wright (2005); Bezanson (2006); Russo and Young (2011)
Tamarins *Saguinus oedipus*	C	D	King et al. (1974); Snowdon and Soini (1988)
Howler monkeys *Alouatta caraya, A. guariba, A. palliata, A. seniculus*	N	D, DQ	Mack (1979); Clarke (1990); Miranda et al. (2005); Bezanson (2006, 2012); Podgaiski and de Assis Jardim (2009); Raguet-Schofield and Pavé (2014)
Tarsiers *Tarsius bancanus, T. tarsier*	N	D	Kirkwood and Stathatos (1992); Gursky (2007)
Mouse lemurs *Microcebus murinus, M. rufus*	C, N	D	Kirkwood and Stathatos (1992); Atsalis (2015)

Table 11.1 *(cont.)*

Species	Environment	Data	References
Ruffed lemurs *Varecia variegata, V. rubra*	C, N	D	Morland (1990); Kirkwood and Stathatos (1992); Baden et al. (2013); Vasey et al. (2018)
Ring-tailed lemurs *Lemur catta*	N	D, DQ	Sussman (1977); Gould (1990); Wolf (2016)
Brown lemurs *Eulemur fulvus*	N	D	Sussman (1977)
Sifakas *Propithecus coquereli, P. diadema, P. verrauxi*	C, SN, N	D, B	Richard (1976); Eaglen and Boskoff (1978); Wright (1995); Demes et al. (1999); Lawler (2006); Ross and Lehman (2016); Wolf (2016)
Galagos *Galago senegalensis, G. moholi, Otolemur crassicaudatus*	C, N	D, DQ	Ehrlich (1974); Doyle (1979); Crompton (1983); Kirkwood and Stathatos (1992)
Lorises *Loris lydekkerianus, Perodicticus potto*	N	D	Ehrlich (1974); Charles-Dominique (1977); Kirkwood and Stathatos (1992); Nekaris (2003); Radhakrishna and Singh (2004)

Notes: Environment: C, captive; SN, semi-naturalistic (i.e., in an outdoor enclosure); N, naturalistic. Data: D, descriptive; DQ, descriptive–quantitative; B, biomechanical (i.e., data on locomotor kinematics and kinetics).

Findings on captive chimpanzees are generally similar to those above in that as animals age, the frequency of suspensory locomotion decreases while that of quadrupedalism increases. These behavioral changes were matched anatomically by a decrease in the brachial index (radius length/humerus length) over ontogeny. Having a longer radius is beneficial for climbing taxa as it increases the mechanical advantage of the forelimb flexor muscles (Keith, 1919; Fleagle, 1977). Some studies of captive chimpanzees find that infants less than 2 years of age do not engage in any more suspensory activities than adults (Schwandt, 2001, 2002; Schwandt and Marzke, 2003), unlike the findings on wild chimpanzees (Doran, 1992). Furthermore, young captive chimpanzees engage solely in knuckle-walking and do not show the palmigrade to knuckle-walking transition observed in wild chimpanzees.

Locomotor development in bonobos (*Pan paniscus*) closely parallels that of its sister species (i.e., *P. troglodytes*), in that the frequency of suspensory locomotion decreases while quadrupedal locomotion becomes more common. However, the decrease in suspensory behaviors is not as extensive as among common chimpanzees. As a result, the overall locomotor pattern of adult bonobos most closely resembles 2–5-year-old common chimpanzees, prompting Doran (1992) to refer to the former as a "locomotor paedomorph" of the latter.

11.2.1.3 Gorillas (e.g., Gorilla*)*

Wild mountain gorillas (*Gorilla berengei*) do not locomote for the first 3 months of life. The first locomotor attempts primarily consist of climbing, either on the mother or on small brush. Bipedal locomotion, with the support of the mother or another substrate, is also common at this time, as in common chimpanzees. By 6 months of age, gorillas are adept at quadrupedalism, with quadrupedal activity constituting more than one-third of all locomotor bouts. As with chimpanzees, initial quadrupedal locomotion is palmigrade. However, knuckle-walking becomes increasingly frequent during ontogeny and by 10 months of age is the predominant posture used in quadrupedal locomotion.

The amount of overall locomotor activity increases during development, reaching its peak during the period from 17 to 21 months. This is also the most arboreal period of life, with infants spending almost a quarter of their total activity time in the trees. The amount of arboreality declines as the animals age, decreasing to only 11% of total activity time by age 2 years and less than 6% among adults. By the middle of the second year, gorillas travel more on their own than with their mother and by 4 years of age, all locomotor behaviors are roughly adult-like in form and frequency.

In general, gorillas follow the same broad developmental trajectory as chimpanzees – a decrease in suspensory locomotion and climbing during ontogeny with a concomitant increase in the amount of terrestrial quadrupedalism. In both chimpanzees and gorillas, palmigrade quadrupedalism precedes knuckle-walking (see also Inouye, 1994). However, the timing of locomotor development differs between the two clades of African ape. Except for very early in ontogeny, when chimpanzees are slightly more precocial than gorillas, gorilla locomotion develops at a faster pace than chimpanzee locomotion. This seems to be caused by an overall higher postnatal growth rate in gorillas, as evidenced by the fact that chimpanzees and gorillas are very similar in their locomotor profiles at equivalent body sizes, but not equivalent ages.

11.2.1.4 Orangutans (e.g., Pongo*)*

Ontogenetic studies of wild orangutans indicate that, overall, orangutans have an extremely slow life history. Weaning occurs around the seventh year and the interbirth interval is greater than nine years. Nonetheless, van Noordwijk and van Schaik (2005) found that locomotor independence in Sumatran orangutans (*Pongo abelii*) is established around the infants' third year, well in advance of chimpanzees and, in some cases, gorillas as well. Similarly, Thorpe and Crompton (2005, 2006) and Manduell et al. (2011) found the locomotor profiles of juvenile Bornean orangutans (*P. pygmaeus*) are nearly identical to those of adults. Smaller-bodied orangutans (including juveniles) tend to travel by climbing and clambering in the middle ranges of the canopy while larger-bodied adult males travel by "tree-swaying" in the lower ranges of the canopy. This is likely due entirely to body size – larger individuals cannot securely climb in the middle canopy, where branches may be too thin to support their weight (see also Myatt and Thorpe, 2011), and smaller individuals do not have sufficient mass to efficiently utilize the tree-swaying method.

11.2.2 Cercopithecines

11.2.2.1 Vervet Monkeys (e.g., Chlorocebus*)*

Vervets (*Chlorocebus aethiops*) can stand quadrupedally by 2 weeks of age (Hurov, 1982). Walking is evident by 1 month. As in many other primate and non-primate taxa, early walking is in infants characterized by excessive flexion, particularly at the more distal joints. There is also a gradual ontogenetic decrease in the number of feet used during support periods (a common pattern during locomotor ontogeny in quadrupedal animals: Shapiro et al., 2014; Cole et al., 2019). By 2 months of age, young vervets are capable of higher-speed galloping gaits. In the wild, they reach full locomotor independence at approximately 3 months of age, near the time their natal black coat changes to the light gray color of adults. The risk of death from predation is particularly strong during this time of initial locomotor independence, with immature vervet monkeys in some populations experiencing up to 17× greater predation risk than adult groupmates (Janson and van Schaik, 1993).

Young vervets have also been shown to engage in "air-stepping," in which suspended infants are able to move their limbs in alternating patterns quite similar to adult-like walking well prior to the development of actual locomotion. Air-stepping is seen in many mammalian taxa, including humans (Thelen and Fisher, 1982) and rats (Vinay et al., 2000b), and reflects the development of flexor muscle control in advance of extensor (i.e., postural) muscles. In vervet monkeys, as in humans, this "stepping reflex" eventually disappears, usually within the first month of life. Whether this disappearance represents the emerging influence of supraspinal centers over spinal-level pattern generators (McGraw, 1945; Forssberg, 1985) or is the outcome of lower-level biomechanical changes, such as weight gain (Thelen and Fisher, 1982), should be addressed in future research.

11.2.2.2 Macaques (e.g., Macaca*)*

Several studies have examined locomotor ontogeny of macaques in both captive and (semi-)natural environments, as might be expected given their ubiquity as biomedical research models. Studies of captive animals have found that newborn macaques are quite precocial. For example, neonatal Japanese macaques (*Macaca fuscata*) can stand immediately after birth or the first day thereafter. They can crawl as newborns, "toddle" unsteadily on flexed and abducted limbs during the first week, walk and run within the first 2–3 weeks, and are adept at trotting and galloping by 2 months of age. Similarly, captive rhesus macaques (*M. mulatta*) are able to climb within the first week of life, walk by the second week of life, and jump by the end of the first month. As in squirrel monkeys, interlimb coordination during walking is quite variable early in development, with infant macaques frequently using lateral sequence gaits rather than the diagonal sequence pattern typical of adult primates.

Data on wild macaques in natural or semi-natural environments (e.g., Cayo Santiago, Puerto Rico) show that the locomotor repertoires of animals throughout development are more varied than observed in captive populations. In general, young infants (0–3 months) show a wide variety of positional behaviors, including galloping, leaping, and hopping in addition to the quadrupedal walking and climbing observed in the laboratory. However, during these behaviors, infants rarely move more than a meter from their mother and spend most of their time clinging to their mother's fur. A mother may encourage their infant to attempt locomotion, by slowly backing away from them and giving a "lip-smacking" sound that entreats the infant to come towards her, a behavior also observed in pig-tailed macaques (*M. nemestrina*). Locomotor independence is achieved around the fifth postnatal month.

In free-ranging bonnet macaques (*M. radiata*), locomotor development is more protracted than that of wild rhesus macaques, or of captive macaques. Infants could stand arboreally and terrestrially within the first month. From 2 to 7 months, locomotion is "unskilled," with poor coordination and excessive dorsoventral movement of the head and trunk (Dunbar and Badam, 1998). More skilled terrestrial and arboreal quadrupedal locomotion, including walking, running, and galloping, is first observed at 6 months of age. These skills continue to mature throughout the end of the first year, when weaning typically occurs.

11.2.2.3 Baboons (e.g., Papio*)*

In the yellow baboon (*Papio cynocephalus*), all infants first cling to their mother's venter, during both travel and rest (Figure 11.1; and see Altmann, 1980). During the second week, infants begin locomotor activity by climbing and moving about on their mother, much as in other primate species. The first locomotor attempts on the ground are clumsy and executed with flexed limb postures, particularly in the hindlimbs. By the end of the first month, infants are able to easily navigate obstacles in their path such as logs or rocks. They begin to climb at 3–4 months. However, infants are still entirely dependent on their mothers for transportation and continue to be carried throughout this period. Climbing ability increases throughout the infant period, and by 8 months infants are able to descend from sleeping trees by themselves. By the end of the first year, when weaning most often takes place, infants are fairly independent and can usually survive the death of their mother. Nevertheless, locomotor independence may be achieved prior to this, as infants are usually able to locomote independently before they could forage independently, a pattern common across primates (Young and Shapiro, 2018). For example, when predators threaten, 10–12-month-old infants frequently flee alongside their mothers, rather than being carried by them.

While in general agreement, two studies of the same troop of free-ranging East African olive baboons (*Papio anubis*) (Rose, 1977; Chalmers, 1980) provide some minor differences in the ontogenetic timing of events, perhaps associated with the durations and sample sizes of the studies. Wild olive baboons are rather precocious in their locomotor development. Animals in both studies could sit and stand within the first week. During initial standing attempts, the limbs are quite abducted and flexed. Crawling – called "squirming" by Rose (1977) – is also seen during the first week. Although Chalmers (1980) records that the infants in his sample could walk during the first week as well, Rose found that coordinated cyclical gaits did not appear until the third week of life. Jumping and hopping was observed between weeks 3 and 4 and infants were able to climb by the end of the first month. Rose (1977) found that locomotor maturation was primarily complete by the fifth or sixth month, at which time postural behaviors and quadrupedal walking are fully developed and adult-like.

Several studies have found that gait selection is quite variable during infant baboons' initial locomotor efforts, with animals frequently switching between the diagonal sequence footfall pattern that is common to most adult primates and the lateral sequence pattern that is typical of most non-primate quadrupeds. We review possible biomechanical explanations for this phenomenon in Section 11.4 (also see Figure 11.1).

11.2.3 Colobines

11.2.3.1 Langurs (e.g., Pygathrix, Semnopithecus, Trachypithecus*)*

Captive Hanuman langur (*Semnopithecus entellus*) infants do not begin to locomote until their second postnatal week. The first locomotor attempts are primarily quadrupedal crawling and walking, with climbing efforts not appearing until week 5. All locomotor behaviors show a steady increase in frequency from this point on with the infant becoming relatively independent by 3 months of age. In free-ranging langurs, the full range of adult locomotor behaviors is typically evident by 6 months of age (although agility is still not as practiced as among adults). Juveniles nonetheless express a comparatively wide array of positional behaviors, with the number of positional behaviors nearly double that of their mature adult groupmates (Workman and Covert, 2005).

11.2.3.2 Colobus Monkeys (e.g., Colobus*)*

Data on developing king colobus monkeys (*Colobus polykomos*) indicate that behavioral development is quite slow. Independent locomotion is not seen until the third week of life. Not until the fifth week can infants restrain other females from engaging in "aunting behavior" (Mitani and Watts, 1997) and removing them from their mothers. Observations of single infants in captivity and the wild indicate that climbing is achieved by the end of the first month and jumping during the third month, with infants able to leap a meter or more by the time they are 3 months old. In wild black and white colobus monkeys (*Colobus angolensis palliatus*) juveniles devote significantly more of their activity budget towards locomotion than adults and are able to move on significantly smaller substrates.

11.2.4 Platyrrhines

11.2.4.1 Squirrel Monkeys (e.g., Saimiri*)*

In the wild, squirrel monkeys are primarily transported on their mothers' backs during the first 4 months of life. They begin initial locomotor efforts at approximately 1 month of age and gradually reach locomotor independence by 8 months. Captive studies have also examined the normative development of crawling, walking, and climbing abilities of squirrel monkeys in greater detail. However, the results of these studies are quite variable. For example, while Schusterman and Sjoberg (1968) and King et al. (1974) found that squirrel monkey infants could not crawl until 36 and 42 days after birth, respectively, Elias (1977) found that her sample could crawl only 17 days after birth and were walking by 22 days. Similarly, King et al. (1974) and Elias (1977) found that young squirrel monkeys cannot climb prior to 54 and 40 days of life, respectively, while Schusterman and Sjoberg (1968) report regular climbing by 12 days. Many of these differences are likely due to differences in testing procedures and skill definitions. Small sample sizes likely also biased the results, as the pace of locomotor development is known to be quite variable across individuals. Overall, these data illustrate that the development of locomotion in squirrel monkeys takes place at a much slower rate than observed in the strepsirrhines and tarsiers, although not as slowly as in other platyrrhines (i.e., capuchin monkeys – see below).

Rosenblum (1968) and Schusterman and Sjoberg (1968) reported that infant squirrel monkeys have a functional prehensile tail (i.e., are able to support their body weight) during the first month of life, subsequently losing this ability. This is an interesting case of non-human primates displaying an ephemeral locomotor skill, a well-known phenomenon in human motor development (e.g., quadrupedal crawling and upright cruising).

In studies of growing *Saimiri boliviensis* (aged 4–10 months), Young (2009, 2012a) found that having relatively short limbs and habitually using more extended elbow postures permitted juvenile squirrel monkeys to reduce joint loading at the youngest ages, perhaps compensating for the reduced muscle strength characteristic of an immature musculoskeletal system. Additionally, interlimb coordination (i.e., gait selection) varied during squirrel monkey growth, and was significantly associated with variation in the biomechanical properties of the young monkeys' bodies (i.e., center of mass position and relative limb length).

11.2.4.2 Owl Monkeys (e.g., Aotus*)*

Although infant owl monkeys cling to their mother's hip and thigh during the first few days of life, fathers are primarily responsible for infant care at all later ages, with infants spending 93% of the time with their fathers by 2 months of age. Infants begin locomotor efforts at 4 weeks of age, can leap up to half a meter within 1.5 months, and are generally independent by 3 months.

11.2.4.3 *Capuchin Monkeys (e.g.,* Cebus*)*

Several studies have investigated the development of locomotion in brown capuchin monkeys (*Cebus apella*). In captivity, newborn infants show low muscle tone, although flexors clearly dominate over extensors (a common pattern in infant primates; see Section 11.3). Whereas flexor dominance facilitates grasping onto the mother's venter during travel and rest, it makes it impossible for young infants to execute any postural or locomotor activities on their own. This flexor dominance gradually disappears by the second week and infants are able to stand on their own by the fifth week. Attempts at crawling are first seen by the sixth week and walking is evident after 8 weeks. At this point, infants gradually begin spending increasing amounts of time away from their mother. By the twelfth week, infants begin going on long excursions of social activity and play by themselves. Adult-like postural facility is evident by the seventh month of life. Captive infant white-fronted capuchin monkeys (*Cebus albifrons*) show a similar developmental time course in captivity, with infants beginning to climb by 31 days, crawl by 44 days, and walk by 52 days.

As in captivity, wild capuchin monkeys begin moving independently at around 2 months of age, although they still prefer to be carried during almost all travel. The incidence of infant carrying decreases over the first 6 months of life, at which time juvenile locomotor abilities generally match adults, although in some species significant differences remain. For instance, Wright (2005) found that brown capuchin juveniles walked less frequently than adults but tended to leap, bound, and climb more. This is most likely related to smaller body size of the juveniles, as smaller primates in general tend to leap more than larger taxa (Fleagle and Mittermeier, 1980) and are generally more agile.

Prehensile tail use begins early in wild capuchin monkeys, although prior to 2 months of age the tail is only used as another means of securely grasping the mother. Prehensile tail use increased with age during infancy and juvenescence, but declines in adults, perhaps as a result of their larger body mass and generally reduced locomotor repertoires. Russo and Young (2011) found that morphometric correlates of caudal vertebrae bending strength (i.e., polar section moduli) scale with negative allometry to body mass as capuchin monkeys age, suggesting that developmental changes in caudal vertebral robusticity track this generalized decline in tail suspensory behaviors during the transition to adulthood.

11.2.4.4 *Tamarins (e.g.,* Saguinus*)*

Captive cotton-top tamarins are reported to crawl consistently within 14 days after birth and are consistently walking within 20 days. Consistent climbing was observed by 18 days after birth. Nevertheless, infant tamarins primarily travel by being transported on their parents' backs during the first 1–2 months.

11.2.4.5 *Howler Monkeys (e.g.,* Alouatta*)*

As with many primate species, wild newborn howler monkeys are able to cling to their mother's venter with their hands and feet. Red howler (*Alouatta seniculus*) infants are carried ventrally during the first 2 months of life and continue to be carried ventrally and dorsally for the first 4 months, and are sometimes carried up until 9 months when crossing large gaps in the canopy. Similarly, infant brown howler monkeys (*A. guariba*) and mantled howler monkeys (*A. palliata*) are carried ventrally, and then dorsally, during the first 5–6 months of life. Black howler monkeys (*A. caraya*) develop more slowly and are carried into the ninth month of life. Slow and awkward independent locomotion on branches generally begins in the first to second month, although Bezanson (2006) did not observe independent locomotion in infants younger than 3 months of age. Prehensile tail use also begins early, with the infant able to hang entirely by its tail and use it to bridge gaps in the canopy by the third month. As in capuchin monkeys, infant – and especially juvenile – howler monkeys use their prehensile tails more often than adults, likely as a result of their smaller size and more athletic locomotor repertoire. Jumps and leaps are seen in the fourth month. By the end of this month, infants had usually attained the full range of positional behaviors seen in adults, although they did not necessarily display the same facility. Infant carrying is generally terminated by the sixth to seventh month, suggesting that locomotor independence is achieved by this age.

11.2.5 Tarsiers

11.2.5.1 Tarsiers (e.g., Tarsier*)*

In captivity, young tarsiers are quite precocial. They are able to crawl within the first 2 days after birth and are leaping within 1–2 weeks. Once leaping emerges, this quickly becomes the predominant mode of locomotion, even for very young animals. Prior to this age, quadrupedalism is the primary mode of transport.

Infant development has been studied in wild spectral tarsiers (*Tarsier tarsier*). Although spectral tarsiers are able to strongly grasp twigs and small branches at birth, they are always transported orally. However, oral transport constitutes a significant energetic cost to the tarsier mothers, and this cost likely explains the need for infant parking during the mother's active period. Wild infant tarsiers begin exploring their environment at about 3 weeks of age, typically by moving about the site at which they are parked. Vertical clinging and leaping (VCL) is first attempted at 1 month of age. As in captivity, arboreal quadrupedalism is the predominant locomotor mode prior to the emergency of VCL.

11.2.6 Strepsirrhines

11.2.6.1 Mouse Lemurs (e.g., Microcebus*)*

Mouse lemurs do not open their eyes for the first 2–4 days after birth. In captivity, infants make attempts to crawl and climb at around 7 days although they cannot completely raise themselves off the ground for the first 2 weeks. Locomotor development proceeds rapidly from that point on. Infants begin jumping at 1 month and are independent by 3 months.

11.2.6.2 Ruffed Lemurs (e.g., Varecia*)*

Infants are usually born into litters of 2–4. Although data from captivity indicate that infants cling to their mothers following birth, wild ruffed lemur infants are typically transported orally and parked during the daily activity period (Morland, 1990; Vasey et al., 2018). Ruffed lemurs are selective in choosing nesting sites in which to park infants, preferentially using trees that are taller, wider, and surrounded by a high number of lianas. Such sites provide refuge from predators and more secure substrates for initial locomotor efforts. Nevertheless, infant mortality is high during the first year, with the predominant cause being falls from nesting sites.

In captivity, infants are able to support themselves and show some walking at 8 days of age. In the wild, infants begin to cling to small branches and walk at 1 month. At 2–3 months, infants are adept enough to follow their mothers and travel up to 100 meters in a single bout. Adult-like locomotor facility is achieved within 3–4 months.

11.2.6.3 Ring-Tailed Lemurs (e.g., Lemur*)*

Infants cling to their mother's ventral surface immediately after birth, transitioning to dorsal riding within a week of birth. This sequence of infant carriage – initial ventral contact followed by dorsal contact – is common to many primates (Figure 11.1). Within 3 days, infant ring-tailed lemurs actively move around their mother's body, and by 2 weeks make short forays away from her, often in order to climb on another female groupmate, logs, bushes, or low branches in the environment. Over the first few months of life, infants spend increasing amounts of time away from their mothers (i.e., away from her ~30% of the time by 6–7 weeks after birth) and become independent at approximately 2.5–3.5 months of age.

11.2.6.4 Brown Lemurs (e.g., Eulemur*)*

Infant development in *Eulemur fulvus* is comparatively slower than in ring-tailed lemurs. Infants only begin to move freely about their mothers at 2 weeks, an age at which *L. catta* are already moving off their mothers. Similarly, the transition from ventral to dorsal riding takes place at 1 month of age, substantial forays of independent locomotion do not begin until about 3 months of age, and locomotor independence is achieved at round 3.5–4 months of age.

11.2.6.5 Sifakas (e.g., Propithecus*)*

Infants begin climbing on their mothers immediately following birth. For their first month, infants are entirely transported by clinging onto their mother's venter, transitioning to dorsal riding at around 2 weeks of age in captivity and between 3 and 7 weeks of age in the wild. Independent locomotion is first observed between weeks 3 and 5 in *P. coquereli*, and somewhat later at 8 weeks of age in *P. verreauxi*. By 9 weeks, infants are regularly exploring within a 1-meter radius of their mothers. They begin taking short leaps at 8–10 weeks and attempt longer ones at 11–14 weeks. Adult-like locomotion becomes evident by the third month, although infants will still cling to their mother's back if the distance to be traveled is beyond their capacity. They reach locomotor independence between 5 and 6 months of age. By 1 year of age, juvenile positional behavior is indistinguishable from that of adults. Demes et al. (1999), using a customized force pole, found that a juvenile sifaka frequently overcompensated when leaping from one support to the next (i.e., take-off and subsequent landing forces are much greater than required to safely span the gap between substrates). This indicates that even after adult-like facility has been attained, growing animals are still learning to gauge their locomotor efforts to the necessities of the task at hand.

11.2.6.6 Galagos (e.g., Galago, Otolemur*)*

In captivity, lesser galagos (i.e., *Galago senegalensis* and *G. moholi*) are rather precocial. They begin walking by 8–10 days after birth and can run and jump by 2 weeks of age. Climbing is evident at 3 weeks and infants are independent by 4–6 weeks. In accordance with their larger body size, captive greater galagos (i.e., *Otolemur*) develop at a slightly slower rate than lesser galagos. Infants begin to crawl at 8–12 days and begin climbing at 14–16 days. Leaping is not evident until 31 days. In general, the attainment of locomotor independence is associated with the full appearance of leaping abilities.

In the wild, the locomotor behavior of juvenile lesser and greater galagos differs from their adult counterparts in many respects. Immature galagos utilize much more arboreal quadrupedalism and less leaping than adults. Climbing and hopping short distances (rather than leaping) are also more common during the immature period. Overall, these data show that although galagos are fairly precocious, leaping (particularly vertical clinging and leaping) is a relatively difficult skill for developing animals to master.

11.2.6.7 Lorises (e.g., Loris, Perodicticus*)*

Gray slender lorises (i.e., *L. lydekkerianus*, also known as Mysore slender lorises) are usually born into twin litters. Although infants are able to open their eyes at birth, they predominantly keep them closed during the first week of life. Infants are able to cling to their mother's fur immediately upon birth, and travel with their mothers constantly during the first few weeks of life. Mothers begin parking their infants at 3–4 weeks of age. The first attempts at locomotion are also seen around this time. Weaning takes place around the fourth or fifth month, at which time locomotor independence is also achieved. Nekaris (2003) notes that this behavioral profile largely also describes locomotor development in *L. tardigradus*, a close relative of *L. lydekkerianus*.

Pottos (*Perodicticus*) are able to cling to their mother immediately after birth. They ride on their mother's back for the first 3–8 days after birth and are parked during maternal activity periods after this. Independence is achieved slightly later than in slender lorises; infants begin following their mothers during the night at 3–4 months and become independent at 6 months.

11.3 Trends in Primate Locomotor Development

In general, the most intensive studies of primate locomotor development have been focused on catarrhine taxa (i.e., macaques, baboons, and great apes), whereas only patchy accounts are available for most platyrrhines and prosimians, and even these are primarily based on very small sample sizes (even a single animal). This bias is understandable. Many of the most frequently studied taxa are (1) terrestrial, and thus more easily observed in the wild (e.g., *Macaca* sp., *Papio* sp.), (2) common laboratory models, making them easier to access for kinesiological research, and/or (3) particularly charismatic for anthropologists (e.g., great apes). However, data drawn from these species should not necessarily be considered to be representative of primates in general. Clearly, more comprehensive data sets need to be collected on prosimians, platyrrhines, and arboreal taxa in general. Given these caveats, we will only highlight general trends and discuss some links between these patterns and associated aspects of somatic growth and development.

Whereas newborn primates are not able to locomote independently (although tarsiers, mouse lemurs, and macaques can do so soon thereafter), nearly all primates are born with well-developed sensory systems (i.e., eyes open, and responsive to visual and auditory stimuli) and most newborn primates can immediately grasp onto a caregiver following birth. Even newborn human infants show a strong "grasping reflex" soon after birth, well before independent locomotion is first attempted (Twitchell, 1965). Infant primates are also morphologically well-adapted to elicit a strong grasping response, perhaps even more so than adult conspecifics. Dissection studies by Grand (1977, 1983), Hamada (1983), and Atzeva et al. (2007) have shown that in several primates – including lemurs, tarsiers, and macaques – extrinsic digital flexor muscles are particularly massive in infants and proportionally decrease relative to overall limb muscle mass during postnatal development (Figure 11.2) (see also Turnquist and Wells, 1994; Raichlen, 2005; Druelle et al., 2016). Additionally, young primates are characterized by relatively large hands and feet for their body size (Lumer and Schultz, 1947; Grand, 1977; Jungers and Fleagle, 1980; Ford and Corruccini, 1985; Jungers, 1985; Ravosa et al., 1993; Turnquist and Wells, 1994; Turner et al., 1997; Raichlen, 2005; Lawler, 2006; Young, 2009; Druelle et al., 2018), and have relatively

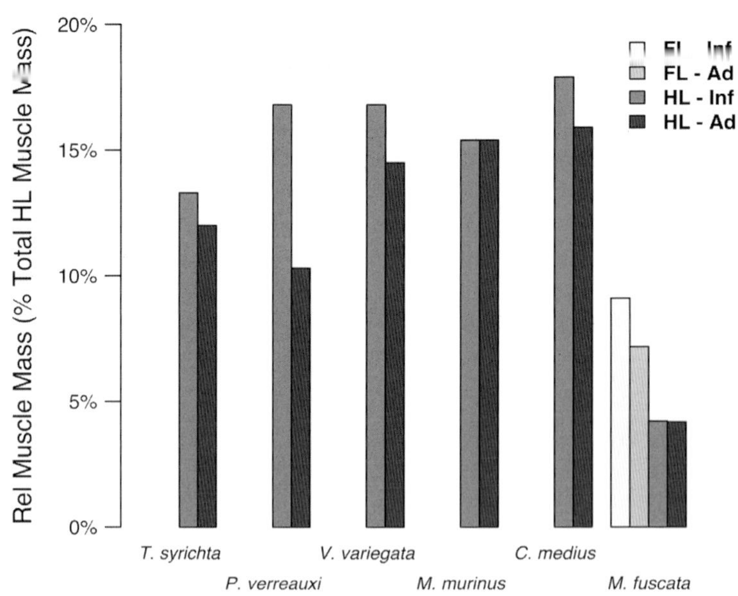

Figure 11.2 Developmental changes in the relative mass of extrinsic digital flexor muscles in *Tarsius syrichta, Propithecus verreauxi, Variecia variegata, Microcebus murinus, Cheirogaleus medius*, and *Macaca fuscata* (expressed as a percentage of total limb muscle mass). Infants tend to have greater digital flexor mass than adults, both in the fore- and hindlimbs. Inf, Infant; Ad, Adult; FL, forelimb; HL, hindlimb. Data from Hamada (1983) and Atzeva et al. (2007).

long fingers and toes (Young and Heard-Booth, 2016; Druelle et al., 2018). Together, these aspects of neonatal primate anatomy likely foster improved grasping ability in infants clinging to caregivers, as well as in juvenile individuals making their initial locomotor attempts on precarious arboreal substrates.

Further development from a passively transported, grasping infant to an independently mobile juvenile is predicated on postural development (Kernell, 1998; Massion, 1998). For example, in their study of the locomotor development of captive Japanese macaques, Negayama et al. (1983) found that postural skills (such as standing) always emerged prior to related locomotor skills (such as walking). A similar pattern characterizes human development (Adolph and Franchak, 2016). Insufficient postural development also explains why the kinematics of air-stepping, in which postural stability is not required, are grossly more similar to the kinematics of mature adults than is actual infant locomotion (Thelen and Fisher, 1982; Vilensky and Gankiewicz, 1986, 1989; Vilensky et al., 1989).

In most primates, once infants begin to take their first steps, limb postures are characterized by pronounced flexor bias and an inability to fully extend the limb joints. Flexor dominance is undoubtedly due to immaturity in both the neuromotor and musculoskeletal systems. In mammals, including primates, the neural tracts necessary to control quadrupedal locomotion continue maturing into postnatal ontogeny, particularly those that control the hindlimbs and maintain extensor muscle tonus (Lawrence and Kuypers, 1965; Kuypers, 1982; Kolb and Whishaw, 1996; Brocard et al., 1999; Cabana, 2000; Vinay et al., 2000a, 2000b; Bewick et al., 2004). Because the relatively large brains of primates take a long time to grow (Barrickman et al., 2008), these processes are likely to be even more protracted in primate infants, leading to poor postural control and a weak antigravity response during early locomotion.

Newborn primates are generally less muscular than adults. For example, Grand (1977) found that muscle accounts for approximately 43% of body mass in adult macaques (*Macaca mulatta*), but only 25% of body mass in late fetal specimens. Histological research in several mammals, including primates, has also shown that due to an initial dearth of Type I muscle fibers, postural muscles do not have the stamina required for tonic isometric activation (Tomanek, 1975; Rubinstein and Kelly, 1978; Jouffroy and Medina, 1996; Bewick et al., 2004). Perhaps in compensation for weak and underdeveloped muscles, studies of both platyrrhine and catarrhine primates have found that infants are often characterized by increased extensor muscle leverage (i.e., greater limb joint mechanical advantage) early in life (Young, 2005; Fellmann, 2011; Figure 11.3). Increased extensor muscle leverage permits young primates to achieve greater output forces for a given muscle force, functionally compensating for proportionally weaker muscles.

The onset of locomotor activity appears to consistently begin earlier in some primate groups than in others. Strepsirrhines tend to begin locomotor activity earlier than monkeys, who in turn tend to begin locomoting more quickly than apes. However, average body mass also tends to increase along these same lines (from strepsirrhines to monkeys to apes), and larger organisms tend to develop more slowly (Harvey and Clutton-Brock, 1985). We therefore used phylogenetic generalized least-squares regression (PGLS; Nunn, 2011) to test for any phylogenetic or ecological influences on age at locomotor onset, after taking variation in body size into account.

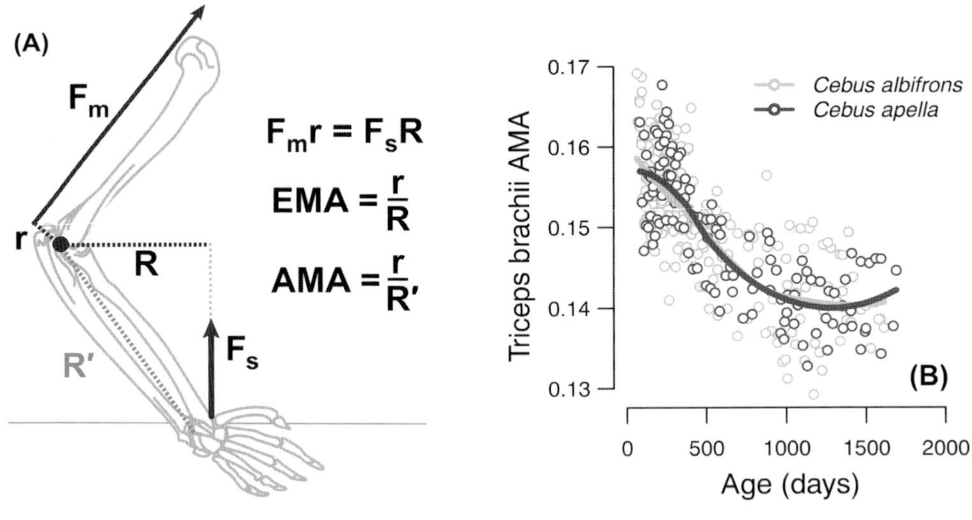

Figure 11.3 (A) Static determinants of mechanical advantage (i.e., muscle leverage) at a limb joint. Due to torque balance about the joint, the magnitude of the output force (F_s) produced for a given muscle force (F_m) is proportional to the ratio of the muscle's lever arm length (r) and the load arm length of the substrate reaction force resulting from pushing on the locomotor substrate (R). The ratio of r and R (i.e., r/R) defines the *effective mechanical advantage* (EMA) of the limb joint in this specific postural configuration. Note that the magnitude of R can be altered via either morphological changes in limb length or behavioral changes in limb posture. R′ indicates skeletal limb length distal to the joint in question, the proxy load arm length used to calculate *anatomical mechanical advantage* (i.e., AMA, r/R′), a morphological estimate of EMA. (B) Ontogenetic changes in triceps brachii AMA in two species of capuchin monkey (*Cebus albifrons* and *Sapajus apella*; original data from Young, 2005). Trends lines represent LOESS (locally estimated scatterplot smoothing) fits to the data for each species. Note that triceps brachii AMA peaks near birth and declines during the first 3 years of life. Ontogenetic declines in limb muscle mechanical advantage have also been observed in several catarrhine primates (i.e., *Macaca fasicularis, M. fuscata, M. mulatta*, and *Papio cynocephalus*) (Fellmann, 2011). Figures modified from Young and Shapiro (2018).

From the studies summarized in Section 11.2, we collated a data set of ages at locomotor onset for 26 species of primates distributed across the extant primate radiation, and including strepsirrhines, tarsiers, platyrrhines, cercopithecines, colobines, and hominoids (see phylogram in Figure 11.4A). Overall, age at locomotor onset shows strong phylogenetic signal in primates. Pagel's λ was estimated to be 0.924 (95% confidence interval = 0.018, 1.00), where $\lambda = 1$ would indicate a Brownian motion process of phenotypic change, such that trait values vary directly as a function of divergence time across taxa. If a given species displays a relatively early (or late) age at locomotor onset, it is highly likely that its close relatives will do the same.

As with many aspects of life history (see Chapter 9), age at locomotor onset was significantly positively correlated with body mass, although nonetheless highly variable (Figure 11.4B; PGLS regression: $p = 0.023$, $R^2 = 0.215$). Larger species tend to begin locomotor activity at older ages than do smaller species. Nevertheless, some taxonomic

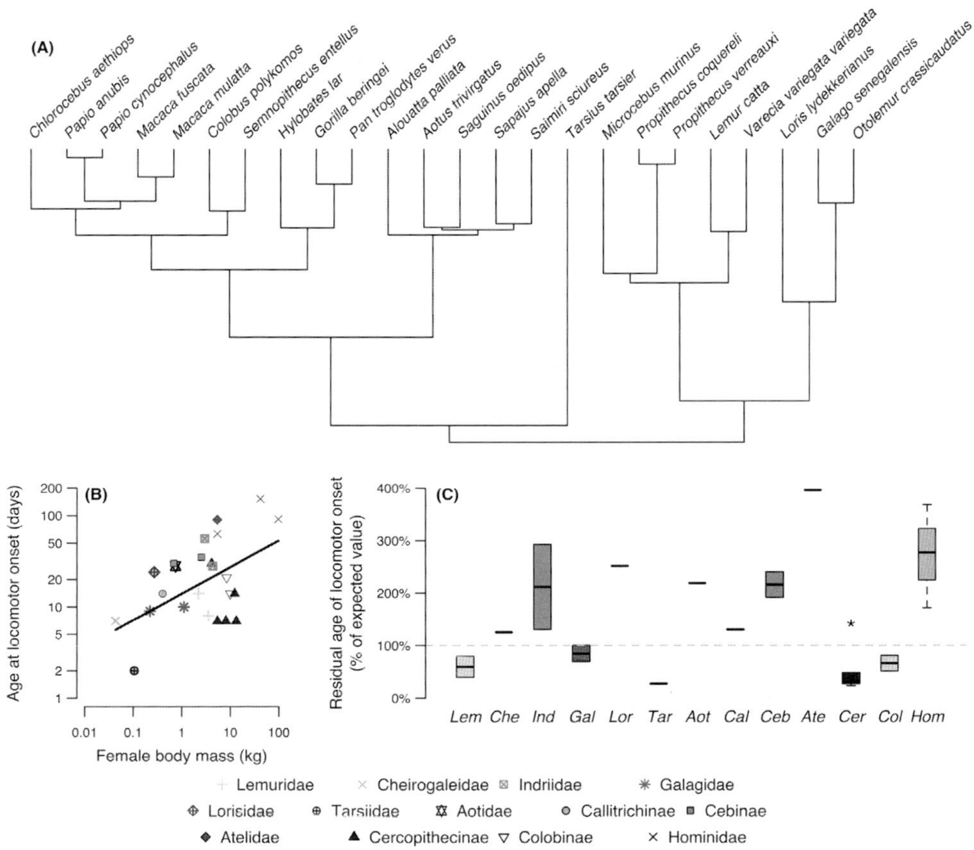

Figure 11.4 (A) Chronogram of primate species used in phylogenetic least-squares (PGLS) analyses of age at locomotor onset. Branch lengths are proportional to times since divergence from common ancestor. Phylogeny is sampled from that of Arnold et al. (2010). (B) Association between age at locomotor onset and average adult female body mass, adjusted for phylogenetic relatedness among taxa. Line indicates PGLS regression fit. (C) Residual age at locomotor onset (after controlling for the effects of adult female body mass), expressed as a percentage of the expected (i.e., fitted) value, grouped by taxonomic family. Values near 100% indicate taxa that begin locomotor activity at the expected age for their body size. Values less than or greater than 100% indicate taxa that begin locomotor activity at relatively early or late ages. Abbreviations: Lem, Lemuridae; Che, Cheirogaleidae; Ind, Indridae; Gal, Galagidae; Lor, Lorisidae; Tar, Tarsiidae; Aot, Aotidae; Cal, Callitrichinae; Ceb, Cebinae; Ate, Atelidae; Cer, Cercopithecinae; Col, Colobinae; Hom, Hominidae.

groups reach locomotor maturity at relatively early, or late, ages compared to their mass-specific expectation (Figure 11.4C). For instance, within strepsirrhines, lemurids (i.e., *Lemur catta* and *Varecia variegata*) begin locomotor activity at relatively young ages for their size, whereas indriids (i.e., *Propithecus coquereli* and *P. verreauxi*) begin locomotion at relatively later ages. Interestingly, this difference runs counter to other aspects of behavioral development, where indriids appear relatively accelerated versus lemurids. In particular, indriids are characterized by an unusually rapid pace of dental development, allowing young sifakas to process hard fruits on their own and become nutritionally independent sooner (Godfrey et al., 2001, 2004, 2006; Schwartz et al., 2002, 2005). Similarly, sifakas are characterized by adult-like locomotor repertoires at early ages, whereas lemurid locomotion is not adult-like until later in life (Sussman, 1977; Lawler, 2006; Wolf, 2016). Perhaps beginning locomotor activity at relatively later ages allows indriids to be somatically more mature at locomotor onset, compressing the time course of locomotor ontogeny and facilitating an early age of locomotor maturity.

Among anthropoids, Old World monkeys (i.e., cercopithecines and colobines) tend to begin locomotion at relatively young ages. Old World monkeys have long been recognized as behaviorally and morphologically distinct from other anthropoids, most likely associated with a long history of selection for increased terrestriality (Di Fiore and Rendall, 1994; McCrossin et al., 1998; Fleagle, 2013). It could be that the unique challenges of locomoting in an arboreal environment may select for a more protracted time course of locomotor development. The arboreal habitat is discontinuous, multidimensional, and frequently unstable. Developing arboreal infants therefore face a host of locomotor challenges not encountered when moving terrestrially – as attested by the frequency of long bone traumas due to falling in free-ranging arboreal primates (Lovell, 1991). Studies of locomotor development corroborate these dangers. Morland (1990) found that falling from arboreal perches was the most common cause of infant mortality in ruffed lemurs (*Varecia variegata*). Field studies of spectral tarsiers (*Tarsier tarsier*) and blue monkeys (*Cercopithecus mitis*) have noted that infants often fall, requiring either the infants themselves or a caregiver to expend the energy to descend to the forest floor and climb back to a safe arboreal location (Förster and Cords, 2002; Gursky, 2007). Finally, Sussman (1977) noted that wild *Lemur catta* tend to progress through locomotor development at a faster pace than sympatric *Eulemur fulvus*, a difference he attributed to the increased arboreality of the latter. To test whether increased arboreality may be associated with more protracted locomotor development across primates, we grouped the species residuals from the PGLS regression presented in Figure 11.4B into arboreal and terrestrial groups (following the habitat coding of Nunn and Barton, 2001). Overall, residual ages at locomotor onset did not differ between arboreal and terrestrial taxa (Figure 11.5; $p = 0.35$). However, among the terrestrial taxa, the highly encephalized and slow-growing African apes (i.e., *Pan troglodytes* and *Gorilla berengei*) appeared as particularly late-developing outliers relative to other terrestrial primates. Among non-hominoid primates in the sample (i.e., strepsirrhines, tarsiers, platyrrhines, and cercopithecoids), arboreal taxa have significantly later mass-specific ages at locomotor onset than terrestrial taxa ($p = 0.01$).

Some of the taxa that were marked by particularly late ages of locomotor onset (e.g., cebine monkeys and hominoids) are also notable for being highly encephalized. Previous studies have found a negative association between relative brain size and the pace of locomotor development, such that more encephalized species begin walking at later ages. For example, Garwicz et al. (2009) found that absolute brain size was the best predictor of age at walking onset across 24 mammals representing 11 orders (including four primate species), even after controlling for the influence of body size. Why should the pace of locomotor development be correlated with brain size? Independent locomotion is predicated on sufficient development of the motor control system, including outgrowth of relevant spinal tracts from the central nervous system,

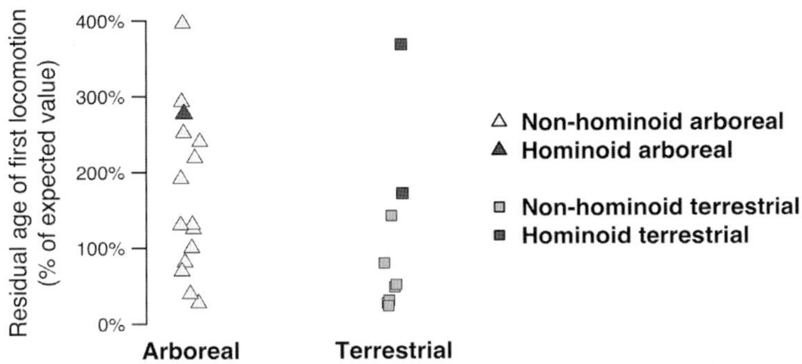

Figure 11.5 Residual age at locomotor onset (after controlling for the influence of body mass), expressed as a percentage of the fitted value from the PGLS regression represented in Figure 11.4B, grouped according to whether the species is primarily arboreal or terrestrial during active foraging and travel (following the coding of Nunn and Barton, 2001). On average, non-hominoid arboreal species begin locomotion at later ages than non-hominoid terrestrial taxa ($p = 0.01$).

the pruning and refinement of effector neurons in the peripheral nervous system, and the maturation of higher-level control centers, such as the cerebellum (Kuypers, 1962; Eyre et al., 2000; ten Donkalaar, 2000). Because the rate of brain development is similar across mammals (Passingham, 1985), species with larger brains should exhibit protracted durations of behavioral development, as growing infants wait for relevant neural substrates to sufficiently mature. This may be particularly true for primates, where locomotion has been shown to depend on supraspinal (i.e., cortical) control to a much greater degree than in other mammals (Vilensky and Larson, 1989; Courtine et al., 2005). Additionally, growing a large brain is energetically expensive, and may come at the cost of slower somatic development, including slower development of the musculoskeletal substrates required for mature locomotion (but see Deaner et al., 2003; Barrickman and Lin, 2010). Interestingly, however, endocranial volume (i.e., ECV, a validated isometric proxy for brain size in primates: Isler et al., 2008) was not significantly correlated with age at locomotor onset in the sample of primates included here ($p = 0.117$), despite evidence from a previous study that ECV is the best predictor of age at locomotor maturity across primates (Young and Shapiro, 2018). It may be that due to the protracted timescale of primate brain growth and behavioral development, much of which occurs postnatally, correlations between brain size and developmental pace are only evident for metrics of skill maturation, not skill onset per se.

Once locomotion has begun, the early locomotor movements of all are generally limited to slow, creeping quadrupedalism, even among taxa for which quadrupedalism is not the primary mode of locomotion in adults (e.g., galagos and tarsiers). Only later do more specialized forms of locomotion, such as VCL, or brachiation, enter the expanding behavioral repertoire. The explanation for this general trend is likely multifactorial. First, complex locomotor skills require more intricate movement trajectories and greater precision of hand and foot placement, and therefore implicate greater cortical control over locomotion. In macaques – and perhaps other primates as well – the relevant corticospinal tracts that permit such precision of limb placements do not functionally mature for several months after birth (Kuypers, 1962; Lawrence and Hopkins, 1976; Armstrong, 1988; Drew et al., 2002; MacKay-Lyons, 2002; Bauby and Kuo, 2000). The slow development of the necessary corticospinal tracts may account for the comparatively late emergence of climbing, leaping, and other complex locomotor modes, as these movements require more precise placement of the extremities on the substrate.

It is also important to consider how ontogenetic changes in lower-level biomechanical parameters, often resulting from allometric patterns of somatic growth, may influence the sequential acquisition of locomotor skills during primate development. For instance, the intermembral index (i.e., IMI; forelimb length as a percentage of hindlimb length) is typically greater in neonates than in older individuals (Figure 11.6). Across primates, IMI has been shown to be broadly predictive of predominant locomotor mode, with low values (i.e., < 100, indicating longer hindlimbs) associated with a propensity for leaping locomotion, high values (i.e., > 100, indicating longer forelimbs) associated with suspensory locomotion, and intermediate values (i.e., ~100, indicating comparable limb lengths) associated with a more exclusive focus on quadrupedalism (Fleagle, 2013). A PGLS regression across a broad sample of strepsirrhine and haplorrhine primates (Figure 11.6A) found that neonatal IMI scales to adult IMI with a slope of 0.69 (95% CI: 0.645, 0.804) and an intercept of 31 (95% CI: 20.7, 59.7) (Figure 11.6B; R^2 = 0.57). This scaling relationship indicates that (1) newborns have significantly longer forelimbs than adults but that (2) this difference is mitigated in the upper range of IMI (i.e., among suspensory primates, such as the great apes). Functionally, given that most non-hominoid adult primates have relatively longer hindlimbs (i.e., IMI < 100), the higher IMI of neonatal infants places them closer to the "quadrupedalism" configuration of approximately equal fore- and hindlimb lengths. As most primates age, differential scaling of fore- and hindlimb lengths tends to drive IMI toward lower values, such that juvenile primates closely approximate the adult configuration – juvenile IMI scales to adult IMI with a slope of 0.89 (95% CI: 0.739, 0.947) and an intercept of 11 (95% CI: 4.8, 23.7) (Figure 11.6C; R^2 = 0.94). These values are very close to those predicted under a scaling relationship of identity (i.e., a slope of 1 and an intercept of 0, indicating equality between the x and y variables). In summary, ontogenetic changes in IMI broadly track developmental changes in locomotor modes. In non-hominoid species, quadrupedalism predominates initially, followed by more specialized forms (such as leaping) as limb proportions become more adult-like. By contrast, chimpanzees and gorillas, who are more suspensory at young ages than during adulthood (although climbing remains an important behavior throughout life: Doran, 1992, 1997; Sarringhaus et al., 2014), have IMI > 100 at birth and maintain similar values throughout growth. Recent studies of other postcranial regions have identified other measures in which variation in newborn anatomy appears to already reflect variation in adult positional behavior. For instance, Zaleski et al. (2018) noted mediolaterally broader thoracic shape in orthograde primate species (e.g., *Tarsius, Galago*) compared to other species at birth.

Finally, allometric changes in limb proportions, whole-body center of mass position, and other aspects of body shape also influence the biomechanics of quadrupedal locomotion during primate growth. We will only briefly touch on these trends – interested readers will find a more comprehensive discussion of these topics in Young

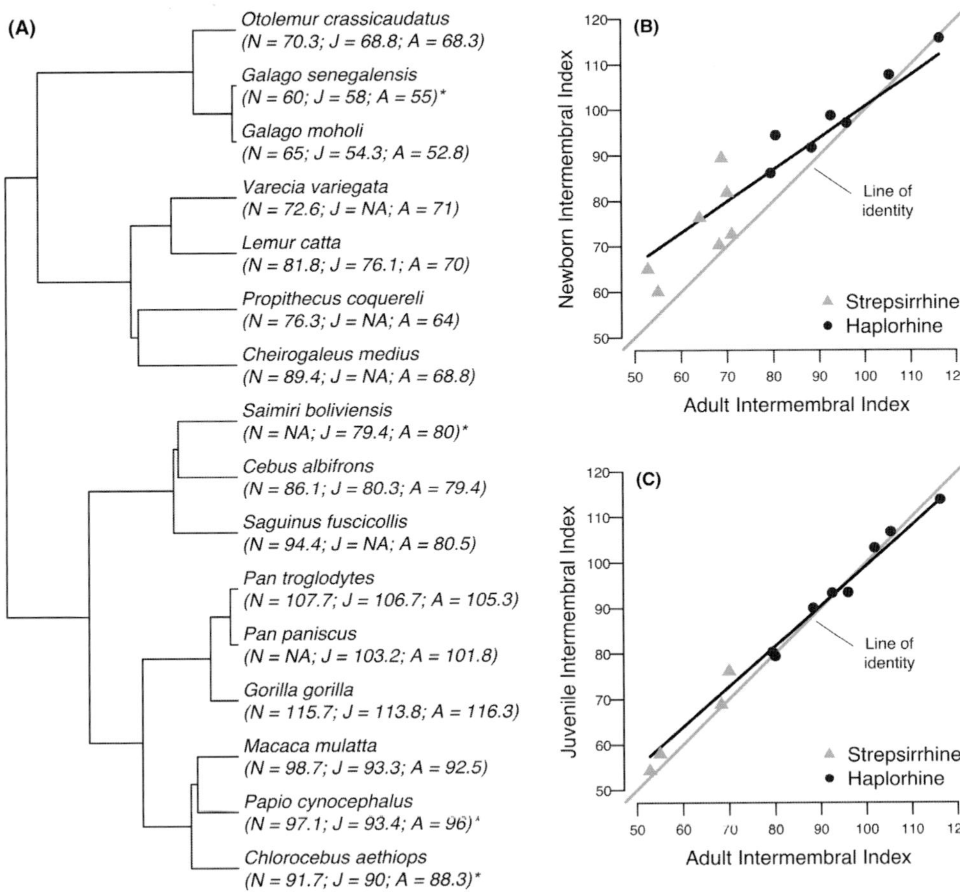

Figure 11.6 (A) Chronogram of primate species used in comparative analyses of developmental changes in intermembral index (IMI). Branch lengths are proportional to times since divergence from common ancestor. Phylogeny is sampled from that of Arnold et al. (2010). Values next to species names are IMI values for newborns (N), juveniles (J), and adults (A). NA signifies that the indicated measurement was not available, whereas asterisks indicate that IMI values are based on external measurements of live animals or cadavers (rather than osseous measurements of limb bones). Data sources include: Fleagle and Samonds (1975); Shea (1981); Glassman (1983); Anemone (1993); Turnquist and Wells (1994); Turner et al. (1997); Shapiro and Raichlen (2006); Schaefer and Nash (2007); Young (2008); Fleagle (2013); and unpublished data from the authors. (B) Scaling of neonatal versus adult IMI. Black line indicates PGLS regression fit, whereas gray line indicates a line of identify (i.e., slope = 1 and intercept = 0). (C) Scaling of juvenile versus adult IMI. Trend lines are drawn as in panel B.

and Shapiro (2018). Ever since the pioneering work of Muybridge (1887), researchers interested in mammalian locomotion have recognized that adult quadrupedal primates use a different footfall sequence during walking than most other mammals. Most quadrupedal mammals use lateral sequence (LS) walking gaits, in which hind foot touchdowns are followed by the touchdown of an ipsilateral forefoot. In contrast, most quadrupedal primates use diagonal sequence (DS) walking gaits, in which hind foot touchdowns are followed by the touchdown of a contralateral forefoot (Hildebrand, 1967; Cartmill et al., 2002). Gait selection in immature primates is more variable. Infants and juveniles frequently use LS gaits either exclusively or in addition to DS (Hildebrand, 1967; Rose, 1977; Rollinson and Martin, 1981; Hurov, 1982; Kimura, 1987; Vilensky and Gankiewicz, 1989; Nakano, 1996; Dunbar and Badam, 1998; Shapiro and Raichlen, 2005, 2006; Young, 2012a; Druelle et al., 2017b). Whereas some researchers have attributed ontogenetic changes in gait selection to the general neurological and musculoskeletal immaturity of infant primates (Hildebrand, 1967; Prost, 1969; Hurov, 1982; Vilensky and Gankiewicz, 1989; Nakano, 1996; Druelle et al., 2017b), citing theoretical arguments that LS gaits potentially confer greater locomotor stability, other studies have found more direct associations between functional aspects of somatic growth and gait preference. For example, as noted in our discussion of developmental changes in IMI, limb proportions can change dramatically during primate growth. The use of DS gaits could be particularly problematic during periods of rapid limb growth. This is because in a DS gait, hindlimb touchdowns coincide with maximal retraction of the ipsilateral forelimb, increasing the potential that the two limbs will physically collide. Indeed, research on squirrel monkeys and baboons shows that LS gaits are preferred during periods when growing animals are relatively long-limbed (Shapiro and Raichlen, 2006; Young, 2012a). Additionally, transitions between preferential use of LS versus DS gaits parallels developmental transitions in whole-body center of mass (CoM) position. Infant primates tend to have relatively large heads and reduced hindlimb mass compared to adults (Grand, 1977, 1983; Druelle et al., 2017a), resulting in a net caudal transition in CoM position as body shape matures (Young, 2012a, 2012b; Druelle et al., 2017a). Under assumptions of static stability (i.e., where balance is achieved as long as the CoM falls within the base of support provided by the supporting limbs), LS gaits are potentially more stable when the CoM is in an anterior location (Cartmill et al., 2002) – perhaps accounting for the more frequent use of LS gaits in heavy-headed infants.

11.4 Conclusions: A Life-History Perspective on Locomotor Development

Name a fitness-critical function, from foraging for or hunting food resources, to securing access to potential mates, to evading predators, and it is likely that safe, efficient locomotion will be somehow central to success. An animal that is unable to feed, mate, or evade predators – or one that spends too much metabolic energy or sustains major injuries while doing so – will not be able to survive, reproduce, and maintain Darwinian fitness. From a developmental perspective, alongside nutritional independence (i.e., weaning), locomotor independence is a primary determinant of when immature animals make the transition from dependent infants to ecologically independent juveniles (Altmann, 1980; Altmann and Samuels, 1992; Young and Shapiro, 2018). In summary, locomotion is central to the biology and ecology of most organisms.

Given this central place in animal biology, it is not surprising that locomotor behavior and the underlying morphological structures that promote locomotion are evolutionary labile and shaped by the ecological pressures placed upon animals across their lifespan (Biewener, 2003). In this vein, locomotor development can be viewed as a fundamental life-history parameter, responding to the same classes of evolutionary pressures that have been shown to be fundamental in previous studies of primate life-history variation (e.g., predation pressures, resource access, overall body size, encephalization). As detailed in Chapter 9, comparative studies of primate life history typically focus on parameters that are tightly constrained (i.e., average body mass, gestation length, growth rates) or only tangentially related to behavior (i.e., age of first reproduction, interbirth interval). However, broadscale variation in primate life-history parameters is ultimately predicated on the daily actions of individuals. Where to forage, how to avoid predators, with whom to mate, and myriad other choices all cascade to determine ecological – and reproductive – success over an individual's lifetime. The pace of developmental changes in locomotor behavior fundamentally determines when a growing infant can become ecologically independent, and how well independent juveniles are able to access and exploit resources. Consideration of locomotor development as a life-history parameter could provide focus for future integrative research. For example, we presented evidence here that the age at locomotor onset occurs earlier in terrestrial primates than in arboreal primates (excluding hominoids). What determines this difference in developmental pacing? Biomechanical factors (i.e., the precariousness of arboreal versus terrestrial locomotion), ecological factors (greater risk of predation in terrestrial environments), or energetic factors (i.e., longer daily path lengths of terrestrial primates) could all play a role. To address these questions, and similar ones, future work on primate locomotor development should seek to move beyond captive laboratory research and study growing primates in their natural environments. New and developing techniques permit the remote, non-invasive study of both somatic growth (Rothman et al., 2008) and locomotor kinematics (Dunham et al., 2018) in arboreal primates. Additionally, future collections-based studies should seek to characterize functional aspects of musculoskeletal growth – such as mechanical advantage (Young, 2005) and bone strength (Young et al., 2010, 2019) – in a variety of strepsirrhine and haplorhine primates, with an eye toward linking the functional capacity and limitations of the growing skeleton to concomitant changes in positional behavior and biomechanics (e.g., Sarringhaus et al., 2016). Only a unified framework of investigation will be able to parse out the broad proximate and ultimate determinants of variation in the tempo and mode of primate locomotor development.

References

Adams LG, Singer FJ, Dale BW. 1995. Caribou calf mortality in Denali National Park, Alaska. *J Wildlife Management* 59:584–594.

Adolph KE, Franchak JM. 2016. The development of motor behavior. *Wiley Interdiscip Rev Cogn Sci* 8:e1430.

Altmann J. 1980. *Baboon Mothers and Infants*. Chicago: University of Chicago Press.

Altmann J, Samuels A. 1992. Costs of maternal care: Infant carrying in baboons. *Behav Ecol Sociobiol* 29:391–398.

Anemone RL. 1993. The functional anatomy of the hip and thigh in primates. In: Gebo DL, editor. *Postcranial Adaptation in Nonhuman Primates*. Dekalb: Northern Illinois University Press, pp. 150–174.

Anemone RL, Watts ES, Swindler DR. 1991. Dental development of known-age chimpanzees, *Pan troglodytes* (Primates, Pongidae). *Am J Phys Anthropol* 86:229–241.

Anemone RL, Mooney MP, Siegel MI. 1996. Longitudinal study of dental development in chimpanzees of known chronological age: Implications for understanding age at death of Plio-Pleistocene hominids. *Am J Phys Anthropol* 99:119–133.

Ankel-Simons F. 2007. *Primate Anatomy: An Introduction*. New York: Academic Press.

Armstrong DM. 1988. The supraspinal control of mammalian locomotion. *J Physiol* 405:1–37.

Arnold C, Matthews LJ, Nunn CL. 2010. The 10kTrees website: A new online resource for primate phylogeny. *Evol Anthropol* 19:114–118.

Arnott SA, Chiba S, Conover DO. 2006. Evolution of intrinsic growth rate: Metabolic costs drive trade-offs between growth and swimming performance in *Menidia menidia*. *Evolution* 60:1269–1278.

Asher RJ, Lehmann T. 2008. Dental eruption in afrotherian mammals. *BMC Biol* 6:14.

Ashley-Montagu MF. 1933. The anthropological significance of the pterion in the Primates. *Am J Phys Anthropol* 18: 159–336.

Ashley Montagu MF. 1943. The mesethmoid–presphenoid relationship in the primates. *Am J Phys Anthropol* 1:129–141.

Atsalis S. 2015. *A Natural History of the Brown Mouse Lemur*. London: Routledge.

Atzeva M, Demes B, Kirkbride ML, Burrows AM, Smith TD. 2007. Comparison of hind limb muscle mass in neonate and adult prosimian primates. *J Hum Evol* 52:231–242.

Avery JK. 2002. *Oral Development and Histology*. 3rd ed. New York: Thieme Medical Publishers.

Baden AL, Wright PC, Louis EE, Bradley BJ. 2013. Communal nesting, kinship, and maternal success in a social primate. *Behav Ecol Sociobio* 67:1939–1950.

Barrickman N, Lin M. 2010. Encephalization, expensive tissues, and energetics: An examination of the relative costs of brain size in strepsirrhines. *Am J Phys Anthropol* 143:579–590.

Barrickman NL, Bastian ML, Isler K, Van Schaik CP. 2008. Life history costs and benefits of encephalization: A comparative test using data from long-term studies of primates in the wild. *J Hum Evol* 54:568–590.

Barros A. 2013. Ontogeny, phylogeny and functional morphology of the hominoid shoulder girdle. PhD dissertation. University College London.

Bauby CE, Kuo AD. 2000. Active control of lateral balance in human walking. *J Biomech* 33:1433–1440.

Baume LJ. 1953. The postnatal growth of the mandible of *Macaca mulatta*, a metricroentgenographic and histologic study. *Am J Ortho* 39:228–229.

Baume LJ. 1968. Patterns of cephalofacial growth and development. A comparative study of the basicranial growth centers in rat and man. *Int Dent J* 13:489–513.

Beecher RM. 1977. Function and fusion at the mandibular symphysis. *Am J Phys Anthropol* 47:325–335.

Bewick GS, Reid B, Jawaid S, Hatcher T, Shanley L. 2004. Postnatal emergence of mature release properties in terminals of rat fast- and slow-twitch muscles. *Eur J Neurosci* 19:2967–2976.

Beynon AD, Dean MC. 1988. Distinct dental development patterns in early fossil hominids. *Nature* 335:509–514

Bezanson M. 2006. Ontogenetic patterns of positional behavior in *Cebus capucinus* and *Alouatta palliata*. PhD dissertation, University of Arizona.

Bezanson M. 2012. The ontogeny of prehensile tail use in *Cebus capucinus* and *Alouatta palliata*. *Am J Primatol* 74:770–782.

Bezanson M. 2017. Primate positional behavior development and evolution. *Annu Rev Anthrop* 46:279–298.

Bezanson M, Morbeck ME. 2013. Future adults or old children? Integrating life history frameworks for understanding primate positional patterns. In: Clancy KBH, Rutherford JN, Hinde K, editors. *Building Babies*. New York: Springer, pp. 435–458.

Biewener AA. 2003. *Animal Locomotion*. Oxford: Oxford University Press.

Blomquist GE, Williams LE. 2013. Quantitative genetics of costly neonatal sexual size dimorphism in squirrel monkeys (*Saimiri boliviensis*). *J Evol Biol* 26:756–765.

Bogin B. 1999a. Evolutionary perspective on human growth. *Annu Rev Anthropol* 28:109–153.

Bogin B. 1999b. *Patterns of Human Growth*. Cambridge: Cambridge University Press.

Boinski S, Fragaszy DM. 1989. The ontogeny of foraging in squirrel monkeys, *Saimiri oerstedi*. *Anim Behav* 37:415–428.

Bolk L, Goeppert E, Kallius E, Lubosch W. 1931–1939. *Handbuch der vergleichenden Anatomie der Wirbeltiere*, vols. 1–6. Berlin: Urban & Schwarzenberg.

Bolter DR, Zihlman AL. 2003. Morphometric analysis of growth and development in wild-collected vervet monkeys (*Cercopithecus aethiops*), with implications for growth patterns in Old World monkeys, apes and humans. *J Zool, Lond* 260:99–110.

Bolter D, Zihlman A. 2007. Primate growth and development. In: Campbell CJ et al., editors. *Primates in Perspective*. New York: Oxford University Press, pp. 408–422.

Born G. 1883. Die Plattenmodeliermethode. *Archiv Mikroskopische Anatomie* 22: 584–599.

Borries C, Lu A, Ossi-Lupo K, Larney E, Koenig A. 2011. Primate life histories and dietary adaptations: A comparison of Asian colobines and macaques. *Am J Phys Anthropol* 144:286–299.

Bosma JF. 1986. *Anatomy of the Infant Head*. Baltimore: Johns Hopkins University Press.

Boughner JC. 2011. Making space for permanent molars in growing baboon (*Papio anubis*) and great ape (*Pan paniscus* and *P. troglodytes*) mandibles: Possible ontogenetic strategies and solution. *Anat Res Int* 2011:1–16.

Boughner J, Dean MC. 2004. Does space in the jaw influence the timing of molar crown initiation? A model using baboons (*Papio anubis*) and great apes (*Pan troglodytes, Pan paniscus*). *J Hum Evol* 46:255–277.

Bouvier M, Hylander WL. 1981. The relationship between split-line orientation and in vivo bone strain in galago (*G. crassicaudatus*) and macaque (*Macaca mulatta* and *M. fascicularis*) mandibles. *Am J Phys Anthropol* 56:147–156.

Bouvier M, Hylander WL. 1982. The effect of dietary consistency on morphology of the mandibular condylar cartilage in young macaques (*Macaca mulatta*). In: Dixon AD, Sarnat BG, editors. *Factors and Mechanisms Influencing Bone Growth*. New York: Alan R. Liss, pp. 569–579.

Bouvier M, Hylander WL. 1996. The mechanical or metabolic function of secondary osteonal bone in the monkey *Macaca fascicularis*. *Arch Oral Biol* 41:941–950.

Bowler M, Bodmer RE. 2011. Diet and food choice in Peruvian red uakaris (*Cacajao calvus ucayalii*): Selective or opportunistic seed predation? *Int J Primatol* 32:1109–1122.

Bray J, Thompson ME, Muller MN, Wrangham RW, Machanda ZP. 2018. The development of feeding behavior in wild chimpanzees (*Pan troglodytes schweinfurthii*). *Am J Phys Anthropol* 165:34–46.

Brocard F, Vinay L, Clarac F. 1999. Gradual development of the ventral funiculus input to lumbar motoneurons in the neonatal rat. *Neuroscience* 90:1543–1554.

Bromage TG, Janal MN. 2014. The Havers–Halberg oscillation regulates primate tissue and organ masses across the life-history continuum. *Biol J Linn Soc* 112:649–656

Bromage TG, Hogg RT, Lacruz RS, Hou C. 2012. Primate enamel evinces long period biological timing and regulation of life history. *J Theor Biol* 305:131–144.

Burr DB, Organ JM. 2017. Postcranial skeletal development and its evolutionary implications. In: Percival CJ, Richtmeier JT, editors. *Building Bones*. Cambridge: Cambridge University Press, pp. 148–174.

Butler H. 1983. The embryology of the lesser galago (*Galago sengalensis*). In Szalay FS, editor. *Contributions to Primatology*. Berlin: Karger.

Cabana T. 2000. The development of mammalian motor systems: The opossum *Monodelphis domestica* as a model. *Brain Res Bull* 53:615–626.

Cachel S. 1984. Growth and allometry in primate masticatory muscles. *Archs Oral Biol* 29:287–293.

Cardini A, Elton S. 2008. Does the skull carry a phylogenetic signal? Evolution and modularity in the guenons. *Biol J Linn Soc Lond* 93:813–834.

Carlson BM. 2014. *Human Embryology and Developmental Biology*. 5th ed. Philadelphia: Elsevier-Saunders.

Cartmill M. 1980. Morphology, function, and evolution of the anthropoid postorbital septum. In: Ciochon RL, Chiarelli AB, editors. *Evolutionary Biology of the New World Monkeys and Continental Drift*. New York: Plenum Press, pp. 243–274.

Cartmill M, Lemelin P, Schmitt D. 2002. Support polygons and symmetrical gaits in mammals. *Zool J Linn Soc* 136:401–420.

Castell R, Sackett G. 1972. Motor behaviors of neonatal rhesus monkeys: Measurement techniques and early development. *Dev Psychobiol* 6:191–202.

Castellana C, Kosa F. 1999. Morphology of the cervical vertebrae in the fetal–neonatal human skeleton. *J Anat* 194:147–152.

Chalk J, Wright BW, Lucas PW, et al. 2016. Age-related variation in the mechanical properties of foods processed by *Sapajus libidinosus*. *Am J Phys Anthropol* 159:199–209.

Chalk-Wilayto J, Ossi-Lupo K, Raguet-Schofield M. 2016. Growing up tough: Comparing the effects of food toughness on juvenile feeding in *Sapajus libidinosus* and *Trachypithecus phayrei crepusculus*. *J Hum Evol* 98:76–89.

Chalmers NR. 1980. Developmental relationships among social, manipulatory, postural and locomotor behaviours in olive baboons, *Papio anubis*. *Behaviour* 74:22–37.

Charles-Dominique P. 1977. *Ecology and Behaviour of Nocturnal Primates: Prosimians of Equatorial West Africa*. New York: Columbia University Press.

Chase JE, Cooper RW. 1969. *Saguinus nigricollis* – Physical growth and dental eruption in a small population of captive-born individuals. *Am J Phys Anthropol* 30:111–116.

Cheverud JM. 1996. Developmental integration and the evolution of pleiotropy. *Amer Zool* 36:44–50.

Clarke MR. 1990. Behavioral development and socialization of infants in a free-ranging group of howling monkeys (*Alouatta palliata*). *Folia Primatol (Basel)* 54:1–15.

Cofran Z, Walker CS. 2017. Dental development in *Homo naledi*. *Biol Lett* 13(8): 20170339. doi: 10.1098/rsbl.2017.0339

Cole TM. 1992. Postnatal heterochrony of the masticatory apparatus in *Cebus apella* and *Cebus albifrons*. *J Hum Evol* 23:253–282.

Cole WG, Vereijken B, Young JW, Robinson SR, Adolph KE. 2019. Use it or lose it? Effects of age, experience, and disuse on crawling. *Dev Psychobiol* 61:29–42.

Conroy GC. 1982. A study of cerebral vascular evolution in primates. In: Armstrong E, Falk D, editors. *Primate Brain Evolution: Methods and Concepts*. New York: Plenum Press, pp. 247–262.

Cooper KL, Oh S, Sung Y, et al. 2013. Multiple phases of chondrocyte enlargement underlie differences in skeletal proportions. *Nature* 495:375–378.

Corruccini RS, Beecher RM. 1982. Occlusal variation related to soft diet in a nonhuman primate. *Science* 218:74–76.

Corruccini RS, Beecher RM. 1984. Occlusofacial morphological integration lowered in baboons raised on soft diet. *J Craniofac Genet Develop Biol* 4:135–142.

Courtine G, Roy RR, Hodgson H, et al. 2005. Kinematic and EMG determinants in quadrupedal locomotion of a non-human primate (Rhesus). *J Neurophysiol* 93:3127–3145.

Crompton RH. 1983. Age differences in locomotion of two subtropical Galaginae. *Primates* 24:241–259.

Cummings JR, Muchlinski MN, Kirk EC, et al. 2012. Eye size at birth in prosimian primates: Life history correlates and growth patterns. *PLoS ONE* 7(5):e36097. doi:10.1371/journal.pone.0036097.

Cunningham C, Scheuer L, Black S. 2016. *Developmental Juvenile Osteology*. London: Elsevier Academic Press.

Cupero TM, Middleton CE, Silva AB. 2001. Effects of functional septoplasty on the facial growth of ferrets. *Arch Otolaryngol Head Neck Surg* 127:1367–1369.

Daegling DJ, Granatosky MC, McGraw WS. 2013. Ontogeny of material stiffness heterogeneity in the macaque mandibular corpus. *Am J Phys Anthropol* 153:297–304.

Darwin C. 1902. *Origin of Species*. 6th ed. New York: PF Collier & Son.

Dayeh AA, Herring SW. 2014. Cellular proliferation in the nasal septal cartilage in juvenile minipigs. *J Anat* 225:603–613.

Deaner RO, Barton RA, van Schaik CP. 2003. Primate brains and life histories: renewing the connection. In: Kappeler PM, Pereira M, editors. *Primate Life Histories and Socioecology*. Chicago: University of Chicago Press, pp. 233–265.

de Beer GR. 1937. *The Development of the Vertebrate Skull*. Chicago: Chicago University Press.

Dechow PC, Carlson DS. 1990. Occlusal force and craniofacial biomechanics during growth in Rhesus monkeys. *Am J Phys Anthropol* 83:219–237.

DeLeon VB, Smith TD. 2014. Mapping the nasal airways: Using histology to enhance CT-based three-dimensional reconstruction. *Anat Rec* 297:2113–2120.

DeLeon VB, Smith TD. 2018. My, what big teeth you have! A structural model for relative midfacial length in primates. *Am J Phys Anth* 165(S66):64.

DeLeon VB, Smith TD, Rosenberger AL. 2016. Ontogeny of the postorbital region in tarsiers and other primates. *Anat Rec* 199:1631–1645.

DeLeon VB, Rosenberger AL, Smith TD. 2017. Unique ontogenetic patterns of postorbital septation in tarsiers and the issue of trait homology. In: Percival CJ, Richtsmeier JT, editors. *Building Bone: Bone Formation and Development in Anthropology*. Cambridge: Cambridge University Press, pp. 79–103.

Demes B, Fleagle JG, Jungers WL. 1999. Takeoff and landing forces of leaping strepsirrhine primates. *J Hum Evol* 37:279–292.

Derrickson EM. 1992. Comparative reproductive strategies of altricial and precocial mammals. *Funct Ecol* 6:57–65.

DeSilva JM, Rosenberg KR. 2017. Anatomy, development, and function of the human pelvis. *Anat Rec* 300:628–632.

Dickinson E, Fitton LC, Kupczik K. 2018. Ontogenetic changes to muscle architectural properties within the jaw-adductor musculature of *Macaca fascicularis*. *Am J Phys Anthropol* 167:291–310.

Di Fiore A, Rendall D. 1994. Evolution of social organization: A reappraisal for primates using phylogenetic methods. *PNAS* 91:9941–9945.

Diogo R, Pastor JF, Hartstone-Rose A, Muchlinski MN. 2015. *Baby Gorilla: Photographic and Descriptive Atlas of Skeleton, Muscles, and Internal Organs*. Boca Raton: CRC Press.

Dolhinow P, Murphy G. 1982. Langur monkey (*Presbytis entellus*) development – The first three months of life. *Folia Primat* 39:305–331.

Doran DM. 1992. The ontogeny of chimpanzee and pygmy chimpanzee locomotor behavior: A case study of paedomorphism and its behavioral correlates. *J Hum Evol* 23:139–157.

Doran DM. 1997. Ontogeny of locomotion in mountain gorillas and chimpanzees. *J Hum Evol* 32:323–344.

Doyle GA. 1979. Development of behavior in prosimians with special reference to the lesser bushbaby, *Galago senegalensis moholi*. In: Doyle GA, Martin RD, editors. *The Study of Prosimian Behavior*. New York: Academic Press, pp. 157–206.

Drew T, Jiang W, Widajewicz W. 2002. Contributions of the motor cortex to the control of the hindlimbs during locomotion in the cat. *Brain Res Rev* 40(1–3): 178–191.

Druelle F, Aerts P, Berillon G. 2016. Effect of body mass distribution on the ontogeny of positional behaviors in non-human primates: Longitudinal follow-up of infant captive olive baboons (*Papio anubis*). *Am J Primatol* 78:1201–1221.

Druelle F, Aerts P, D'Aoû K, Moulin V, Berillon G. 2017a. Segmental morphometrics of the olive baboon (*Papio anubis*): A longitudinal study from birth to adulthood. *J Anat* 230:805–819.

Druelle F, Berillon G, Aerts P. 2017b. Intrinsic limb morpho-dynamics and the early development of interlimb coordination of walking in a quadrupedal primate. *J Zool* 301:235–247.

Druelle F, Young J, Berillon G. 2018. Behavioral implications of ontogenetic changes in intrinsic hand and foot proportions in olive baboons (*Papio anubis*). *Am J Phys Anthropol* 165:65–76.

Dunbar DC, Badam GL. 1998. Development of posture and locomotion in free-ranging primates. *Neurosci Biobehav Rev* 22:541–546.

Dunham NT. 2015. Ontogeny of positional behavior and support use among *Colobus angolensis palliatus* of the Diani Forest, Kenya. *Primates* 56:1–10.

Dunham NT, McNamara A, Shapiro LJ, Young JW. 2018. A user's guide for the quantitative analysis of substrate characteristics and locomotor kinematics in free-ranging primates. *Am J Phys Anthropol* 167:569–584.

Eadie EC. 2015. Ontogeny of foraging competence in capuchin monkeys (*Cebus capucinus*) for easy versus difficult to acquire fruits: A test of the needing to learn hypothesis. *PLoS ONE* 10:e0138001.

Eaglen RH. 1985. Behavioral correlates of tooth eruption in Madagascar lemurs. *Am J Phys Anthropol* 66:307–315.

Eaglen RH, Boskoff KJ. 1978. The birth and early development of a captive sifaka, *Propithecus verrauxi coquereli*. *Folia Primatol* 30:206–219.

Ehrlich A. 1974. Infant development in two prosimian species: Greater galago and slow loris. *Dev Psychobio* 7:439–454.

Elad D, Kozlovsky P, Blum O, et al. 2014. Biomechanics of milk extraction during breast-feeding. *PNAS* 111:5230–5235.

Elias MF. 1977. Relative maturing of cebus and squirrel monkeys at birth and during infancy. *Dev Psychobiol* 10:519–528.

Enlow DR. 1990. *Facial Growth*. Philadelphia: WB Saunders.

Escobar-Páramo P. 1989. The development of the wild black-capped capuchin (*Cebus apella*) at La Macarena, Colombia. *Field Studies of New World Monkeys, La Macarena, Colombia* 2:45–56.

Eyre JA, Miller S, Clowry GJ, Conway EA, Watts C. 2000. Functional corticospinal projections are established prenatally in the human foetus permitting involvment in the development of spinal motor centres. *Brain* 123:51–64.

Fairbanks LA. 1993. Juvenile vervet monkeys: Establishing relationships and practicing skill for the future. In: Pereira ME, Fairbanks LA, editors. *Juvenile Primates: Life History, Development and Behavior*. New York: Oxford University Press.

Fan P, Scott MB, Fei H, Ma C. 2012. Locomotion behavior of Cao Vit gibbon (*Nomascus nasutus*) living in karst forest in Bangliang Nature Reserve, Guangxi, China. *Integr Zool* 8:356–364.

Farnum CE, Tinsley M, Hermanson JW. 2008. Forelimb versus hindlimb skeletal development in the big brown bat, *Eptesicus fuscus*: Functional divergence is reflected in chondrocytic performance in autopodial growth plates. *Cells Tiss Org* 187:35–47.

Fei H, Ma C, Bartlett TQ, et al. 2015. Feeding postures of Cao Vit gibbons (*Nomascus nasutus*) living in a low-canopy karst forest. *Int J Primatol* 36:1036–1054.

Fellmann CD. 2011. Limb ontogeny, function, and locomotor development in catarrhines. PhD thesis. Department of Anthropology, New York University.

Fietz J, Dausmann KH. 2003. Costs and potential benefits of parental care in the nocturnal fat-tailed dwarf lemur (*Cheirogaleus medius*). *Folia Primatol* 74: 246–258.

Fleagle JG. 1977. Locomotor behavior and skeletal anatomy of sympatric Malaysian leaf-monkeys (*Presbytis obscura* and *Presbytis melalophos*). *Yearb Phys Anthropol* 20:440–453.

Fleagle JG. 2013. *Primate Evolution and Adaptation*. San Diego: Academic Press.

Fleagle JG, Mittermeier RA. 1980. Locomotor behavior, body size, and comparative ecology of seven Surinam monkeys. *Am J Phys Anthropol* 52:301–314.

Fleagle JG, Samonds K. 1975. Physical growth of cebus monkeys (*Cebus albifrons*) during the first year of life. *Growth* 39:35–52.

Fleagle JG, Gilbert CC, Baden AL. 2010. Primate cranial diversity. *Am J Phys Anthropol* 142:565–578.

Fooden J, Izor RJ. 1983. Growth curves, dental emergence norms, and supplementary morphological observations in known-age captive orangutans. *Am J Primatol* 5:285–301.

Ford SM, Corruccini RS. 1985. Intraspecific, interspecific, metabolic, and phylogenetic scaling in platyrrhine primates. In: Jungers WL, editor. *Size and Scaling in Primate Biology*. New York: Plenum Press, pp. 401–435.

Forssberg H. 1985. Ontogeny of human locomotor control. I. Infant stepping, supported locomotion, and transition to independent locomotion. *Exp Brain Res* 57:480–493.

Förster S, Cords M. 2002. Development of mother–infant relationships and infant behaviors in wild blue monkeys (*Cercopithecus mitis stuhlmanni*). In: Glenn ME, Cords M, editors. *The Guenons: Diversity and Adaptation in African Monkeys*. New York: Kluwer Academic/Plenum Publishers, pp. 245–272.

Foster A, Holton N. 2016. Variation in the developmental and morphological interaction between the nasal septum and facial skeleton. *Anat Rec* 299:730–740.

Fragaszy DM. 1989. Activity states and motor activity in an infant capuchin monkey (*Cebus apella*) from birth through eleven weeks. *Dev Psychobiol* 22:141–157.

Fragaszy DM. 1990. Early behavioral development in capuchins (*Cebus*). *Folia Primatol (Basel)* 54:119–128.

Fragaszy D, Visalberghi E, Fedigan L. 2004. *The Complete Capuchin*. Cambridge: Cambridge University Press.

Franzen JL, Gingerich PD, Habersetzer J, et al. 2009. Complete primate skeleton from the Middle Eocene of Messel in Germany: Morphology and paleobiology. *PLoS ONE* 4(5):e5723.

Frazier SB, Roodhouse KA, Hourcade DE, Zhang L. 2008. The quantification of glycosaminoglycans: A comparison of HPLC, carbazole, and Alcian blue methods. *Open Glycosci* 1:31–99.

Fricano EEI. 2018. The primate ectotympanic tube: Correlates of structure, function, and development. PhD dissertation. Johns Hopkins University, Baltimore.

Fröbisch NB. 2008. Ossification patterns in the tetrapod limb – Conservation and divergence from morphogenetic events. *Biol Rev* 83:571–600.

Galliari CA. 1988. A study of postnatal appendicular skeletal maturation in captive-born squirrel monkeys (*Saimiri boliviensi*). *Am J Primatol* 16:51–61.

Gans C, Northcutt RG. 1983. Neural crest and the origin of vertebrates: A new head. *Science* 220:268–273.

Garbino GST, de Aquino CC. 2016. Evolutionary significance of the entepicondylar foramen of the humerus in New World monkeys (Platyrrhini). *J Mamm Evol* 25:141–151.

Garn SM, Lewis AB, Polacheck DL. 1959. Variability of tooth formation. *J Dent Res* 38:135–148.

Garwicz M, Christensson M, Psouni E. 2009. A unifying model for timing of walking onset in humans and other mammals. *PNAS* 106:21889–21893.

Gebo DL. 2014. *Primate Comparative Anatomy*. Baltimore: Johns Hopkins University Press.

German RZ, Crompton AW. 2000. The ontogeny of feeding in mammals. In: Schwenck K, editor. *Feeding*. New York: Academic Press, pp. 449–457.

German RZ, Crompton AW, Levitch LC, Thexton A. 1992. The mechanism of suckling in two species of infant mammal: Miniature pigs and long-tailed macaques. *J Exp Zool* 261:322–330.

Gilbert SF. 2014. *Developmental Biology*. 10th ed. Sunderland, MA: Sinauer Associates.

Glander KE. 1980. Reproduction and population growth in free-ranging mantled howling monkeys. *Am J Phys Anthropol* 53:25–36.

Glaser D. 1970. Über die Ossifikation der Extremitäten bei neugeborenen Primaten (Mammalia). *Z Morphol Tiere* 68:127–139.

Glassman DM. 1983. Growth and development in the saddle-back tamarin: The sequence and timing of dental eruption and epiphyseal union. *Am J Primatol* 5:51–59.

Godfrey LR, Samonds KE, Jungers WL, Sutherland MR. 2001. Teeth, brains, and primate life histories. *Am J Phys Anthropol* 114:192–214.

Godfrey LR, Samonds KE, Jungers WL, Sutherland MR. 2003. Dental development and primate life histories. In: Kappeler PM, Pereira ME, editors. *Primate Life Histories and Socioecology*. Chicago: University of Chicago Press, pp. 177–203.

Godfrey LR, Samonds KE, Jungers WJ, Sutherland MR, Irwin MT. 2004. Ontogenetic correlates of diet in Malagasy lemurs. *Am J Phys Anthropol* 123:250–276.

Godfrey LR, Samonds KE, Wright PC, King SJ. 2005. Schultz's unruly rule: Dental developmental sequences and schedules in small-bodied, folivorous lemurs. *Folia Primatol* 76:77–99.

Godfrey LR, Schwartz GT, Samonds KE, Jungers WL, Catlett KK. 2006. The secrets of lemur teeth. *Evol Anthropol* 15:142–154.

Gonzalez PN, Kristensen, Morck DW, Boyd S, Hallgrimsson B. 2013. Effects of growth hormone on the ontogenetic allometry of craniofacial bones. *Evol Dev* 15:133–145.

Goswami A, Polly PD. 2010. The influence of modularity on cranial morphological disparity in Carnivora and Primates (Mammalia). *PLoS ONE* 5(3):e9517. doi:10.1371/journal.pone.0009517

Gould L. 1990. The social development of free-ranging infant *Lemur catta* at Berenty Reserve, Madagascar. *Int J Primatol* 11:297–318.

Gould SJ. 1977. *Ontogeny and Phylogeny*. Cambridge, MA: Belknap Press.

Grand TI. 1977. Body weight: Its relation to tissue composition, segment distribution, and motor function. I. Interspecific comparisons. *Am J Phys Anthropol* 47:211–239.

Grand TI. 1983. The anatomy of growth and its relation to locomotor capacity in *Macaca*. In: Eisenberg JF, Kleiman DG, editors. *Advances in the Study of Mammalian Behavior*. Shippensburg, PA: American Society of Mammalogists, pp. 5–23.

Grand TI. 1992. Altricial and precocial mammals: a model of neural and muscular development. *Zool Biol* 11:3–15.

Gregory WK. 1934. A half century of trituberculy: The Cope-Osborn theory of dental evolution with a revised summary of molar evolution from fish to man. *Proc Am Philos Soc* 73:169–317.

Großmann M, Sánchez-Villagra MR, Maier W. 2002. On the development of the shoulder girdle in *Crocidura russula* (Soricidae) and other placental mammals: Evolutionary and function aspects. *J Anat* 201:371–381.

Gruss LT, Gruss R, Schmitt D. 2017. Pelvic breadth and locomotor kinematics in human evolution. *Anat Rec* 300:739–751.

Gunst N, Leca JB, Boinski S, Fragaszy DM. 2010. The ontogeny of handling hard-to-process food in wild brown capuchins (*Cebus apella apella*): Evidence from foraging on the fruit of *Maximiliana maripa*. *Behaviour* 145:195–229.

Gursky SL. 2007. *The Spectral Tarsier*. Upper Saddle River: Pearson Prentice Hall.

Guthrie EH, Frost SR. 2011. Pattern and pace of dental eruption in *Tarsius*. *Am J Phys Anthropol* 145: 446–451.

Hall BK. 1995. Homology and embryological development. In: Hecht MK, Macintyre RJ, Clegg MT, editors. *Evolutionary Biology*, vol. 28. Boston: Springer, pp. 1–37.

Hall BK. 2015. *Bones and Cartilage*. 2nd ed. New York: Academic Press.

Halley AC. 2017. Minimal variation in eutherian brain growth rates during fetal neurogenesis. *Proc Biol Sci* 284(1854).

Hamada Y. 1983. Growth of the locomotor apparatus of the Japanese macaque: Based on analysis of muscular weights and limb segments. *J Anthropol Soc Nippon* 91:347–362.

Harvey PH, Clutton-Brock TH. 1985. Life history variation in primates. *Evolution* 39:559–581.

Harvey PH, Martin RD, Clutton-Brock TH. 1987. Life histories in comparative perspective. In: Smuts BH, Cheney DL, Seyfarth RM, Wrangham RW, Struhsaker TT, editors. *Primate Societies*. Chicago: University of Chicago Press, pp. 181–196.

Hautier L, Weisbecker V, Sánchez-Villagra M, Goswami A, Asher RJ. 2010. Skeletal development in sloths and the evolution of mammalian vertebral patterning. *PNAS* 107:18903–18908.

Henderson E. 2007. Platyrrhine dental eruption sequences. *Am J Phys Anthropol* 134:226–239.

Hendrickx AG. 1971. *Embryology of the Baboon*. Chicago: University of Chicago Press.

Hershkovitz P. 1977. *Living New World Monkeys (Platyrrhini)*. Vol. 1. Chicago: University of Chicago Press.

Hertwig O. 1888. *Lehrbuch der Entwicklungsgeschichte des Menschen und der Wirbelthiere*. Jena: Gustav Fischer.

Hertwig O. 1906. *Handbuch der vergleichenden und experimentellen Entwickelungslehre der Wirbeltiere*. Vols. I–III. Jena: Fischer.

Hiiemae KM. 2000. Feeding in mammals. In: Schwenck K, editor. *Feeding*. New York: Academic Press, pp. 411–448.

Hildebrand M. 1967. Symmetrical gaits of primates. *Am J Phys Anthropol* 26:119–130.

Hofmann MI, Schradin C, Geissmann T. 2007. Radiographic evaluation of neonatal skeletal development in *Callimico goeldii* reveals closer similarity to *Callithrix jacchus* than to *Saguinus oedipus*. *Am J Primatol* 69:420–433.

Hohl TH. 1983. Masticatory muscle transposition in primates: Effects on craniofacial growth. *J Maxillofac Surg* 11:149–156.

Hughes RL, Hall LS. 1998. Early development and embryology of the platypus. *Phil Trans R Soc Lond B* 353:1101–1114.

Hunt K, Cant J, Gebo D, Rose MD, Walker S, Youlatos D. 1996. Standardized descriptions of primate locomotor and postural modes. *Primates* 37(4):363–387.

Hunziker EB. 1994. Mechanism of longitudinal bone growth and its regulation by growth plate chondrocytes. *Microsc Res Techn* 28:505–519.

Hunziker EB, Schenk RK, Cruz-Orive LM. 1987. Quantitation of chondrocyte performance in growth-plate cartilage during longitudinal bone growth. *J Bone Joint Surg Am* 69: 1621–173.

Hurov JR. 1982. Diagonal walking in captive infant vervet monkeys. *Am J Primatol* 2:211–213.

Hylander WL, Ravosa MJ, Ross CF, Johnson KR. 1998. Mandibular corpus strain in primates: Further evidence for a functional link between symphyseal fusion and jaw-adductor muscle force. *Am J Phys Anthropol* 107:257–271.

Hylander WL, Ravosa MJ, Ross CF, Wall CE, Johnson KR. 2000. Symphyseal fusion and jaw-adductor muscle force: An EMG study. *Am J Phys Anthropol* 112:469–492.

Hylander WL, Vinyard CJ, Wall CE, Williams SH, Johnson KR. 2011. Functional and evolutionary significance of the recruitment and firing patterns of the jaw adductors during chewing in Verreauxi's sifaka (*Propithecus verreauxi*). *Am J Phys Anthropol* 145:531–547.

Inouye SE. 1994. Ontogeny of knuckle-walking hand postures in African apes. *J Hum Evol* 26:459–485.

Isler KI, van Shaik CP. 2009. The expensive brain: A framework for explaining changes in brain size. *J Hum Evol*: 392–400.

Isler K, Kirk EC, Miller J. 2008. Endocranial volumes of primate species: Scaling analyses using a comprehensive and reliable data set. *J Hum Evol* 55: 967–978.

Janson CH, Van Schaik CP. 1993. Ecological risk aversion in juvenile primates: slow and steady wins the race. In: Pereira ME, Fairbanks LA, editors. *Juvenile Primates. Life History, Development, and Behavior*. New York: Oxford University Press, pp. 57–76.

Jeffery N, Davies K, Kockenberger W, Williams S. 2007. Craniofacial growth in fetal *Tarsius bancanus*: brains, eyes and nasal septa. *J Anat* 210:703–722.

Jerison HJ. 1975. Fossil evidence of the evolution of the human brain. *Ann Rev Anthropol* 4: 27–58.

Jinkins JR. 2000. *Atlas of Neuroradioloic Embryology, Anatomy, and Variants.* Philadelphia: Lippincott Williams and Wilkins.

Johnson SE, Bock J. 2004. Trade-offs in skill acquisition and time allocation among juvenile Chacma baboons. *Hum Nat* 15:45–62.

Johnston GW, Dreizen S, Levy M. 1970. Dental development in the cotton ear marmoset (*Callithrix jacchus*). *Am J Phys Anthropol* 33:41–48.

Jouffroy FK, Medina MF. 1996. Developmental changes in the fibre composition of elbow, knee, and ankle extensor muscles in cercopithecid monkeys. *Folia Primatol (Basel)* 66:55–67.

Jungers WL. 1985. Body size and scaling of limb proportions in primates. In: Jungers WL, editor. *Size and Scaling in Primate Biology*. New York, NY: Plenum Press, pp. 345–381.

Jungers WL, Fleagle JG. 1980. Postnatal growth allometry of the extremities in *Cebus albifrons* and *Cebus apella*: A longitudinal and comparative study. *Am J Phys Anthropol* 53:471–478.

Kaessmann H, Pääbo S. 2002. The genetical history of humans and the great apes. *J Intern Med* 251:1–18.

Kappeler PM. 1998. Nests, tree holes, and the evolution of primate life histories. *Am J Primatol* 46:7–33.

Kappeler PM, Pereira ME. 2003. *Primate Life Histories and Socioecology*. Chicago: University of Chicago Press.

Karsenty G, Oury F. 2010. The central regulation of bone mass, the first link between bone remodeling and energy metabolism. *J Clin Endocrinol Metab* 95:4795–4801.

Keith A. 1919. *The Engines of the Human Body*. London: Williams and Norgate.

Kernell D. 1998. The final common pathway in postural control – Developmental perspective. *Neurosci Biobehav Rev* 22:479–484.

Keyte AL, Smith KK. 2010. Developmental origins of precocial forelimbs in marsupial neonates. *Development* 137:4283–4294.

Kilborn SH, Trudel G, Uhtoff H. 2002. Review of growth plate closure compared with age at sexual maturity and lifespan in laboratory animals. *Contemp Top Lab Anim Sci* 41:21–26.

Kimes KR, Siegel MI, Mooney MP, Todhunter J. 1988. Relative contributions of the nasal septum and airways to total nasal capsule volume in normal and cleft lip and palate fetal specimens. *Cleft Palate J* 25(3):282–287.

Kimura T. 1987. Development of chimpanzee locomotion on level surfaces. *Hum Evol* 2:107–119.

Kimura T. 2000. Development of quadrupedal locomotion on level surfaces in Japanese macaques. *Folia Primatol (Basel)* 71:323–333.

King JE, Fobes JT, Fobes JL. 1974. Development of early behaviors in neonatal squirrel monkeys and cotton-top tamarins. *Dev Psychobiol* 7:97–109.

Kirk EC. 2006. Effects of activity pattern on eye size and orbital aperture size in primates. *J Hum Evol* 51:159–170.

Kirkwood JK, Kember NF. 1993. Comparative quantitative histology of mammalian growth plates. *J Zool* 231:543–562.

Kirkwood JK, Stathatos K. 1992. *Biology, Rearing, and Care of Young Primates*. Oxford: Oxford University Press.

Kivell TL. 2016. The primate wrist. In: Kivell TL, Lemelin P, Richmond BG, Schmitt D, editors. *The Evolution of the Primate Hand: Anatomical, Developmental, Functional, and Paleontological Evidence*. New York: Springer, pp. 17–54.

Kivell TL, Begun DR. 2007. Frequency and timing of scaphoid–centrale fusion in hominoids. *J Hum Evol* 52:321–340.

Kivell TL, Schmitt D. 2009. Independent evolution of knuckle-walking in African apes shows that humans did not evolve from a knucklewalking ancestor. *PNAS* 106:14241–14246.

Kjosness KM, Reno PL. 2017. Using comparisons between species and anatomical locations to discover mechanisms of growth plate patterning and differential growth. In: Percival CJ, Richtmeier JT, editors. *Building Bones*. Cambridge: Cambridge University Press, pp. 205–232.

Kjosness KM, Hines JE, Lovejoy CO, Reno PL. 2014. The pisiform growth plate is lost in humans and supports a role for *Hox* in growth plate formation. *J Anat* 225:527–538.

Kohler IV, Preston SH, Lackey LB. 2006. Comparative mortality levels among selected species of captive animals. *Demogr Res* 15:413–434.

Kolb B, Whishaw IQ. 1996. *Fundamentals of Human Neuropsychology*. New York: W.H. Freeman and Company.

Kölliker A. 1879. *Grundriss der Entwickelungsgeschichte des Menschen und der höheren Tiere*. 2nd ed. Leipzig: Engelmann.

Kraus BS, Jordan R. 1965. *The Human Dentition Before Birth*. Philadelphia: Lea and Febiger.

Krstic RV. 1984. *Illustrated Encyclopedia of Human Histology*. Berlin: Springer.

Kuykendall KL, Maloney CJ, Conroy GC. 1992. Probit and survival analysis of tooth emergence in a mixed-longitudinal sample of chimpanzees (*Pan troglodytes*). *Am J Phys Anthropol* 89:379–399.

Kuypers HGJM. 1962. Corticospinal connections: Postnatal development in the rhesus monkey. *Science* 138:678–680.

Kuypers HGJM. 1982. A new look at the organization of the motor system. *Prog Brain Res* 57:381–403.

Lampl M, Monge JM, Mann AE. 1993. Further observations on a method for estimating hominoid dental developmental patterns. *Am J Phys Anthropol* 90:113–127.

Lancaster JB. 1971. Play-mothering: The relations between juvenile females and young infants among free-ranging vervet monkeys (*Cercopithecus aethiops*). *Folia Primatol (Basel)* 15:161–182.

Latham RA. 1970. Maxillary development and growth: The septopremaxillary ligament. *J Anat* 107:471–478.

Lawler RR. 2006. Sifaka positional behavior: Ontogenetic and quantitative genetic approaches. *Am J Phys Anthropol* 131:261–271.

Lawrence D, Hopkins D. 1976. The development of motor control in the rhesus monkey: Evidence concerning the role of corticomotoneuronal connections. *Brain* 99(2):235–254.

Lawrence DG, Kuypers HGJM. 1965. Pyramidal and non-pyramidal pathways in monkeys: Anatomical and functional correlation. *Science* 148:973–975.

LeDouarin N. 1982. *The Neural Crest*. Cambridge: Cambridge University Press.

Lee PC. 1996. The meanings of weaning: Growth, lactation, and life history. *Evol Anthropol* 5:87–98.

Lee PC. 1999. Comparative ecology of postnatal growth and weaning among haplorhine primates. In Lee PC, editor. *Comparative Primate Socioecology*. Cambridge: Cambridge University Press, pp. 111–139.

Leigh SR. 1994. Ontogenetic correlates of diet in anthropoid primates. *Am J Phys Anthropol* 94:499–522.

Leigh SR. 2004. Brain growth, life history, and cognition in primate and human evolution. *Am J Primatol* 62:139–164.

Leigh SR, Bernstein RM. 2006. Ontogeny, life history, and maternal investment in baboons. In Swedell L, Leigh SR, editors. *Reproduction and Fitness in Baboons*. New York: Springer, pp. 225–256.

Leigh SR, Blomquist GE. 2007. Life history. In: MacKinnon KC, Panger M, Bearder S, editors. *Primates in Perspective*. Oxford: Oxford University Press, pp. 396–407.

Leigh SR, Park PB. 1998. Evolution of human growth prolongation. *Am J Phys Anthropol* 107:331–350.

Lieberman DE, Ross CF, Ravosa MJ. 2000. The primate cranial base: Ontogeny, function, and integration. *Yearb Phys Anthropol* 43:117–169.

Lieberman P. 2002. On the nature and evolution of the neural bases of human language. *Am J Phys Anthropol* 119:36–62.

Llera C, Smith T, Young J, DeLeon V. 2016. Pattern of ossification in tarsals reflects locomotor specialization in primates. *Am J Phys Anthropol* 159(S62):211.

Loty S, Foll C, Forest N, Sautier J-M. 2000. Association of enhanced expression of gap junctions with *in vitro* chondrogenic differentiation of rat nasal septal cartilage – Released cells following their dedifferentiation and redifferentiation. *Arch Oral Biol* 45:843–856.

Lovell NC. 1991. An evolutionary framework for assessing illness and injury in nonhuman primates. *Yearb Phys Anthropol* 34:117–155.

Luckett WP. 1993. An ontogenetic assessment of dental homologies in therian mammals. In: Szalay FS, Novacek MJ, McKenna MC, editors. *Mammal Phylogeny. Mesozoic Differentiation, Multituberculates, Monotremes, Early Therians, and Marsupials*. New York: Springer-Verlag, pp. 182–204.

Luckett WP, Maier W. 1982. Development of deciduous and permanent dentition in *Tarsius* and its phylogenetic significance. *Folia Primatol* 37:1–36.

Lui JC, Brown J. 2011. Mechanisms limiting body growth in mammals. *Endocr Rev* 32:422–440.

Lumer H, Schultz A. 1947. Relative growth of the limb segments and tail in *Ateles geoffroyi* and *Cebus capucinus*. *Hum Biol* 19:53–67.

Mack D. 1979. Growth and development of infant red howling monkeys (*Alouatta seniculus*) in a free-ranging population. In: Eisenberg JF, editor. *Vertebrate Ecology in the Northern Neotropics*. Washington, DC: Smithsonian Institution Press, pp. 127–136.

MacKay-Lyons M. 2002. Central pattern generation of locomotion: A review of the evidence. *Phys Ther* 82(1):69–83.

Mackeprang M, Hay S. 1972. Cleft lip and palate mortality study. *Cleft Palate J* 9:51–63.

MacPhee RDE. 1979. Entotympanics, ontogeny, and primates. *Folia Primatol* 31:23–47.

Maestripieri D. 1995. First steps in the macaque world: Do rhesus mothers encourage their infants' independent locomotion? *Anim Behav* 49:1541–1549.

Maier W. 1980. Nasal structures in Old and New World primates. In: Ciochon RL, Chiarelli AB, editors. *Evolutionary Biology of the New World Monkeys and Continental Drift*. New York: Plenum Press, pp. 219–241.

Maier W. 1983. Morphology of the interorbital region of *Saimiri sciureus*. *Folia Primatol* 41:277–303.

Maier W. 1987. The ontogenetic development of the orbitotemporal region in the skull of *Monodelphis domestica* (Didelphidae, Marsupilaia), and the problem of the mammalian alisphenoid. In: Kuhn H-J, Zeller U, editors. *Morphogenesis of the Mammalian Skull*. Hamburg: Verlag, pp. 71–90.

Maier W. 1993. Cranial morphology of the therian common ancestor, as suggested by the adaptations of neonate marsupials. In: Szalay F, Novacek MJ, McKenna MC, editors. *Mammal Phylogeny*. New York: Springer-Verlag, pp. 165–181.

Maier W. 2000. Ontogeny of the nasal capsule in cercopithecoids: A contribution to the comparative and evolutionary morphology of catarrhines. In: Whitehead PF, Jolly CJ, editors. *Old World Monkeys*. Cambridge: Cambridge University Press, pp. 99–132.

Maier W, Ruf I. 2014. Morphology of the nasal capsule of primates – With special reference to *Daubentonia* and *Homo*. *Anat Rec* 297:2018–2030.

Manduell KL, Morrogh-Bernard HC, Thorpe SKS. 2011. Locomotor behavior of wild orangutans (*Pongo pygmaeus wurmbii*) in disturbed peat swamp forest, Sabangau, Central Kalimantan, Indonesia. *Am J Phys Anthropol* 145:348–359.

Mann A. 1988. The nature of Taung dental maturation. *Nature* 333:123.

Mann A, Weiss M. 1996. Hominoid phylogeny and taxonomy: A consideration of the molecular and fossil evidence in a historical perspective. *Mol Phylogenet Evol* 5:169–181.

Martell MC, Smith TD, Rossie JB, DeLeon VB. 2016. Heterochrony of nasal turbinal development in primates. *Am J Phys Anthropol* 159(S62):220.

Martin RD. 1981. Relative brain size and metabolic rate in terrestrial vertebrates. *Nature* 293:57–60.

Martin RD. 1990. *Primate Origins and Evolution. A Phylogenetic Reconstruction*. Princeton: Princeton University Press.

Massion J. 1998. Postural control systems in developmental perspective. *Neurosci Biobehav Rev* 21:465–472.

McBratney-Owen B, Iseki S, Bamforth SD, Olsen BR, Morriss-Kay GM. 2008. Development and tissue origins of the mammalian cranial base. *Dev Biol* 322:121–132.

McCrossin ML, Benefit BR, Gitau SN, Palmer AK, Blue KT. 1998. Fossil evidence for the origins of terrestriality among Old World higher primates. In: Strasser E, Fleagle J, Rosenberger AL, McHenry H, editors. *Primate Locomotion*. New York, NY: Springer, pp. 353–396.

McGraw M. 1945. *The Neuromuscular Maturation of the Human Infant*. New York: Columbia University Press.

McGraw WS, Vick AE, Daegling DJ. 2011. Sex and age differences in the diet and ingestive behaviors of sooty mangabeys (*Cercocebus atys*) in the Tai Forest, Ivory Coast. *Am J Phys Anthropol* 144:140–153.

Melsen B. 1974. The cranial base: The postnatal development of the cranial base studied histologically on human autopsy material. *Acta Odont Scand Suppl* 62:1–126.

Miranda JMD, Aguiar LM, Ludwig G, Moro-Rios RF, Passos FC. 2005. The first seven months of an infant of *Alouatta guariba* (Humboldt) (Primates, Atelidae): Interactions and the development of behavioral patterns. *Revista Brasileira de Zoologia* 22:1191–1195.

Mitani JC, Watts DP. 1997. The evolution of non-maternal caretaking among anthropoid primates: Do helpers help? *Behav Ecol Sociobiol* 40:213–220.

Mitteroecker P, Gunz P, Bernhard M, Schaefer K, Bookstein FL. 2004. Comparison of cranial ontogenetic trajectories among great apes and humans. *J Hum Evol* 46:679–698.

Mooney MP, Siegel MI, Kimes KR, Todhunter J. 1989. A test of two midfacial growth models using path analysis of normal human fetal material. *Cleft Palate J* 26:93–99.

Mooney MP, Siegel MI, Kimes KR, Todhunter J. 1991. Premaxillary development in normal and cleft lip and palate human fetuses using three-dimensional computer reconstruction. *Cleft Palate Craniofac J* 28:49–54.

Monson TA, Coleman JL, Hlusko LJ. 2019. Craniodental allometry, prenatal growth rates, and the evolutionary loss of the third molars in New World monkeys. *Anat Rec* 302:1419–1433.

Morgado E, Ocqueteau C, Cury M, et al. 1990. Three dimensional morphometry of mammalian cells II. Areas, volumes, and area-volume ratios. *Arch Biol Med Exp* 23:21–27.

Morimoto N. 2018. What could hominoid fetuses tell us about human evolution? *Anat Rec* 301:970–972.

Morland HS. 1990. Parental behavior and infant development in ruffed lemurs (*Varecia variegata*) in a northeast Madagascar rain forest. *Am J Primatol* 20:253–265.

Moss M, Greenberg S. 1967. Functional cranial analysis of the human maxillary bone. *Ang Orthod* 37:151–164.

Moss ML, Young RW. 1960. A functional approach to craniology. *Am J Phys Anthropol* 18:281–292.

Mundy GT 1999. Bone remodeling. In: Favus MJ, editor. *Primer on The Metabolic Bone Diseases and Disorders of Mineral Metabolism*. Philadelphia: Lippincott, Williams and Wilkins, pp. 30–38.

Muybridge E. 1887. *Animal Locomotion*. New York: Dover.

Myatt JP, Thorpe SKS. 2011. Postural strategies employed by orangutans (*Pongo abelii*) during feeding in the terminal branch niche. *Am J Phys Anthropol* 146:73–82.

Nakano Y. 1996. Footfall patterns in the early development of the quadrupedal walking of Japanese macaques. *Folia Primatol (Basel)* 66:113–125.

Nanci A. 2007. *Ten Cate's Oral Histology: Development, Structure, and Function*. 9th ed. St. Louis: Elsevier.

Napier JP, Napier P. 1967. *A Handbook of Living Primates*. London: Academic Press.

Napier JR, Walker AC. 1967. Vertical clinging and leaping – A newly recognized category of locomotor behavior of primates. *Folia Primatol* 6:204–219.

Neaux D, Sansalone C, Ledogar JA, Ledogar SH, Luk THY, Wroe S. 2018. Basicranium and face: Assessing the impact of morphological integration on primate evolution. *J Hum Evol* 118:43–55.

Negayama K, Kondo K, Itoigawa N. 1983. Development of locomotor behavior in infant Japanese macaques (*Macaca fuscata*). *Annal Sci Natur: Zool* 5:169–180.

Nekaris KAI. 2003. Observations of mating, birthing and parental behaviour in three subspecies of slender loris (*Loris tardigradus* and *Loris lydekkerianus*) in India and Sri Lanka. *Folia Primatol (Basel)* 74:312–336.

Newell-Morris L, Tarrant LH. 1978. Ossification in the hand and foot of the macacque (*Macaca nemistrina*). I. General features. *Am J Phys Anthropol* 48:441–454.

Newell-Morris L, Tarrant LH, Fahrenbruch CE, Burbacher TM, Sackett GP. 1980. Ossification in the hand and foot of the pigtail macacque (*Macaca nemistrina*). II. Order of appearance of centers and variability in sequence. *Am J Phys Anthropol* 53:423–439.

Newman LM, Hendrickx AG. 1984. Fetal development in the normal thick-tailed bushbaby (*Galago crassicaudatus panganiensis*). *Am J Primatol* 6:337–355.

Nicolson NA. 1984. Infants, mothers, and other females. In: Smuts BA, Cheney DL, Seyfarth RM, Wrangham RW, Struhsaker TT, editors. *Primate Societies*. Chicago: University of Chicago Press, pp. 330–342.

Niemitz C, Sprankel H. 1974. Frühe postnatale Ossifikation bei *Tarsius bancanus* Horsfield, 1821 (Primates) und ihr Bezug zur Nestflüchter-Nesthocker-Hypothese. *Zeitschr Morphol Tiere* 79:155–163.

Nissen HW, Riesen AH. 1949. Onset of ossification in the epiphyses and short bones of the extremeties in chimpanzee. *Growth* 13:45–70.

Nolte MJ, Hockman D, Cretekos CJ, Behringer RR, Rasweiler JJ. 2009. Embryonic staging system for the black mastiff bat, *Molossus rufus* (Molossidae), correlated with structure–function relationships in the adult. *Anat Rec* 92:155–168.

Norconk MA, Wright B, Conklin-Brittain NL, Vinyard CJ. 2009. Mechanical and nutritional properties of food as factors in platyrrhine dietary adaptations. In Garber PA, Estrada A, Bicca-Marques JC, Heymann E, Strier K, editors. *South American Primates: Testing New Theories in the Study of Primate Behavior, Ecology, and Conservation*. New York: Springer, pp. 279–319.

Nowell AA, Fletcher AW. 2008. The development of feeding behaviour in wild western lowland gorillas (*Gorilla gorilla gorilla*). *Behaviour* 145:171–193.

Nunn CL. 2011. *The Comparative Approach in Evolutionary Anthropology and Biology*. Chicago: University of Chicago Press.

Nunn CL, Barton RA. 2001. Comparative methods for studying primate adaptation and allometry. *Evol Anthropol* 10:81–98.

Ogden JA, Phillips SB. 1983. Radiology of postnatal skeletal development. VII. The scapula. *Skel Radiol* 9: 157–169.

Oka SW, Kraus BS. 1969. The circumnatal status of molar crown maturation among the Hominoidea. *Archives Oral Biol* 14:639–659.

Ooë T. 1979. Development of human first and second permanent molar, with special reference to the distal portion of the dental lamina. *Anat Embryol* 155:221–240.

O'Rahilly R, Müller F. 1987. *Developmental Stages in Human Embryos, Including a Revision of Streeter's 'Horizons' and a Survey of the Carnegie Collection*. Washington, DC: Carnegie Institution of Washington.

O'Rahilly R, Müller F. 2010. Developmental stages in human embryos: Revised and new measurements. *Cells Tiss Org* 192:73–84.

Osborn JW. 1981. *Dental Anatomy and Embryology*. Oxford: Blackwell Scientific.

Oyen OJ, Walker AC, Rice RW. 1979. Craniofacial growth in olive baboons (*Papio cynocephalus anubis*): Browridge formation. *Growth* 43:174–187.

Paddock K, Zeigler L, Harvey B, et al. In press. Comparative dental anatomy in newborn primates: Cusp mineralization. *Anat Rec.*

Pagel MD, Harvey PH. 1988. Recent developments in the analysis of comparative data. *Quart Rev Biol* 63:413–440.

Parsons TE, Downey CM, Jirik FR, Hallgrimsson B, Jamniczky. 2015. Mind the gap: Genetic manipulation of basicranial growth within synchondroses modulates calvarial and facial shape in mice through epigenetic interactions. *PLoS ONE* 10(2):e0118355. doi:10.1371/journal.pone.0118355

Passingham RE. 1985. Rates of brain development in mammals including man. *Brain Behav Evol* 26:167–175.

Pereira ME, Altman J. 1985. Development of social behavior in free-living nonhuman primates. In: Watts ES, editor. *Nonhuman Primate Models for Human Growth and Development*. New York: Alan R Liss, pp. 217–309.

Pereira ME, Fairbanks LA. 1993. *Juvenile Primates. Life History, Development, and Behavior*. New York: Oxford University Press.

Perlyn CA, DeLeon VB, Babbs C, et al. 2006. The craniofacial phenotype of the Crouzon mouse: Analysis of a model for syndromic craniosynostosis using 3D microCT. *Cleft Palate Craniofac J* 43:740–748.

Phillips IR. 1976. Skeletal development in the foetal and neonatal marmoset (*Callithrix jacchus*). *Lab Anim* 10:317–333.

Plavcan JM, Daegling DJ. 2006. Interspecific and intraspecific relationships between tooth size and jaw size in primates. *J Hum Evol* 51:171–184.

Pocock RI. 1918. On the external characters of the lemurs and *Tarsius*. *Proc Zool Soc Lond* 1918:19–53.

Podgaiski LR, de Assis Jardim MM. 2009. Early behavioral development of a free-ranging howler monkey infant (*Alouatta guariba clamitans*) in Southern Brazil. *Neotropical Primates* 16:27–31.

Portmann A. 1939. Nesthocker und nestflüchter als entwicklungszustände von verschiedener wertigkeit bei vögeln und säugern. *Rev Suisse Zool* 46:385–390.

Portmann A. 1941. Die Tragzeiten der Primaten und die Dauer der Schwangerschaft beim Menschen: Ein Problem der vergleichenden Biologie. *Rev Suisse Zool* 48:511–518.

Presley R, Steel FLD. 1976. On the homology of the alisphenoid. *J Anat* 121:441–459.

Promislow DEL, Harvey PH. 1990. Living fast and dying young: A comparative analysis of life history variation among mammals. *J Zool Lond* 220:417–437.

Prost JH. 1969. A replication study on monkey gaits. *Am J Phys Anthropol* 30:203–208.

Radhakrishna S, Singh M. 2004. Infant development in the slender loris (*Loris lydekkerianus lydekkerianus*). *Current Science* 86:1121–1127.

Rae TC, Koppe T. 2004. Holes in the head. Evolutionary interpretations of the paranasal sinuses in catarrhines. *Evol Anthropol* 13:211–223.

Raguet-Schofield M. 2010. The ontogeny of feeding behavior of Nicaraguan mantled howler monkeys (*Alouatta palliata*). PhD thesis. University of Illinois, Urbana-Champaign.

Raguet-Schofield M, Pavé R. 2014. An ontogenetic framework for *Alouatta*: infant development and evaluating models of life history. In: *Howler Monkeys*. New York: Springer, pp. 289–316.

Raichlen DA. 2005. Ontogeny of limb mass distribution in infant baboons (*Papio cynocephalus*). *J Hum Evol* 49:452–467.

Ramaswami LS. 1957. The development of the skull in the slender loris, *Loris tardigradus lydekkerianus cabr*. *Acta Zool* 38:27–68.

Randall FE. 1944. The skeletal and dental development and variability of the gorilla. *Hum Biol* 15–16:236–254, 307–337.

Rapaport LG, Brown GR. 2008. Social influences on foraging behavior in young nonhuman primates: Learning what, where, and how to eat. *Evol Anthropol* 17:189–201.

Ravosa MJ. 1992. Allometry and heterochrony in extant and extinct Malagasy primates. *J Hum Evol* 23:197–217.

Ravosa MJ. 1999. Anthropoid origins and the modern symphysis. *Folia Primatol* 70:65–78.

Ravosa MJ. 2007. Cranial ontogeny, diet, and ecogeographic variation in African lorises. *Am J Primatol* 69:59–73.

Ravosa MJ, Ross CF. 1994. Craniodental allometry and heterochrony in two howler monkeys: *Alouatta seniculus* and *A. palliata*. *Am J Primatol* 33:277–299.

Ravosa MJ, Simons EL. 1994. Mandibular growth and function in *Archaeolemur*. *Am J Phys Anthropol* 95:63–76.

Ravosa MJ, Meyers DM, Glander KE. 1993. Relative growth of the limbs and trunk in sifakas: Heterochronic, ecological, and functional considerations. *Am J Phys Anthropol* 92:499–520.

Rawlins RG. 1976. Locomotor ontogeny in *Macaca mulatta*: I. Behavioral strategies and tactics. *Am J Phys Anthropol* 44:201.

Reed DA, Ross CF. 2010. The influence of food material properties on jaw kinematics in the primate *Cebus*. *Arch Oral Biol* 55:946–952.

Rehorek SJ, Smith TD, Beeching SC. 2016. Early development and differentiation of the Laysan albatross (*Phoebastria immutabilis* Rothschild, 1893): Procellariiformes). *J Morphol* 277:1231–1249.

Reid DJ, Schwartz GT, Dean C, Chandrasekera MS. 1998. A histological reconstruction of dental development in the common chimpanzee, *Pan troglodytes*. *J Hum Evol* 35:427–448.

Reinholt LE, Burrows AM, Eiting TP, Dumont ER, Smith TD. 2009. Brief communication: Histology and microCT as methods for assessing facial suture patency. *Am J Phys Anthropol* 138:499–506.

Richard AF. 1976. Preliminary observations on the birth and development of *Propithecus verreauxi* to the age of six months. *Primates* 17:357–366.

Richardson MK, Hanken J, Gooneratne ML, et al. 1997. There is no highly conserved embryonic stage in the vertebrates: Implications for current theories of evolution and development. *Anat Embryol* 196:91–106.

Robinow M, Richards TW, Anderson M. 1942. The organization of the deciduous teeth. *Growth* 6:127–133.

Robl N, 2008. Ontogenetic resource partitioning in white faced sakis (*Pithecia pithecia*). MA thesis, Kent State University.

Rolian C. 2008. Developmental basis of limb length in rodents: Evidence for multiple divisions of labor in mechanisms of endochondral bone growth. *Evolution & Development* 10:12–28.

Rollinson J, Martin RD. 1981. Comparative aspects of primate locomotion, with special reference to arboreal cercopithecines. *Symp Zool Soc London* 48:377–427.

Rolls G, Farmer N. 2008. *Artifacts in Histological and Cytological Preparations*. Wetzlar: Leica Biosystems.

Rose MD. 1977. Positional behavior of olive baboons (*Papio anubis*) and its relationship to maintenance and social activities. *Primates* 18:59–116.

Rosenberg KR. 1992. The evolution of modern human childbirth. *Yearb Phys Anthropol* 35:89–124.

Rosenberg KR, Trevathan WR. 2015. Are human infants precocial? *Am J Phys Anthropol* 156(Suppl 60):271.

Rosenberger AL. 2010. The skull of *Tarsius*: Functional morphology, eyeballs, and the nonpursuit predatory lifestyle. *Int J Primatol* 31:1032–1054.

Rosenberger AL, Smith TD, DeLeon VB, et al. 2016. Eye size and set in small-bodied fossil primates: A three-dimensional method. *Anat Rec* 299:1671–1689.

Rosenblum LA. 1968. Mother–infant relations and early behavioral development in the squirrel monkey. In: Rosenblum LA, Cooper RW, editors. *The Squirrel Monkey*. New York: Academic Press, pp. 207–233.

Ross AC, Lehman SM. 2016. Infant transport and mother-infant contact from 1 to 26 weeks postnatal in Coquerel's sifaka (*Propithecus coquereli*) in northwestern Madagascar. *Am J Primatol* 78:646–658.

Ross CF. 1995. Allometric and functional influences on primate orbit orientation and the origins of Anthropoidea. *J Hum Evol* 29:201–227.

Ross C. 2001. Park or ride? Evolution of infant carrying in primates. *Int J Primatol* 22:749–771.

Ross C. 2003a. Life history, infant care strategies, and brain size in primates. In: Kappeler PM, Pereira ME, editors. *Primate Life Histories and Socioecology*. Chicago: University of Chicago Press, pp. 266–284.

Ross C. 2003b. Primate life histories. *Evol Anthropol* 6:54–63.

Ross C, Jones KE. 1999. Socioecology and the evolution of primate reproductive rates. In: Lee PC, editor. *Comparative Primate Socioecology*. Vol. 22. Cambridge: Cambridge University Press, pp. 73–110.

Rossie J. 2006. Ontogeny and homology of the paranasal sinuses in Platyrrhini (Mammalia: Primates). *J Morphol* 267:1–40.

Rossie JB, Smith TD. 2007. Ontogeny of the nasolacrimal duct in primates: Functional and phylogenetic implications. *J Anat* 210:195–208.

Rot C, Stern T, Blecher R, Friesem B, Zelcher E. 2014. A mechanical jack-like mechanism drives spontaneous fracture healing in neonatal mice. *Devel Cell* 31:159–170.

Rothman J, Chapman C, Twinomugisha D, et al. 2008. Measuring physical traits of primates remotely: The use of parallel lasers. *Am J Primatol* 70:1191–1195.

Rubinstein NA, Kelly AM. 1978. Myogenic and neurogenic contributions to the development of fast and slow twitch muscles in rats. *Developmental Biology* 62:473–485.

Russo GA, Young JW. 2011. Tail growth tracks the ontogeny of prehensile tail use in capuchin monkeys (*Cebus albifrons* and *C. apella*). *Am J Phys Anthropol* 146:465–473.

Ryan TM, Krovitz GE. 2006. Trabecular bone ontogeny in the human proximal femur. *J Hum Evol* 51:591–602.

Sacher GA, Staffeldt EF. 1974. Relation of gestation time to brain weight for placental mammals: Implications for the theory of vertebrate growth. *Amer Nat* 108:593–615.

Sánchez-Villagra MR, Maier W. 2002. Ontogenetic data and the evolutionary origin of the mammalian scapula. *Naturwissenschaften* 89:459–461.

Sarnat BG, Wexler MR. 1966. Growth of the face and jaws after resection of the septal cartilage in the rabbit. *Am J Anat* 118:755–767.

Sarringhaus LA, MacLatchy LM, Mitani JC. 2014. Locomotor and postural development of wild chimpanzees. *J Hum Evol* 66:29–38.

Sarringhaus LA, Maclatchy LM, Mitani JC. 2016. Long bone cross-sectional properties reflect changes in locomotor behavior in developing chimpanzees. *Am J Phys Anthropol* 160:16–29.

Savage VM, Allen AP, Brown JH, et al. 2007. Scaling of number, size, and and metabolic rate of cells with body size in mammals. *PNAS* 104:4718–4723.

Schaefer MS, Nash LT. 2007. Limb growth in captive *Galago senegalensis*: Getting in shape to be an adult. *Am J Primatol* 69:103–111.

Scheuer L, Black S. 2004. *The Juvenile Skeleton*. London: Elsevier Academic Press.

Schiel N, Huber L. 2006. Social influences on the development of foraging behavior in free-living common marmosets (*Callithrix jacchus*). *Am J Primatol* 68:1–11.

Schultz AH. 1926. Fetal growth in man and other primates. *Quart Rev Biol* 1:465–521.

Schultz AH. 1935. Eruption and decay of the permanent teeth in primates. *Am J Phys Anthropol* 19:489–581.

Schultz AH. 1940. Growth and development of the chimpanzee. *Carnegie Inst Wash Pub 518, Contrib Embryol* 28:1–63.

Schultz AH. 1941. Growth and development of the orangutan. *Carnegie Inst Wash Pub XXX, Contrib Embryol* 29:57–110.

Schultz AH. 1944. Age changes and variability in gibbons. *Am J Phys Anthropol* 2:1–129.

Schultz AH. 1949. Sex differences in the pelves of primates. *Am J Phys Anthropol* 7:401–423.

Schultz AH. 1960. Age changes in primates and their modification in man. In: Tanner JM, editor. *Human Growth*. Oxford: Pergamon, pp. 1–20.

Schultz AH. 1969. *Life of Primates*. New York: Universe Books.

Schultz AH. 1973. Age changes, variability and generic differences in body proportions of recent hominoids. *Folia Primatol* 19:338–359.

Schusterman RJ, Sjoberg A. 1968. Early behavior patterns of squirrel monkeys (*Saimiri sciureus*). In: Carpenter CA, editor. *Proceedings of the Second International Congress of Primatology*. Volume 1: Behavior. New York: Karger, pp. 194–203.

Schwandt ML. 2001. The ontogeny of positional behavior in captive chimpanzees: A mixed cross-sectional/longitudinal study. *Am J Phys Anthropol Supplement* 32:133.

Schwandt ML. 2002. The ontogeny of positional behavior in captive chimpanzees. PhD thesis. University of Arizona.

Schwandt ML, Marzke M. 2003. Body size, limb proportions, and positional behavior during ontogeny in captive chimpanzees (*Pan troglodytes*). *Am J Phys Anthropol Supplement* 36:187.

Schwartz GT, Samonds KE, Godfrey LR, Jungers WL, Simons EL. 2002. Dental microstructure and life history in subfossil Malagasy lemurs. *PNAS* 99:6124–6129.

Schwartz GT, Mahoney P, Godfrey LR, et al. 2005. Dental development in *Megaladapis edwardsi* (Primates, Lemuriformes): Implications for understanding life history variation in subfossil lemurs. *J Hum Evol* 49:702–721.

Schwartz JH. 2003. How close are the similarities between *Tarsius* and other primates? In: Wright PC, Simons EL, Gursky S, editors. *Tarsiers: Past, Present, and Future*. New Jersey: Rutgers University Press, pp. 50–96.

Schwartz JH. 2007. *Skeleton Keys*. New York: Oxford University Press.

Scott JE. 2012. Molar size and diet in the Strepsirrhini: Implications for size-adjustment in studies of primate dental adaptation. *J Hum Evol* 63:796–804.

Scott JH. 1958. The cranial base. *Am J Phys Anthropol* 16:319–348.

Seibert JR, Swindler DR. 1991. Perinatal dental development in the chimpanzee (*Pan troglodytes*). *Am J Phys Anthropol* 86: 287–294.

Serrat MA. 2013. Allen's rule revisited: Temperature influences bone elongation during a critical period of postnatal development. *Anat Rec* 296:1534–1545.

Shapiro LJ, Raichlen DA. 2005. Lateral sequence walking in infant *Papio cynocephalus*: Implications for the evolution of diagonal sequence walking in primates. *Am J Phys Anthropol* 126:205–213.

Shapiro LJ, Raichlen DA. 2006. Limb proportions and the ontogeny of quadrupedal walking in infant baboons (*Papio cynocephalus*). *J Zool* 269:191–203.

Shapiro LJ, Young JW, VandeBerg JL. 2014. Body size and the small branch niche: Using marsupial ontogeny to model primate locomotor evolution. *J Hum Evol* 68:14–31.

Shea BT. 1981. Relative growth of the limbs and trunk in African apes. *Am J Phys Anthropol* 56:179–201.

Shea BT. 1989. Heterochrony in human evolution: The case for neoteny reconsidered. *Yearb Phys Anthropol* 32:69–101.

Siegel MI, Sadler D. 1981. Nasal septum resection and craniofacial growth in a chimpanzee animal model: Implications for cleft palate surgery. *Plast Reconstr Surg* 68:849–853.

Simons EL, Rasmussen DT. 1989. Cranial anatomy of *Aegyptopithecus* and *Tarsius* and the question of the tarsier–anthropoidean clade. *Am J Phys Anthropol* 79:1–23.

Simpson SW, Lovejoy CO, Meindl RS. 1990. Hominoid dental maturation. *J Hum Evol* 19:285–297.

Sirianni JE, Swindler DR. 1985. *Growth and Development of the Pigtailed Macaque*. Boca Raton: CRC Press.

Smith BH. 1989. Dental development as a measure of life history in primates. *Evolution* 43:683–688.

Smith BH. 1992. Life history and the evolution of human maturation. *Evol Anthropol* 1:134–142.

Smith BH. 1994. Patterns of dental development in *Homo, Australopithecus, Pan*, and *Gorilla*. *Am J Phys Anthropol* 94:307–325.

Smith BH. 2000. 'Schultz's Rule' and the evolution of tooth emergence and replacement patterns in primates and ungulates. In: Teaford M, Smith MM, Ferguson MWJ, editors. *Development, Function and Evolution of Teeth*. Cambridge: Cambridge University Press, pp. 218–228.

Smith BH, Crummett, TL, Brandt KL. 1994. Ages of eruption of primate teeth: A compendium foraging individuals and comparing life histories. *Yearb Phys Anthropol* 37:177–231.

Smith KK, Keyte AL. in press. Adaptations of the marsupial newborn: Birth as an extreme environment. *Anat Rec*.

Smith TD, Rossie JB. 2008. The nasal fossa of mouse and dwarf lemurs (Primates, Cheirogaleidae). *Anat Rec* 291:895–915.

Smith TD, Rossie JB, Bhatnagar KP. 2007. Evolution of the nose and nasal skeleton in primates. *Evol Anthropol* 16:132–146.

Smith TD, Rossie JB, Docherty BA, Cooper GM, Bonar CJ, Silverio AL, Burrows AM. 2008. Fate of the nasal capsular cartilages in prenatal and perinatal tamarins (*Saguinus geoffroyi*) and extent of secondary pneumatization of maxillary and frontal sinuses. *Anat Rec* 291:1397–1413.

Smith TD, Rossie JB, Cooper GM, et al. 2010. The maxillary sinus in three genera of New World monkeys: Factors that constrain secondary pneumatization. *Anat Rec* 293:91–107.

Smith TD, Rossie JB, Cooper GM, et al. 2011. Comparative micro CT and histological study of maxillary pneumatization in four species of New World monkeys: The perinatal period. *Am J Phys Anthropol* 144:392–410.

Smith TD, Rossie JB, Cooper GM, et al. 2012. Microanatomical variation of the nasal capsular cartilage in newborn primates. *Anat Rec* 295:950–960.

Smith TD, DeLeon VB, Rosenberger AL. 2013. At birth, tarsiers lack a postorbital bar or septum. *Anat Rec* 296:365–377.

Smith TD, Kentzel ES, Cunningham JM, et al. 2014. Mapping bone cell distributions to assess ontogenetic origin of primate midfacial form. *Am J Phys Anthropol* 154:424–435.

Smith TD, Muchlinski MN, Jankord KD, et al. 2015. Dental maturation, eruption, and gingival emergence in the upper jaw of newborn primates. *Anat Rec* 298:2098–2131.

Smith TD, Martell MC, Rossie JB, Bonar CJ, DeLeon VB. 2016. Ontogeny and microanatomy of the nasal turbinals in lemuriformes. *Anat Rec* 299:1492–1510.

Smith TD, Muchlinski MN, Bucher WR, et al. 2017a. Relative tooth size at birth in primates: Life history correlates. *Am J Phys Anthropol* 164:623–634.

Smith TD, McMahon MJ, Millen ME, et al. 2017b. Growth and development at the sphenoethmoidal junction in perinatal primates. *Anat Rec* 300:2115–2137.

Snowdon CT, Soini P. 1988. The tarmarins, genus *Saguinus*. In: Mittermeier RA, Rylands AB, Coimbra-Filho A, Fonseca GAB, editors. *Ecology and Behavior of Neotropical Primates: Volume 2*. Washington, DC: World Wildlife Fund, pp. 223–298.

Spatz W. 1964. Beitrag zur Kenntnis der Ontogenese des Cranium von *Tupaia glis* (Diard 1820). *Gegenbaurs Morphol Jahrb* 106:321–416.

Sperber G. 2000. *Craniofacial Development*. Hamilton: B.C. Decker.

Sprankel H. 1965. Untersuchungen an *Tarsius*. I. Morphologie des Schwanzes nebst ethologischen Bemerkungen. *Folia Primatol* 3:153–188.

Starck D. 1967. Le crâne des mammifères. In: Grassé P-P, editor. *Traité de Zoologie XVI*. Paris: Masson, pp. 405–549.

Starck D. 1975. The development of the chondrocranium in primates. In: Luckett WP, Szalay FS, editors. *Phylogeny of the Primates*. New York: Plenum Press, pp. 127–155.

Stone AI. 2006. Foraging ontogeny is not linked to delayed maturation in squirrel monkeys (*Saimiri sciureus*). *Ethology* 112:105–115.

Streeter GL. 1920. Weight, sitting height, head size, foot length, and menstrual age of the human embryo. *Carnegie Institute of Washington Publication* 11:142–170.

Sugardjito J, van Hooff JARAM. 1986. Age–sex class-differences in the positional behavior of the Sumatran orangutan (*Pongo pygmaeus abelii*) in the Gunung Leuser National Park, Indonesia. *Folia Primatol* 47:14–25.

Sussman RW. 1977. Socialization, social structure, and ecology of two sympatric species of *Lemur*. In: Chevalier-Skolnikoff S, Poirer FE, editors. *Primate Bio-Social Development*. New York: Garland Publishing, pp. 515–528.

Swarts JD. 1988. Deciduous dentition: Implications for hominoid phylogeny. In: Schwatz JH, editor. *Orang-utan Biology*. New York: Oxford University Press, pp. 263–270.

Swindler DR. 2002. *Primate Dentition. An Introduction to the Teeth of Non-human Primates*. New York: Cambridge University Press.

Swindler DR, McCoy HA. 1964. Calcification of deciduous teeth in rhesus monkeys. *Science* 144:1243–1244.

Swindler DR, Orlosky FJ, Hendrickx AG. 1968. Calcification of the deciduous molars in baboons (*Papio anubis*) and other primates. *J Dent Res* 47:167–170.

Symington J. 1889. The vertebral column of a young gorilla. *J Anat Physiol* 24:42–51.

Tague RG. 1991. Commonalities in dimorphism and variability in the anthropoid pelvis, with implications for the fossil record. *J Hum Evol* 21:153–176.

Taniguchi H. 2015. How the physical properties of food influence its selection by infant Japanese macaques inhabiting a snow-covered area. *Am J Primatol* 77:285–295.

Tanner JM. 1990. *Fetus into Man*. Cambridge: Harvard University Press.

Tarnaud L. 2004. Ontogeny of feeding behavior of *Eulemur fulvus* in the dry forest of Mayotte. *Int J Primatol* 25:803–824.

Tarrant LH, Swindler DR. 1973. Prenatal dental development in the black howler monkey (*Alouatta caraya*). *Am J Phys Anthropol* 38:255–260.

Tattersall I, Schwartz JH. 1974. Craniodental morphology and the systematics of the Malagasy lemurs (Primates, Prosimii). *Anthrop Papers Amer Mus Nat Hist* 52:139–192.

Taylor AB. 2002. Masticatory form and function in African apes. *Am J Phys Anthropol* 117:133–156.

Taylor AB, Vinyard CJ. 2013. The relationships among jaw-muscle fiber architecture, jaw morphology, and feeding behavior in extant apes and modern humans. *Am J Phys Anthropol* 151:120–134.

Tecot SR, Baden AL, Romine NK, Kamilar JM. 2012. Infant parking and nesting, not allomaternal care, influence Malagasy primate life histories. *Behav Ecol Sociobio* 66:1375–1386.

ten Donkalaar HJ. 2000. Development and regenerative capacity of descending supraspinal pathways in tetrapods: A comparative approach. *Adv Anat Embryo Cell Biol* 154:1–145.

Thelen E, Fisher DM. 1982. Newborn stepping: An explanation for a "disappearing reflex". *Dev Psychol* 18:760–775.

Thorogood PV, Hinchliffe JR. 1975. An analysis of the condensation process during chondrogenesis in the embryonic chick hind limb. *J Embryol Exp Morph* 33:581–606.

Thorpe SKS, Crompton RH. 2005. Locomotor ecology of wild orangutans (*Pongo pygmaeus abelii*) in the Gunung Leuser Ecosystem, Sumatra, Indonesia: A multivariate analysis using log-linear modelling. *Am J Phys Anthropol* 127:58–78.

Thorpe SKS, Crompton RH. 2006. Orangutan positional behavior and the nature of arboreal locomotion in Hominoidea. *Am J Phys Anthropol* 131:384–401.

Tomanek RJ. 1975. A histochemical study of postnatal differentiation of skeletal muscle with reference to functional overload. *Dev Biol* 42:305–314.

Torre C, Giacobini G. 1978. Skeletal development of an orang-utan premature newborn: A comparative study with man. *J Hum Evol* 7:143–149.

Trevathan WR. 2015. Primate pelvic anatomy and implications for birth. *Phil Trans R Soc B* 370:20140065.

Turner TR, Anapol F, Jolly CJ. 1997. Growth, development, and sexual dimporphism in vervet monkeys (*Cercopithecus aethiops*) at four sites in Kenya. *Am J Phys Anthropol* 103:19–35.

Turnquist JE, Wells JP. 1994. Ontogeny of locomotion in rhesus macaques (*Macaca mulatta*): I. Early postnatal ontogeny of the musculoskeletal system. *J Hum Evol* 26:487–499.

Twitchell TE. 1965. The automatic grasping responses of infants. *Neuropsychologia* 3:247–259.

van Nievelt AFH, Smith KK. 2005. To replace or not to replace: The significance of reduced tooth replacement in marsupial and placental mammals. *Paleobiology* 31:324–346.

van Noordwijk MA, van Schaik CP. 2005. Development of ecological competence in Sumatran orangutans. *Am J Phys Anthropol* 127:79–94.

van Noordwijk MA, Atmoko SSU, Knott CD, et al. 2018. The slow ape: High infant survival and long interbirth intervals in wild orangutans. *J Hum Evol* 125:38–49.

van Schaik CP, Van Noordwijk MA, Vogel ER. 2009. Ecological sex differences in wild orangutans. In: Wich SA, Atmoko SSU, Setia TM, Van Schaik CP, editors. *Orangutans: Geographic Variation in Behavioral Ecology and Conservation*. Oxford: Oxford University Press, pp. 255–277.

Van Wagenen G, Asling W. 1964. Ossification in the fetal monkey (*Macaca mulatta*). Estimation of age and progress of gestation by roentgenography. *Am J Anat* 114:107–132.

Vasey N, Mogilewsky M, Schatz GE. 2018. Infant nest and stash sites of variegated lemurs (*Varecia rubra*): The extended phenotype. *Am J Primatol* 6:e22911–22915.

Venkataraman VV, Glowacka H, Fritz J, et al. 2014. Effects of dietary fracture toughness and dental wear on chewing efficiency in geladas (*Theropithecus gelada*). *Am J Phys Anthropol* 155:17–32.

Vilensky JA, Gankiewicz E. 1986. Effects of size on vervet (*Cercopithecus aethiops*) gait parameters: A preliminary analysis. *Folia Primatol (Basel)* 46:104–117.

Vilensky JA, Gankiewicz E. 1989. Early development of locomotor behavior in vervet monkeys. *Am J Primatol* 17:11–25.

Vilensky JA, Larson SG. 1989. Primate locomotion: Utilization and control of symmetrical gaits. *Annu Rev Anthrop* 18:17–35.

Vilensky JA, Wilson P, Gankiewicz E, Townsend DW. 1989. An analysis of air-stepping in normal infant vervet monkeys. *J Mot Behav* 21:429–456.

Vinay L, Brocard F, Clarac F. 2000a. Differential maturation of motoneurons innervating ankle flexor and extensor muscles in the neonatal rat. *Eur J Neurosci* 12:4562–4566.

Vinay L, Brocard F, Pflieger JF, Simeoni-Alias J. 2000b. Perinatal development of lumbar motoneurons and their inputs in the rat. *Brain Res Bull* 15:635–647.

Vinyard CJ, Hanna J. 2005. Molar scaling in strepsirrhine primates. *J Hum Evol* 49:241–269.

Vinyard CJ, Ravosa MJ. 1998. Ontogeny, function, and scaling of the mandibular symphysis in papionin primates. *J Morphol* 235:157–175.

Wang Q, Opperman LA, Havill LM, Carlson DS, Dechow PC. 2006. Inheritance of sutural pattern at the pterion in Rhesus monkey skulls. *Anat Rec* 288:1042–1049.

Wang Q, Ashley DW, Dechow PC. 2010. Regional, ontogenetic, and sex-related variations in elastic properties of cortical bone in baboon mandibles. *Am J Phys Anthropol* 141:526–549.

Wasser SK, Barash DP. 1983. Reproductive suppression among female mammals: Implications for biomedicine and sexual selection theory. *Quart Rev Biol* 58:513–538.

Watts DP. 1985. Observations on the ontogeny of feeding behavior in mountain gorillas (*Gorilla gorilla beringei*). *Am J Primatol* 8:1–10.

Watts ES. 1990. A comparative study of neonatal skeletal development in *Cebus* and other primates. *Folia Primatol* 54:217–224.

Watts ES, Gavin JS. 1982. Postnatal growth of nonhuman primates: The problem of the adolescent spurt. *Hum Biol* 54:53–70.

Wealthall RJ, Herring SW. 2006. Endochondral ossification of the mouse nasal septum. *Anat Rec* 288:1163–1172.

Weins F, Zitzmann A. 2003. Social dependence of infant slow lorises to learn diet. *Int J Primatol* 24:1007–1021.

Wells JP, Turnquist JE. 2001. Ontogeny of locomotion in rhesus macaques (*Macaca mulatta*): II. Postural and locomotor behavior and habitat use in a free-ranging colony. *Am J Phys Anthropol* 115:80–94.

Werneburg I, Tzika AC, Hautier L, et al. 2013. Development and embryonic staging in non-model organisms: The case of an Afrotherian mammal. *J Anat* 222:2–18.

Western D. 1979. Size, life history and ecology in mammals. *Afr J Ecol* 17:185–204.

Wich SA, Utami-Atmoko SS, Setia T, et al. 2004. Life history of wild Sumatran orangutans (*Pongo abelii*). *J Hum Evol* 47:385–398.

Winkler LA. 1995. A comparison of radiographic and anatomical evidence of tooth development in infant apes. *Folia Primatol* 65:1–13.

Winkler LA, Schwartz JH, Swindler DR. 1991. Aspects of dental development in the orangutan prior to eruption of the permanent dentition. *Am J Phys Anthropol* 86:255–271.

Wolf SA. 2016. Ontogeny of postcranial morphology and locomotor behavior in *Propithecus verreauxi coquereli* and *Lemur catta*. Master's thesis. James Madison University.

Wolovich CK, Evans S, French JA. 2008. Dads do not pay for sex but do buy the milk: Food sharing and reproduction in owl monkeys (*Aotus* spp.). *Anim Behav* 75:1155–1163.

Workman C, Covert HH. 2005. Learning the ropes: the ontogeny of locomotion in the red-shanked douc (*Pygathrix nemaeus*), Delacour's (*Trachypithecus delacouri*), and Hatinh langurs (*Trachypithecus hatinhensis*) I. Positional behaviors. *Am J Phys Anthropol* 128:371–380.

Wright KA. 2005. Interspecific and ontogenetic variation in locomotor behavior, habitat use, and postcranial morphology in *Cebus apella* and *Cebus olivaceus*. PhD thesis. Northwestern University.

Wright PC. 1995. Demography and life history of free-ranging *Propithecus diadema edwardsi* in Ranomafana National Park, Madagascar. *Int J Primatol* 16:835–854.

Yamashita N, Vinyard CJ, Tan CL. 2009. Food mechanical properties in three sympatric species of *Hapalemur* in Ranomafana National Park, Madagascar. *Am J Phys Anthropol* 139:368–381.

Young JW. 2005. Ontogeny of muscle mechanical advantage in capuchin monkeys (*Cebus albifrons* and *Cebus apella*). *J Zool Lond* 267:351–362.

Young JW. 2008. Ontogeny of locomotion in *Saimiri boliviensis* and *Callithrix jacchus*: Implications for primate locomotor ecology and evolution. Interdepartmental Doctoral Program in Anthropological Sciences, Stony Brook University.

Young JW. 2009. Ontogeny of joint mechanics in squirrel monkeys (*Saimiri boliviensis*): Implications for mammalian limb growth and locomotor development. *J Exp Biol* 212:1576–1591.

Young JW. 2012a. Gait selection and the ontogeny of quadrupedal walking in squirrel monkeys (*Saimiri boliviensis*). *Am J Phys Anthropol* 147:580–592.

Young JW. 2012b. Ontogeny of limb force distribution in squirrel monkeys (*Saimiri boliviensis*): Insights into the mechanical bases of primate hind limb dominance. *J Hum Evol* 62:473–485.

Young JW, Heard-Booth AN. 2016. Grasping primate development: Ontogeny of intrinsic hand and foot proportions in capuchin monkeys (*Cebus albifrons* and *Sapajus apella*). *Am J Phys Anthropol* 161:104–115.

Young JW, Shapiro LJ. 2018. Developments in development: what have we learned from primate locomotor ontogeny? *Yearb Phys Anthropol* S65:37–71.

Young JW, Fernández D, Fleagle JG. 2010. Ontogeny of limb bone geometry in capuchin monkeys (*Cebus albifrons* and *Cebus apella*): Implications for locomotor development and life history. *Biol Lett* 6:197–200.

Young JW, Jankford K, Saunders MM, Smith TD. in press. Getting into shape: Limb bone strength in perinatal *Lemur catta* and *Propithecus coquereli*. *Anat Rec*.

Zaleski SM, Smith TD, DeLeon VB. 2018. Geometric morphometrics of the neonatal thorax in prosimians. *Am J Phys Anthropol* 165(S66):310.

Zehr SM, Roach RG, Haring D, et al. 2014. Life history profiles for 27 strepsirrhine primate taxa generated using captive data from the Duke Lemur Center. *Sci Data* 1:140019.

Index

Note: Locators in italics refer to figures.

Atlas

Contents

(*cont.*)

Plate	Element	Taxa	Views
22	Sternebrae	*Tarsius, Galago, Lemur*	Orientation, ventral
23	Ribs	*Callicebus, Cebuella, Saguinus*	Orientation, right lateral, ventral
24	Ribs	*Tarsius, Galago, Lemur*	Orientation, right lateral, ventral
Forelimb			
25a	Clavicle and scapula	*Callicebus, Cebuella, Saguinus*	Orientation, rostal thorax
25b	Clavicle and scapula	*Callicebus, Cebuella, Saguinus*	Clavicle (ventral), scapula (external)
26a	Clavicle and scapula	*Tarsius, Galago, Lemur*	Orientation, rostal thorax
26b	Clavicle and scapula	*Tarsius, Galago, Lemur*	Clavicle (ventral), scapula (external)
27	Humerus	*Callicebus, Cebuella, Saguinus*	Orientation, right lateral, ventral
28	Humerus	*Tarsius, Galago, Lemur*	Orientation, right lateral, ventral
29	Radius	*Callicebus, Cebuella, Saguinus*	Orientation, right lateral, ventral
30	Radius	*Tarsius, Galago, Lemur*	Orientation, right lateral, ventral
31	Ulna	*Callicebus, Cebuella, Saguinus*	Orientation, right lateral, ventral
32	Ulna	*Tarsius, Galago, Lemur*	Orientation, right lateral, ventral
33a	Wrist and hand	*Callicebus, Cebuella, Saguinus*	Orientation, ulnar (postaxial)
33b	Wrist and hand	*Callicebus, Cebuella, Saguinus*	Dorsal (flattened), palmar (flattened)
34a	Wrist and hand	*Tarsius, Galago, Lemur*	Orientation, ulnar (postaxial)
34b	Wrist and hand	*Tarsius, Galago, Lemur*	Dorsal (flattened), palmar (flattened)
Hindlimb			
35a	Os coxa	*Callicebus, Cebuella, Saguinus*	Orientation, right lateral, right medial
35b	Os coxa	*Callicebus, Cebuella, Saguinus*	Ventral, caudal
36a	Os coxa	*Tarsius, Galago, Lemur*	Orientation, right lateral, right medial
36b	Os coxa	*Tarsius, Galago, Lemur*	Ventral, caudal
37a	Femur	*Callicebus, Cebuella, Saguinus*	Orientation, right lateral, ventral
37b	Femur	*Callicebus, Cebuella, Saguinus*	Dorsal, proximal, distal
38a	Femur	*Tarsius, Galago, Lemur*	Orientation, right lateral, ventral
38b	Femur	*Tarsius, Galago, Lemur*	Dorsal, proximal, distal
39	Tibia	*Callicebus, Cebuella, Saguinus*	Orientation, right lateral, ventral
40	Tibia	*Tarsius, Galago, Lemur*	Orientation, right lateral, ventral

(*cont.*)

Plate	Element	Taxa	Views
41	Fibula	*Callicebus, Cebuella, Saguinus*	Orientation, right lateral, ventral
42	Fibula	*Tarsius, Galago, Lemur*	Orientation, right lateral, ventral
43a	Ankle and foot	*Callicebus, Cebuella, Saguinus*	Orientation, right lateral
43b	Ankle and foot	*Callicebus, Cebuella, Saguinus*	Lateral ankle, dorsal foot (flattened)
44a	Ankle and foot	*Tarsius, Galago, Lemur*	Orientation, right lateral
44b	Ankle and foot	*Tarsius, Galago, Lemur*	Lateral ankle, dorsal foot (flattened)

Specimen Notes

The atlas is intended as both a resource to support the text and a stand-alone comparative atlas of the newborn skeleton. We employ six species of primates to illustrate the lesser-studied taxa, including three monkeys, a tarsier, a bushbaby, and a lemur. Although the perinatal timeframe is a vulnerable period and perinatal deaths are more common than other subadult mortalities, "pristine" specimens are somewhat rare due to trauma that may have led to death, necropsies that may have been conducted to determine the cause of death, or postmortem damage (e.g., necrosis). In addition, primates themselves are relatively rare and slow to reproduce. Thus, for the atlas we took advantage of our larger sample of primate species to select those bodies that were complete, or nearly so, and well-preserved. All were obtained following death by natural causes in zoos or other institutions dedicated to conservation. We thank the Dumond Conservancy for Primates and Tropical Forests (for the *Callicebus* newborn), the Dallas World Aquarium (*Cebuella*), the Gladys Porter Zoo (*Saguinus*), and the Duke Lemur Center (*Tarsius, Galago*, and *Lemur*) for carefully preserving these invaluable remains. All of the atlas specimens either died on the day of birth or were stillborn and of similar size to other newborn specimens of the species.

For each specimen, microCT imaging was performed at Northeast Ohio Medical College (NEOMED) with scan parameters 70 kV and 114 μA. Additional details for each scan are provided below. For all specimens, post-processing of virtual reconstructions was performed at the University of Florida using Amira software (FEI; v. 5.0+). Scan artifacts were removed, including radio-opaque nodules and the tube that held the specimen. For the wrist and hand plate and the foot plate, the digits were reconstructed to align elements in a single plane. Additional details on reconstruction of each specimen are also provided below.

Callicebus cupreus

This *Callicebus cupreus* specimen had necropsy damage that impacted the sternum (the manubrium was sagittally transected and the left half was absent) and had slight displacement of the left thoracic wall. There was no gross damage to the head. A vertebral variant was observed in this specimen and might reflect a developmental segmentation error. We identified seven cervical, 13 thoracic, and six lumbar vertebrae. T1 has a cervical morphology on the transverse processes and reduced ribs and could be interpreted alternatively as an eighth cervical vertebra.

The microCT image volume had reconstructed cubic voxels of 0.025 × 0.025 × 0.025 mm, and volume dimensions 2048 × 2048 × 4178. The skeletal anatomy required no additional reconstruction.

Cebuella pygmaea

This *Cebuella pygmaea* specimen (Cp17) had no apparent necropsy damage.

MicroCT imaging was performed in two scans, both with reconstructed cubic voxels of 0.0205 × 0.0205 × 0.0205 mm. The rostral portion of the body was isolated from the first scan, the caudal part of the body was isolated from the second scan, and the two portions were aligned at the lumbosacral junction. The head was ventrally rotated to mitigate postmortem cervical extension.

Saguinus midas

This *Saguinus midas* specimen (GPZ10672) had the chest plate removed during necropsy, with one cut through the right ribs, and two cuts through the left ribs, resulting in displacement of ribs, sternebrae, and right clavicle, and possible loss of the hyoid. The most distal caudal vertebrae were not present.

The microCT image volume had reconstructed cubic voxels of 0.035 × 0.035 × 0.035 mm, and volume dimensions 2048 x 2048 x 2337. Post-processing repositioned the following elements into anatomic position: right rib plate, two left rib plates, sternebrae, and right clavicle. The right forelimb was rotated to move the elbow out of the reconstructed thoracic cavity.

Tarsius syrichta

This *Tarsius syrichta* specimen (P93, DLC93, or CM116671) had an atlantoaxial dislocation. The shoulders were posteriorly displaced, resulting in deformation (folding) of scapular bodies.

The microCT image volume had reconstructed cubic voxels of 0.025 × 0.025 × 0.025 mm. No additional post-processing was required.

Galago moholi

This *Galago moholi* specimen (P2047, or DLC2047) had an atlantoaxial dislocation. The tail was transected, and the posterior neurocranium showed some flattening.

The microCT image volume had reconstructed cubic voxels of 0.0205 × 0.0205 × 0.0205 mm. No additional post-processing was required.

Lemur catta

This *Lemur catta* specimen (DLC7118) had necropsy damage including decapitation below C2, and additional separation below T4 with damage to the lower left ribs. The manubrium, ventral arch of C1, and pubic bones were all absent and may have been removed in necropsy. The neurocranium was depressed at the vertex.

The microCT image volume had reconstructed cubic voxels of 0.0205 × 0.0205 × 0.0205 mm. Post-processing isolated the anatomic segment including skull and C1–2 vertebrae and aligned this segment with the rest of the body at the C2–C3 articulation.

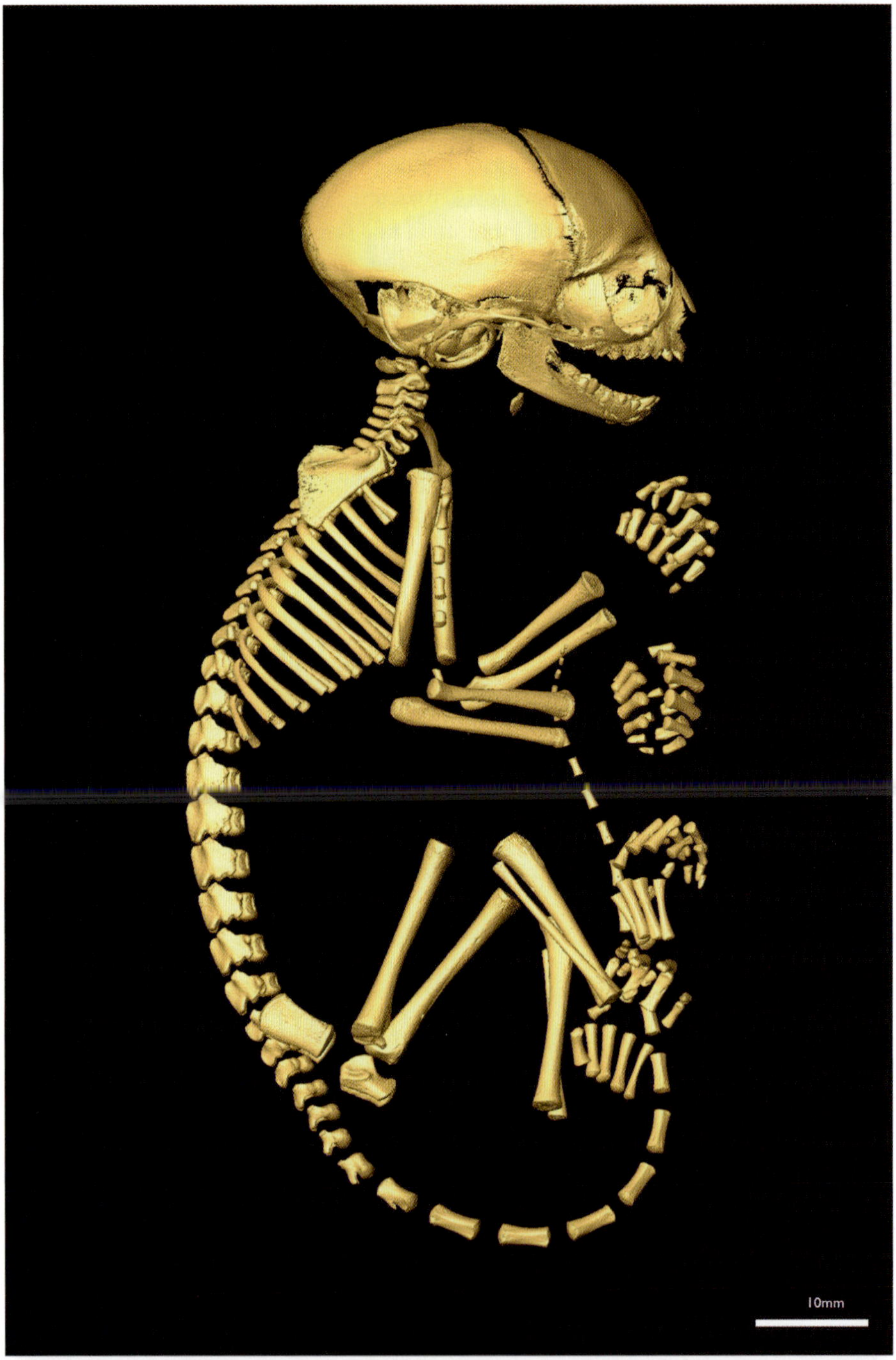

Callicebus cupreus

Plate 1

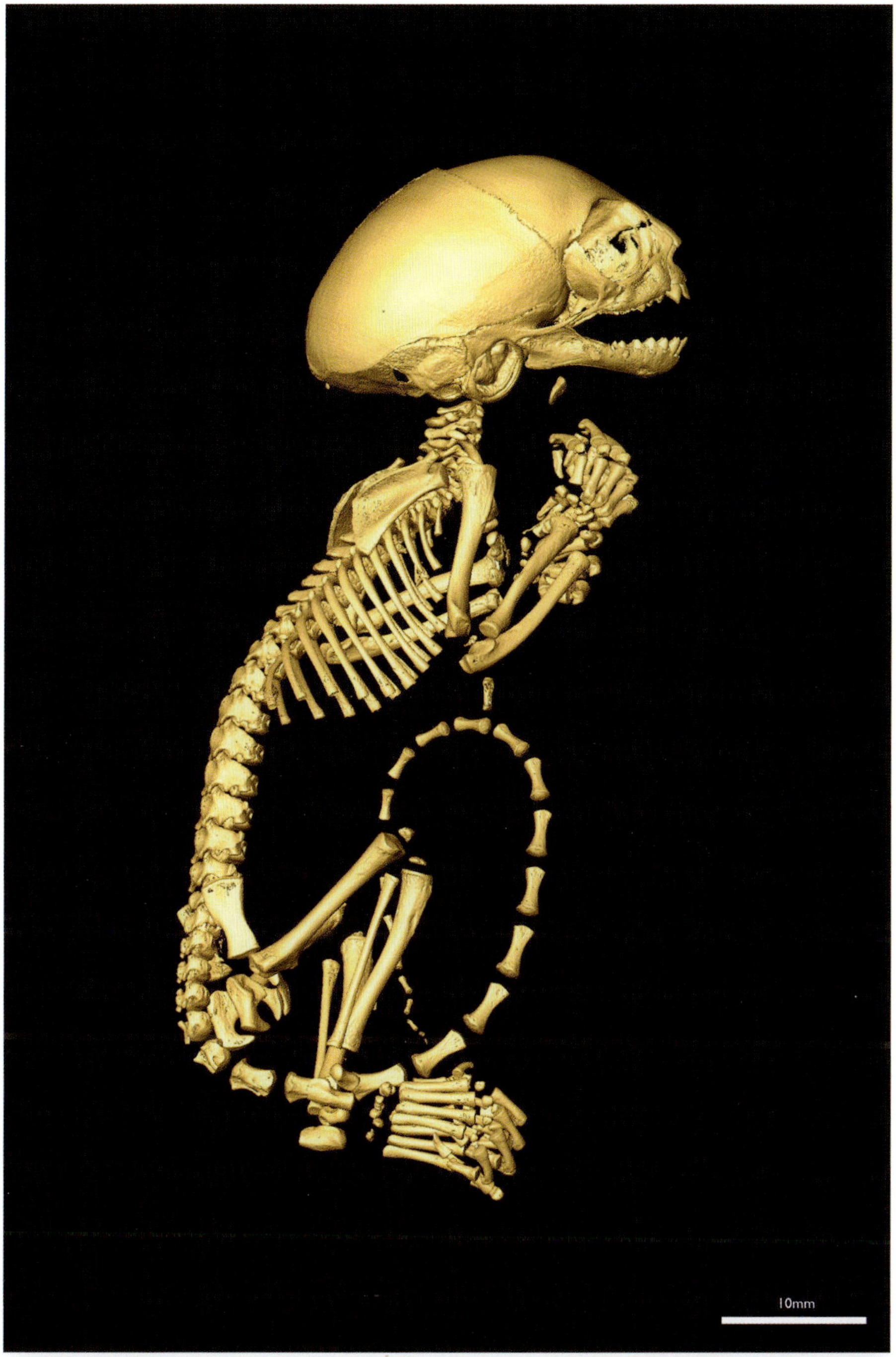

Cebuella pygmaea

Plate 2

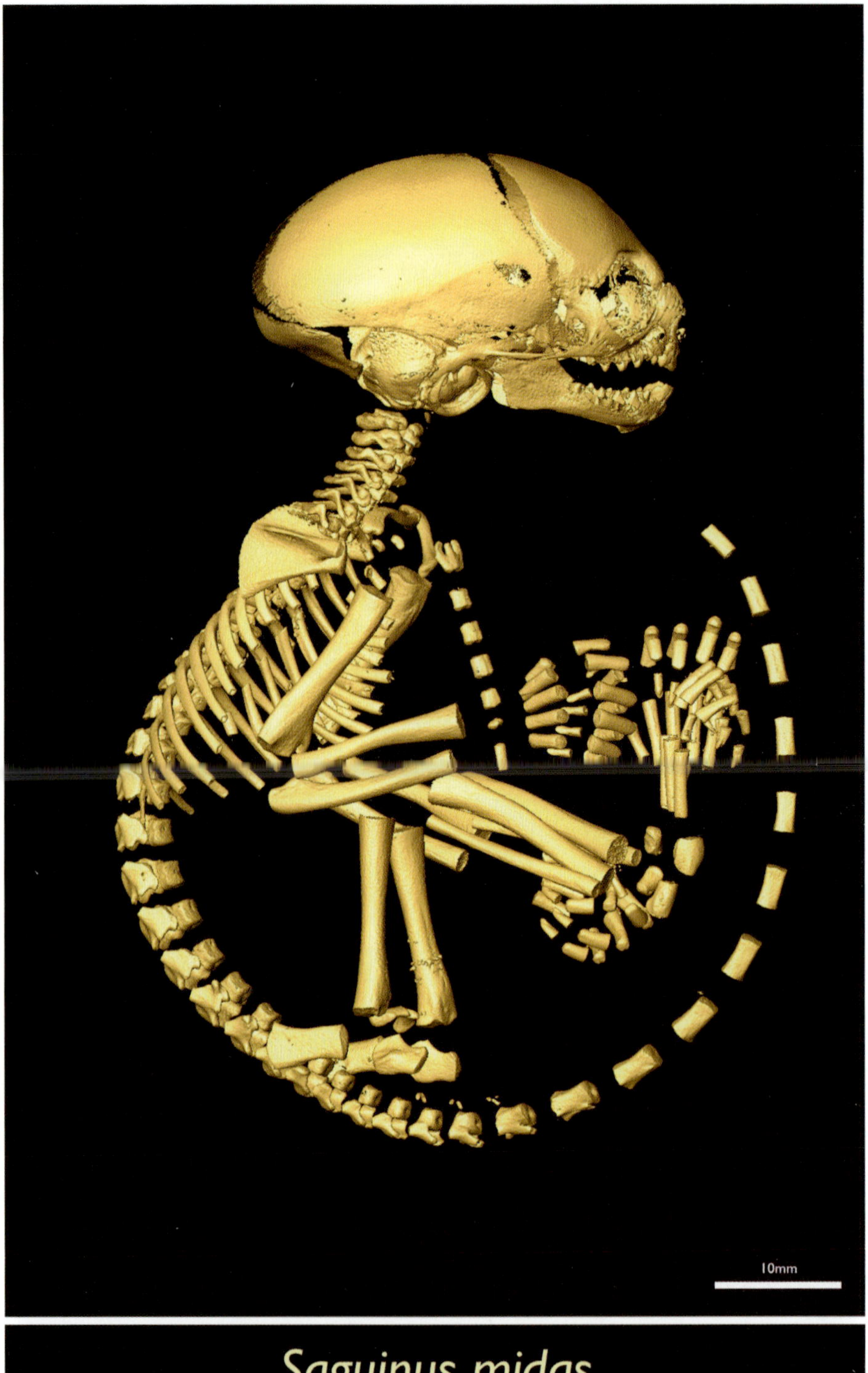

Saguinus midas

Plate 3

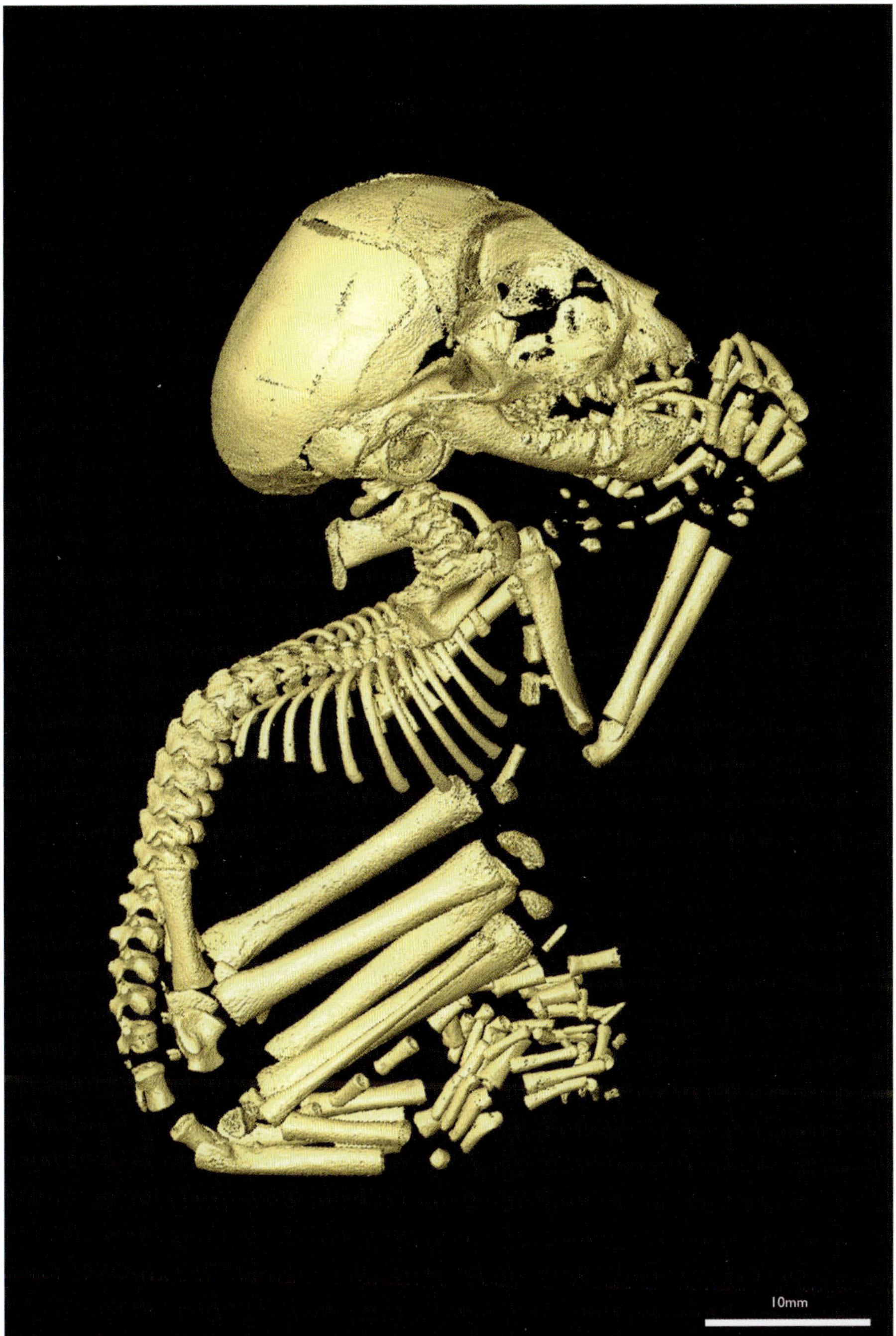

Tarsius syrichta

Plate 4

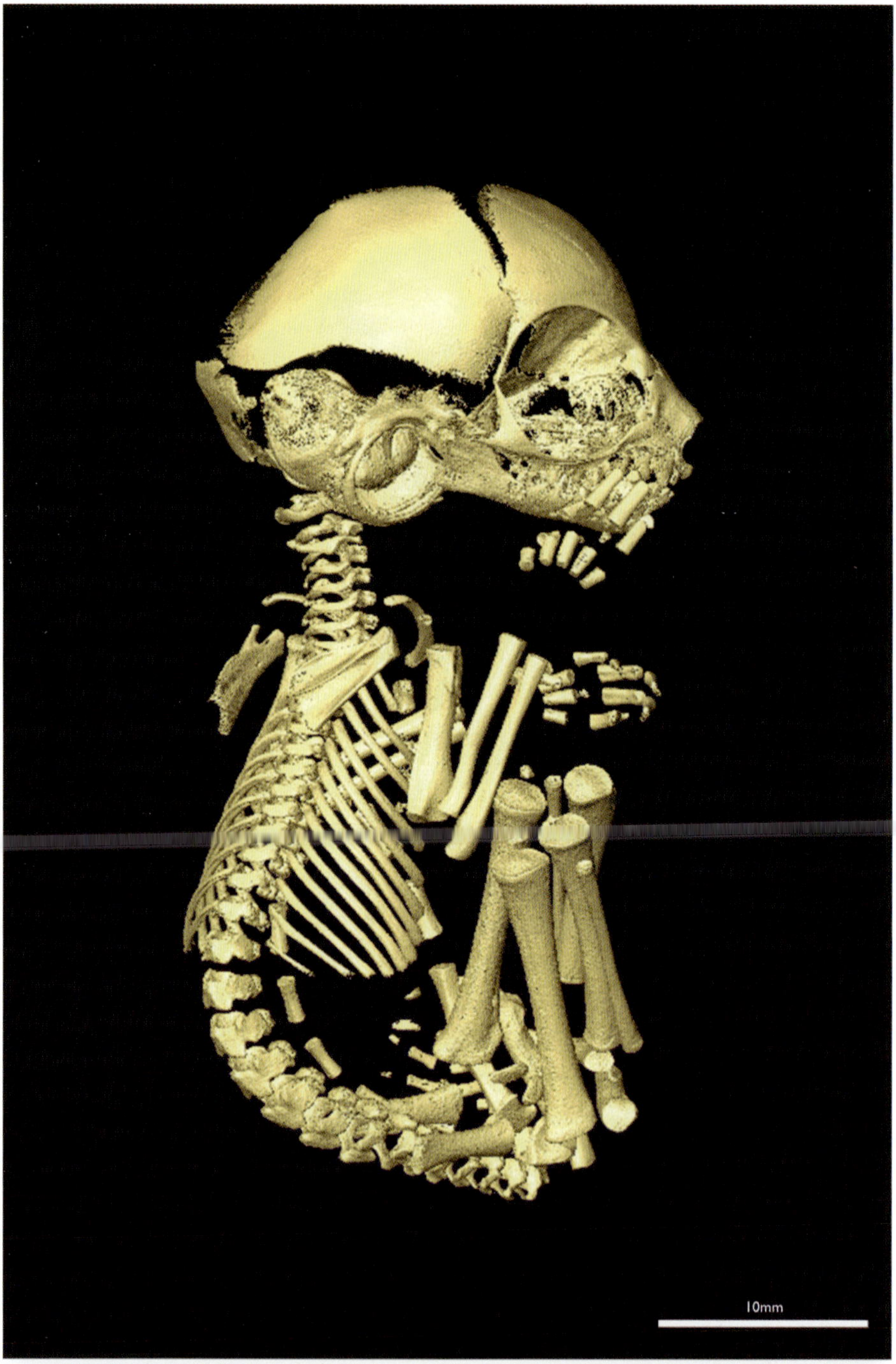

Galago moholi

Plate 5

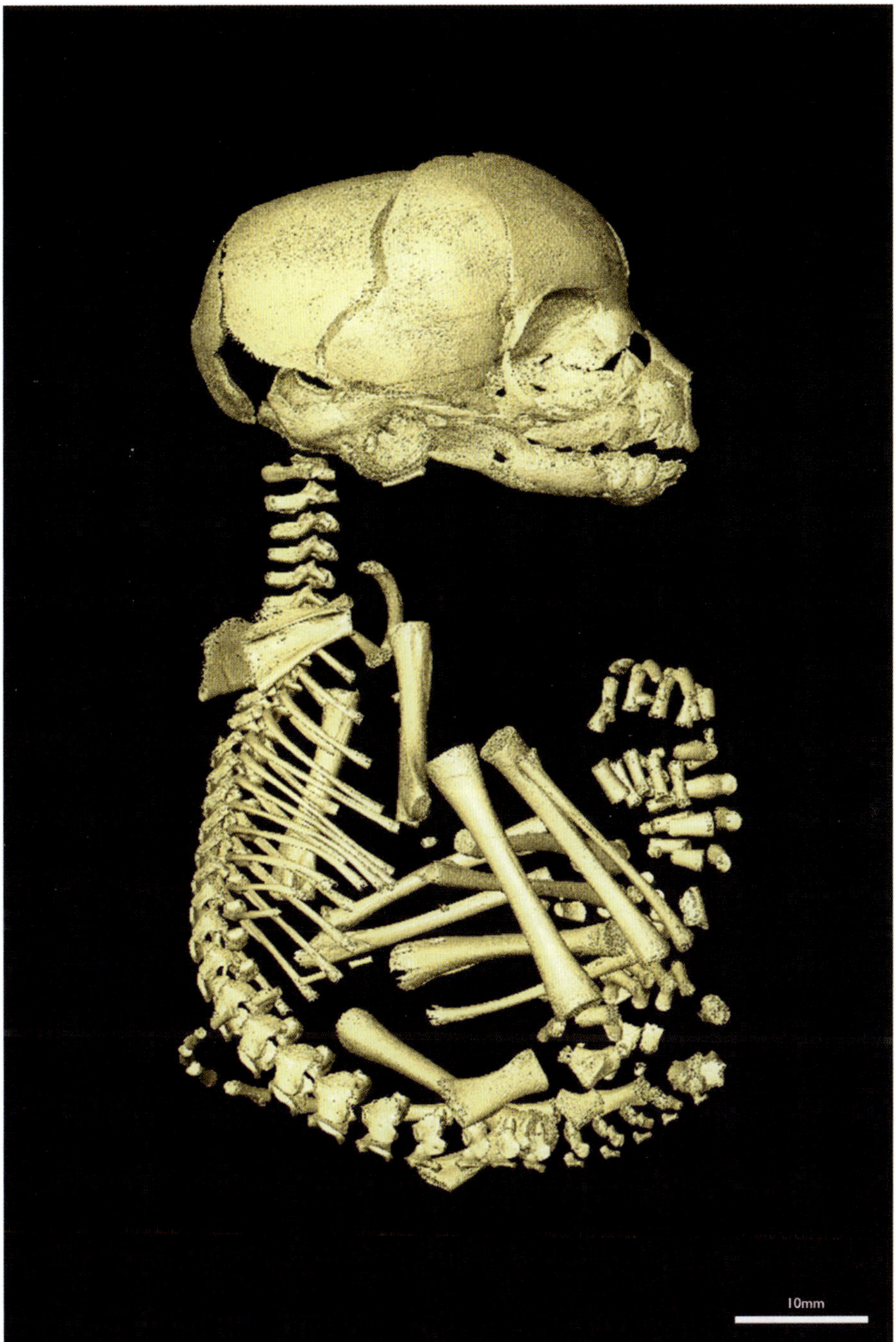

Lemur catta

Plate 6

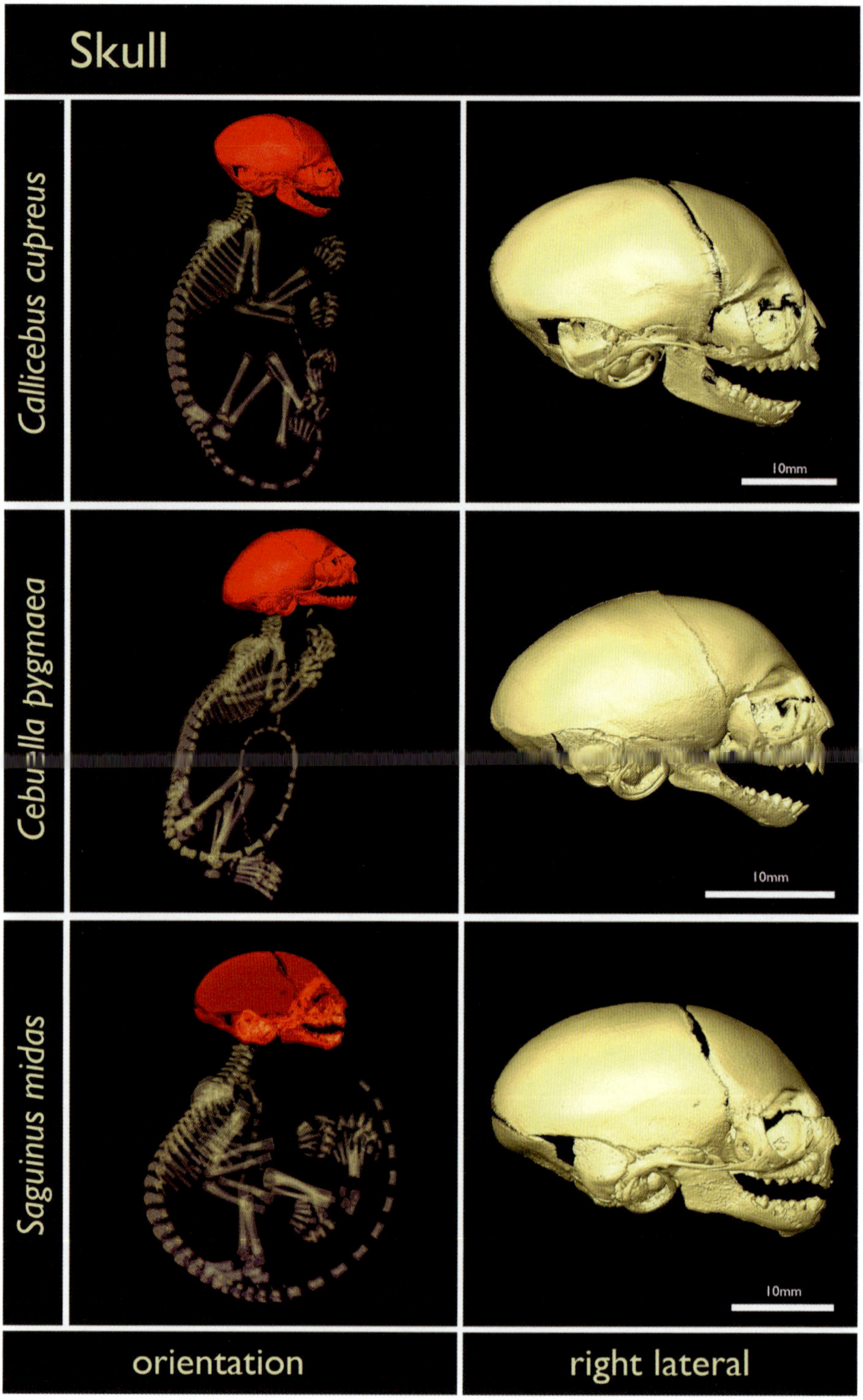
Skull
Callicebus cupreus
10mm
Cebuella pygmaea
10mm
Saguinus midas
10mm
orientation
right lateral

Plate 7a

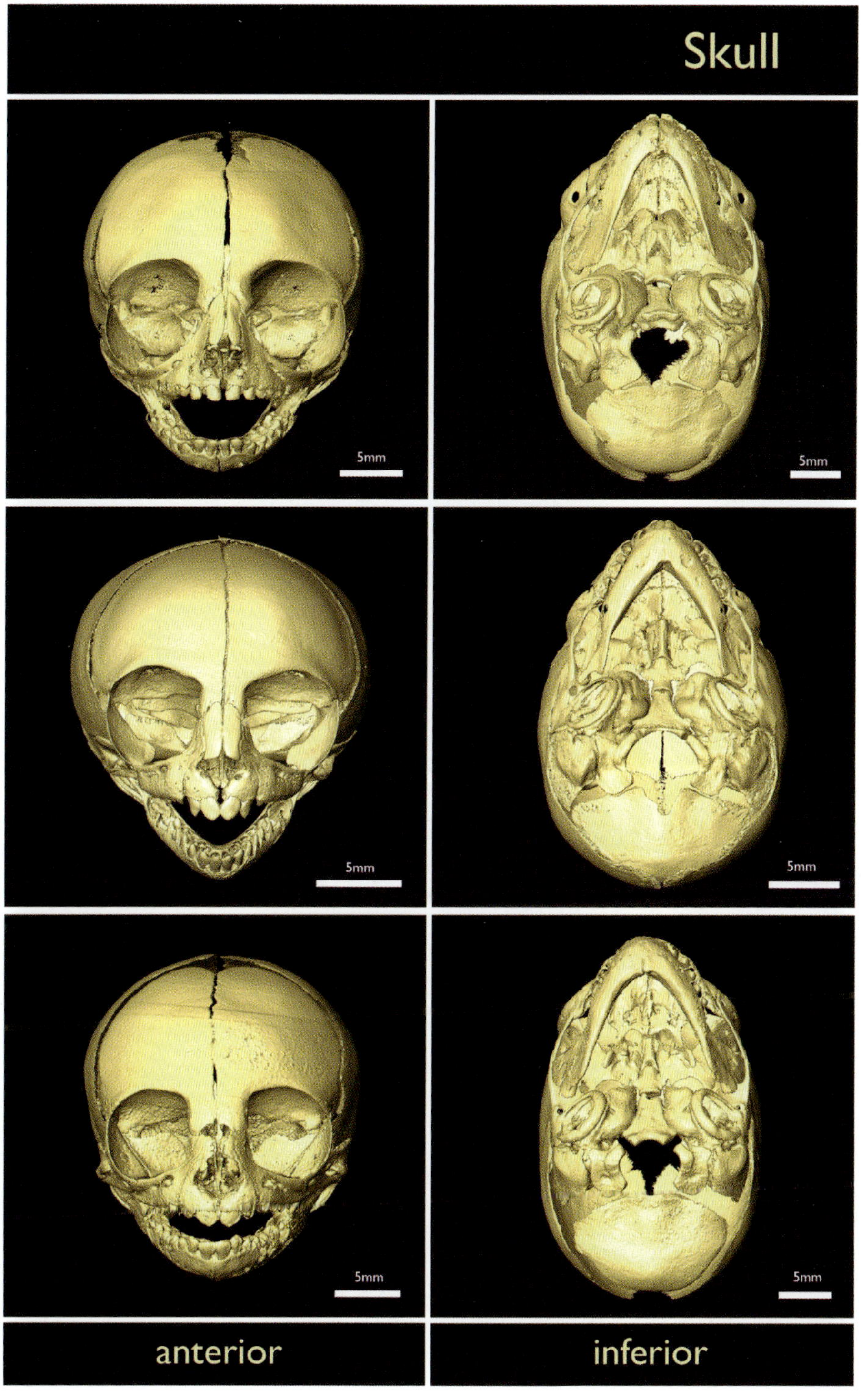

Plate 7b

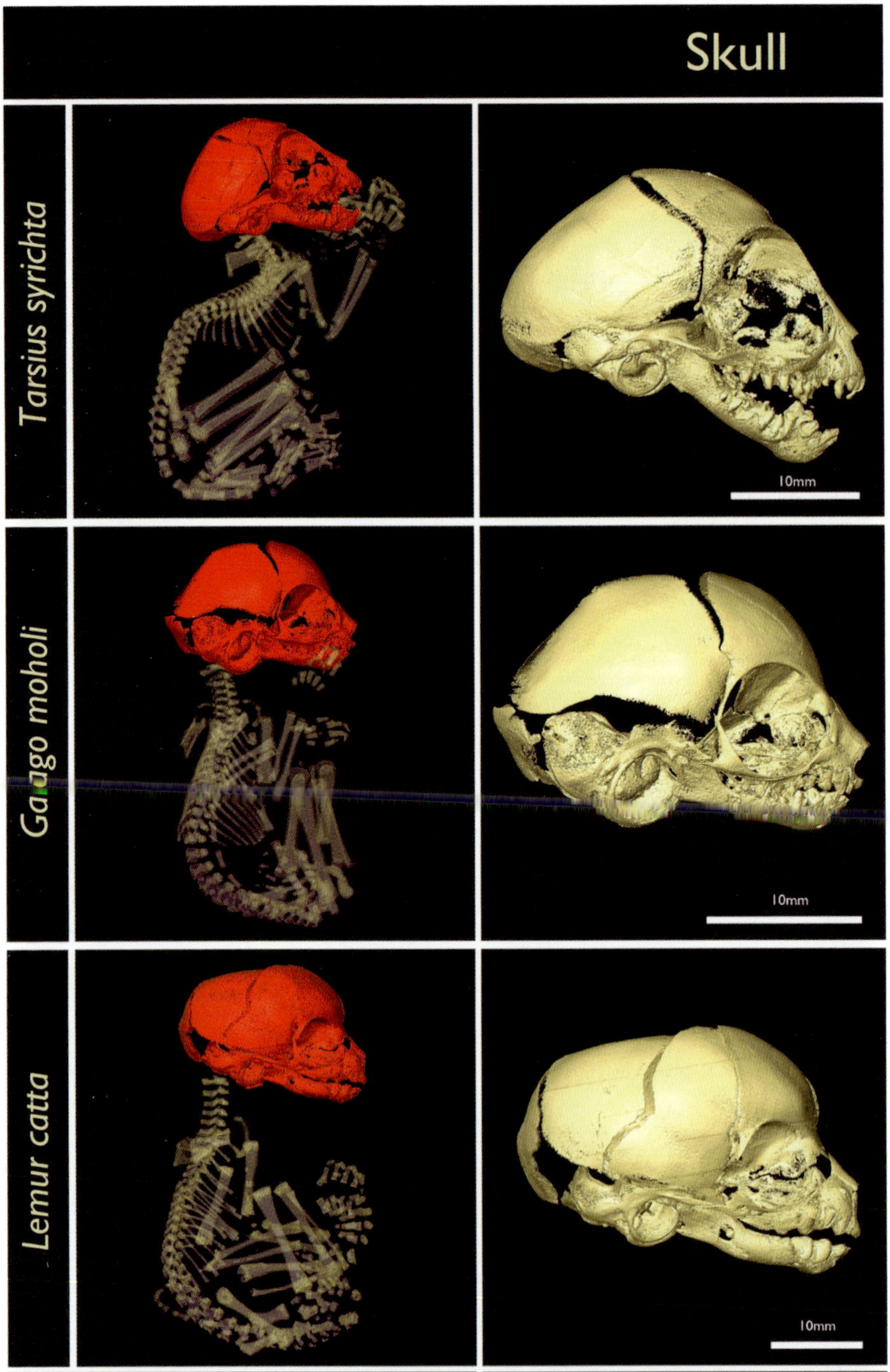
Skull
Tarsius syrichta
10mm
Galago moholi
10mm
Lemur catta
10mm
orientation
right lateral

Plate 8a

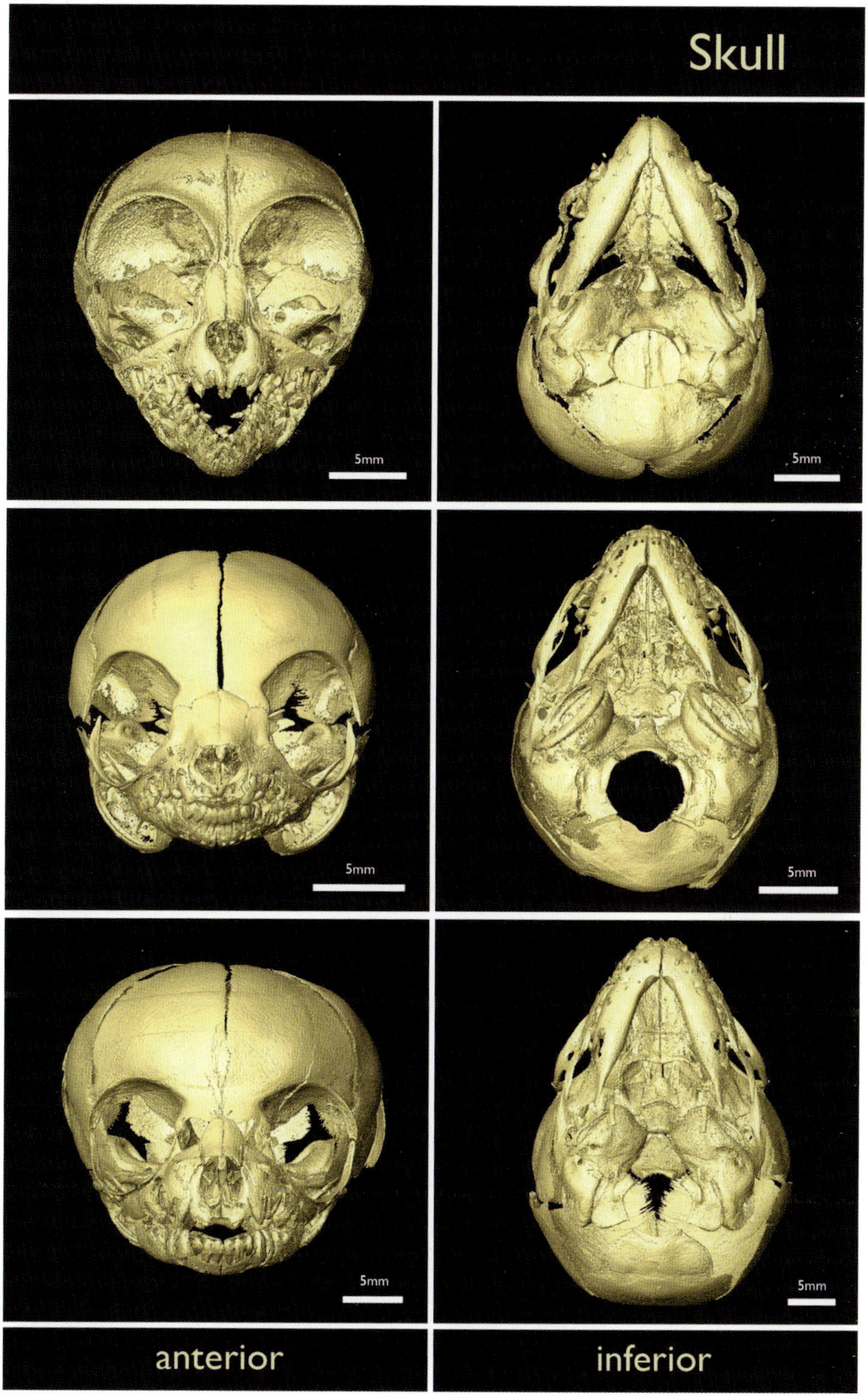

Plate 8b

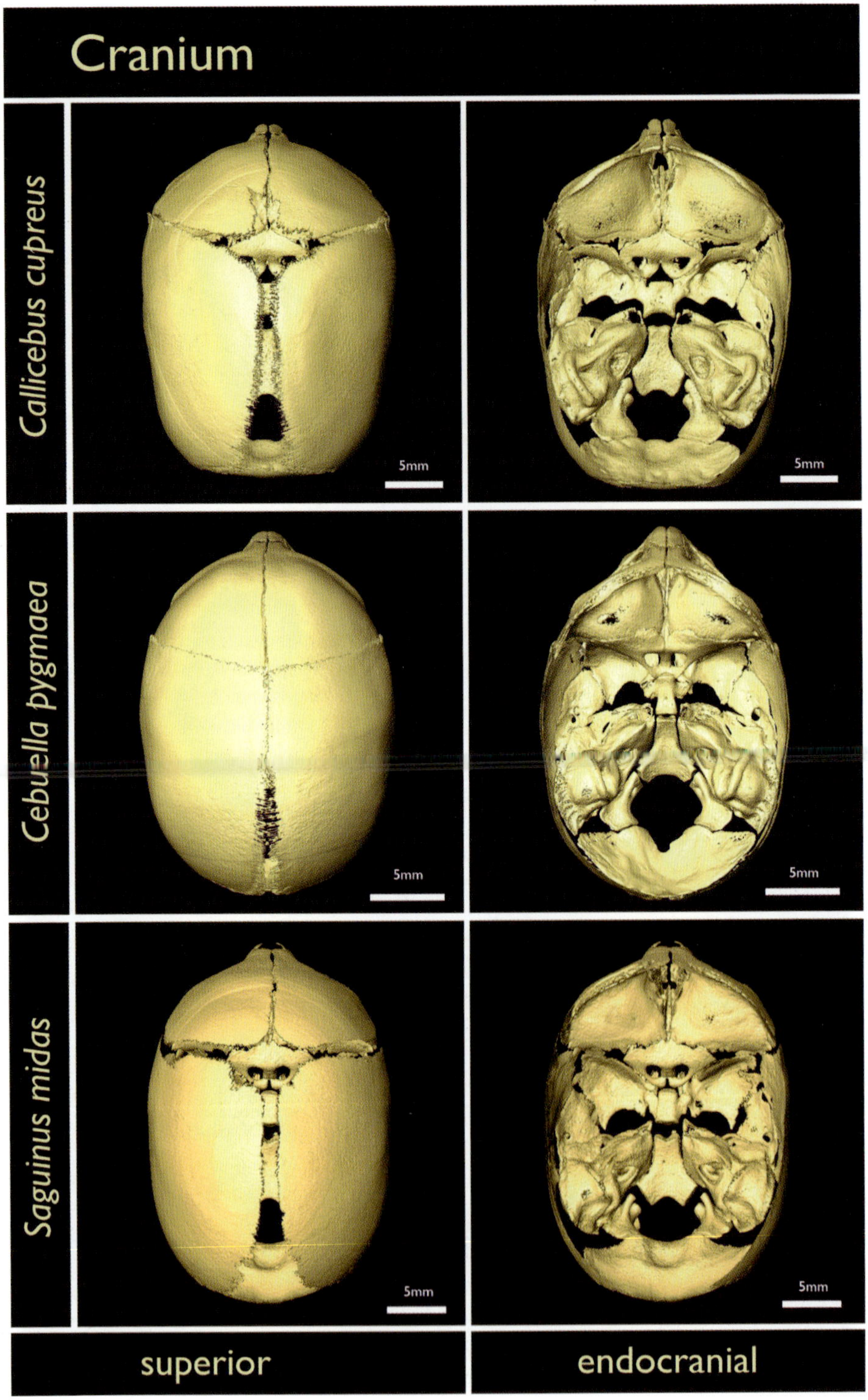
Cranium
Callicebus cupreus
Cebuella pygmaea
Saguinus midas
5mm
superior
endocranial

Plate 9a

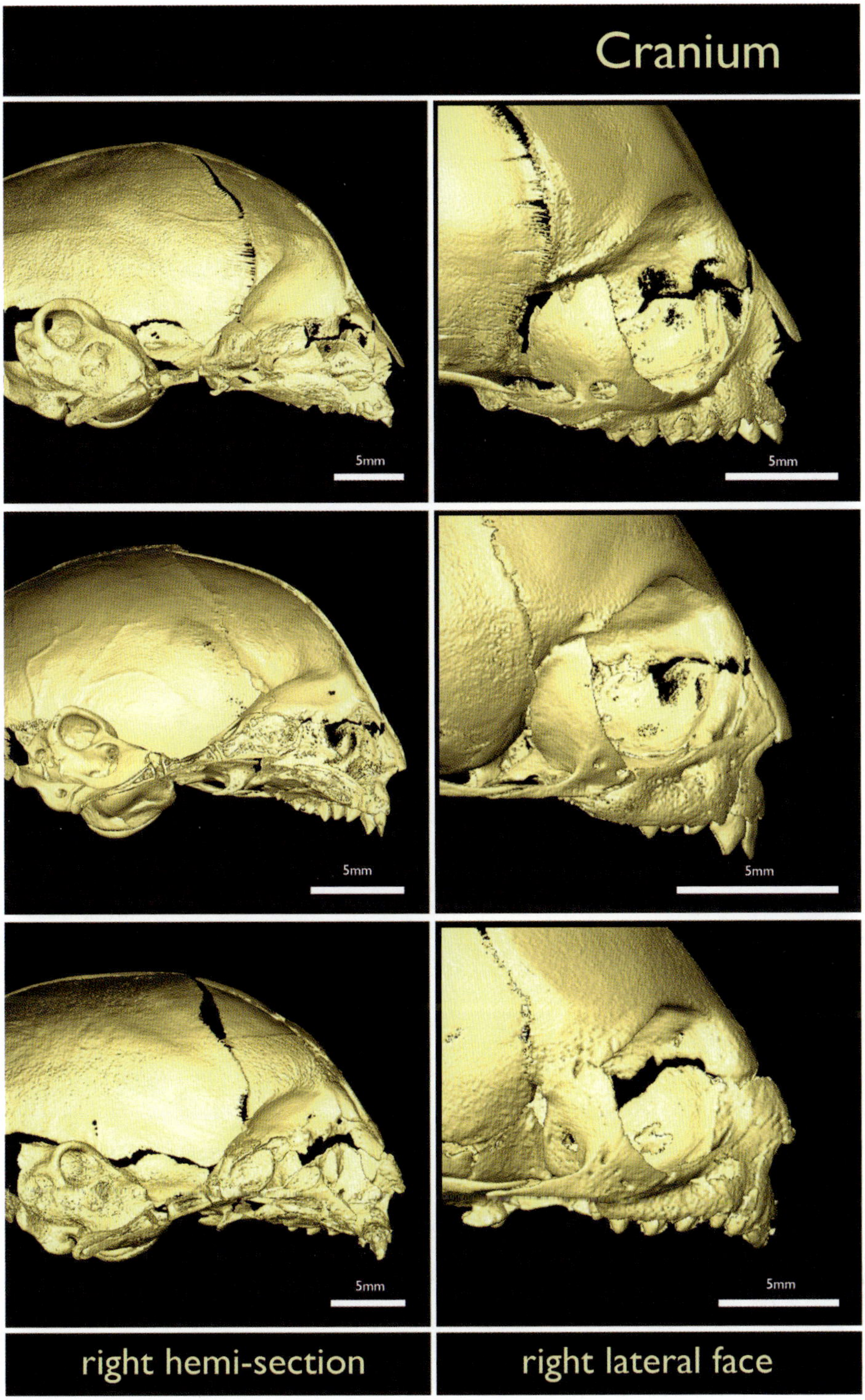
Cranium
5mm
5mm
5mm
5mm
5mm
5mm
right hemi-section
right lateral face

Plate 9b

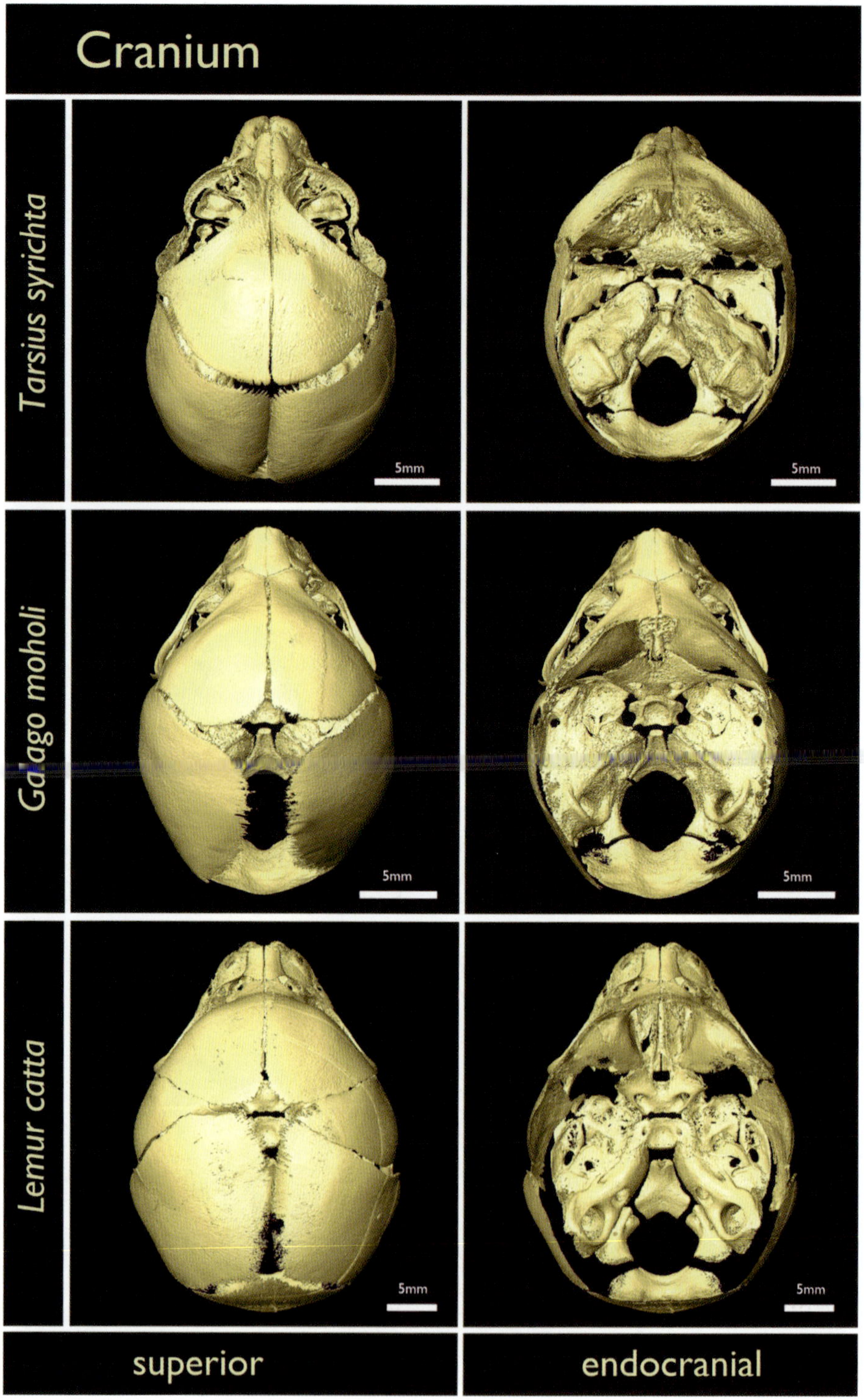
Cranium
Tarsius syrichta
Galago moholi
Lemur catta
5mm
superior
endocranial

Plate 10a

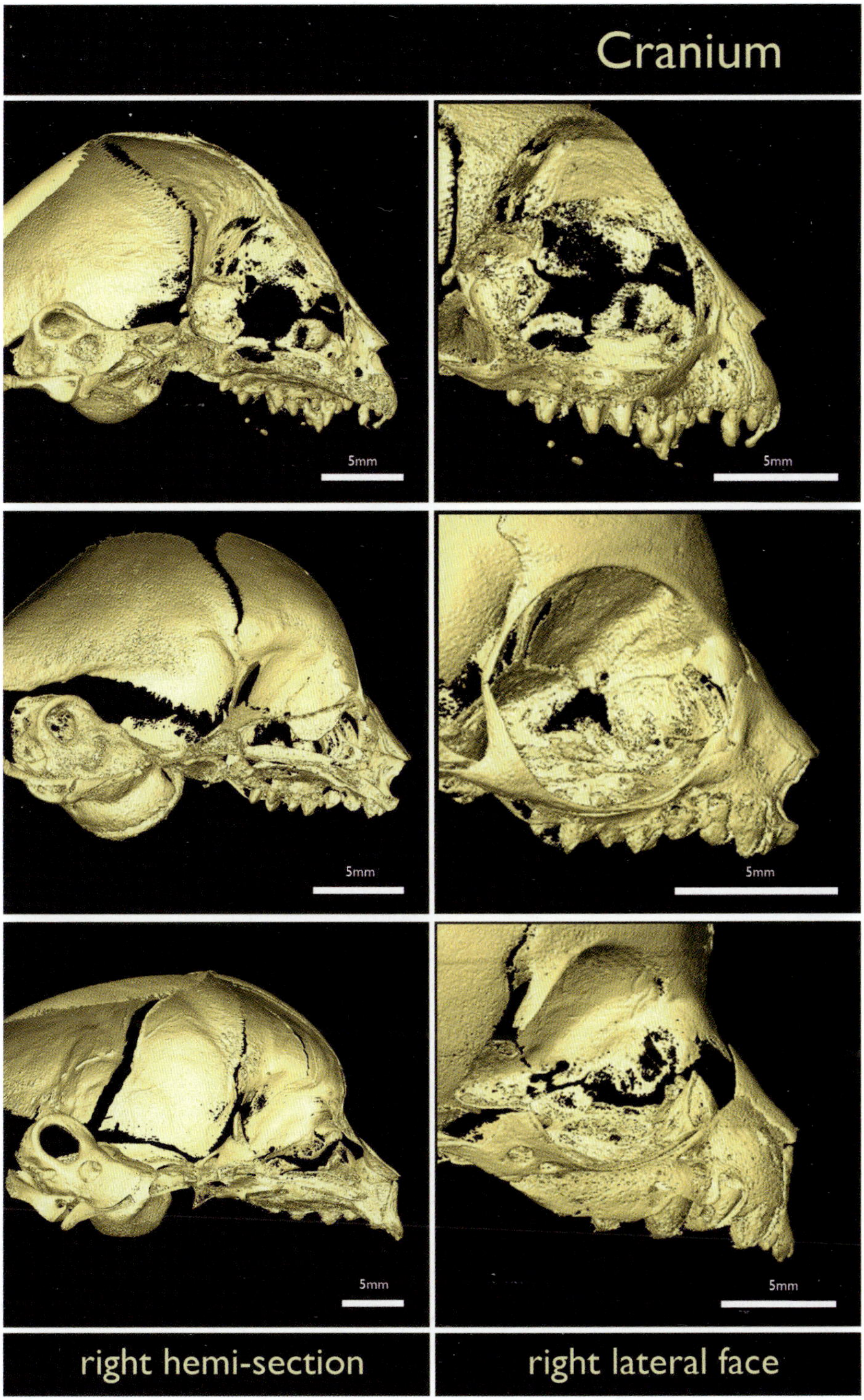
Cranium
5mm
5mm
5mm
5mm
5mm
5mm
right hemi-section
right lateral face

Plate 10b

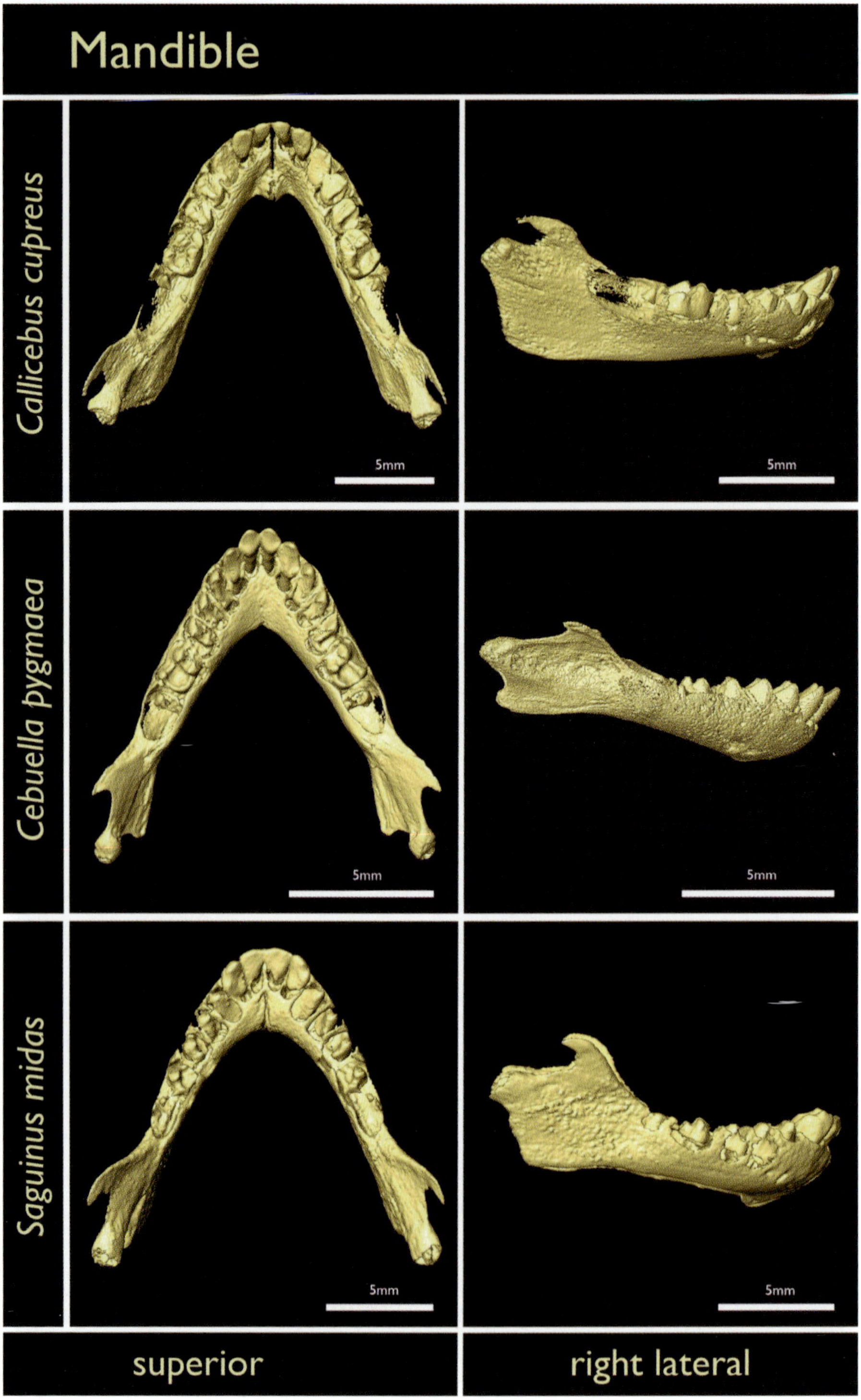
Mandible
Callicebus cupreus
5mm
5mm
Cebuella pygmaea
5mm
5mm
Saguinus midas
5mm
5mm
superior
right lateral

Plate 11

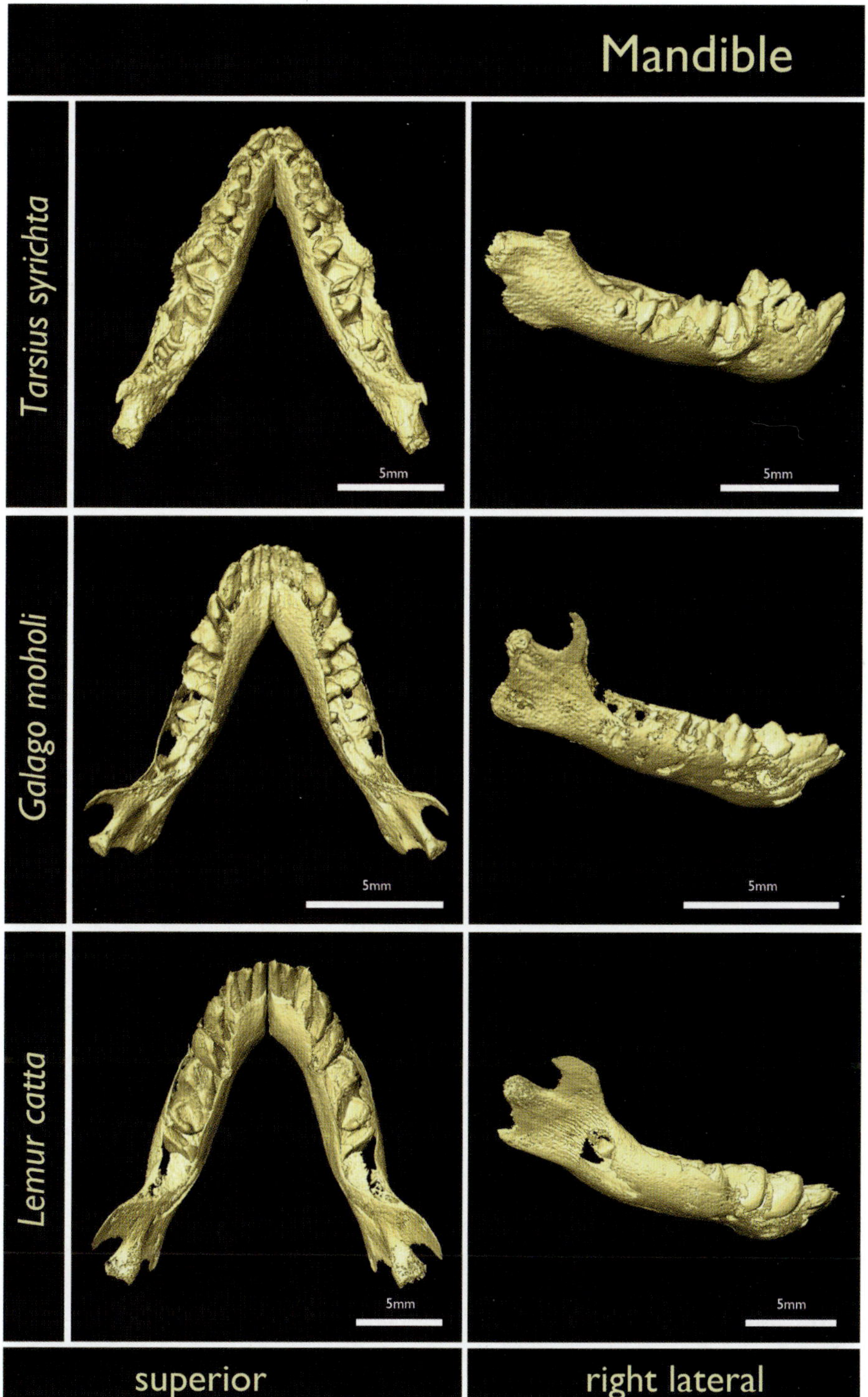
Mandible
Tarsius syrichta
Galago moholi
Lemur catta
5mm
5mm
5mm
5mm
5mm
5mm
superior
right lateral

Plate 12

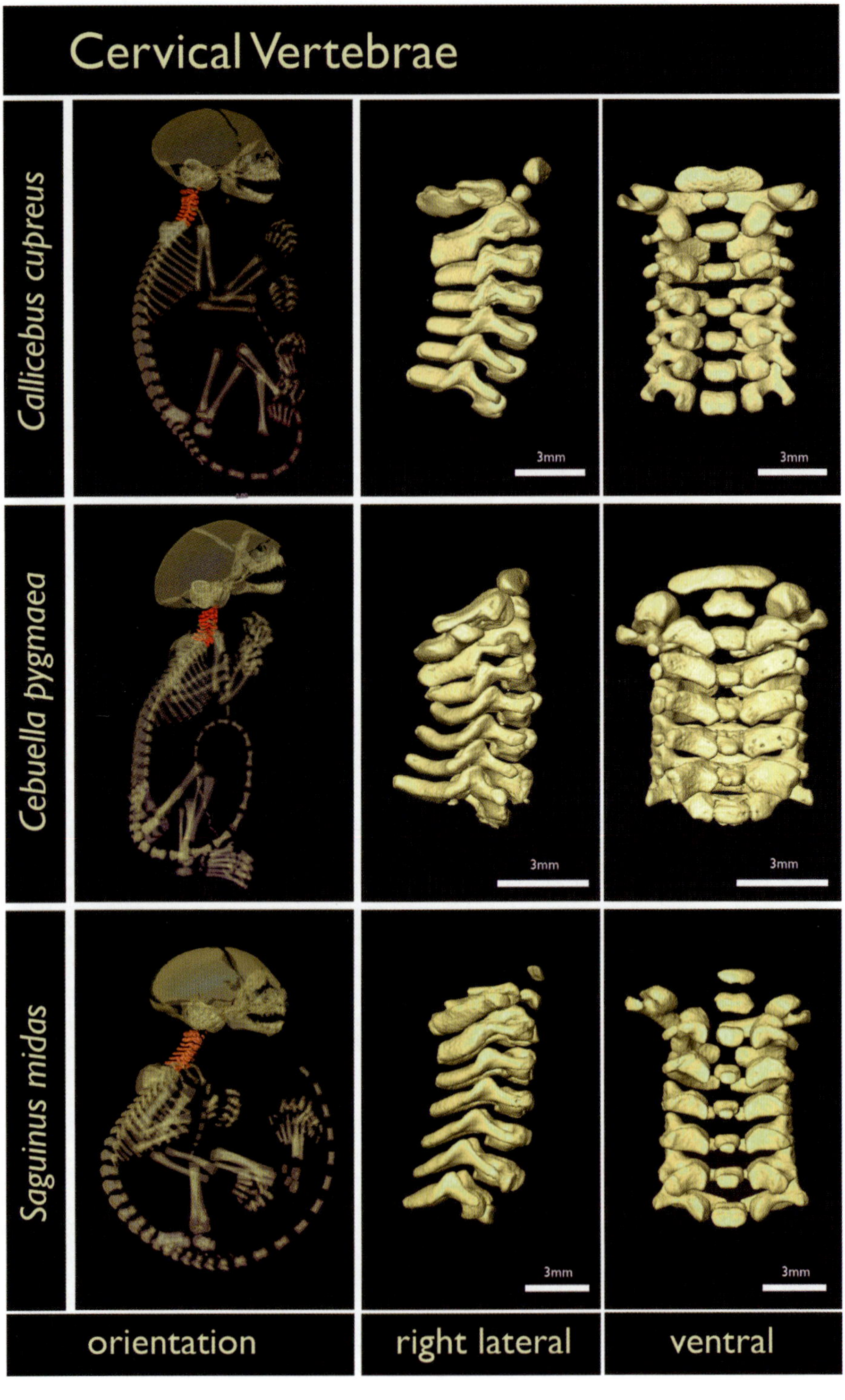
Cervical Vertebrae
Callicebus cupreus
Cebuella pygmaea
Saguinus midas
3mm
orientation
right lateral
ventral

Plate 13

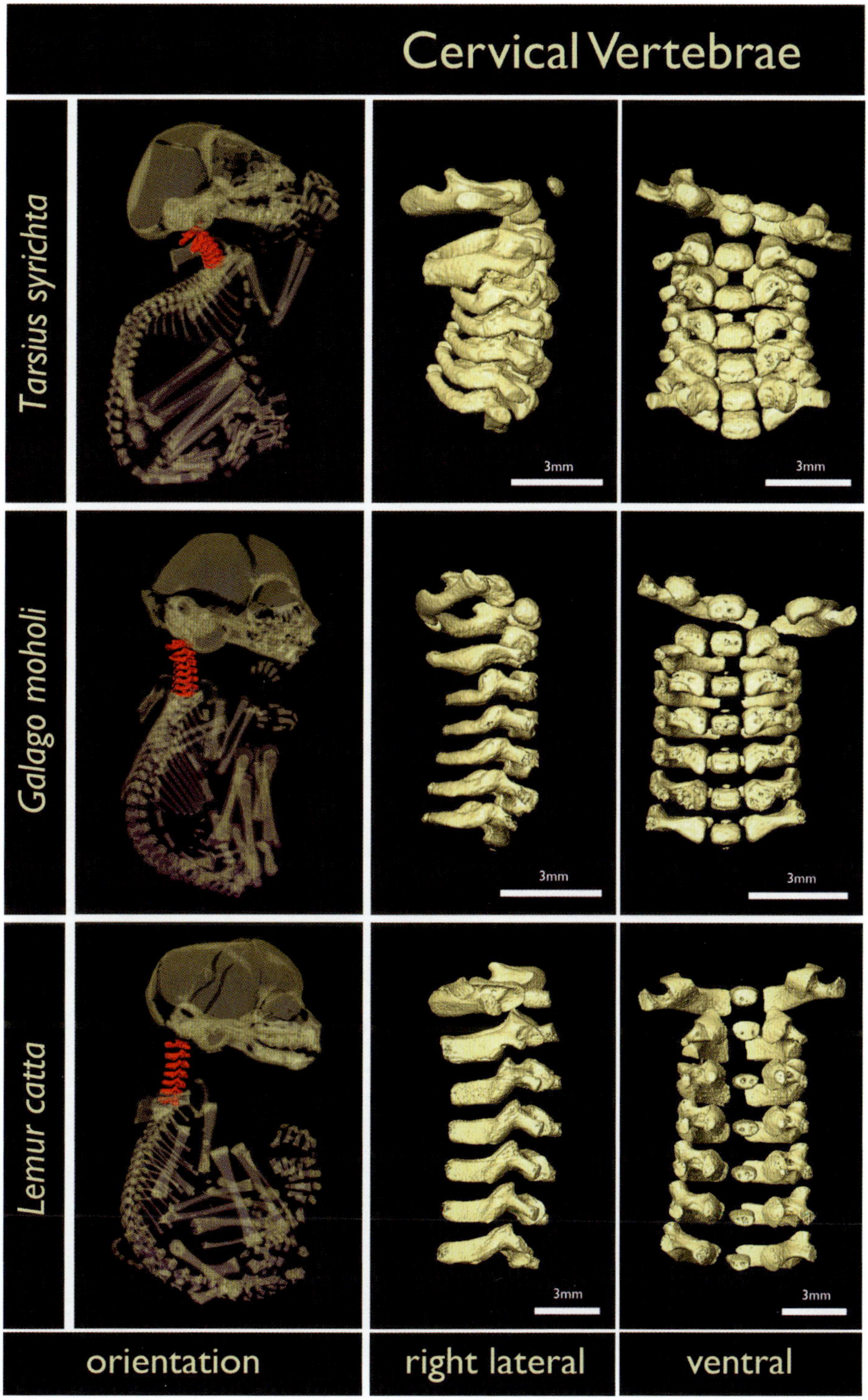
Cervical Vertebrae
Tarsius syrichta
Galago moholi
Lemur catta
3mm
orientation
right lateral
ventral

Plate 14

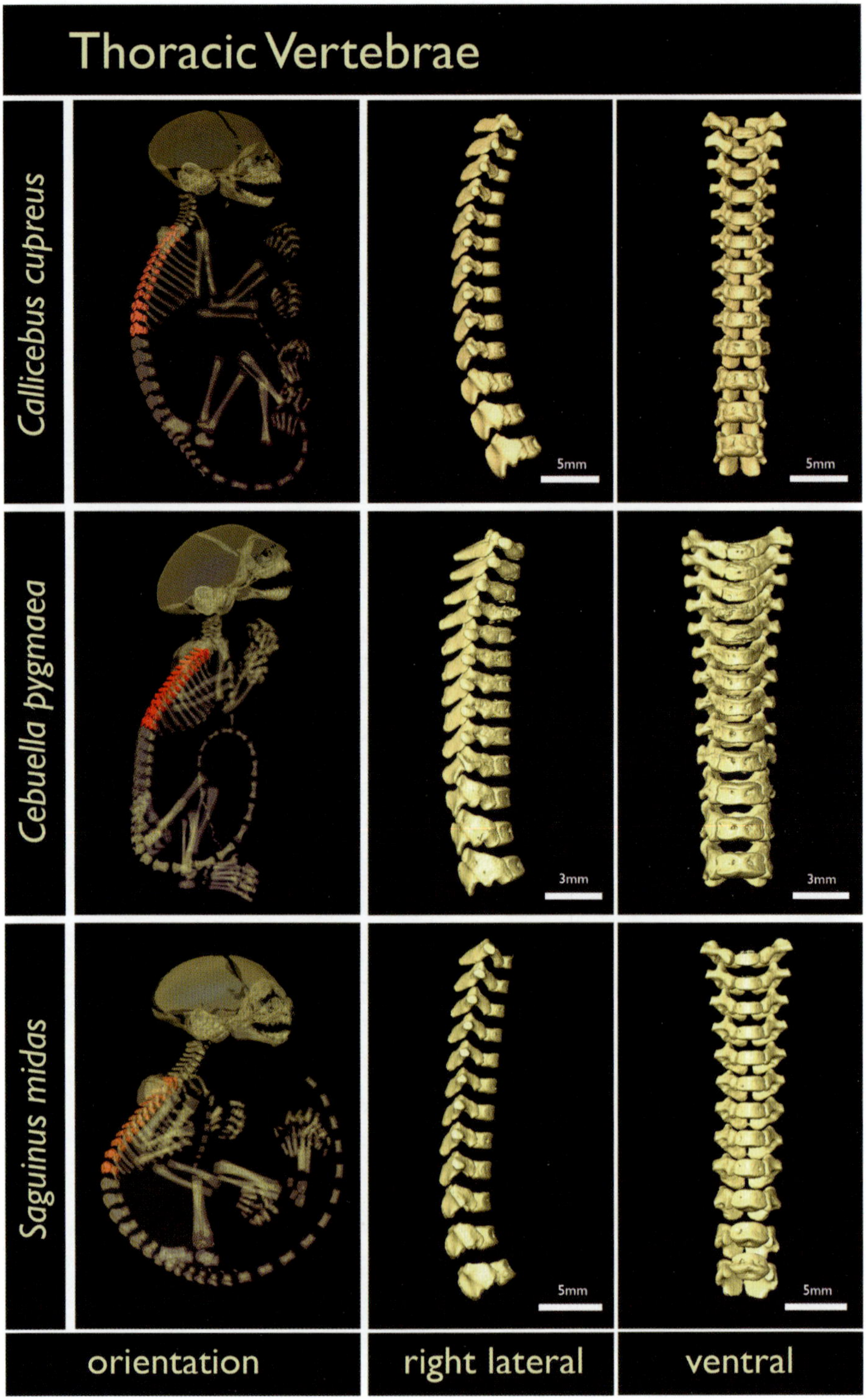
Thoracic Vertebrae
Callicebus cupreus
5mm
5mm
Cebuella pygmaea
3mm
3mm
Saguinus midas
5mm
5mm
orientation
right lateral
ventral

Plate 15

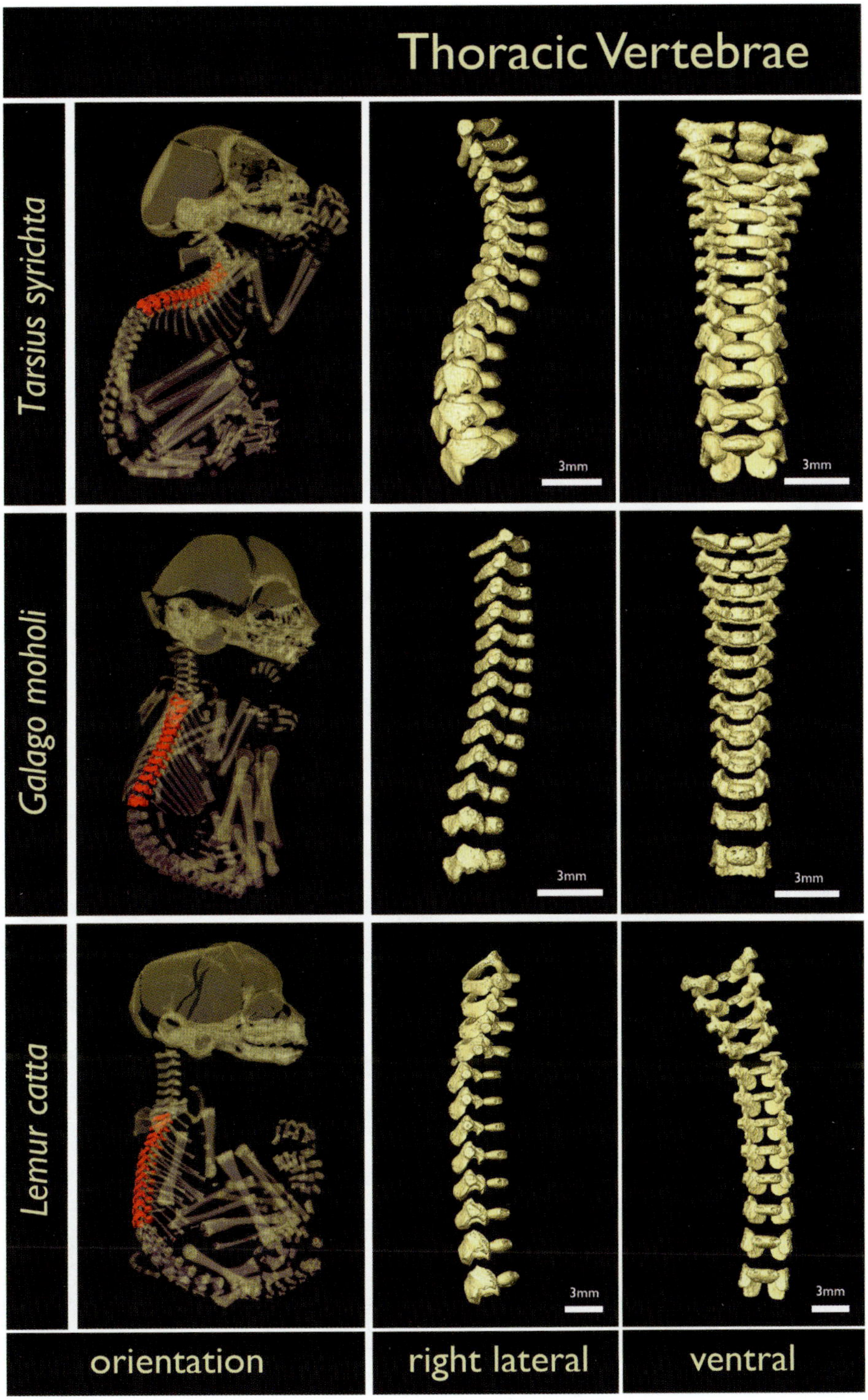
Thoracic Vertebrae
Tarsius syrichta
3mm
3mm
Galago moholi
3mm
3mm
Lemur catta
3mm
3mm
orientation
right lateral
ventral

Plate 16

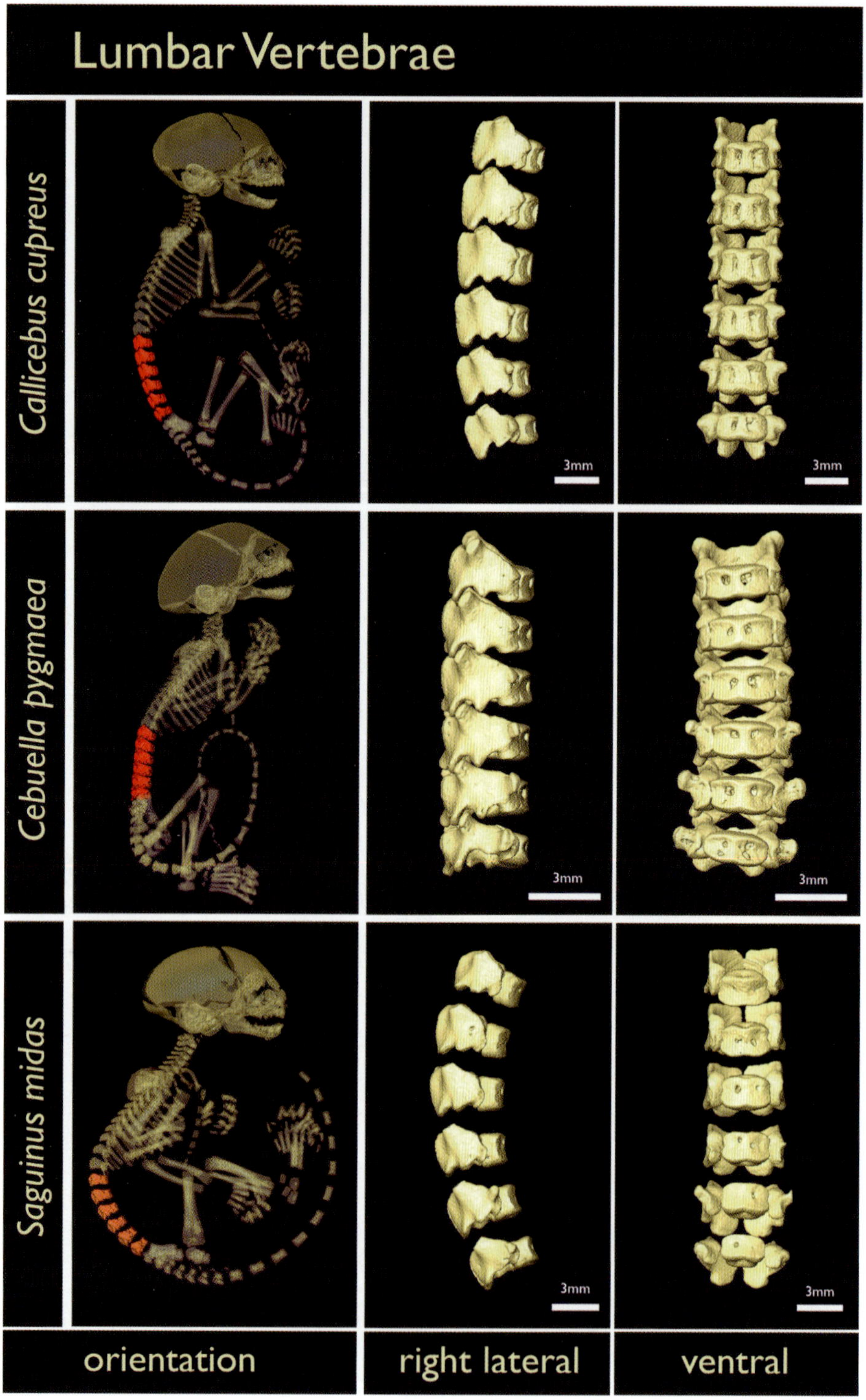

Lumbar Vertebrae
Callicebus cupreus
Cebuella pygmaea
Saguinus midas
3mm
3mm
3mm
3mm
3mm
3mm
orientation
right lateral
ventral

Plate 17

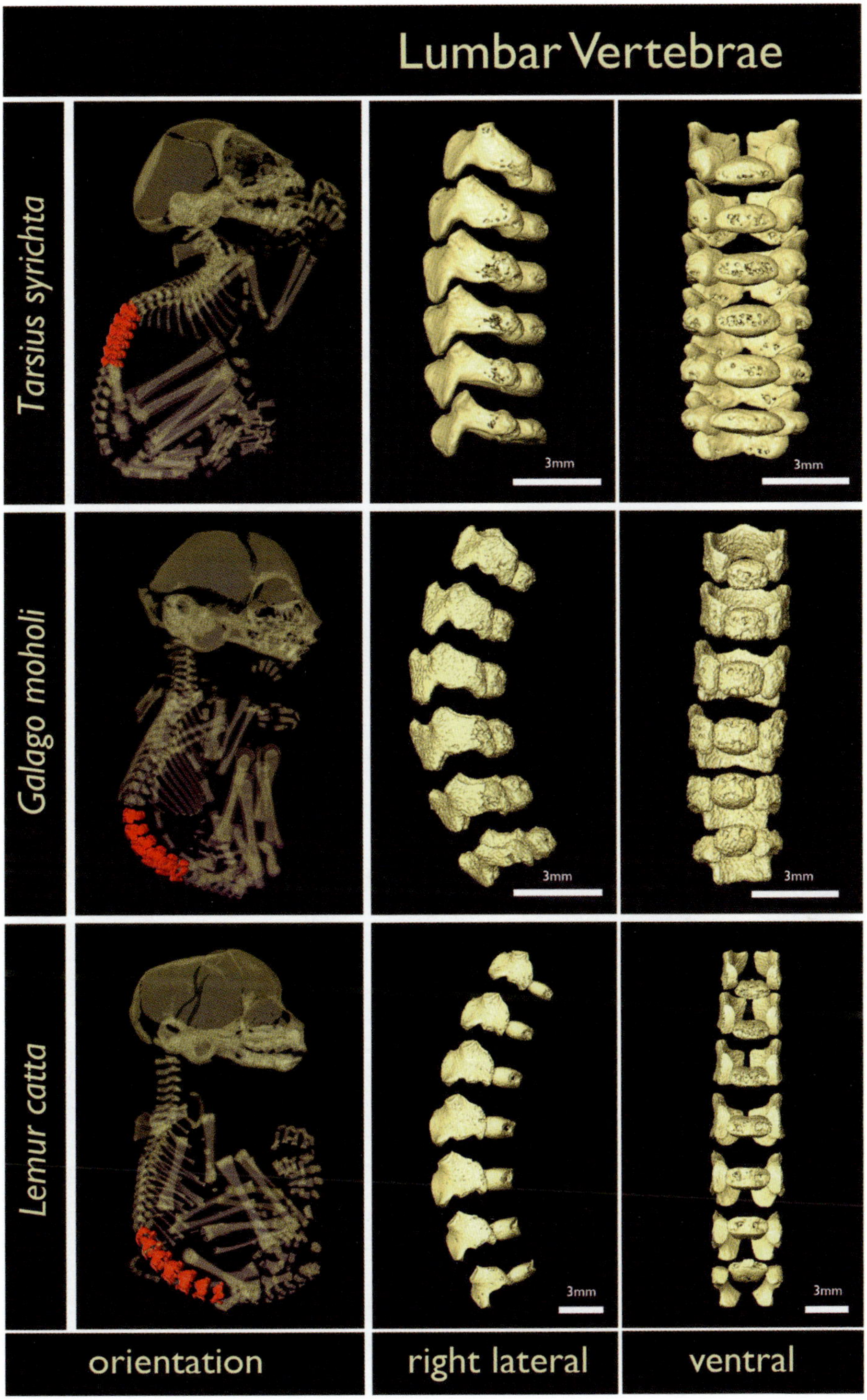
Lumbar Vertebrae
Tarsius syrichta
Galago moholi
Lemur catta
3mm
orientation
right lateral
ventral

Plate 18

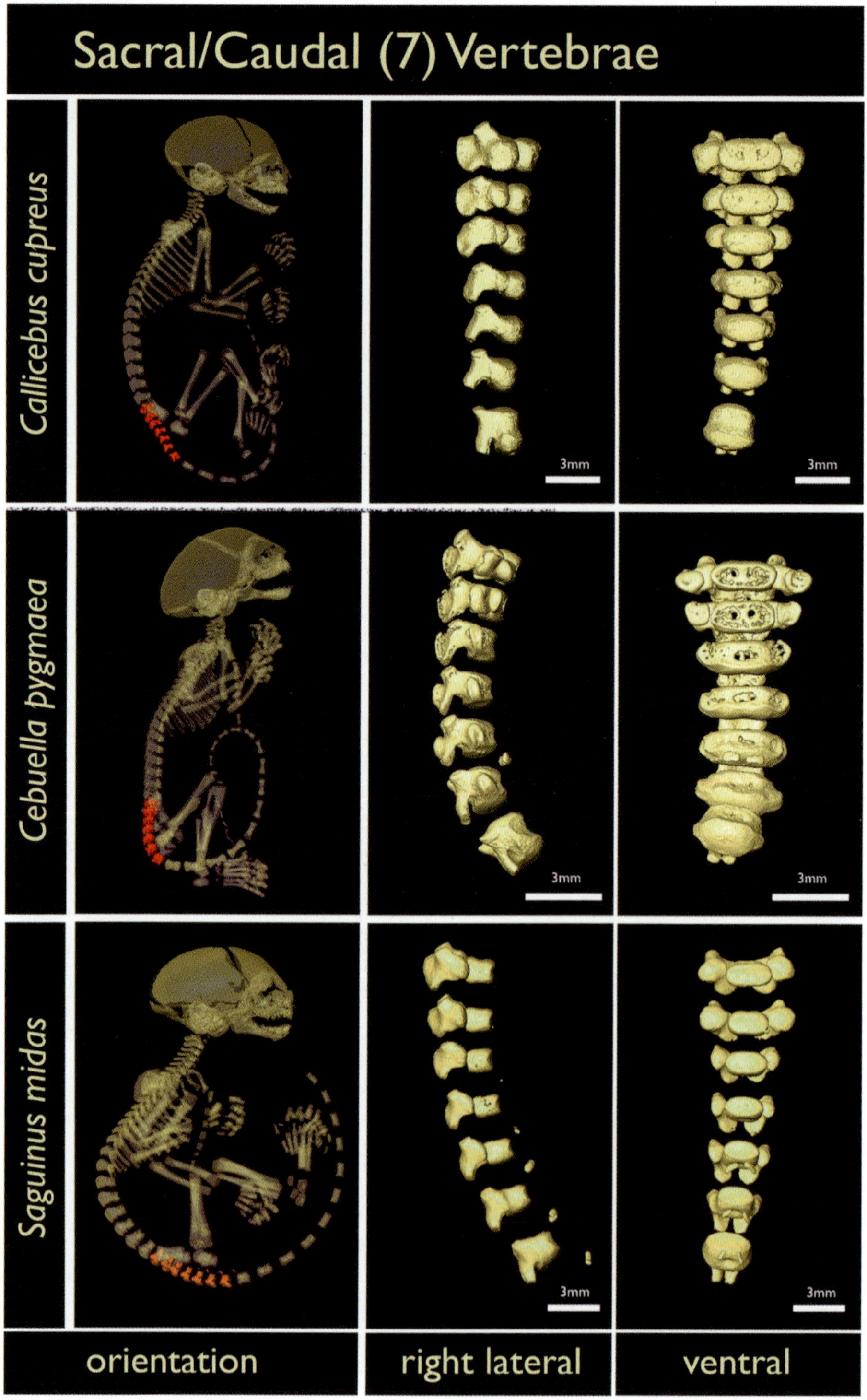

Sacral/Caudal (7) Vertebrae
Callicebus cupreus
Cebuella pygmaea
Saguinus midas
3mm
orientation
right lateral
ventral

Plate 19

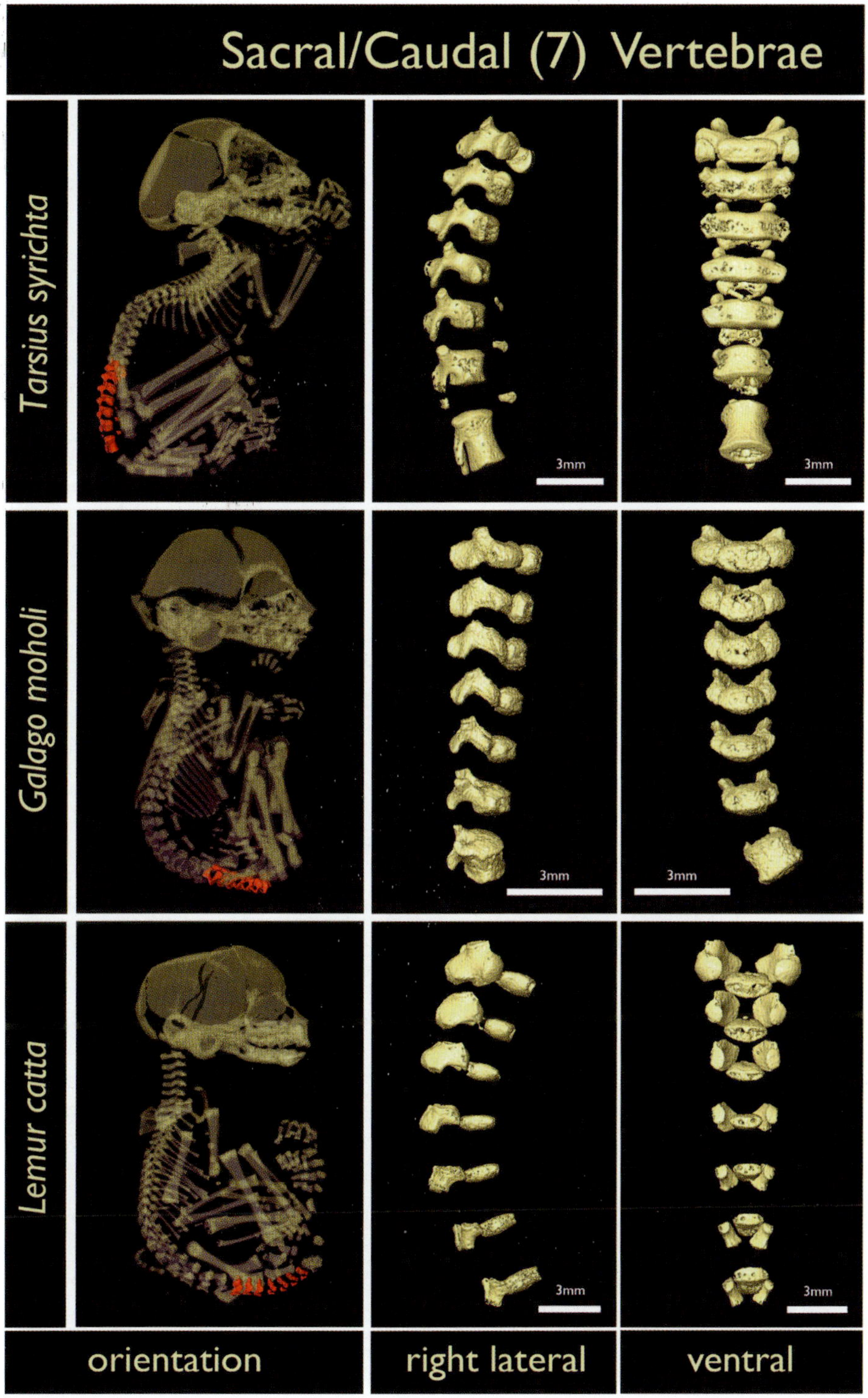
Sacral/Caudal (7) Vertebrae
Tarsius syrichta
Galago moholi
Lemur catta
3mm
orientation
right lateral
ventral

Plate 20

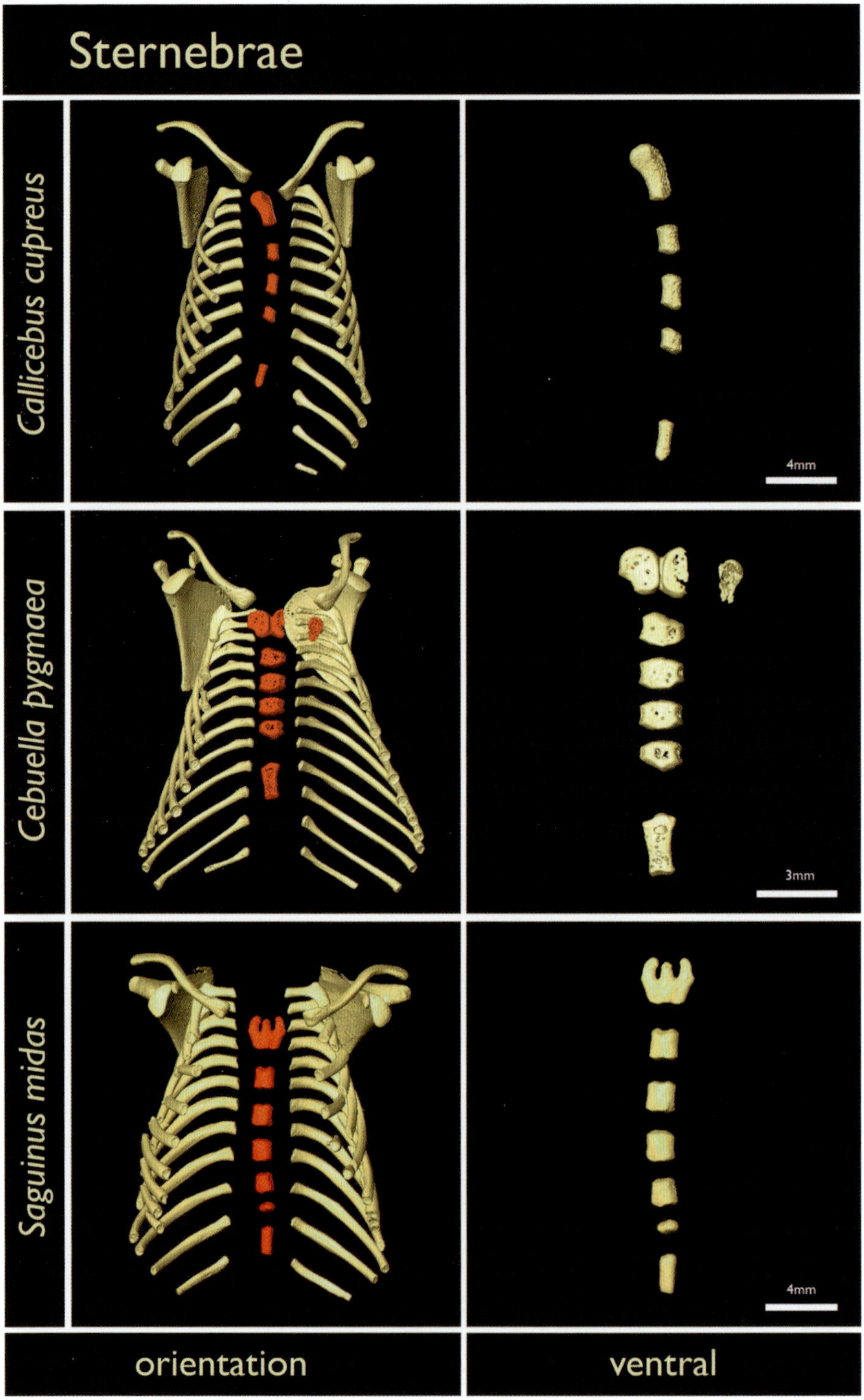
Sternebrae
Callicebus cupreus
4mm
Cebuella pygmaea
3mm
Saguinus midas
4mm
orientation
ventral

Plate 21

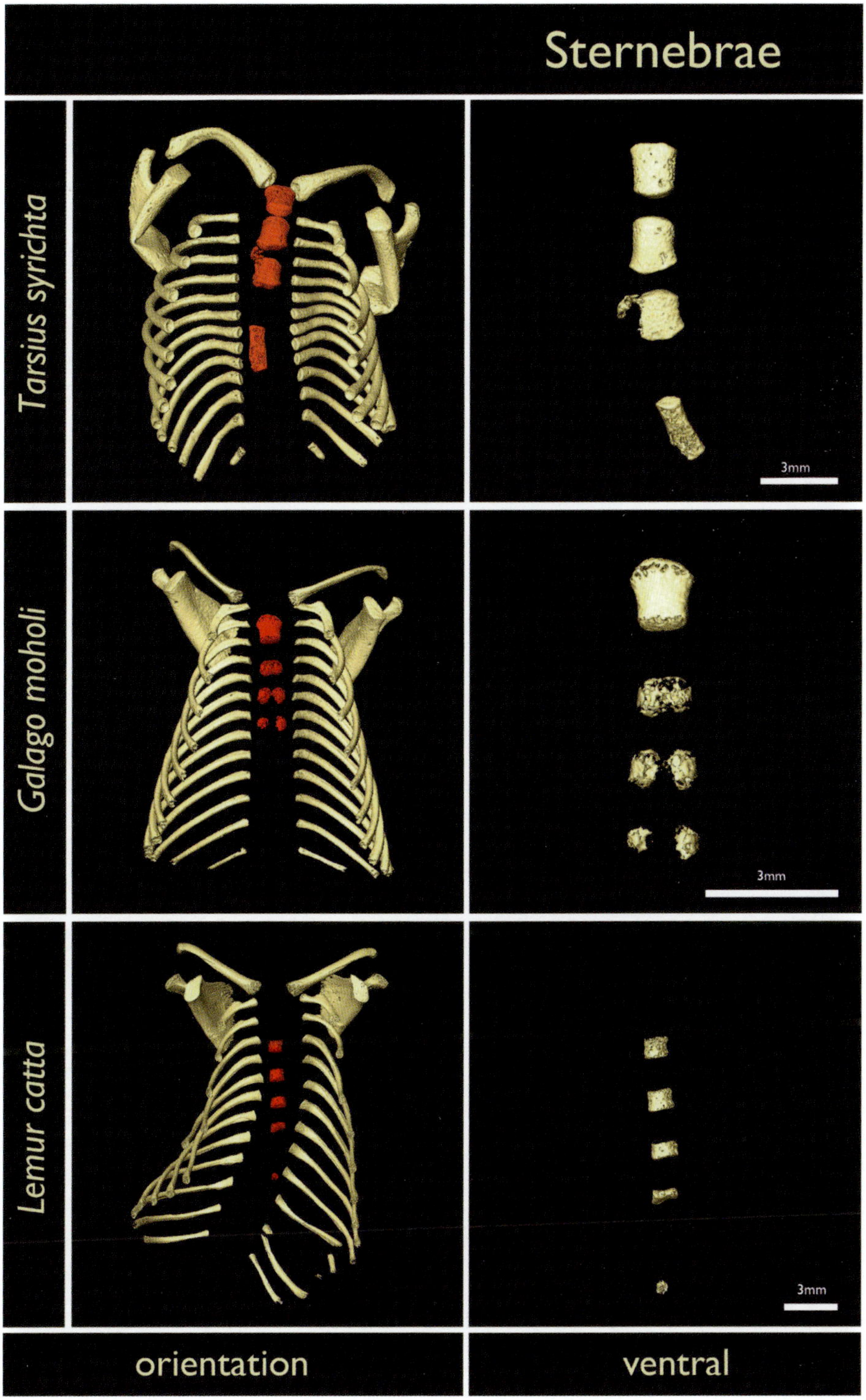

Sternebrae
Tarsius syrichta
3mm
Galago moholi
3mm
Lemur catta
3mm
orientation
ventral

Plate 22

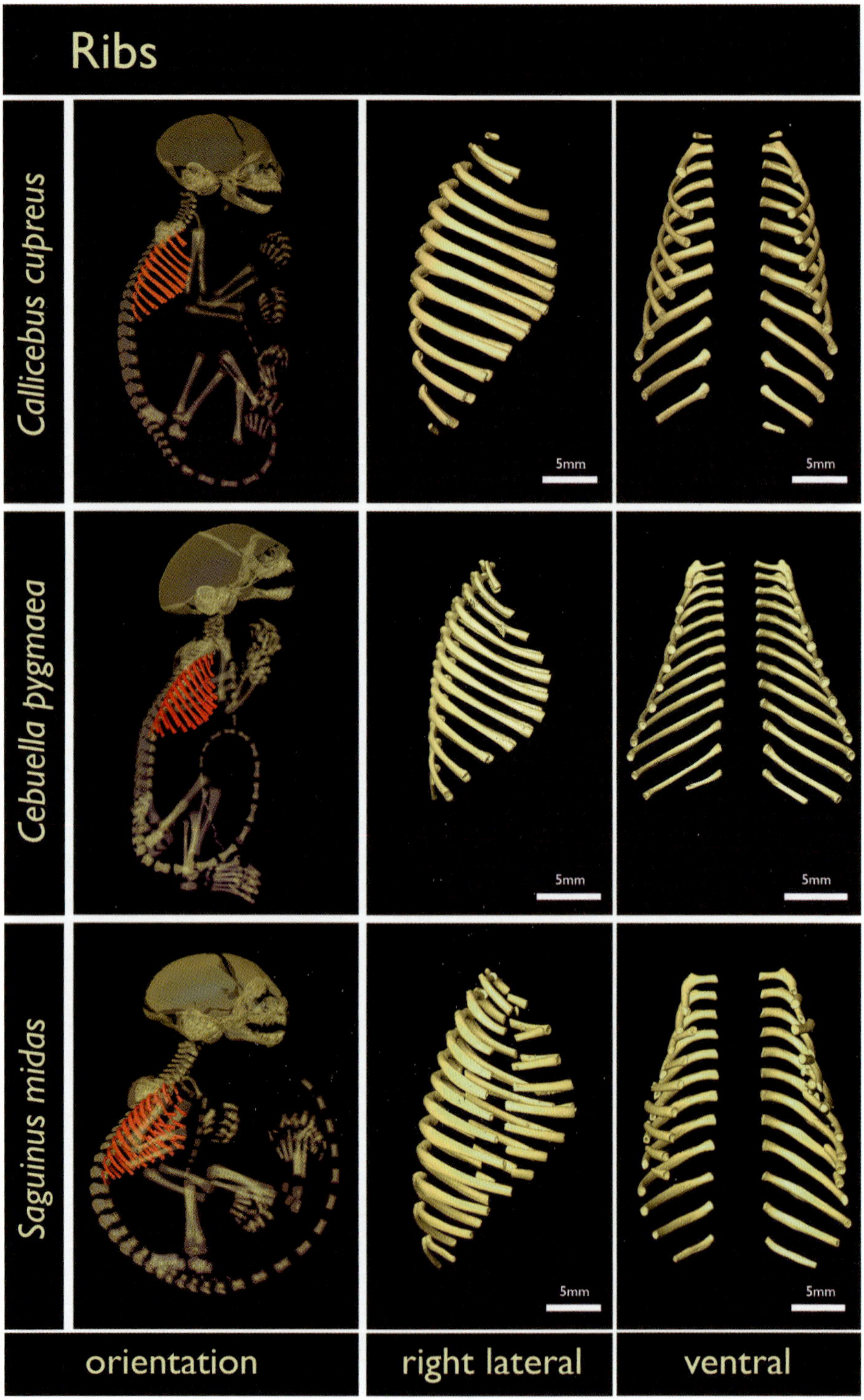
Ribs
Callicebus cupreus
Cebuella pygmaea
Saguinus midas
5mm
5mm
5mm
5mm
5mm
5mm
orientation
right lateral
ventral

Plate 23

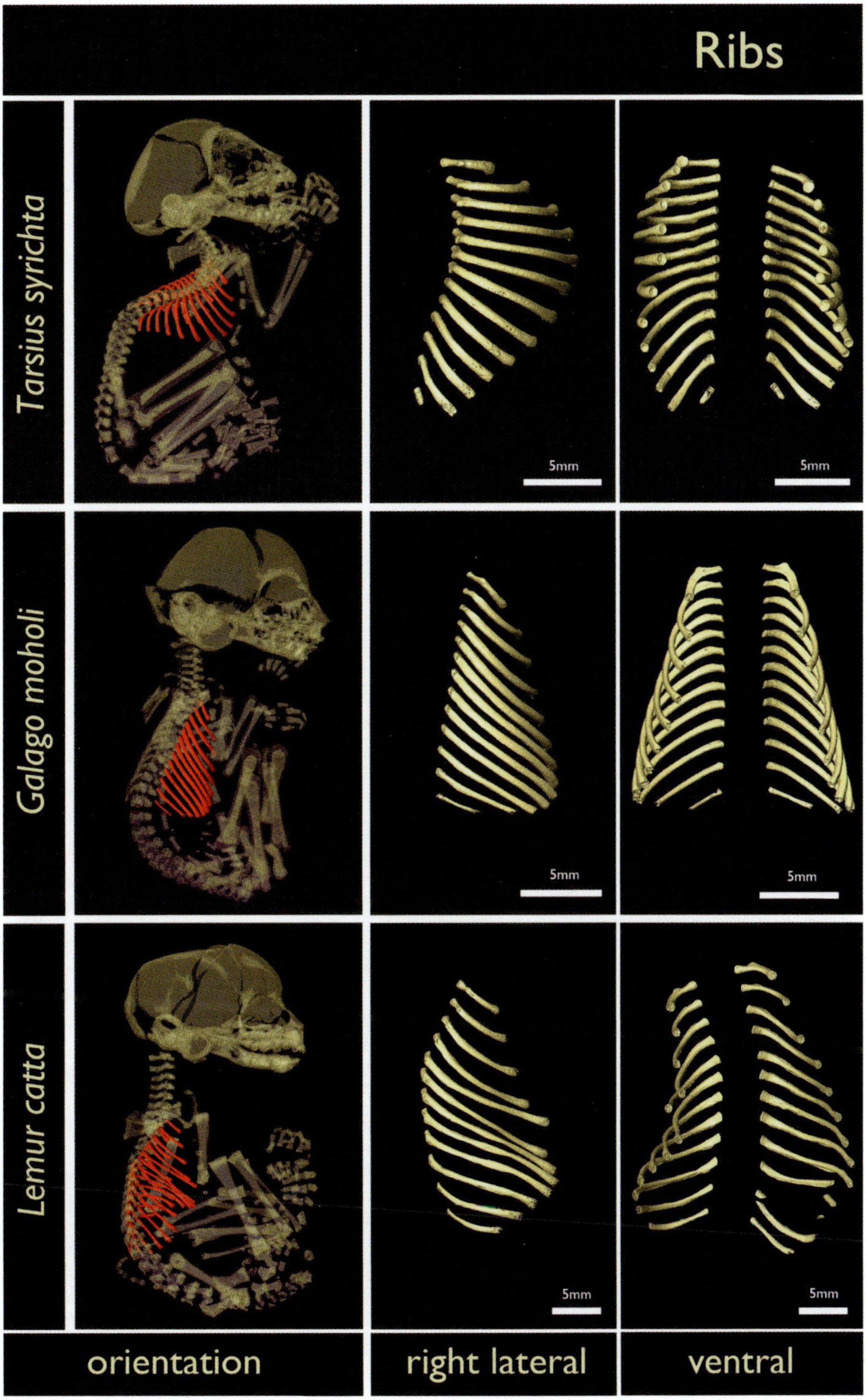
Ribs
Tarsius syrichta
5mm
5mm
Galago moholi
5mm
5mm
Lemur catta
5mm
5mm
orientation
right lateral
ventral

Plate 24

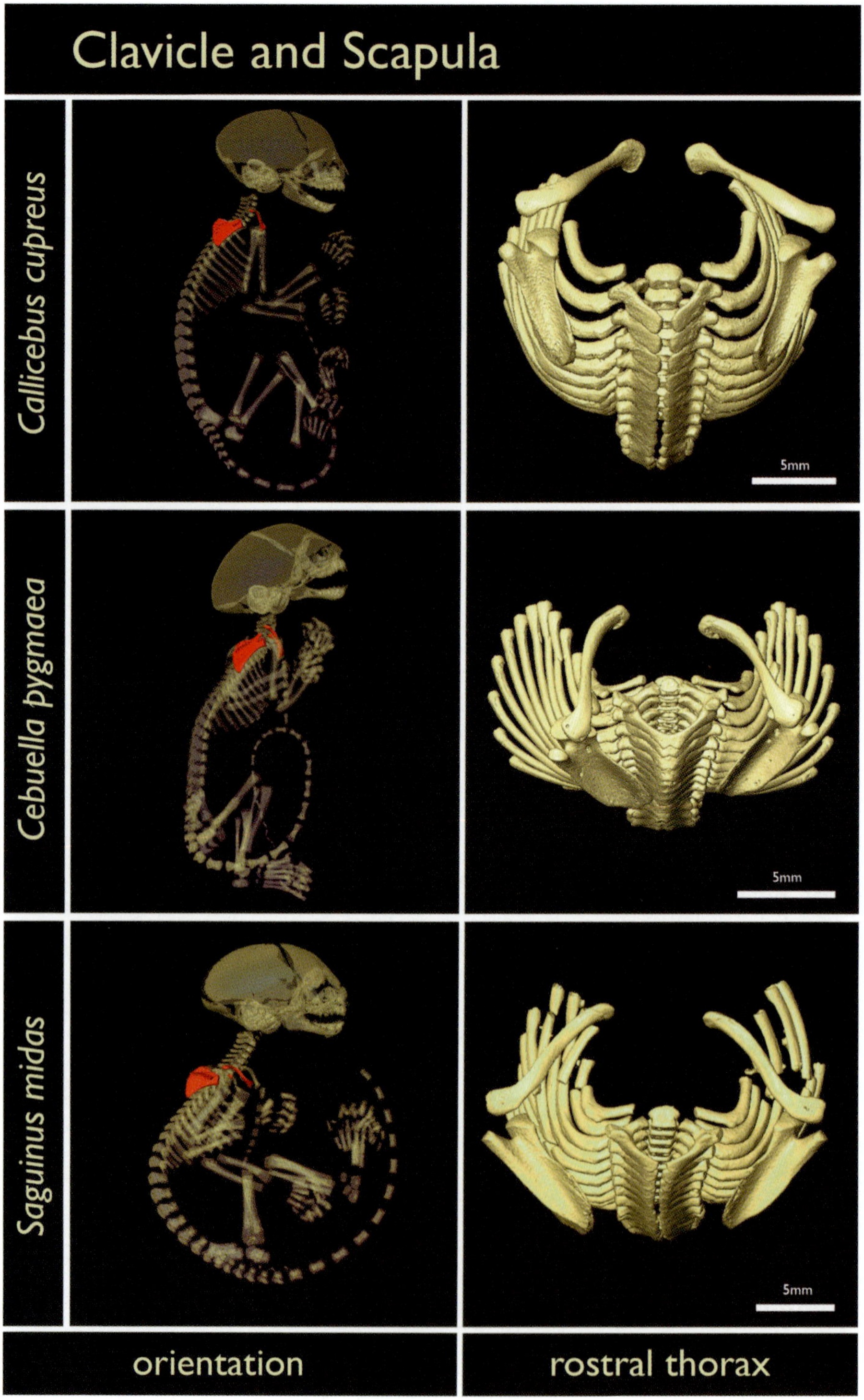
Clavicle and Scapula
Callicebus cupreus
Cebuella pygmaea
Saguinus midas
5mm
5mm
5mm
orientation
rostral thorax

Plate 25a

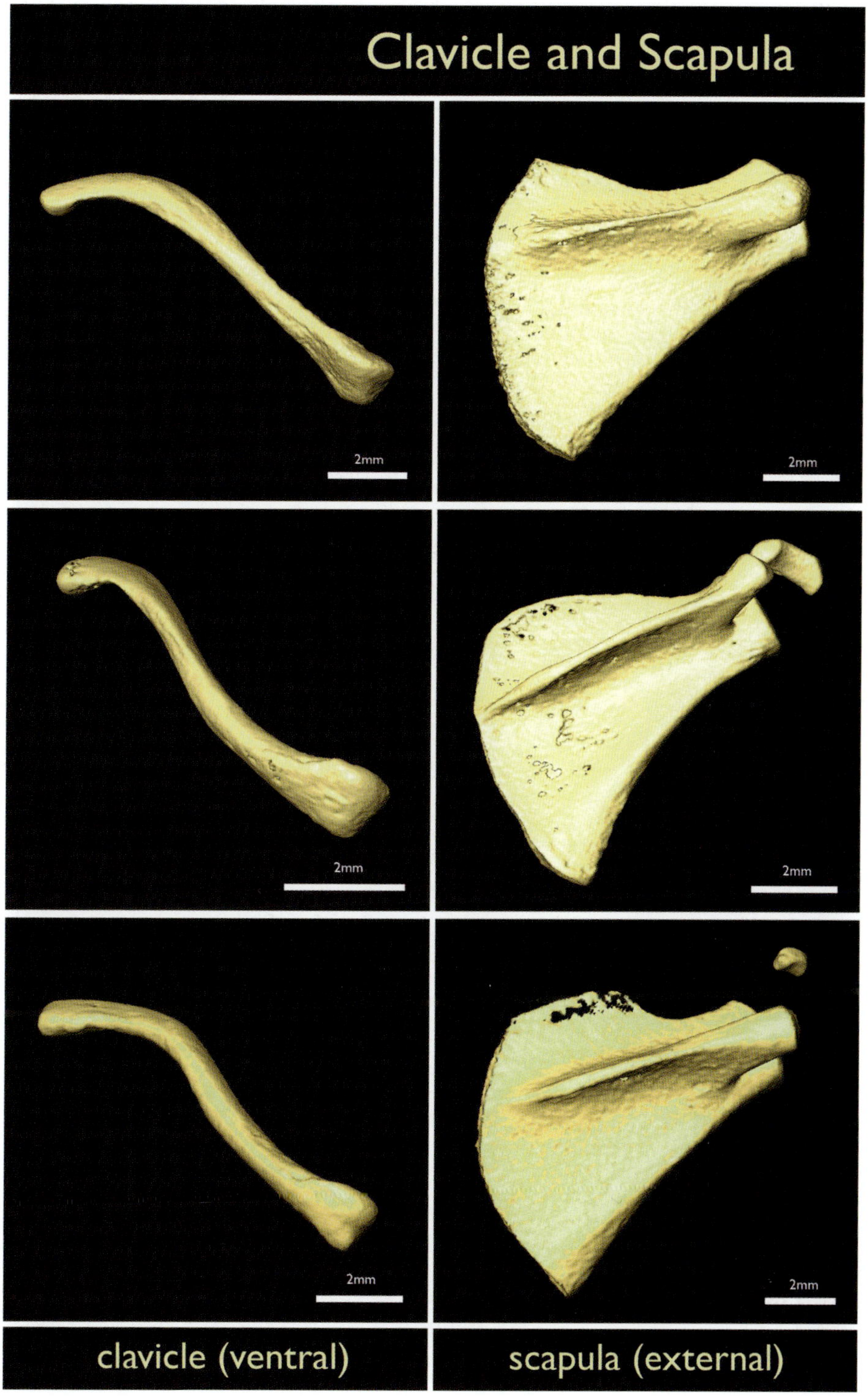
Clavicle and Scapula
2mm
2mm
2mm
2mm
2mm
2mm
clavicle (ventral)
scapula (external)

Plate 25b

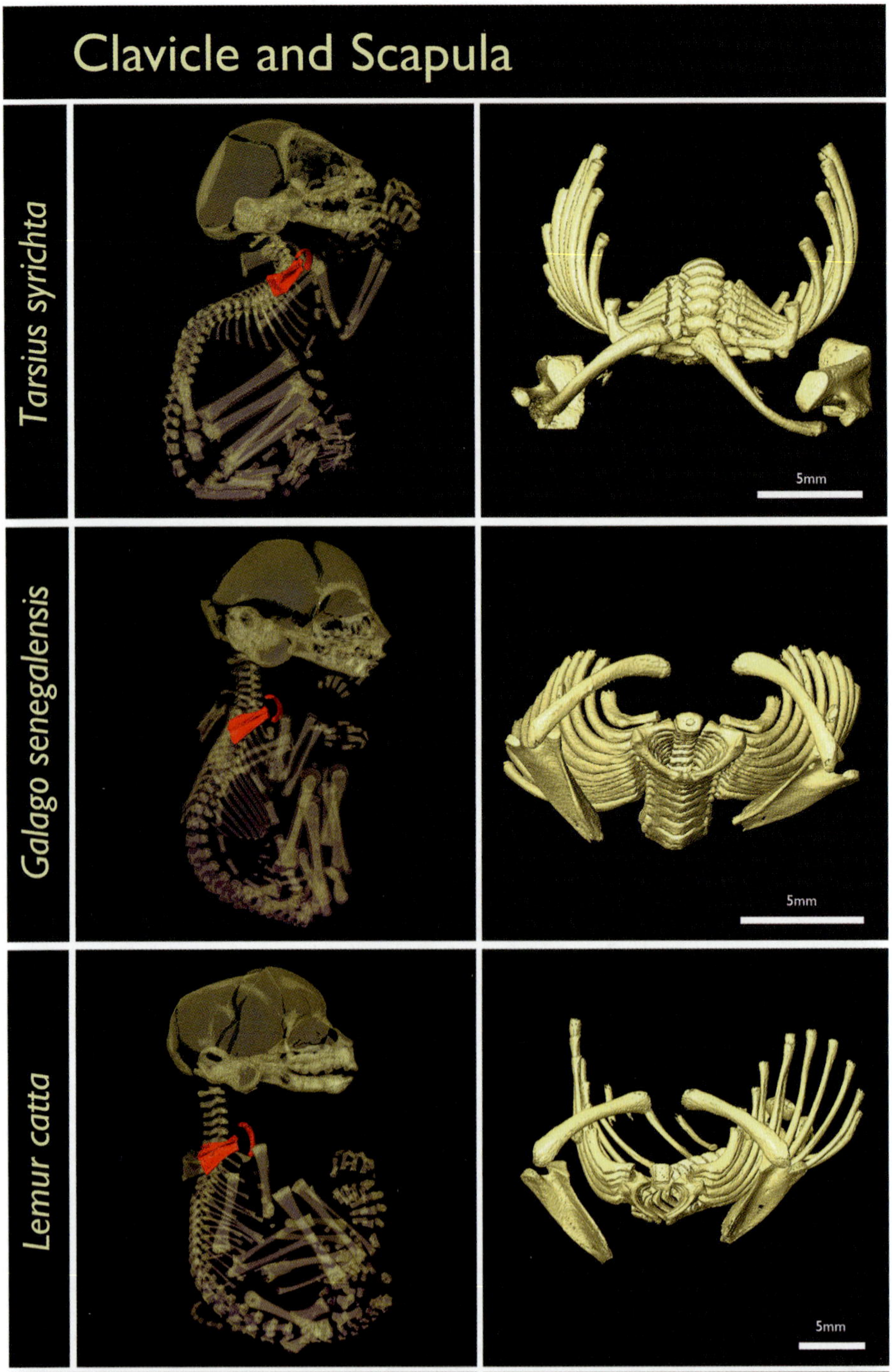
Clavicle and Scapula
Tarsius syrichta
Galago senegalensis
Lemur catta
5mm
5mm
5mm
orientation
rostral thorax

Plate 26a

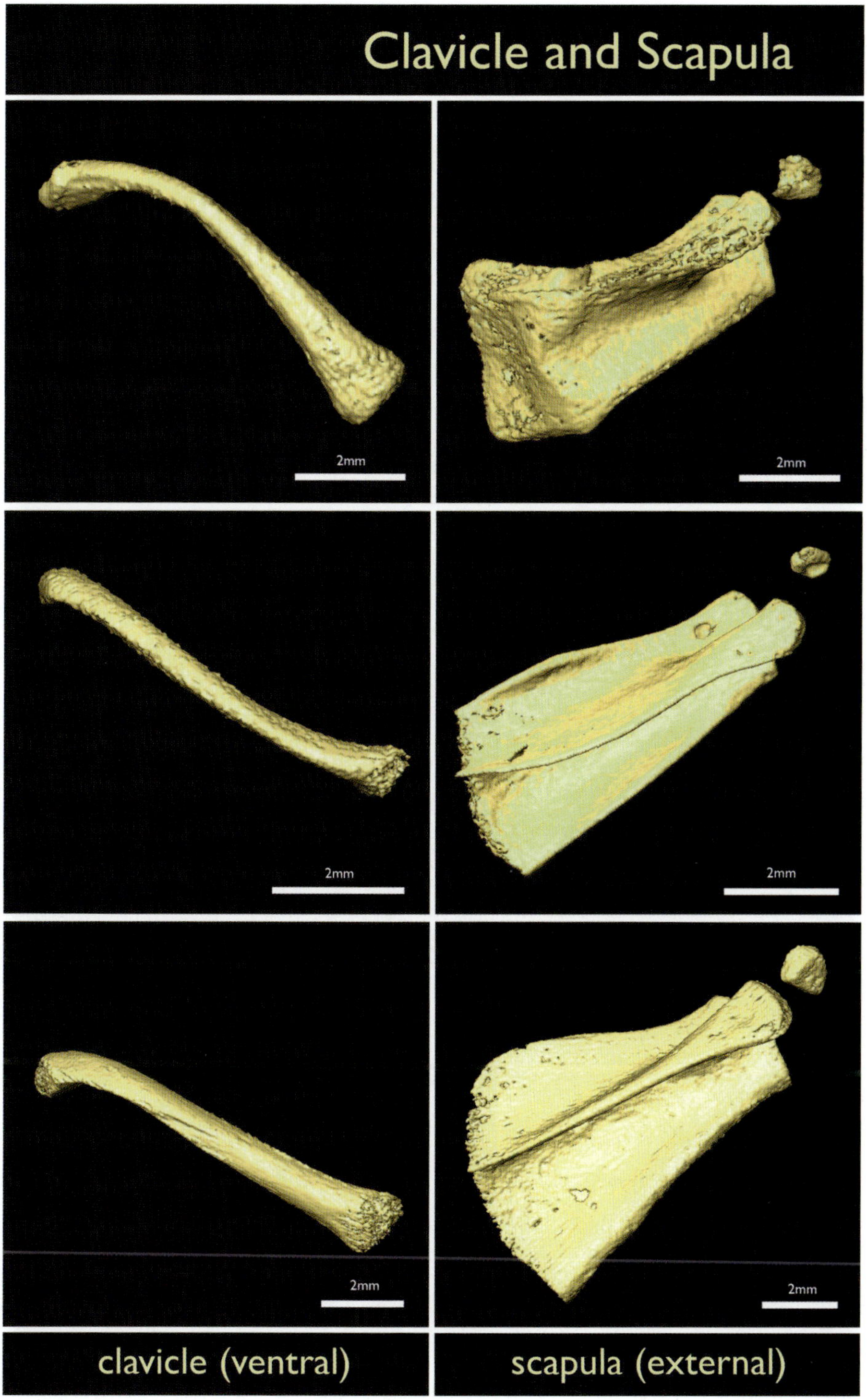
Clavicle and Scapula
2mm
2mm
2mm
2mm
2mm
2mm
clavicle (ventral)
scapula (external)

Plate 26b

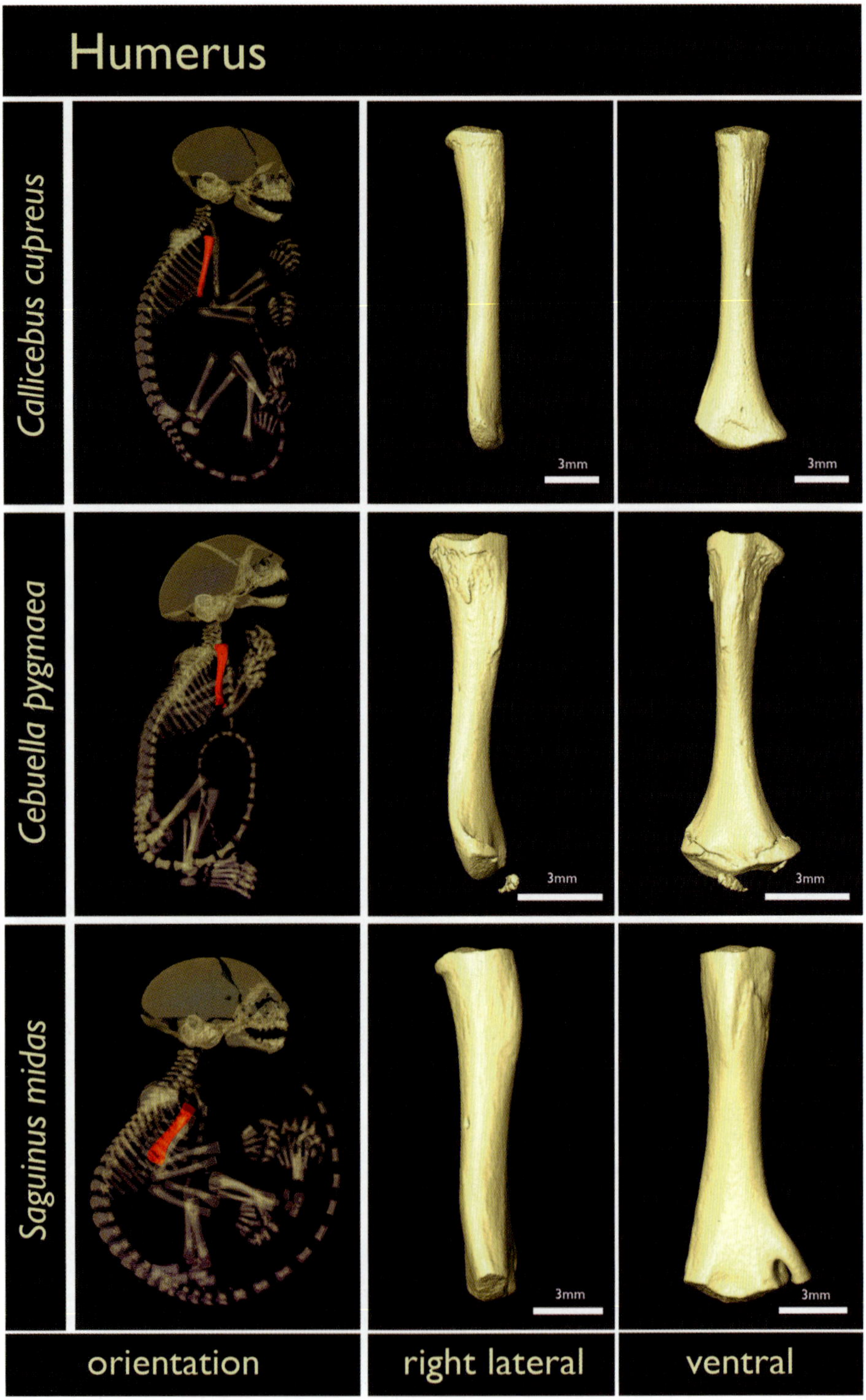
Humerus
Callicebus cupreus
Cebuella pygmaea
Saguinus midas
3mm
orientation
right lateral
ventral

Plate 27

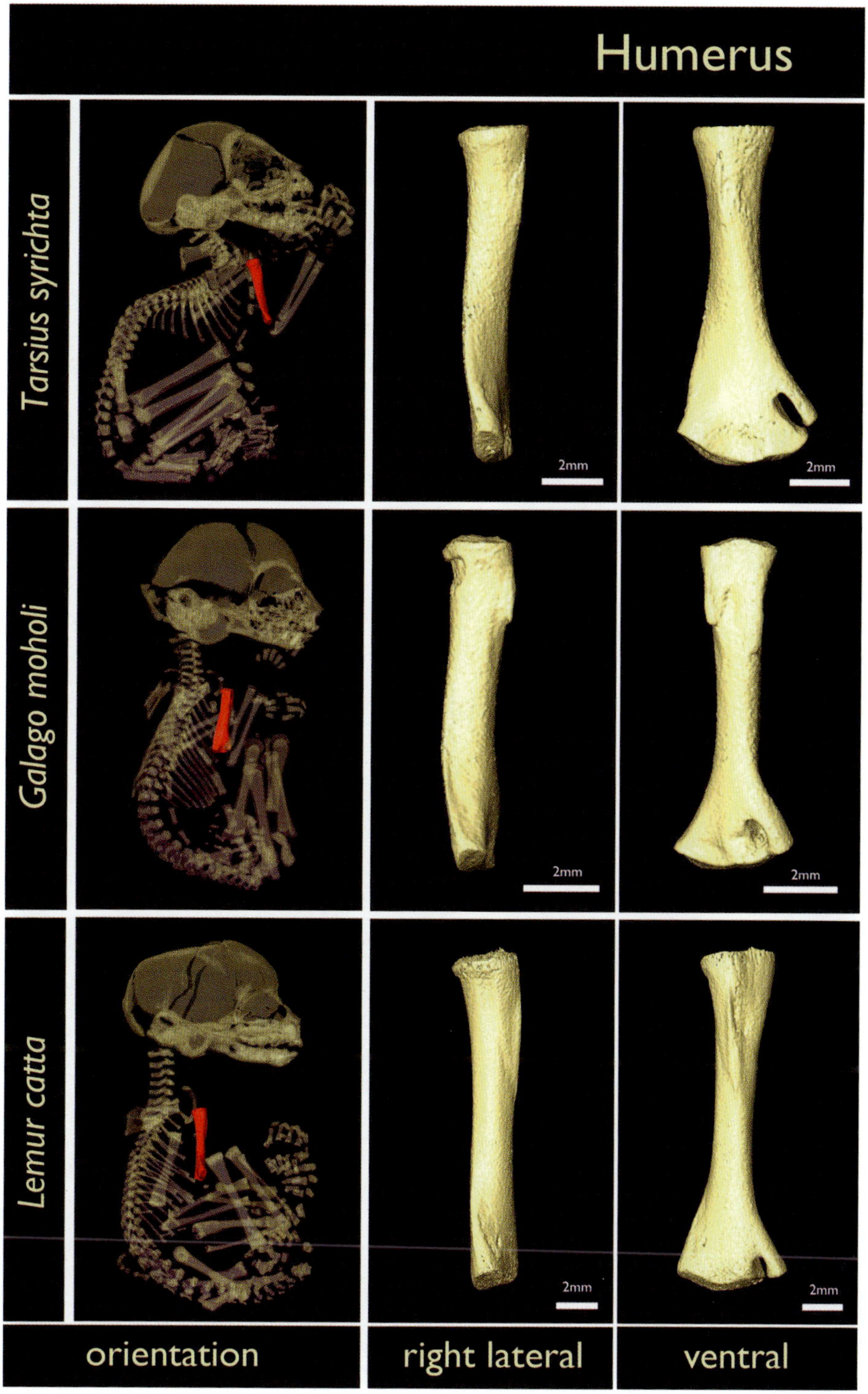
Humerus
Tarsius syrichta
2mm
2mm
Galago moholi
2mm
2mm
Lemur catta
2mm
2mm
orientation
right lateral
ventral

Plate 28

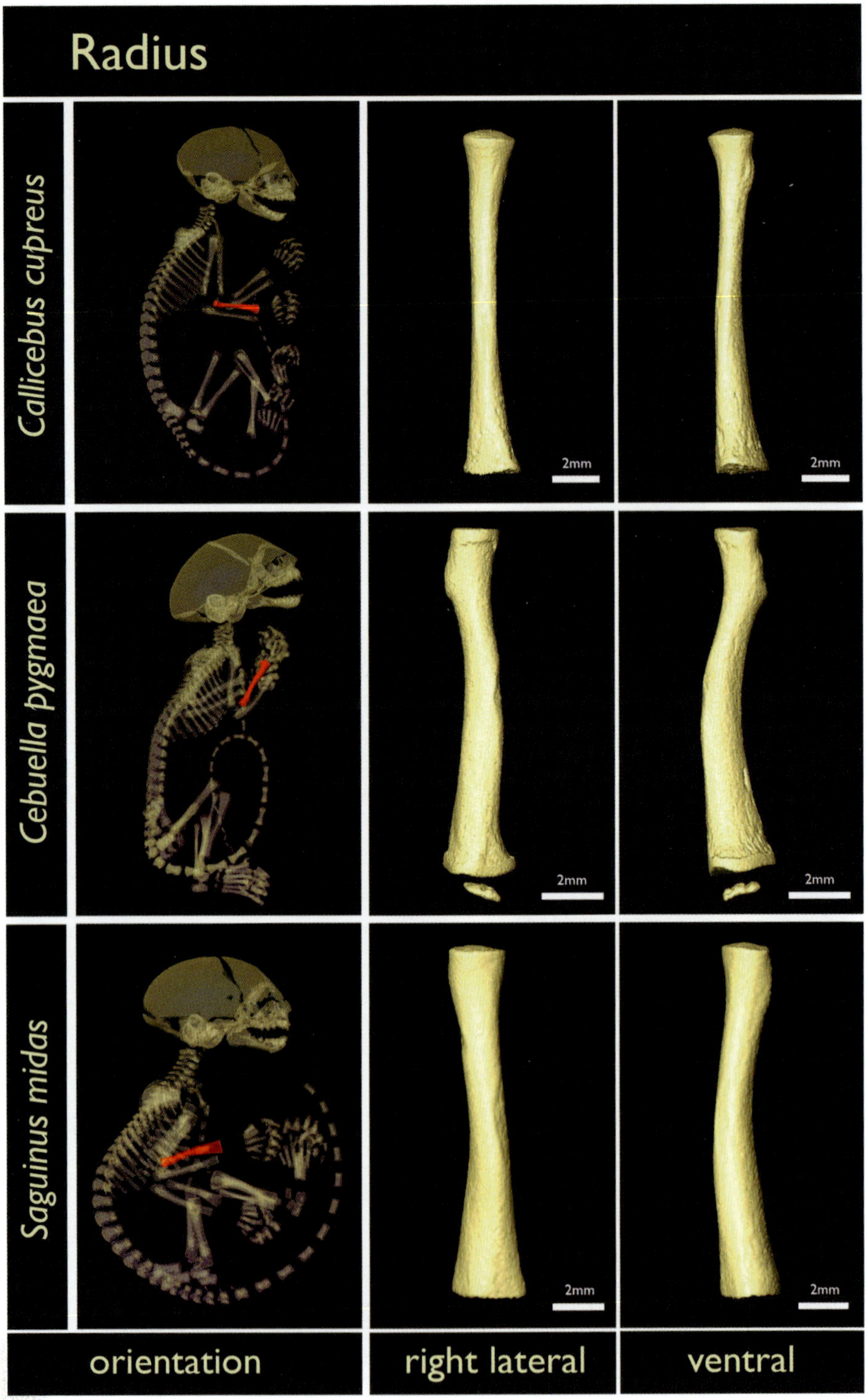
Radius
Callicebus cupreus
Cebuella pygmaea
Saguinus midas
2mm
orientation
right lateral
ventral

Plate 29

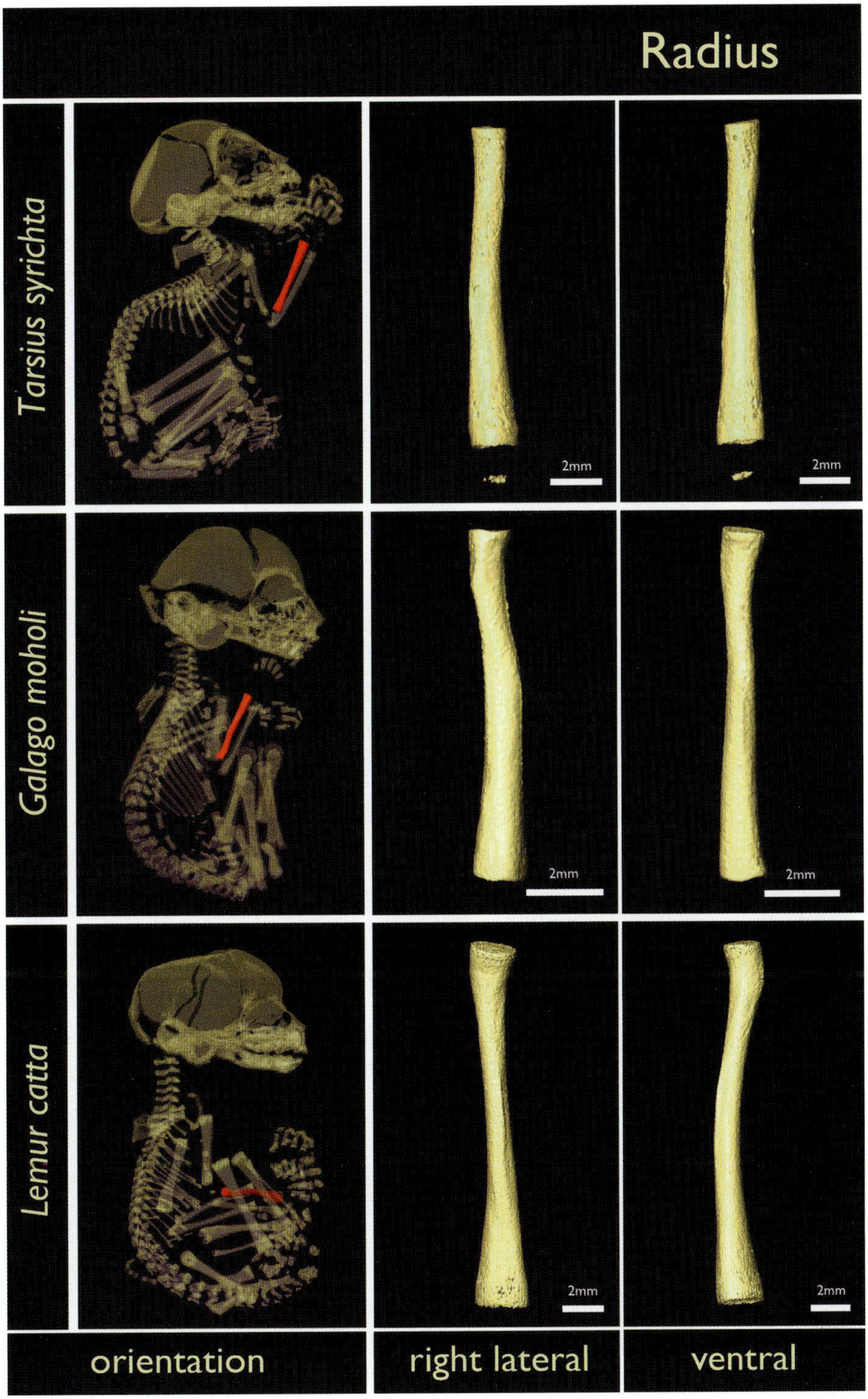
Radius
Tarsius syrichta
2mm
2mm
Galago moholi
2mm
2mm
Lemur catta
2mm
2mm
orientation
right lateral
ventral

Plate 30

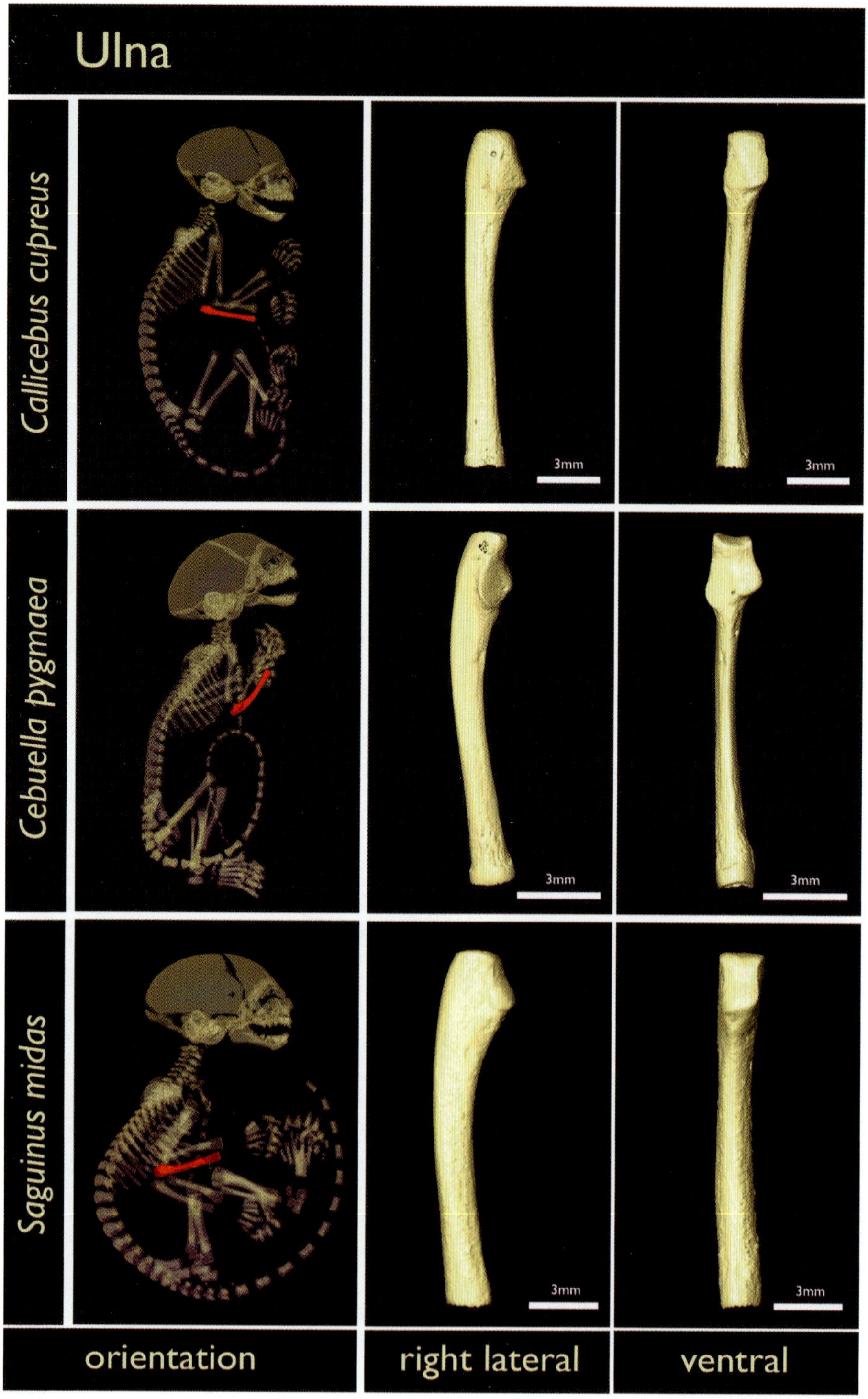
Ulna
Callicebus cupreus
Cebuella pygmaea
Saguinus midas
3mm
orientation
right lateral
ventral

Plate 31

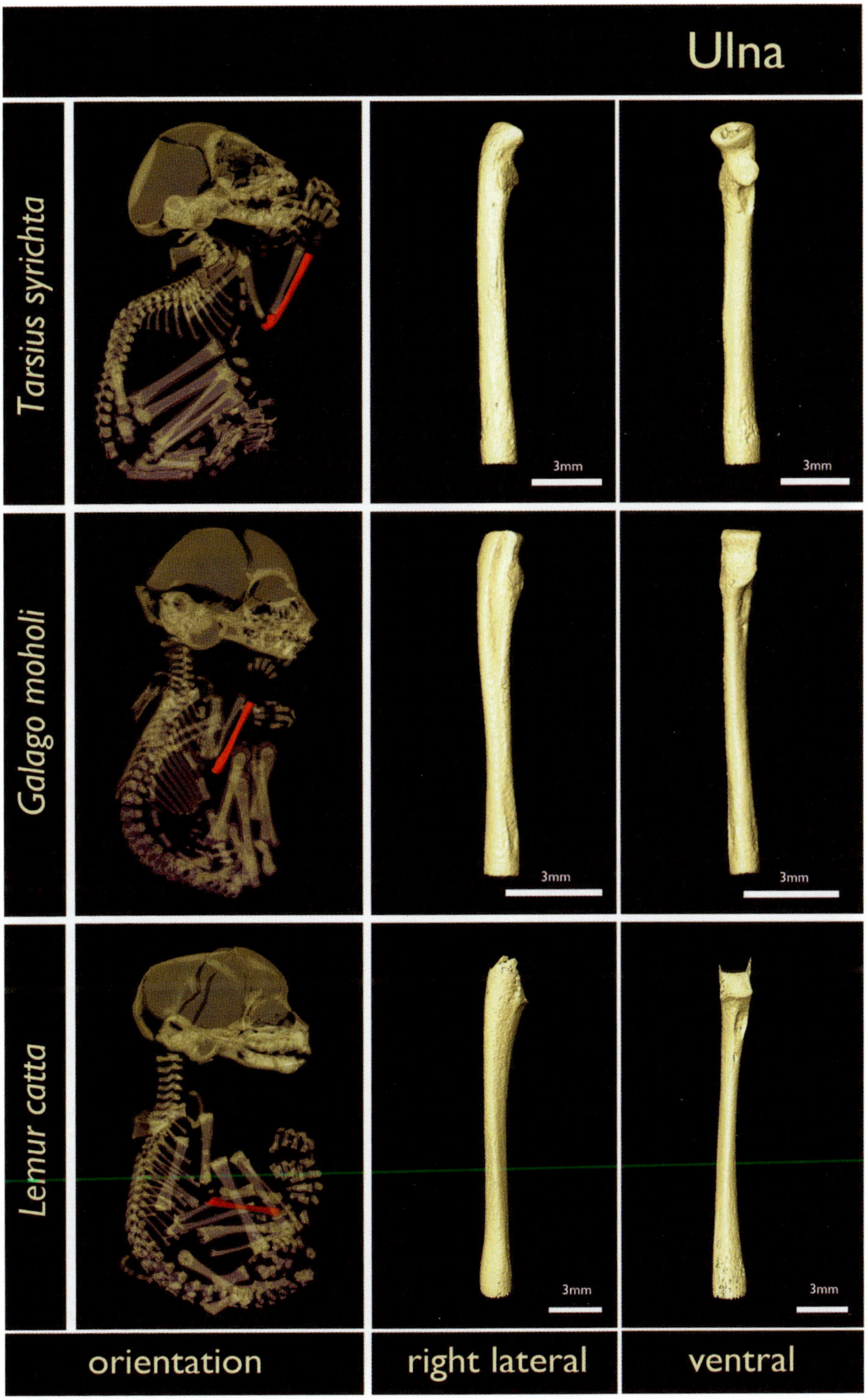
Ulna
Tarsius syrichta
3mm
3mm
Galago moholi
3mm
3mm
Lemur catta
3mm
3mm
orientation
right lateral
ventral

Plate 32

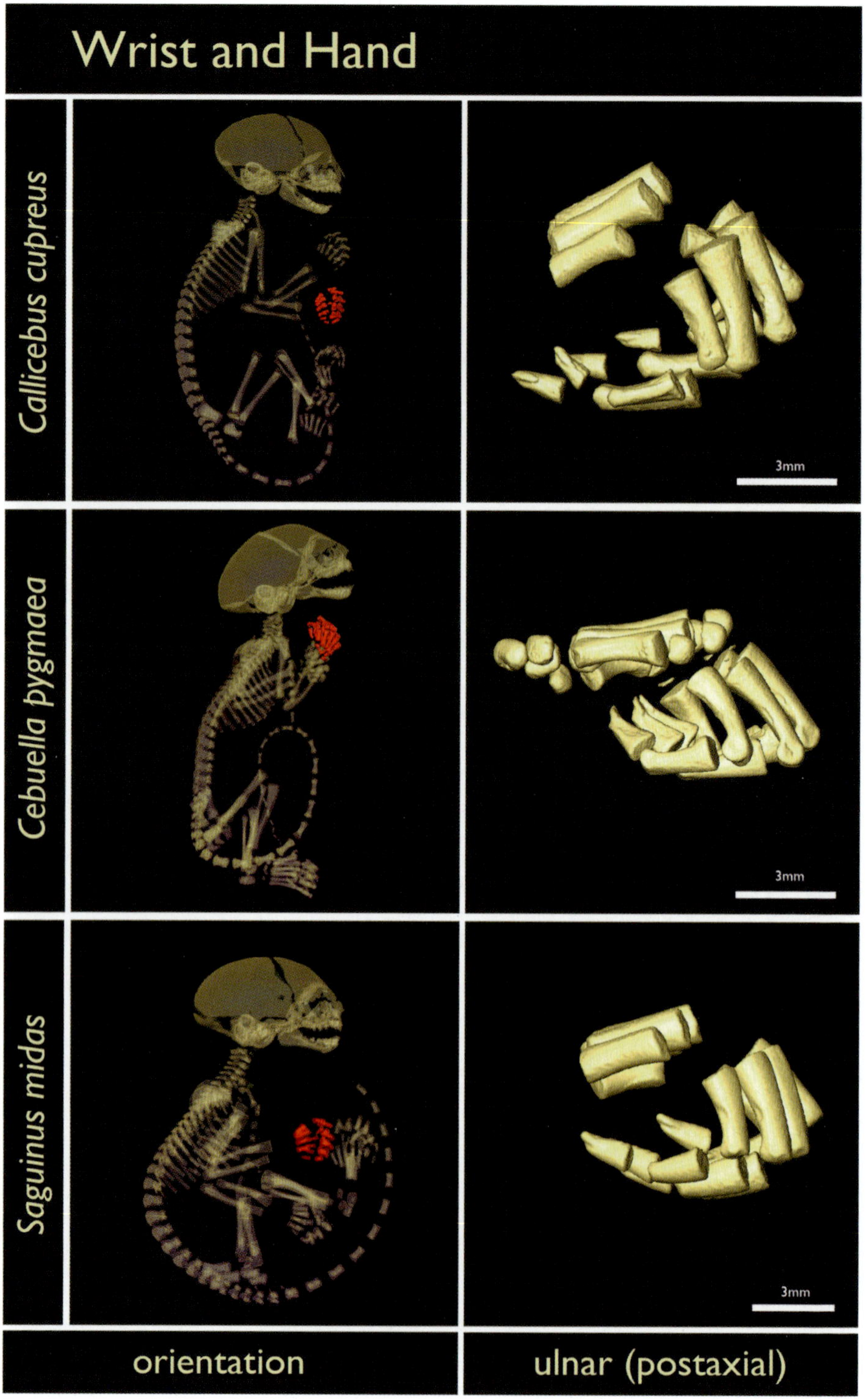
Wrist and Hand
Callicebus cupreus
Cebuella pygmaea
Saguinus midas
3mm
3mm
3mm
orientation
ulnar (postaxial)

Plate 33a

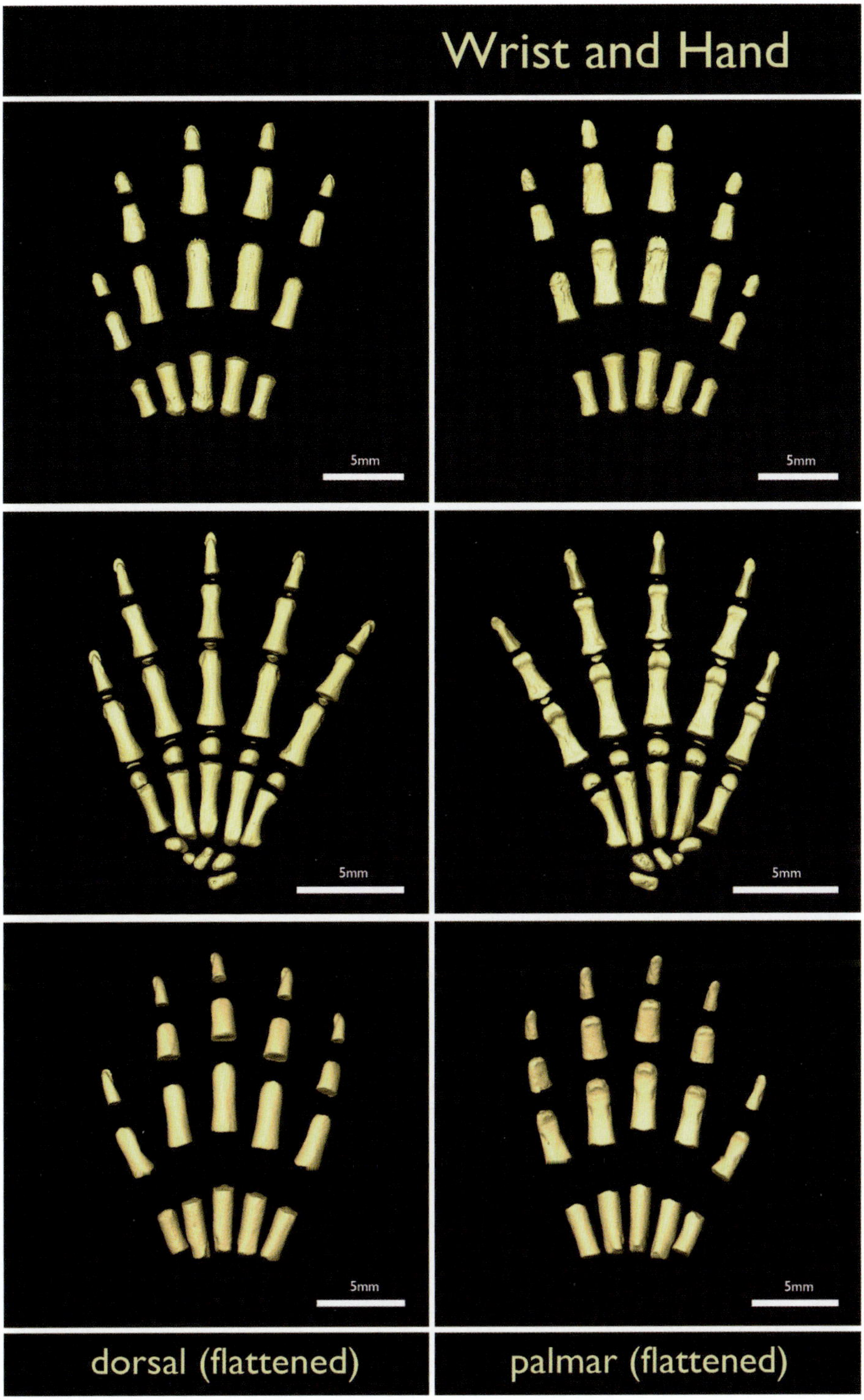
Wrist and Hand
5mm
5mm
5mm
5mm
5mm
5mm
dorsal (flattened)
palmar (flattened)

Plate 33b

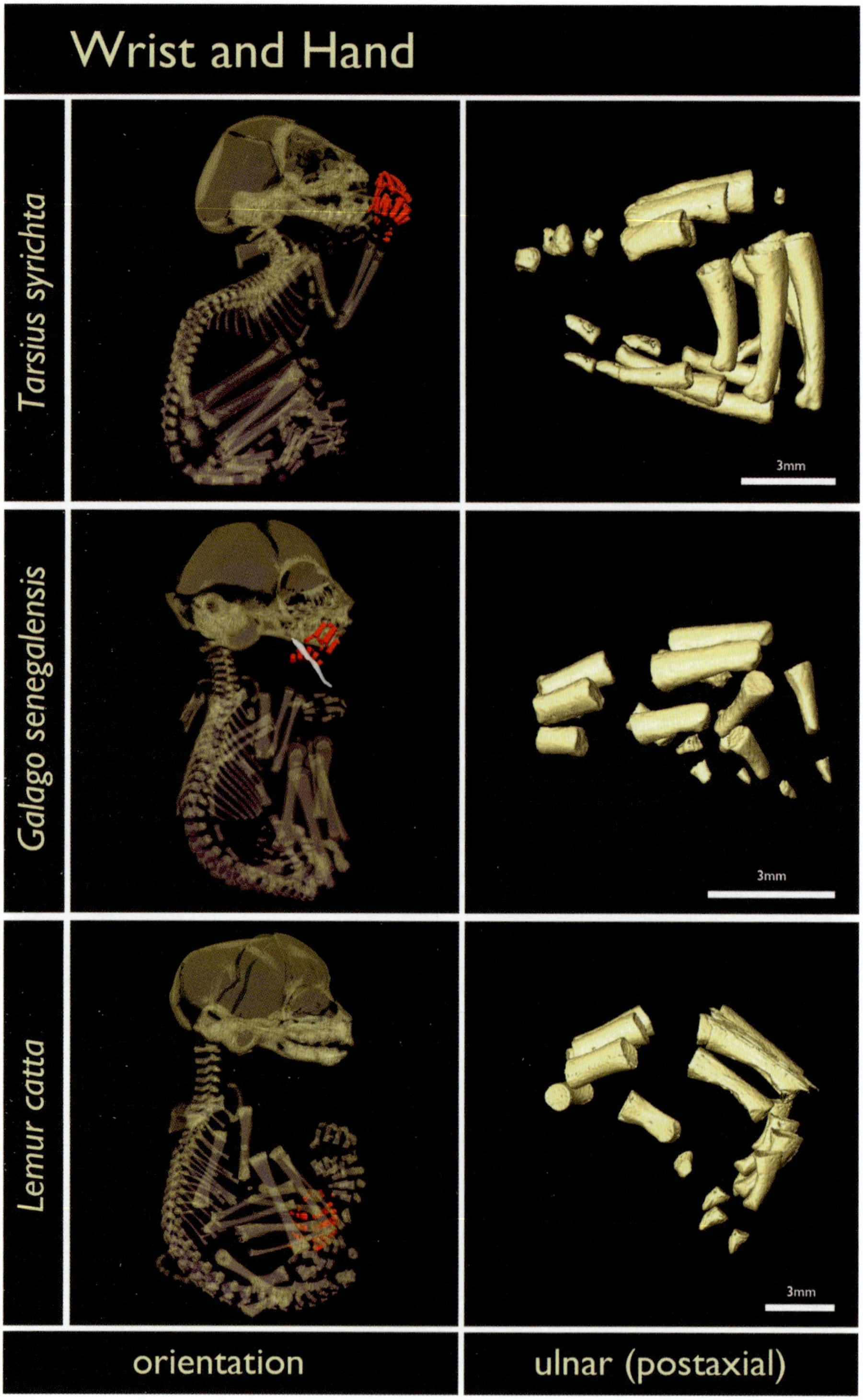

Wrist and Hand
Tarsius syrichta
3mm
Galago senegalensis
3mm
Lemur catta
3mm
orientation
ulnar (postaxial)

Plate 34a

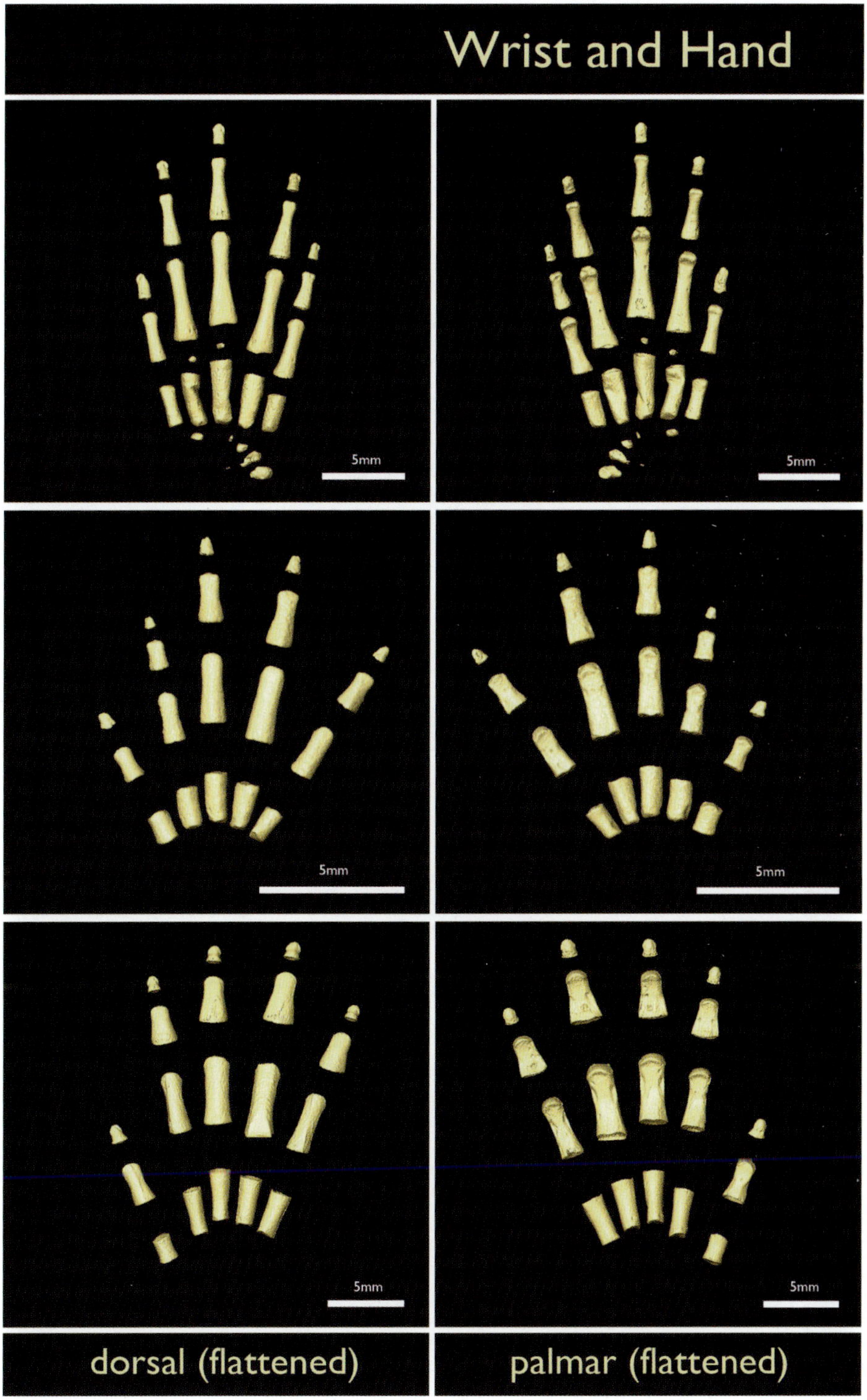
Wrist and Hand
5mm
5mm
5mm
5mm
5mm
5mm
dorsal (flattened)
palmar (flattened)

Plate 34b

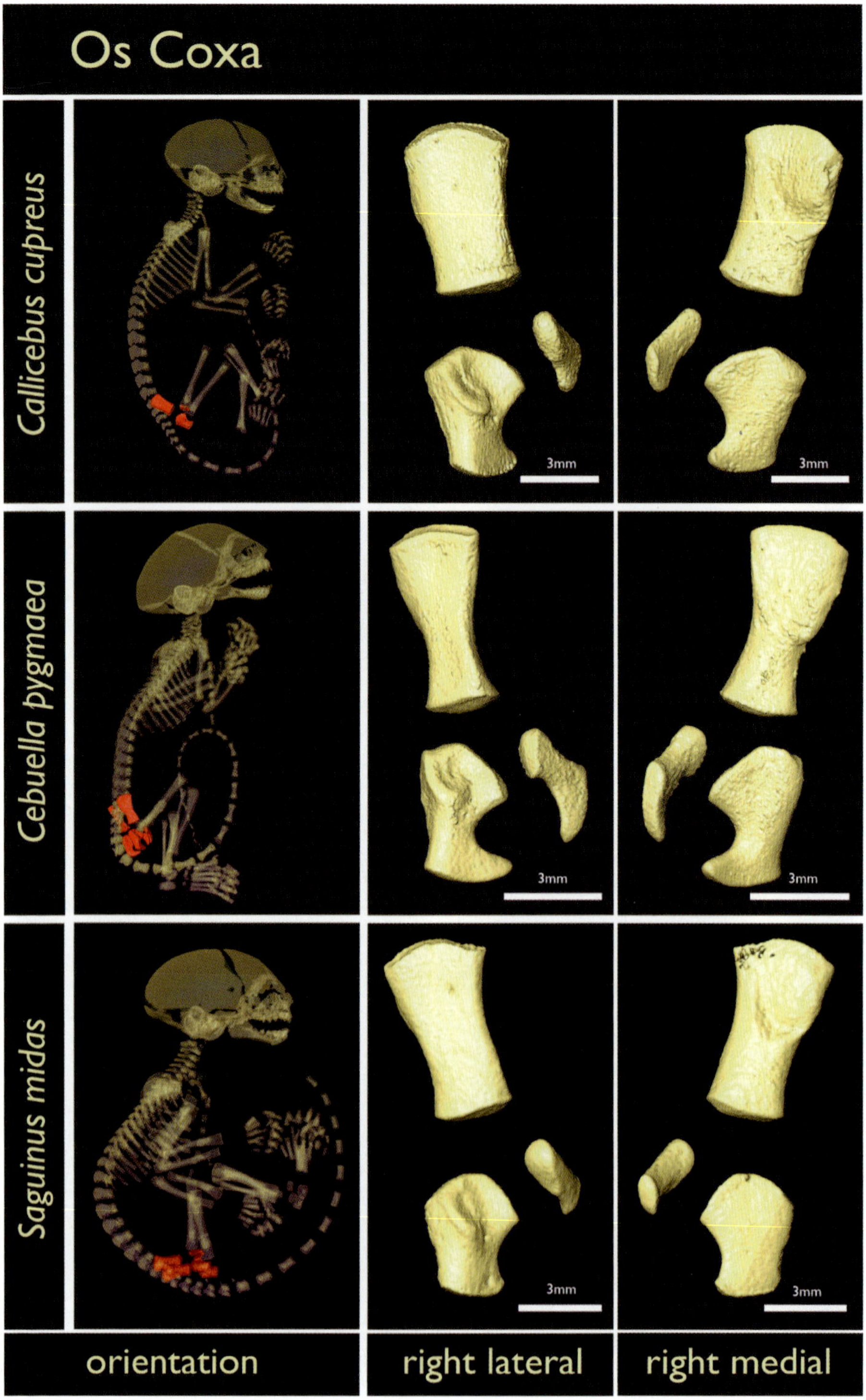
Os Coxa
Callicebus cupreus
Cebuella pygmaea
Saguinus midas
3mm
orientation
right lateral
right medial

Plate 35a

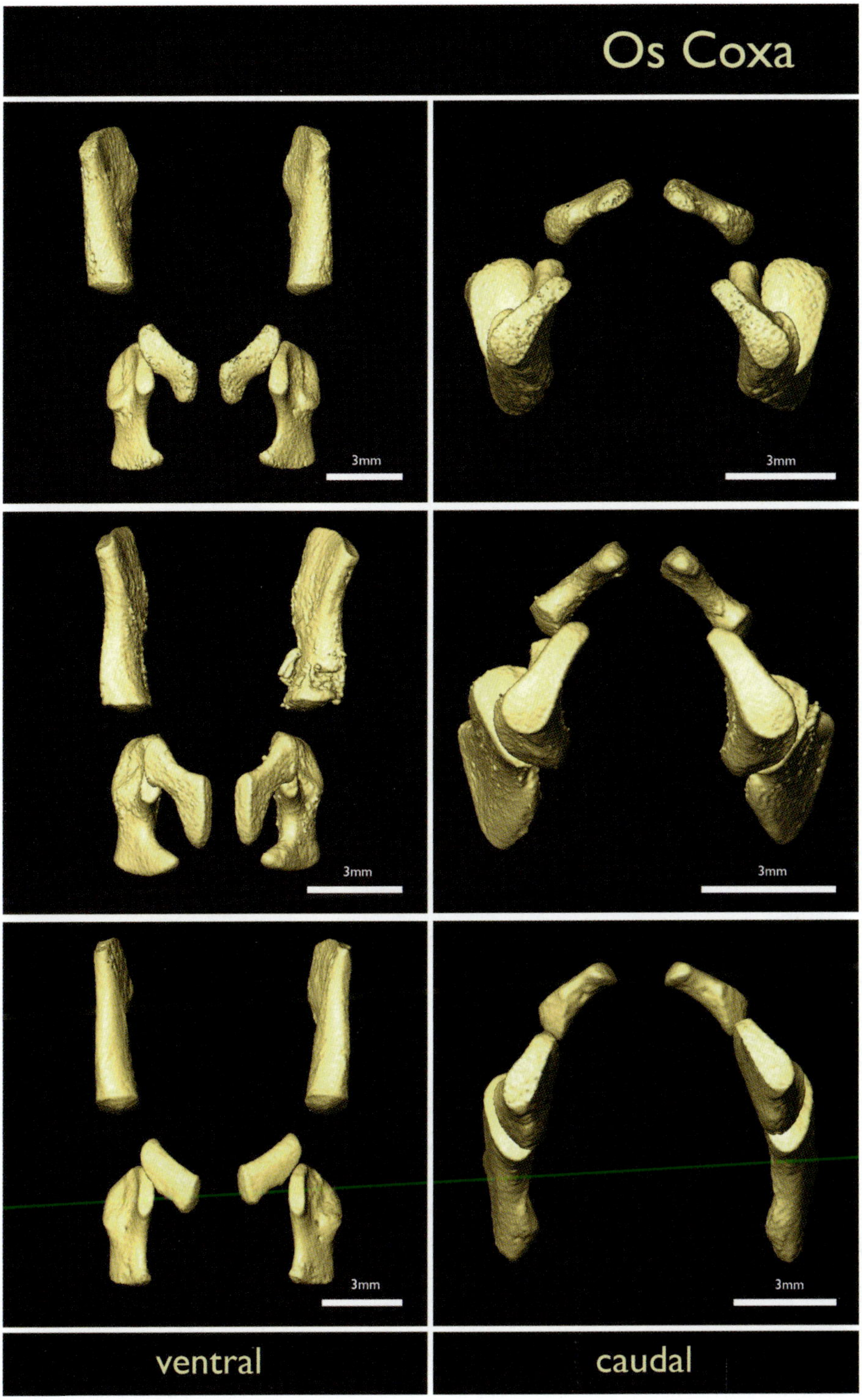

Plate 35b

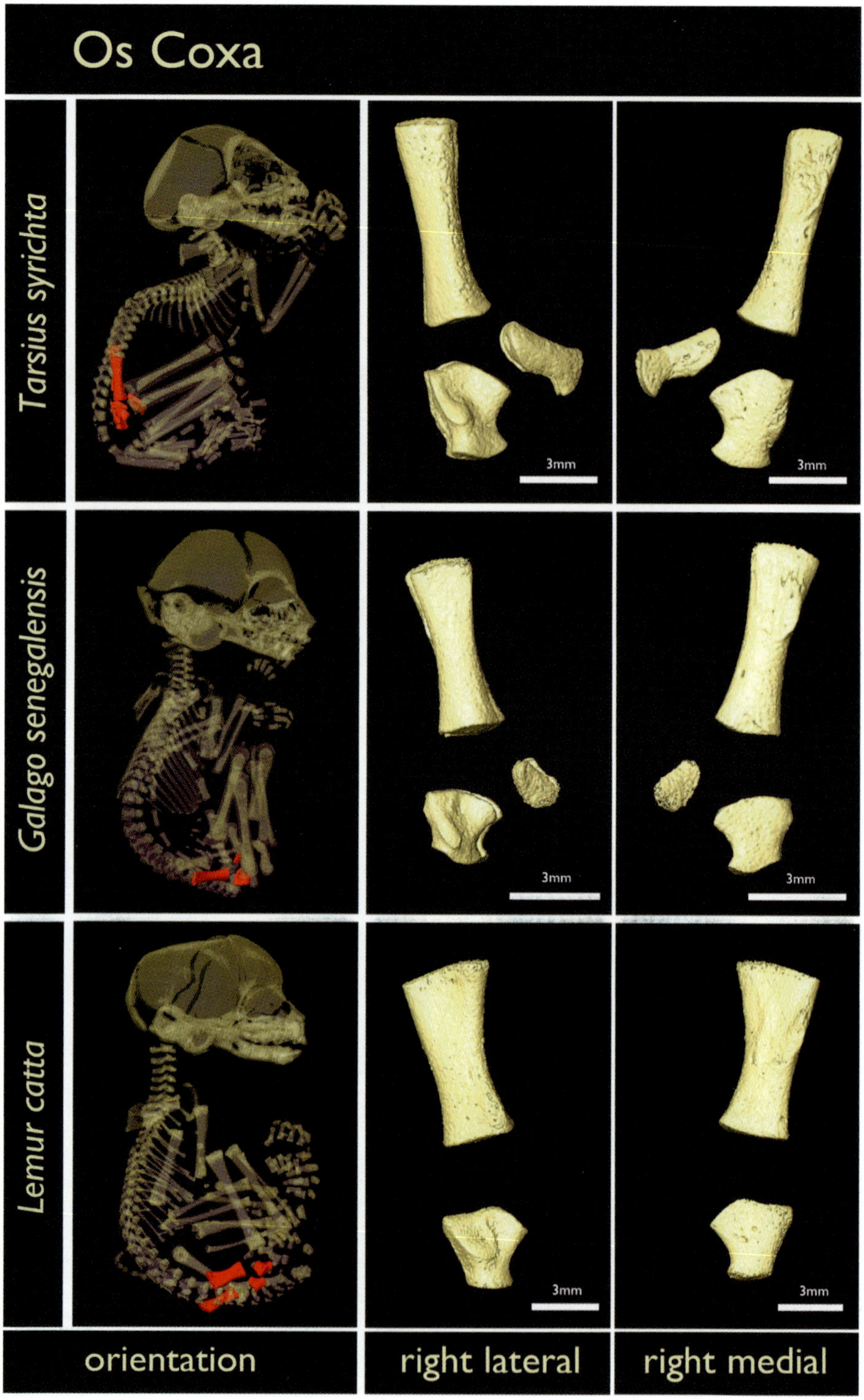
Os Coxa
Tarsius syrichta
Galago senegalensis
Lemur catta
3mm
3mm
3mm
3mm
3mm
3mm
orientation
right lateral
right medial

Plate 36a

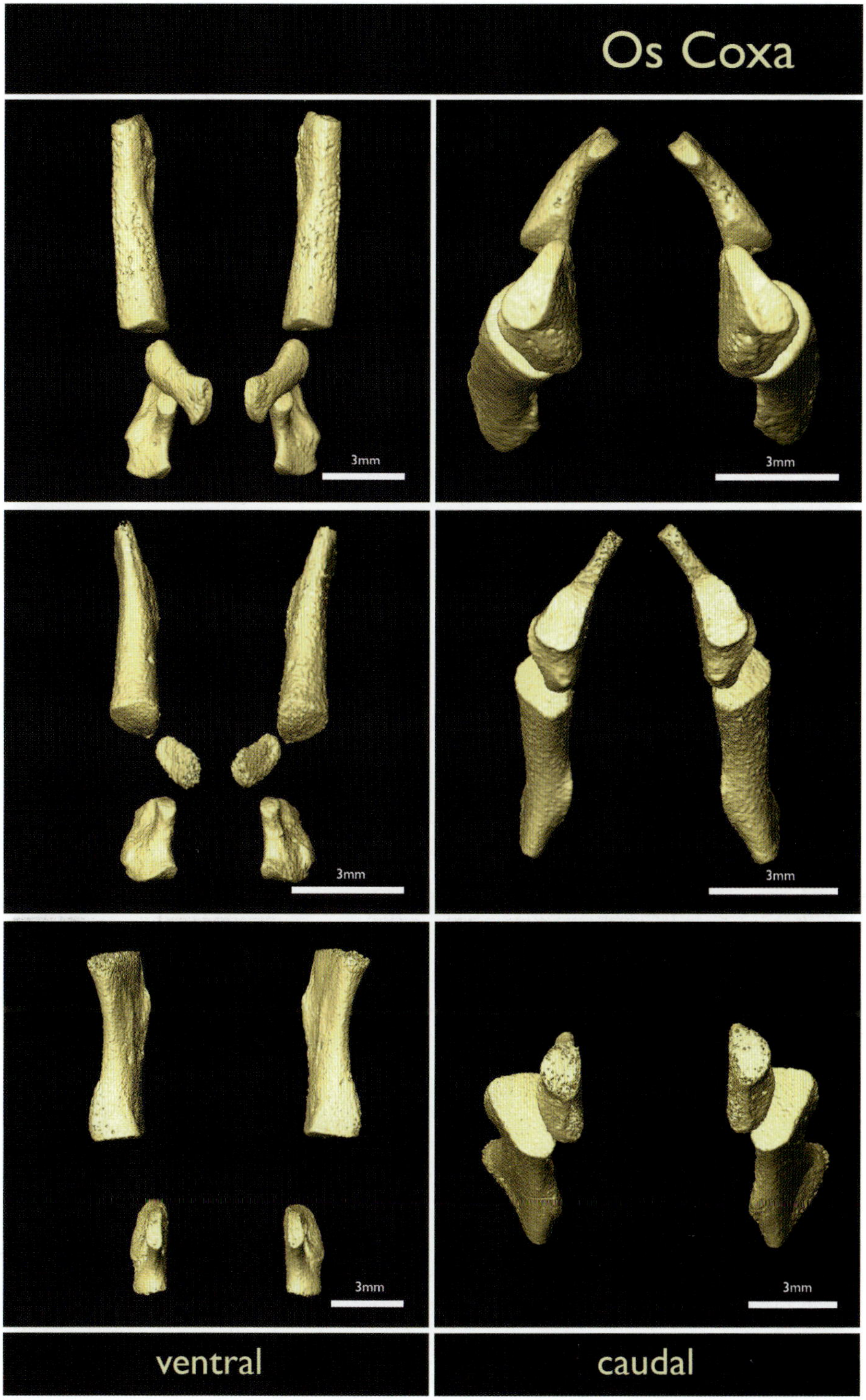

Plate 36b

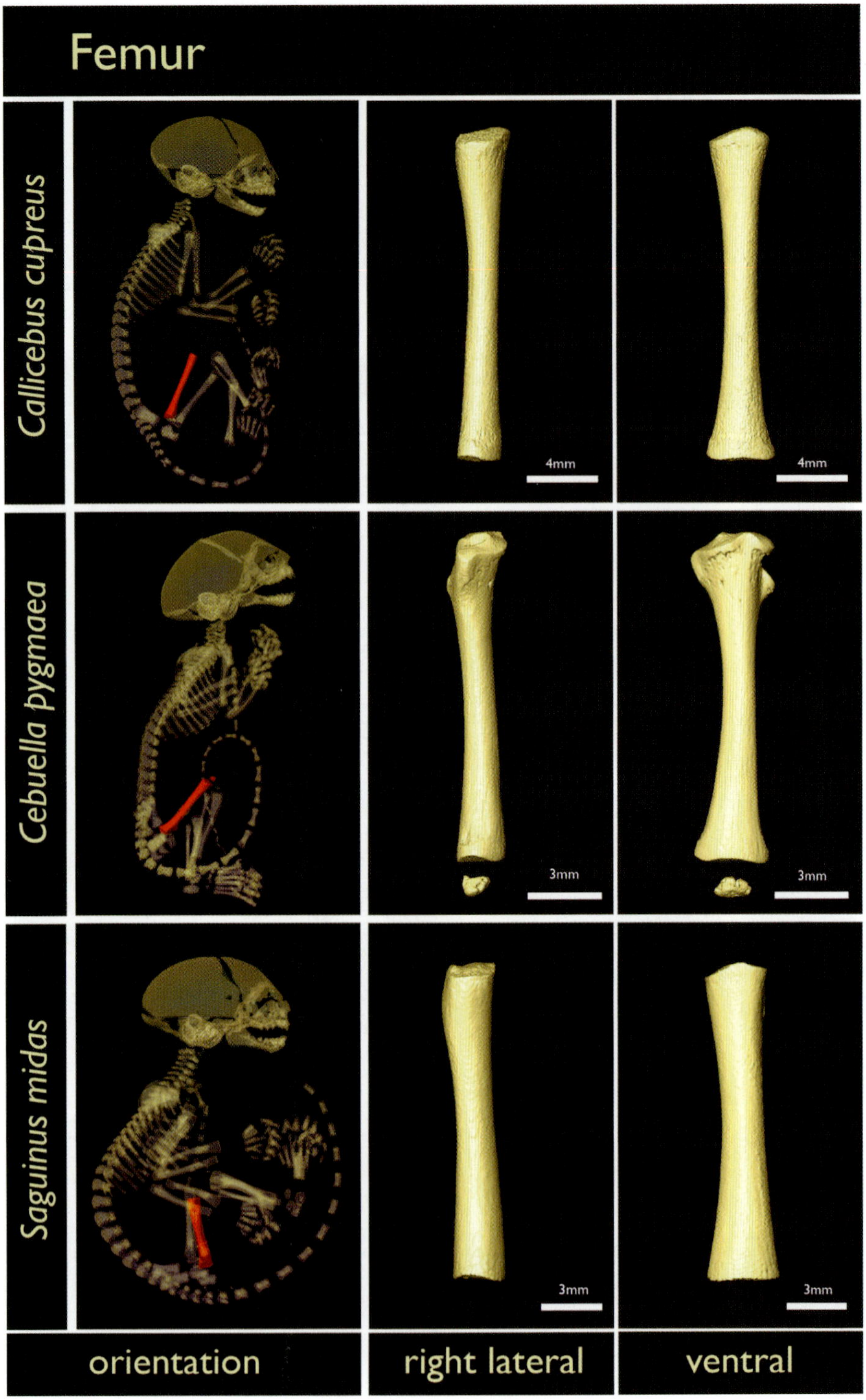
Femur
Callicebus cupreus
Cebuella pygmaea
Saguinus midas
4mm
4mm
3mm
3mm
3mm
3mm
orientation
right lateral
ventral

Plate 37a

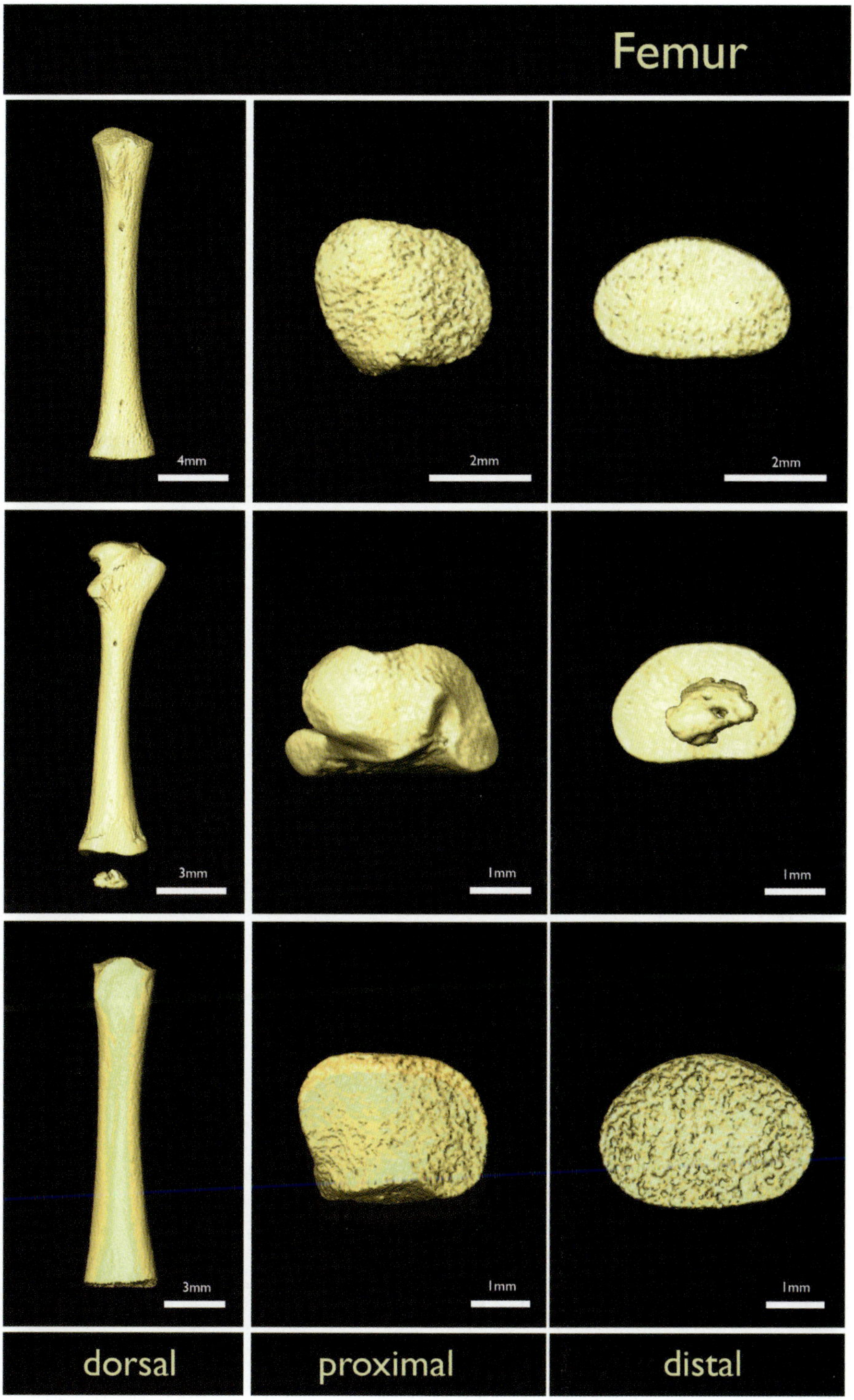
Femur
4mm
2mm
2mm
3mm
1mm
1mm
3mm
1mm
1mm
dorsal
proximal
distal

Plate 37b

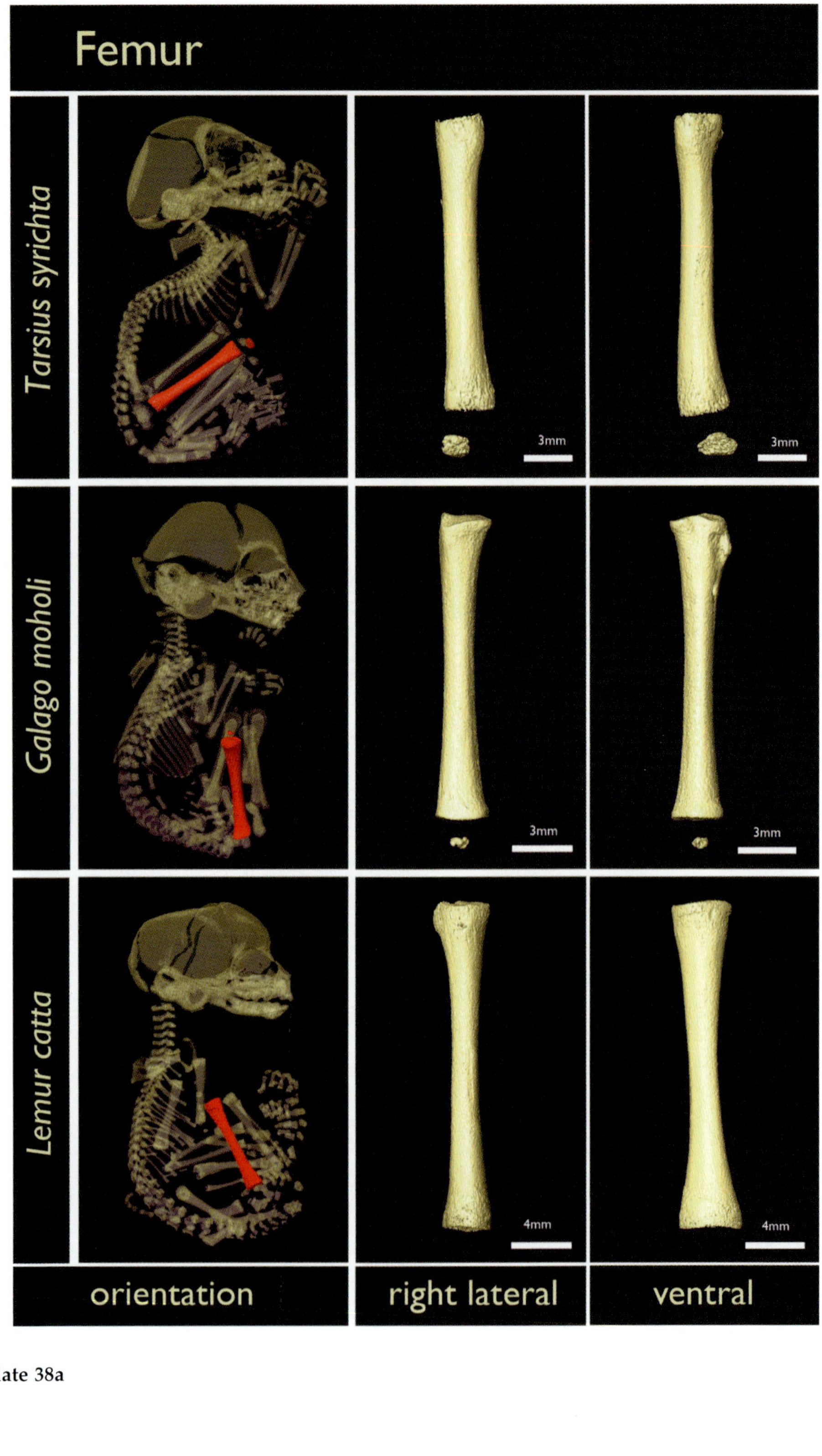
Femur
Tarsius syrichta
3mm
3mm
Galago moholi
3mm
3mm
Lemur catta
4mm
4mm
orientation
right lateral
ventral

Plate 38a

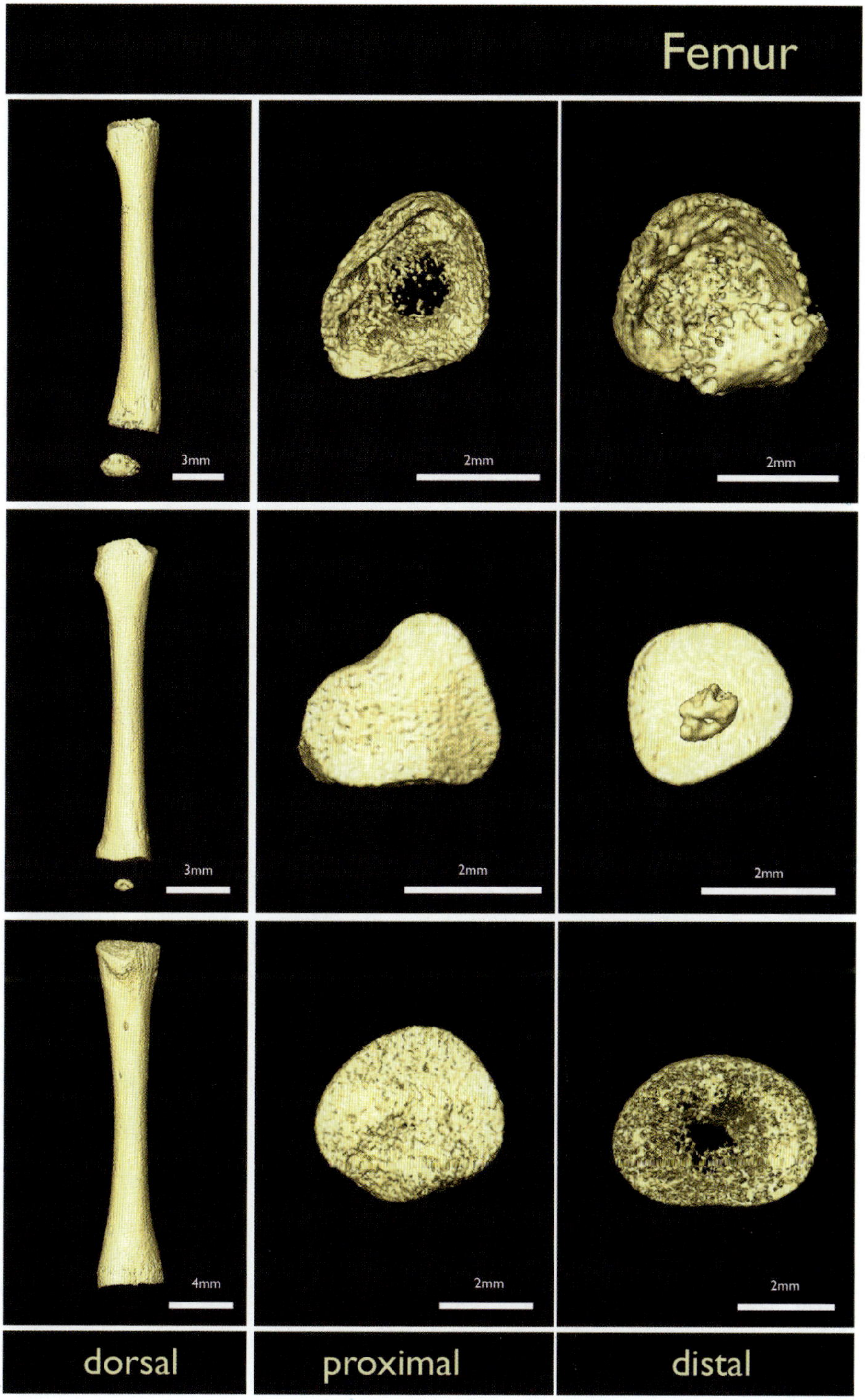
Femur
3mm
2mm
2mm
3mm
2mm
2mm
4mm
2mm
2mm
dorsal
proximal
distal

Plate 38b

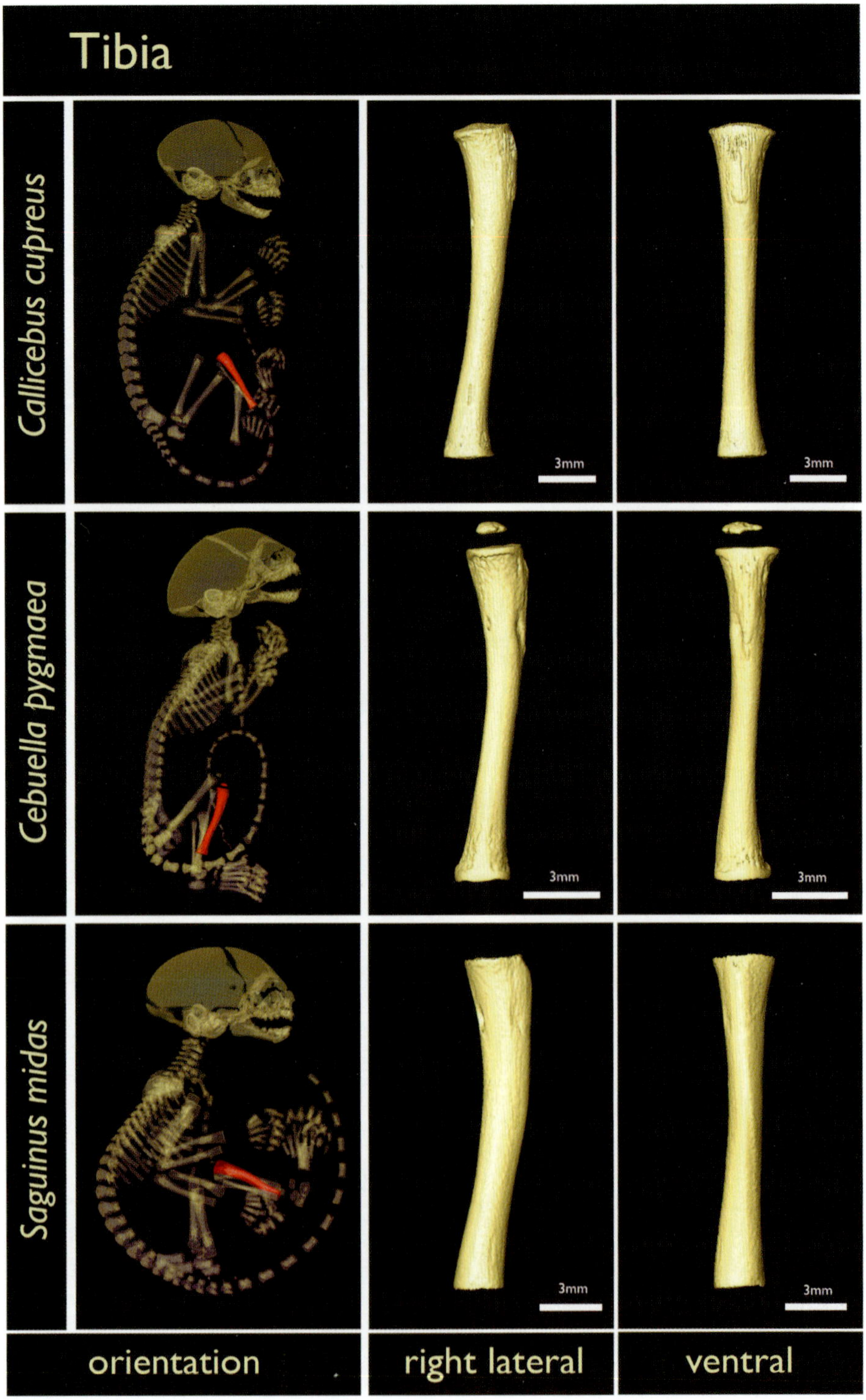
Tibia
Callicebus cupreus
Cebuella pygmaea
Saguinus midas
3mm
3mm
3mm
3mm
3mm
3mm
orientation
right lateral
ventral

Plate 39

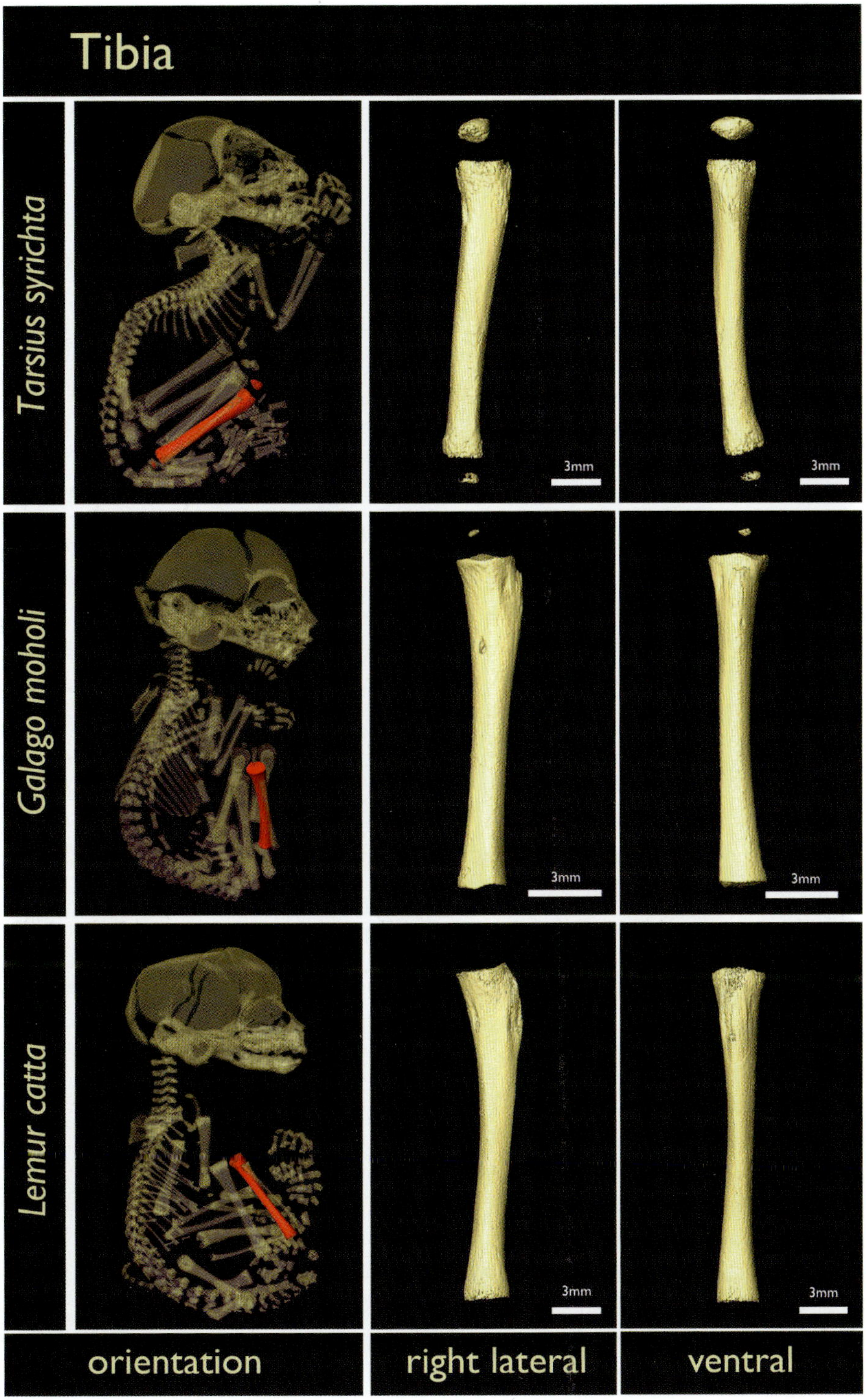
Tibia
Tarsius syrichta
Galago moholi
Lemur catta
3mm
orientation
right lateral
ventral

Plate 40

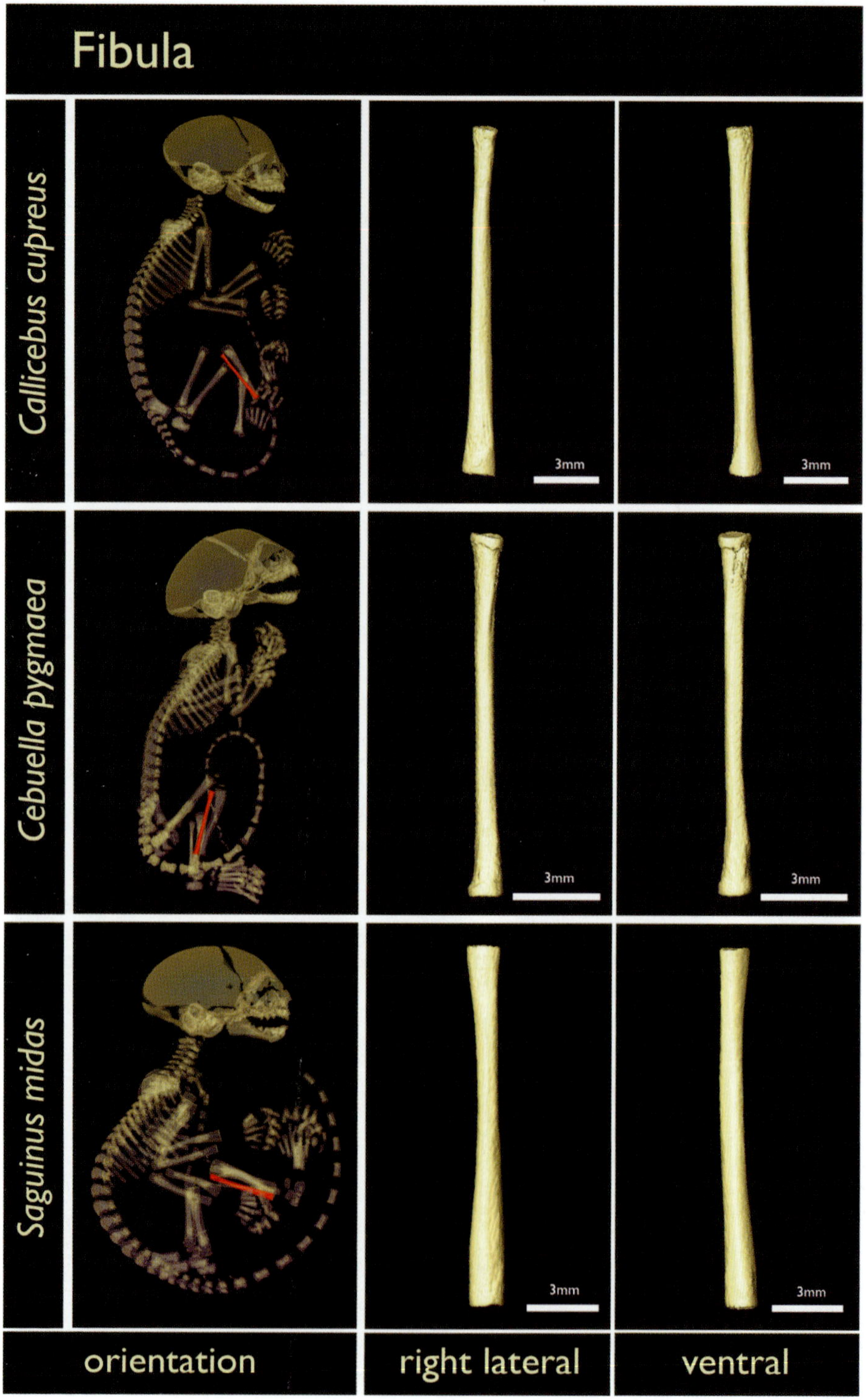
Fibula
Callicebus cupreus
Cebuella pygmaea
Saguinus midas
3mm
3mm
3mm
3mm
3mm
3mm
orientation
right lateral
ventral

Plate 41

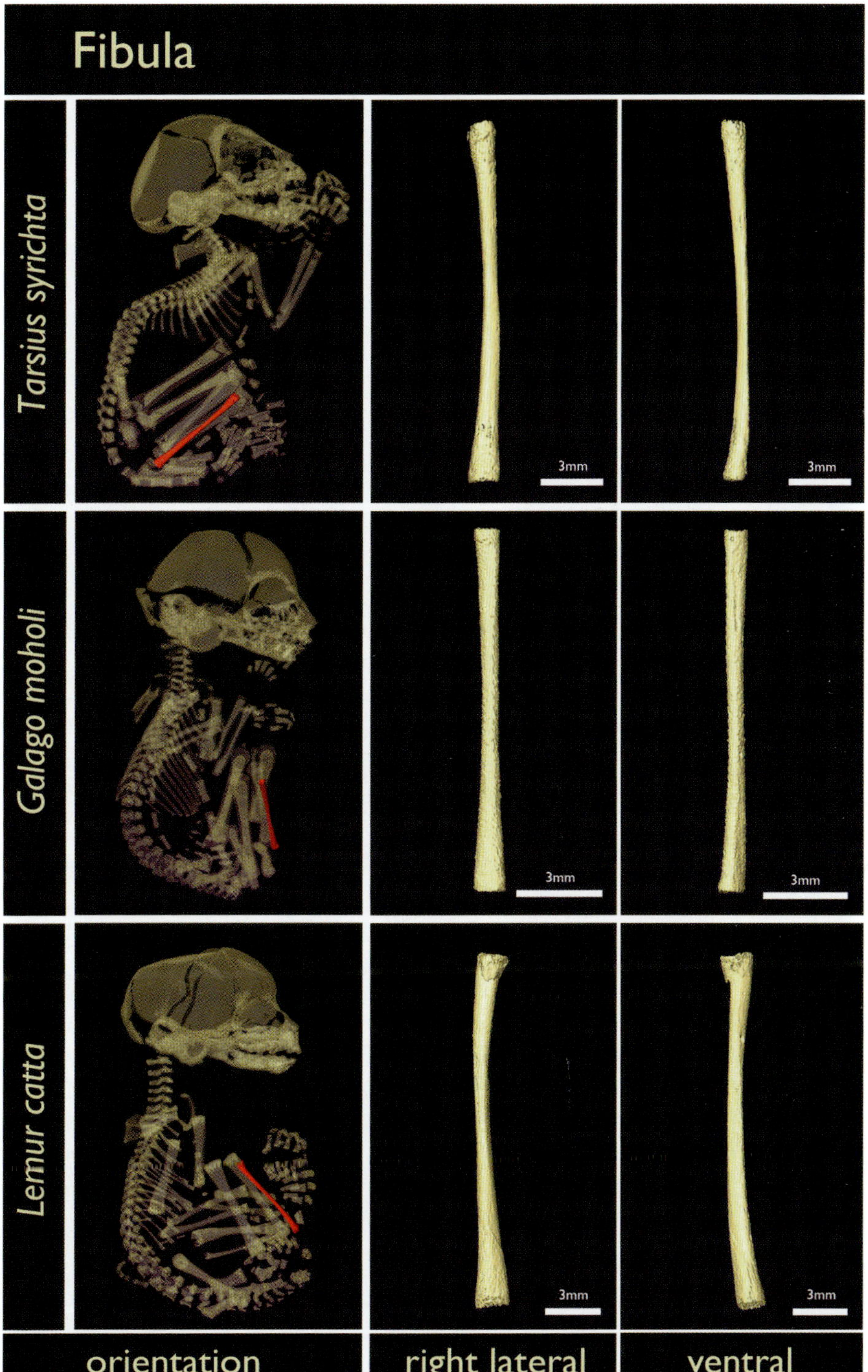
Fibula
Tarsius syrichta
3mm
3mm
Galago moholi
3mm
3mm
Lemur catta
3mm
3mm
orientation
right lateral
ventral

Plate 42

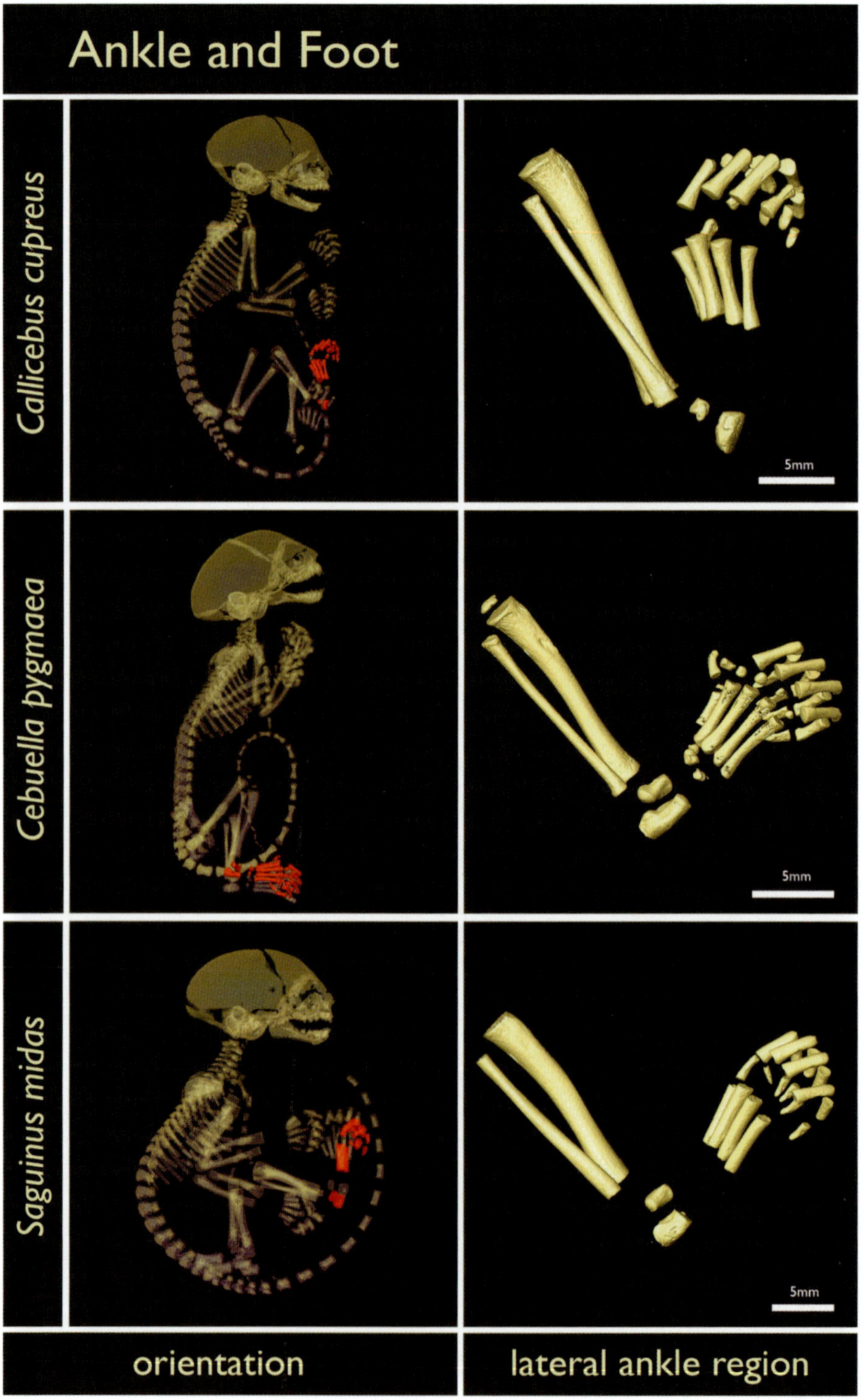
Ankle and Foot
Callicebus cupreus
Cebuella pygmaea
Saguinus midas
5mm
5mm
5mm
orientation
lateral ankle region

Plate 43a

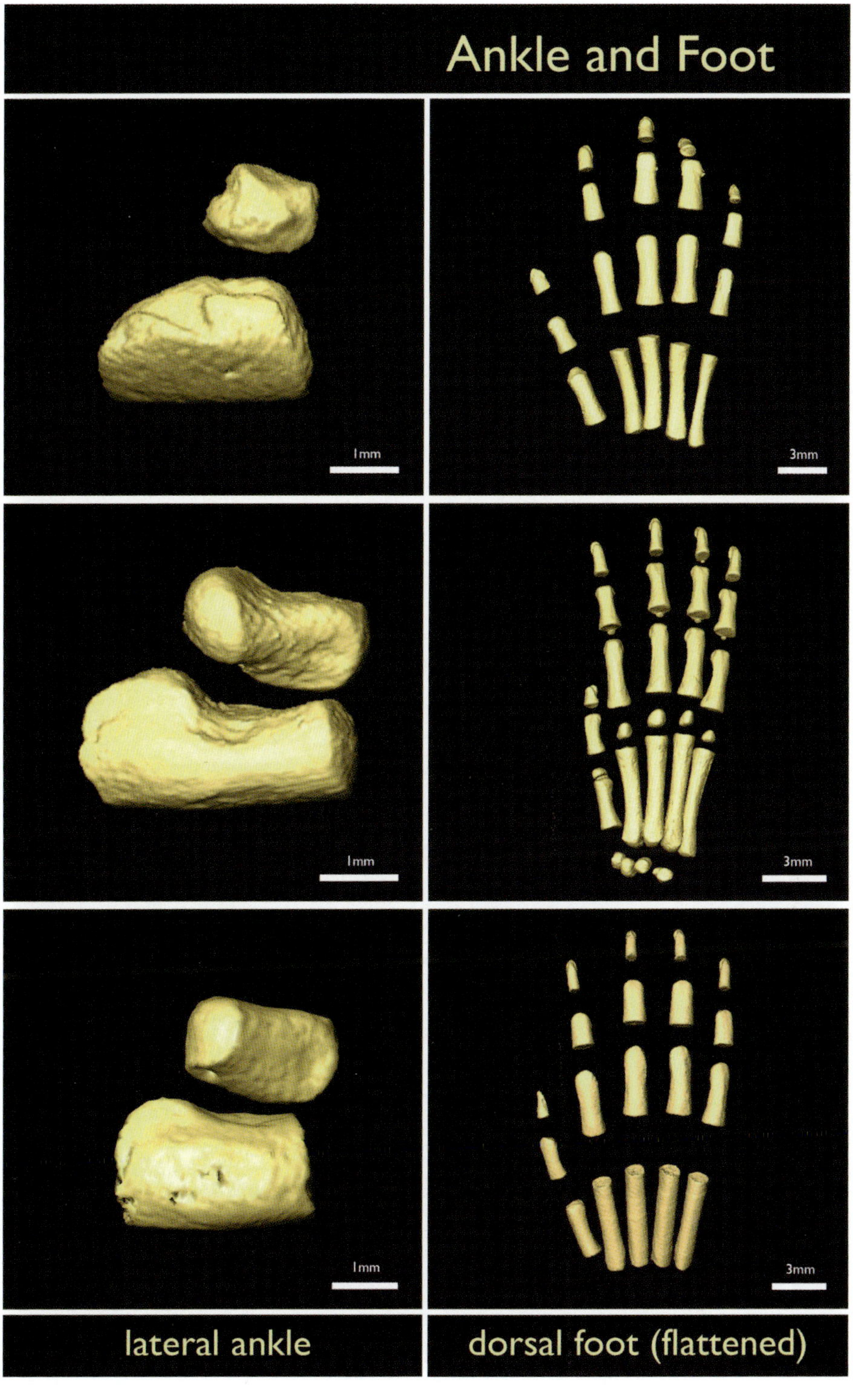
Ankle and Foot
1mm
3mm
1mm
3mm
1mm
3mm
lateral ankle
dorsal foot (flattened)

Plate 43b

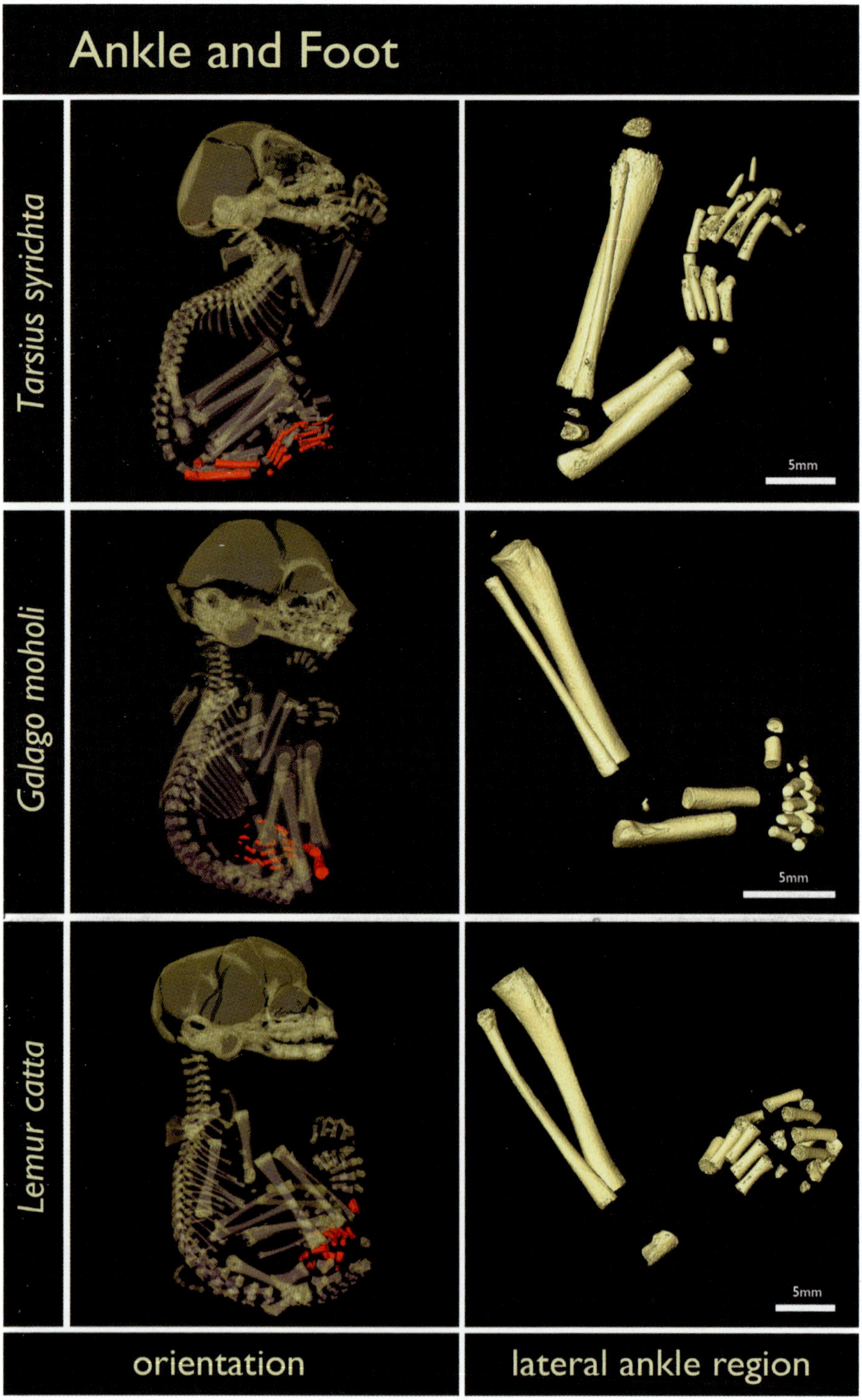
Ankle and Foot
Tarsius syrichta
Galago moholi
Lemur catta
5mm
5mm
5mm
orientation
lateral ankle region

Plate 44a

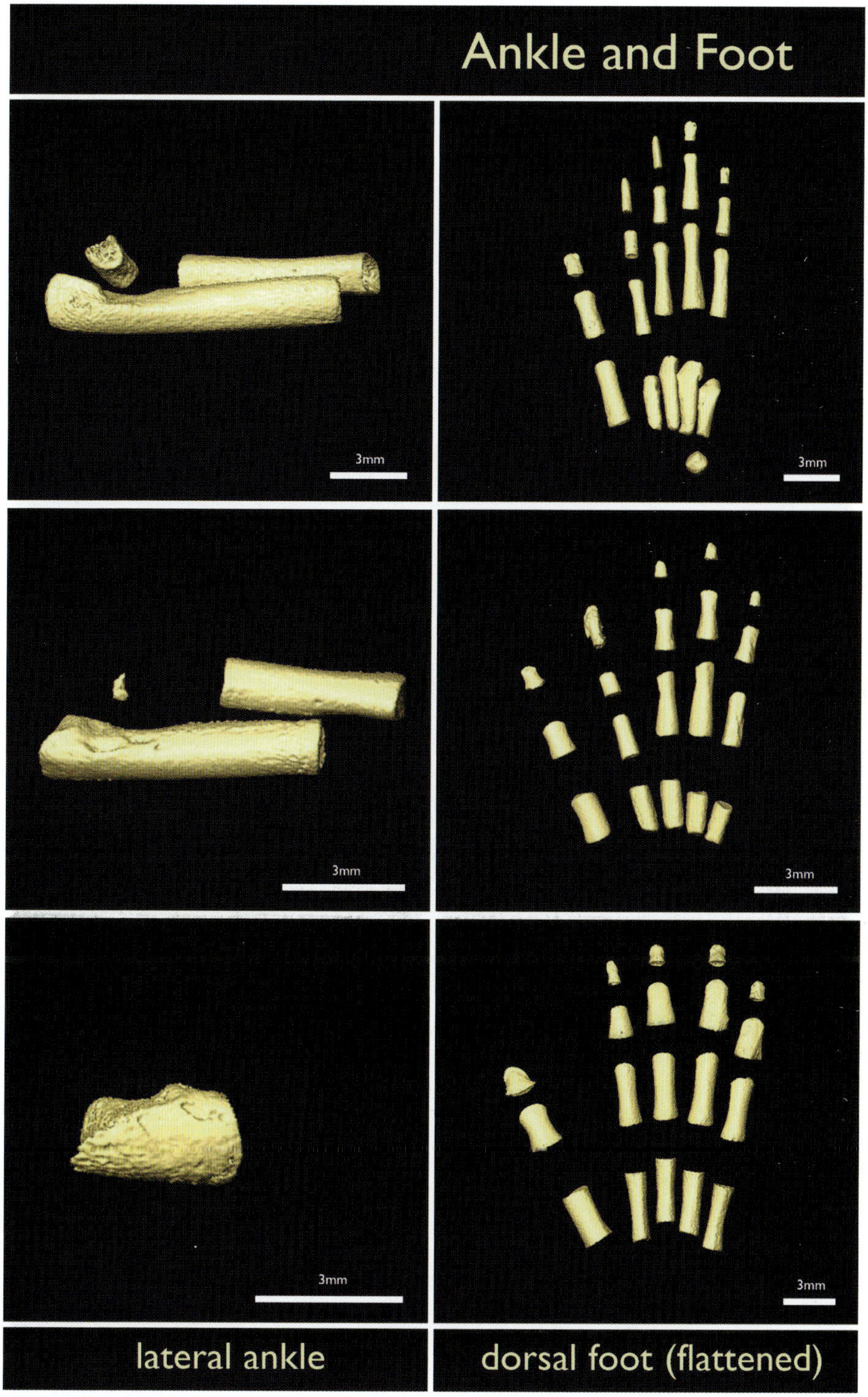
Ankle and Foot
3mm
3mm
3mm
3mm
3mm
3mm
lateral ankle
dorsal foot (flattened)

Plate 44b